D1272962

DATE DUE

Unless Recalled Earlier

HANDBOOK of
ORGANIC
SOLVENTS

HANDBOOK of
ORGANIC
SOLVENTS

David R. Lide

CRC Press

Boca Raton   Ann Arbor   London   Tokyo

Library of Congress Cataloging-in-Publication Data

Catalog record is available from the Library of Congress

INTRODUCTION

A solvent may be defined in rough terms as any liquid that serves as a carrier for another substance, as a medium for conducting a chemical reaction, or as a means of extracting or separating other substances. Whatever the precise definition used, solvents are a ubiquitous feature of both modern industry and everyday life. Solvents serve as carriers for paints, reagents, medications, cleaning agents, and a host of other active ingredients. Many of the most important industrial processes take place in a solvent medium. Solvents find wide application for cleaning and for separations of all kinds. In terms of either production bulk or dollar value, solvents form one of the leading categories of products of the chemical industry.

The most common solvent is water, but next in importance comes a group of organic liquids and their mixtures. This book presents data on 564 organic compounds which are now used (or are potentially useful) as solvents. The selection of a limited set of substances from the extremely large universe of organic compounds must necessarily be somewhat arbitrary. Manufacturers' catalogs and other books on solvents have been consulted, as well as the general chemical literature. In some cases, the complete set of isomers of a compound has been covered, even though not all of the isomers find current use. This has been done in the hope that new ideas for replacement solvents may emerge. Emphasis has also been given to environmentally benign compounds which are finding increasing use as replacements for more hazardous solvents such as chlorinated hydrocarbons.

In practice, many solvents are mixtures rather than pure compounds. It would be a formidable task to compile data on all the mixtures of industrial significance. However, the pure compound data in this book form a basis for estimating properties of mixed solvents. A few useful references to mixture properties and estimation methods are given in the Bibliography (Ref. 8, 10, 35-37).

The properties covered fall into the following categories:
- Identifying information such as name, synonyms, and various registry numbers which are helpful in accessing other reference sources.
- Common physical properties: melting point, normal boiling point, critical constants, density, index of refraction, and solubility in water and other substances.
- Thermodynamic properties, including heat capacity; heats of formation, fusion, and vaporization; surface tension; and vapor pressure as a function of temperature.

- Transport properties: viscosity and thermal conductivity as a function of temperature.
- Electrical properties: dipole moment, dielectric constant, and ionization potential.
- Spectral data: mass spectra, infrared, Raman, ultraviolet, and nuclear magnetic resonance.
- Health and safety information, including flammability, permissible airborne concentration, carcinogenicity, and pertinent information from government regulatory lists concerning use and disposal of the substance. The latter information provides an indication of toxicity and other health hazards of the substance.

The data presented in this book have been taken from evaluated sources whenever possible, rather than the primary literature. Space and format limitations prevent giving detailed references to each piece of data, but the major sources consulted are listed in the bibliography. In addition, reviews and compilations in the *Journal of Physical and Chemical Reference Data* and in other publications of the Standard Reference Data Program of the National Institute of Standards and Technology have been heavily used. Where there were discrepancies between different sources, the author used his judgment on selecting the most likely value.

The compounds are arranged alphabetically by the most frequently used name. Again, the choice of primary name is somewhat arbitrary, but an effort has been made to strike a balance between names that are easily recognizable and names that are chemically informative. A systematic name, generally the Chemical Abstracts Service Index Name, is also given, as well as other synonyms. Each compound is assigned an ID number, which serves as the key in the indexes.

Four indexes are provided to aid in location of a compound. These are the **Name Index**, which includes the primary name, systematic name, and all listed synonyms for each solvent; a **Compound Class Index**, in which the compounds are arranged by chemical class; a **Molecular Formula Index**, and a **CAS Registry Number Index**. There are four additional indexes which list compounds in order of **Normal Boiling Point**, **Melting Point**, **Density**, and **Dielectric Constant**.

Explanation of the Data Fields

The following data items appear for each solvent when the information is available. The principal references used as sources for this book are indicated in brackets.

Syst. name: Generally, the Index Name from the *8th or 9th Collective Index* of Chemical Abstracts Service (CAS). [Ref. 2]

Synonyms: One or more synonyms in common use. [Ref. 2]

CASRN: The Chemical Abstracts Registry Number assigned by CAS as the unique identifier for the specific isomer of the compound. [Ref. 2]

Rel. CASRN's: Related CAS Registry Numbers under which the compound is sometimes listed. In most cases these are "generic" Registry Numbers assigned to a compound without specification of the particular isomer. The number may refer to a mixture of isomers or simply to a substance for which the isomer is unidentified. Such numbers are often used in Government regulatory lists instead of the CASRN's for the individual isomers. [Ref. 31, 32]

Merck No: Monograph Number in *The Merck Index, Eleventh Edition.* It should be noted that this is not a unique identifier for a single compound, since several derivatives or isomers of a compound may be included in the same Monograph. [Ref 3]

DOT No: Registry number assigned by the Department of Transportation for labeling hazardous substances when they are shipped. [Ref. 31]

Beil RN: The Beilstein Registry Number used as a unique identifier in the *Beilstein Database.* [Ref. 9]

Beil Ref: Citation to the printed *Beilstein Handbook of Organic Chemistry.* An entry of 5-18-11-01234, for example, indicates that the compound may be found in the 5th Series, Volume 18, Subvolume 11, page 1234. [Ref. 9]

MF: The molecular formula written in the Hill Order, in which C appears first, H second (if present), and then the other element symbols in alphabetical order. [Ref. 2]

MW: Molecular weight (relative molar mass) as calculated with the 1993 IUPAC Standard Atomic Weights. [Ref. 2]

MP: Normal melting point in $^\circ$C. Although some values are quoted to 0.1°C, uncertainties are typically several degrees Celsius. [Ref. 1-6, 9, 10, 12, 13, 17]

BP: Normal boiling point in $^\circ$C. This is the temperature at which the liquid phase is in equilibrium with the vapor at a pressure of 760 mmHg (101.325 kPa). [Ref. 1-6, 9, 10, 12, 13, 17]

Den: Density (mass per unit volume) in g/cm^3 for the liquid phase. The superscript indicates the temperature in $^\circ$C. All values are true densities, not specific gravities. The number of decimal places gives a rough estimate of the accuracy of the value. [Ref. 1-6, 10, 13, 17]

n_D: Refractive index of the liquid, at the temperature in $^\circ$C indicated by the superscript. All values refer to a wavelength of 589 nm (sodium D line). [Ref. 2, 3]

Vap. pres: Vapor pressure in kPa at the temperature indicated by the superscript (in $^\circ$C). Note that 1 kPa = 7.50 mmHg and 101.325 kPa = 1 atmos. [Ref. 1, 4, 5, 10, 13]

Therm. prop: This field gives the molar enthalpy (heat) of fusion ($\Delta_{fus}H$) at the melting point, enthalpy of vaporization ($\Delta_{vap}H$) at the normal boiling point, and standard enthalpy of formation from the elements in their standard reference states ($\Delta_f H^\circ$). All values are in kJ/mol. [Ref. 1, 4, 5, 11, 12, 14, 15, 17]

C_p **(liq)**: Molar heat capacity of the liquid phase in J/mol $^\circ$C at the temperature indicated by the superscript. [Ref. 1, 4, 5, 10, 14, 16]

γ: Surface tension in mN/m (equivalent to dyn/cm) at the temperature indicated by the superscript. When available, the change in surface tension with temperature, $d\gamma/dT$, is also given. [Ref. 1, 4, 10, 18]

Sol: Solubility on a relative scale: 1 = insoluble; 2 = slightly soluble; 3 = soluble; 4 = very soluble; 5 = miscible; 6 = decomposes. See List of Abbreviations for the solvent abbreviations. [Ref. 2, 3]

Elec. prop: This field includes:
-Electric dipole moment μ in debye units (1 D = 1.33564 X 10^{-30} C m). Values measured in the gas phase are given when available. If there is no gas-phase measurement, a value determined in the liquid phase or in solution is given in parentheses; such values are much less reliable. [Ref. 1, 22, 23]

-Dielectric constant ε (also called relative permittivity) at the temperature in $^{\circ}C$ indicated by the superscript. Information on the temperature dependence of the dielectric constant may be found in the references. [Ref. 1, 24]
-First ionization potential **IP** of the substance. [Ref. 1, 25]

η: Viscosity of the liquid in mPa s (equivalent to centipoise, cP). The temperature is indicated by the superscript. [Ref. 1, 4, 5, 10, 19-21]

k: Thermal conductivity of the liquid in W/m $^{\circ}C$ at the temperature indicated by the superscript. [Ref. 1, 5, 7, 10, 20]

MS: The m/e values of the most abundant peaks in the mass spectrum. Relative intensities are given in parentheses, with the most abundant peak assigned an intensity of (100). [Ref. 2]

IR: Major infrared peaks in cm^{-1}. Values are given to the nearest 10 cm^{-1}. [Ref. 2]

Raman: Most significant peaks in the Raman spectrum, in cm^{-1}. [Ref. 2]

UV: Strongest ultraviolet bands, in nm (1 nm = 10 Å). When available, the molar absorption coefficient follows the wavelength (in parentheses). Solvents are specified in some cases. [Ref. 2]

^{13}C NMR: Chemical shifts in ppm for specific carbon atoms or recognizable groups, referenced to tetramethylsilane (TMS). If known, the solvent in which the spectrum was obtained is given (see List of Abbreviations). [Ref. 2]

^{1}H NMR: Proton chemical shifts in ppm for specific protons or recognizable groups, referenced to tetramethylsilane (TMS). [Ref. 2]

Flammability: This field includes:
-Flash point, which is the minimum temperature at which the vapor pressure of a liquid is sufficient to form an ignitable mixture with air near the surface of the liquid. Flash point is not an intrinsic physical property but depends on the conditions of measurement. See Ref. 29 for details. [Ref. 4, 5, 29]
-Ignition temperature (also called autoignition temperature), which is the minimum temperature required for self-sustained combustion in the absence of an external ignition source. As in the case of flash point, the value depends on specified test conditions. [Ref. 29]
-Flammable limits (often called explosive limits), which specify the range of concentration of the vapor in air (in percent by volume) for which a flame can propagate. Below the lower flammable limit, the gas mixture is too lean to

burn; above the upper flammable limit, the mixture is too rich. Values refer to ambient temperature and pressure and are dependent on the precise test conditions. [Ref. 29]

TLV/TWA: The threshold limit value, expressed as a time weighted average airborne concentration over a normal 8-hour workday and 40-hour workweek, to which most workers can be exposed without adverse effects. The values given here are the recommendations of the American Conference of Governmental Industrial Hygienists (ACGIH), based in part on regulations issued by the Occupational Safety and Health Administration (OSHA). Values conform to the 1993-94 adoptions of ACGIH; intended changes for 1993-94 are not included, since these are still tentative. Changes are issued annually. [Ref. 28, 31]

Reg. lists: The following information has been extracted from Government regulations regarding safe use and disposal of hazardous substances:
-CERCLA indicates the compound is on the list of hazardous substances issued under the Comprehensive Emergency Response, Compensation, and Liability Act of 1980. Any release of such a substance into the environment within a 24-hour period that equals or exceeds the specified Reportable Quantity (RQ) must be reported to the National Response Center (800-424-8802). The RQ value in pounds is given here for each solvent on the CERCLA list. The information is believed to be current as of July 1993. [Ref. 31, 34]
-SARA 313 indicates that the compound appears on the list of toxic chemicals for which reporting standards have been set under Section 313 of the Superfund Amendments and Reauthorization Act of 1986. Anyone using a compound on this list in an amount of 10,000 pounds per year or greater must report the fact to the Environmental Protection Agency. The percentage figure given in parentheses is the threshold concentration of the compound in a mixture below which the reporting requirement is waived. The information is believed to be current as of July 1993. [Ref. 31, 34]
-RCRA indicates that the compound is on the list of hazardous wastes released under Section 3010 of the Resource Conservation and Recovery Act. Hazardous waste containers that contain such a compound must be labeled with the RCRA number given here. A number preceded by "P" indicates a compound that is considered acutely hazardous, while "U" indicates a compound deemed toxic. The information is believed to be current as of July 1993. [Ref. 31, 34]
-If the compound is a confirmed or suspected human carcinogen, this fact is indicated. The information is based on the designations of the International Agency for Research on Cancer (IARC), the National Toxicology Program (NTP), and the Occupational Safety and Health Administration (OSHA); it is believed to be current as of January 1993. All substances so designated by at

least one of the above agencies are indicated as carcinogenic in this book. [Ref. 28, 31, 32, 34]

List of Abbreviations

abs	absolute
ac	acid
ace	acetone
AcOEt	ethyl acetate
alk	alkali
aq	aqueous
Bu	butyl
bz	benzene
CERCLA	Comprehensive Emergency Response, Compensation, and Liability Act of 1980
chl	chloroform
con	concentrated
ctc	carbon tetrachloride
cyhex	cyclohexane
dil	dilute
diox	dioxane
Et	ethyl
eth	ethyl ether
EtOH	ethyl alcohol
HOAc	acetic acid
hp	heptane
hx	hexane
iso	isooctane
lig	ligroin
liq	liquid
MeCN	acetonitrile
MeOH	methyl alcohol
os	organic solvents
peth	petroleum ether
$PhNH_2$	aniline
ppm	parts per million
pres	pressure
PrOH	propyl alcohol
py	pyridine
RCRA	Resource Conservation and Recovery Act
SARA	Superfund Amendments and Reauthorization Act of 1986
sat	saturated, saturation

sulf	sulfuric acid
tol	toluene
TLV	threshold limit value
TMS	tetramethylsilane
TWA	time weighted average

Bibliography

The following references are suggested as sources for more detailed information on the properties covered in this book.

General references covering a variety of properties:

1. Lide, D.R., Editor, *CRC Handbook of Chemistry and Physics, 75th Edition,* CRC Press, Boca Raton, FL, 1994.

2. Lide, D.R., and Milne, G.W.A., Editors, *Handbook of Data on Organic Compounds, Third Edition,* CRC Press, Boca Raton, FL, 1994. (Also available as a CD ROM database.)

3. Budavari, S., Editor, *The Merck Index, Eleventh Edition,* Merck & Co., Rahway, NJ, 1989.

4. Riddick, J.A., Bunger, W.B., and Sakano, T.K., *Organic Solvents, Fourth Edition,* John Wiley & Sons, New York, 1986.

5. .Daubert, T.E., Danner, R.P., Sibul, H.M., and Stebbins, C.C., *Physical and Thermodynamic Properties of Pure Compounds: Data Compilation,* extant 1994 (core with 4 supplements), Taylor & Francis, Bristol, PA.

6. *Physical Constants of Hydrocarbon and Non-Hydrocarbon Compounds, ASTM Data Series DS 4B,* ASTM, Philadelphia, 1988.

7. Marsh, K.N., Editor, *Recommended Reference Materials for the Realization of Physicochemical Properties,* Blackwell Scientific Publications, Oxford, 1987.

8. Beaton, C.F., and Hewitt, G.F., *Physical Property Data for the Design Engineer,* Hemisphere Publishing. Corp., New York, 1989.

9. Luckenbach, R., Editor, *Beilstein Handbook of Organic Chemistry,* Springer-Verlag, Heidelberg. (Also available as an on-line database.)

Thermodynamic properties:

10. Lide, D.R., and Kehiaian, H.V., *CRC Handbook of Thermophysical and Thermochemical Data,* CRC Press, Boca Raton, FL, 1994. (Includes computer disk which permits calculation of properties as a function of temperature.)

11. Pedley, J.B., Naylor, R.D., and Kirby, S.P., *Thermochemical Data of Organic Compounds, Second Edition,* Chapman and Hall, London, 1986.

12. Stevenson, R.M., and Malanowski, S., *Handbook of the Thermodynamics of Organic* Compounds, Elsevier, New York, 1987.

13. *TRC Thermodynamic Tables,* Thermodynamic Research Center, Texas A&M University, College Station, TX. (Also available as an on-line database.)

14. Wagman, D.D., Evans, W.H., Parker, V.B., Schumm, R.H., Halow, I., Bailey, S.M., Churney, K.L., and Nuttall, R.L., *The NBS Tables of Chemical Thermodynamic Properties, J. Phys. Chem. Ref. Data, Vol. 11, Suppl. 2,* 1982.

15. Majer, V., and Svoboda, V., *Enthalpies of Vaporization of Organic* Compounds, Blackwell Scientific Publications, Oxford, 1985.

16. Domalski, E.S., Evans, W.H., and Hearing, E.D., *Heat Capacities and Entropies of Organic Compounds in the Condensed Phase, J. Phys. Chem. Ref. Data, Vol. 13, Suppl. 1, 1984; Vol. 19, No. 4,* 881-1047, 1990.

17. Wilhoit, R.C., and Zwolinski, B.J., *Physical and Thermodynamic Properties of Aliphatic Alcohols, J. Phys. Chem. Ref. Data, Vol. 2, Suppl. 1,* 1973.

18. Jasper, J.J., "The Surface Tension of Pure Liquids", *J. Phys. Chem. Ref. Data 1, 841,* 1972.

Transport properties:

19. Viswanath, D.S., and Natarajan, G., *Data Book on the Viscosity of Liquids,* Hemisphere Publishing Corp., New York, 1989.

20. Liley, P.E., Makita, T., and Tanaka, Y., *Properties of Inorganic and Organic Fluids,* Hemisphere Publishing Corp., New York, 1988.

21. Hellwege, K.H., Editor, *Landolt-Börnstein, Numerical Data and Functional Relationships in Science and Technology, Sixth Edition, II/5a, Transport Phenomena I (Viscosity and Diffusion),* Springer-Verlag, Heidelberg, 1969.

Dipole Moment, dielectric constant, and ionization potential:

22. Nelson, R.D., Lide, D.R., and Maryott, A.A., *Selected Values of Dipole Moments for Molecules in the Gas Phase, Natl. Stand. Ref. Data Ser. - Nat. Bur. Stnds. 10,* 1967.

23. Hellwege, K.H., Editor, *Landolt-Börnstein, Numerical Data and Functional Relationships in Science and Technology, Group II, Vol. 6, 1974; Vol. 14, Subvol. a, 1982; Vol. 14, Subvol. b,* 1983; Springer-Verlag, Heidelberg. (Dipole moments)

24. Madelung, O., Editor, *Landolt-Börnstein, Numerical Data and Functional Relationships in Science and Technology, Group IV, Vol. 6,* Springer-Verlag, Heidelberg, 1991. (Dielectric constants)

25. Lias, S.G., Bartmess, J.E., Liebman, J.F., Holmes, J.L., Levin, R.D., and Mallard, W.G., *Gas-Phase Ion and Neutral Thermochemistry, J. Phys. Chem. Ref. Data, Vol. 17, Suppl. 1,* 1988.

Health and Safety Information:

26. Howard, P.H., Editor, *Handbook of Environmental Fate and Exposure Data for Organic Chemicals, Volume II. Solvents,* Lewis Publishers, Boca Raton, FL, 1990.

27. Lewis, R.J., *Hazardous Chemicals Desk Reference*, Second Edition, Van Nostrand Reinhold, New York, 1991.

28. *Threshold Limit Values for Chemical Substances and Physical Agents 1993-94,* American Conference of Governmental Industrial Hygienists, Cincinnati, OH, 1993.

29. *Fire Protection Guide to Hazardous Materials, 10th Edition,* National Fire Protection Association, Quincy, MA, 1991.

30. *Chemical Safety Data Sheets, Volume 1: Solvents,* Royal Society of Chemistry, Cambridge, 1988.

31. *The Book of Chemical Lists,* Business and Legal Reports, Inc., Madison, CT, 1993.

32. *List of Lists of* Worldwide Hazardous Chemicals and Pollutants, J. B. Lippincott Company, Philadelphia, 1990.

33. *Hazardous Substances Data Bank,* National Library of Medicine, Bethesda, MD.

34. *Code of Federal Regulations,* 40 CRF 261, 40 CRF 302, and 40 CRF 372, Superintendent of Documents, U.S. Government Printing Office, Washington, DC 20402. (Revised periodically.)

Estimation methods, mixture properties:

35. Reid, R.C., Prausnitz, J.M., and Poling, B.E., *The Properties of Gases and Liquids, Fourth Edition,* McGraw-Hill, New York, 1987.

36. Lyman, W.J., Reehl, W.F., and Rosenblatt, D.H., *Handbook of Chemical Property Estimation Methods,* American Chemical Society, Washington, 1990.

37. Wichterle, I., Linek, J., Wagner, Z., and Kehiaian, H.V., *Vapor-Liquid Equilibrium Bibliographic Database,* ELDATA SARL, Montreuil, France, 1993.

1. Acetal

Syst. Name: Ethane, 1,1-diethoxy-
Synonyms: 1,1-Diethoxyethane; Acetylene diethyl
  acetal

CASRN: 105-57-7   **DOT No.**: 1088
Merck No.: 31   **Beil Ref.**: 4-01-00-03103
Beil RN: 1098310
MF: $C_6H_{14}O_2$ **MP[°C]**: -100   **Den.[g/cm³]**: 0.8254^{20}
MW: 118.18   **BP[°C]**: 102.2 n_D: 1.3834^{20}
Sol.: H_2O 3; EtOH 5; eth
  5; ace 4; chl 3
Crit. Const.: T_c = 254°C
Vap. Press. [kPa]: 0.777^0; 3.68^{25}; 13.2^{50}; 38.1^{75}; 93.9^{100}
Therm. Prop.[kJ/mol]: $\Delta_{vap}H$ = 36.3; $\Delta_f H°$(l, 25°C) = -491.4
C_p **(liq.) [J/mol °C]**: 238.0^{25}
γ **[mN/m]**: 20.89^{25}; $d\gamma/dT$ = 0.1030 mN/m °C
Elec. Prop.: μ = (1.4 D); ε = 3.80^{25}; IP = 9.78 eV
MS: 44(100) 43(92) 29(77) 31(76) 45(74) 27(52) 72(48) 73(23) 28(17) 46(15)
IR [cm⁻¹]: 2940 2860 1450 1370 1330 1140 1100 1040 950 850
Raman [cm⁻¹]: 2980 2940 2880 2800 2760 2730 2660 1490 1410 1400 1370
  1340 1280 1170 1140 1100 1060 1030 960 920 860 840 810 660 530 470 450
  360 230
UV [nm]: 286(0) 264(0)
¹³C NMR [ppm]: 15.3 19.9 60.6 99.5 CDCl₃
¹H NMR [ppm]: 1.2 1.3 3.5 3.7 4.7 CDCl₃
Flammability: Flash pt. = -21°C; ign. temp. = 230°C; flam. lim. = 1.6-10.4%

2. Acetaldehyde

Syst. Name: Ethanal

CASRN: 75-07-0   **DOT No.**: 1089
Merck No.: 32   **Beil Ref.**: 4-01-00-03094
Beil RN: 505984
MF: C_2H_4O **MP[°C]**: -123   **Den.[g/cm^3]**: 0.7834[18]
MW: 44.05   **BP[°C]**: 20.1 n_D: 1.3316[20]
Sol.: H_2O 5; EtOH 5; eth
    5; bz 5; chl 2
Crit. Const.: T_c = 193°C; V_c = 154 cm^3
Vap. Press. [kPa]: 12.8[-25]; 44.3[0]; 120[25]
Therm. Prop.[kJ/mol]: $\Delta_{fus}H$ = 3.24; $\Delta_{vap}H$ = 25.8; $\Delta_f H°$(g, 25°C) = -166.2
C_p **(liq.) [J/mol °C]**: 89.0[25] (sat. press.)
γ **[mN/m]**: 21.18[20]; dγ/dT = 0.1360 mN/m °C
Elec. Prop.: μ = 2.750 D; ε = 21.0[18]; IP = 10.23 eV
η **[mPa s]**: 0.224[20]
MS: 29(100) 44(81) 43(33) 42(12) 28(6) 26(6) 41(4) 27(4) 45(2) 25(2)
IR [cm^{-1}]: 3000 2830 2720 1730 1430 1400 1350 1170 1110 860
UV [nm]: 292(10) hp
^{13}C NMR [ppm]: 30.7 199.7 CDCl$_3$
^{1}H NMR [ppm]: 2.2 9.8 CDCl$_3$
Flammability: Flash pt. = -39°C; ign. temp. = 175°C; flam. lim. = 4.0-60%
TLV/TWA: 25 ppm (45 mg/m^3)
Reg. Lists: CERCLA (RQ = 1000 lb.); SARA 313 (0.1%); RCRA U001 (toxic);
    confirmed or suspected carcinogen

3. Acetamide

Syst. Name: Ethanamide

CASRN: 60-35-5   **Beil Ref.**: 4-02-00-00399
Merck No.: 36
Beil RN: 1071207
MF: C_2H_5NO **MP[°C]**: 81   **Den.[g/cm^3]**: 0.9986^{85}
MW: 59.07   **BP[°C]**: 222.0 n_D: 1.4278
Sol.: H_2O 4; EtOH 4
Vap. Press. [kPa]: 0.852^{100}
Therm. Prop.[kJ/mol]: $\Delta_{fus}H$ = 15.71; $\Delta_f H°$(s, 25°C) = -317.0
γ [mN/m]: 38.96^{85}
Elec. Prop.: μ = 3.76 D; ε = 67.6^{91}; IP = 9.65 eV
η [mPa s]: 2.18^{91}
MS: 59(100) 44(79) 43(56) 42(26) 28(14) 41(9) 31(5) 40(4) 60(3) 30(3)
IR [cm^{-1}]: 3330 2940 1670 1640 1470 1410 1150 710
Raman [cm^{-1}]: 3340 3160 3010 2980 2940 1650 1590 1410 1360 1150 880 580
   460 120 90
UV [nm]: 185(8710) H_2O
^{1}H NMR [ppm]: 2.0 D_2O
Reg. Lists: SARA 313 (0.1%)

4. Acetic acid

Syst. Name: Ethanoic acid

CASRN: 64-19-7	**DOT No.**: 2789
Merck No.: 47	**Beil Ref.**: 4-02-00-00094
Beil RN: 506007	

MF: $C_2H_4O_2$	**MP[°C]**: 16.6	**Den.[g/cm^3]**: 1.0492^{20}
MW: 60.05	**BP[°C]**: 117.9	n_D: 1.3720^{20}

Sol.: H_2O 5; EtOH 5; eth
   5; ace 5; bz 5; chl 3;
   os 3; CS_2 3
Crit. Const.: T_c = 319.56°C; P_c = 5.786 MPa; V_c = 171 cm^3
Vap. Press. [kPa]: 2.07^{25}; 7.62^{50}; 22.7^{75}; 57.0^{100}; 125^{125}
Therm. Prop.[kJ/mol]: $\Delta_{fus}H$ = 11.54; $\Delta_{vap}H$ = 23.7; $\Delta_f H°$(l, 25°C) = -484.5
C_p (liq.) [J/mol °C]: 123.6^{25}
γ [mN/m]: 27.10^{25}; $d\gamma/dT$ = 0.0994 mN/m °C
Elec. Prop.: μ = 1.70 D; ϵ = 6.20^{20}; IP = 10.66 eV
η [mPa s]: 1.06^{25}; 0.786^{50}; 0.599^{75}; 0.464^{100}
k [W/m °C]: 0.158^{25}; 0.153^{50}; 0.149^{75}; 0.144^{100}
MS: 43(100) 45(87) 60(57) 15(42) 42(14) 29(13) 14(13) 28(7) 18(6) 16(6)
IR [cm^{-1}]: 3000 2930 2650 1710 1410 1350 1290 940 620
Raman [cm^{-1}]: 2940 2850 2710 1760 1670 1430 1370 1280 1020 940 900 630
   610 4
UV [nm]: 208(32) EtOH
^{13}C NMR [ppm]: 20.6 178.1 CDCl$_3$
^{1}H NMR [ppm]: 2.1 11.4 CDCl$_3$
Flammability: Flash pt. = 39°C; ign. temp. = 463°C; flam. lim. = 4.0-19.9%
TLV/TWA: 10 ppm (25 mg/m^3)
Reg. Lists: CERCLA (RQ = 5000 lb.)

5. Acetic anhydride

Syst. Name: Ethanoic acid, anhydride
Synonyms: Acetyl acetate

CASRN: 108-24-7   **DOT No.**: 1715
Merck No.: 48   **Beil Ref.**: 4-02-00-00386
Beil RN: 385737
MF: $C_4H_6O_3$ **MP[°C]**: -73   **Den.[g/cm^3]**: 1.082^{20}
MW: 102.09   **BP[°C]**: 139.5 n_D: 1.3901^{20}
Sol.: H_2O 4; EtOH 3; eth
   5; bz 3; ctc 2; chl 3
Crit. Const.: $T_c = 333°C$; $P_c = 4.0$ MPa
Vap. Press. [kPa]: 0.680^{25}; 9.98^{75}; 27.4^{100}; 64.4^{125}
Therm. Prop.[kJ/mol]: $\Delta_{vap}H = 38.2$; $\Delta_f H°(l, 25°C) = -624.4$
C_p **(liq.) [J/mol °C]**: 168.2^{30}
γ **[mN/m]**: 31.93^{25}; $d\gamma/dT = 0.1436$ mN/m °C
Elec. Prop.: $\mu = 2.8$ D; $\varepsilon = 22.45^{20}$; IP = 10.00 eV
η **[mPa s]**: 1.24^0; 0.843^{25}; 0.614^{50}; 0.472^{75}; 0.377^{100}
MS: 43(100) 42(35) 45(29) 60(18) 29(9) 41(8) 40(2) 26(2) 87(1) 61(1)
IR [cm^{-1}]: 3030 2940 1820 1750 1430 1370 1220 1120 1040 1000 890 810 780
Raman [cm^{-1}]: 3030 3000 6940 2840 2730 1820 1800 1760 1430 1370 1220
   1120 1050 1020 920 810 780 670 650 600 560 530 430 350 200
UV [nm]: 225(48) iso
^{1}H NMR [ppm]: 2.2 CCl_4
Flammability: Flash pt. = 49°C; ign. temp. = 316°C; flam. lim. = 2.7-10.3%
TLV/TWA: 5 ppm (21 mg/m^3)
Reg. Lists: CERCLA (RQ = 5000 lb.)

6. Acetone

Syst. Name: 2-Propanone
Synonyms: Dimethyl ketone

CASRN: 67-64-1   **DOT No.**: 1090
Merck No.: 58   **Beil Ref.**: 4-01-00-03180
Beil RN: 635680
MF: C_3H_6O **MP[°C]**: -94.8   **Den.[g/cm³]**: 0.7899^{20}
MW: 58.08   **BP[°C]**: 56.0 n_D: 1.3588^{20}
Sol.: H_2O 5; EtOH 5; eth
   5; ace 5; bz 5; chl 5
Crit. Const.: $T_c = 235.0°C$; $P_c = 4.700$ MPa; $V_c = 209$ cm³
Vap. Press. [kPa]: 2.19^{-25}; 9.35^0; 30.8^{25}; 82.0^{50}; 186^{75}; 372^{100}; 676^{125}; 1136^{150}
Therm. Prop.[kJ/mol]: $\Delta_{fus}H = 5.69$; $\Delta_{vap}H = 29.1$; $\Delta_f H°(l, 25°C) = -248.1$
C_p (liq.) [J/mol °C]: 126.6^{25}
γ [mN/m]: 23.46^{25}; $d\gamma/dT = 0.1120$ mN/m °C
Elec. Prop.: $\mu = 2.88$ D; $\varepsilon = 21.01^{20}$; IP = 9.71 eV
η [mPa s]: 0.540^{-25}; 0.395^0; 0.306^{25}; 0.247^{50}
k [W/m °C]: 0.169^0; 0.161^{25}
MS: 43(100) 15(34) 58(23) 27(9) 14(9) 42(8) 26(7) 29(5) 28(5) 39(4)
IR [cm⁻¹]: 3000 1715 1420 1360 1220 1090 900 790 530
Raman [cm⁻¹]: 3010 2930 2850 2700 1740 1710 1430 1360 1220 1060 900 790
   520 490 390
UV [nm]: 270 MeOH
¹³C NMR [ppm]: 30.6 206.0 $CDCl_3$
¹H NMR [ppm]: 2.1 $CDCl_3$
Flammability: Flash pt. = -20°C; ign. temp. = 465°C; flam. lim. = 2.5-12.8%
TLV/TWA: 750 ppm (1780 mg/m³)
Reg. Lists: CERCLA (RQ = 5000 lb.); SARA 313 (1.0%); RCRA U002 (toxic)

7. Acetonitrile

Syst. Name: Ethanenitrile
Synonyms: Methyl cyanide

————≡≡N

CASRN: 75-05-8   **DOT No.**: 1648
Merck No.: 62   **Beil Ref.**: 4-02-00-00419
Beil RN: 741857
MF: C_2H_3N **MP[°C]**: -43.8   **Den.[g/cm^3]**: 0.7857^{20}
MW: 41.05   **BP[°C]**: 81.6 n_D: 1.3442^{30}
Sol.: H_2O 5; EtOH 5; eth
 5; ace 5; bz 5; ctc 5
Crit. Const.: $T_c = 272.4$°C; $P_c = 4.85$ MPa; $V_c = 173$ cm^3
Vap. Press. [kPa]: 11.9^{25}; 33.9^{50}; 82.1^{75}
Therm. Prop.[kJ/mol]: $\Delta_{fus}H = 8.17$; $\Delta_{vap}H = 29.8$; $\Delta_fH°(l, 25°C) = 31.4$
C_p **(liq.) [J/mol °C]**: 91.4^{25}
γ **[mN/m]**: 28.66^{25}; $d\gamma/dT = 0.1263$ mN/m °C
Elec. Prop.: $\mu = 3.924$ D; $\varepsilon = 36.64^{20}$; IP = 12.19 eV
η **[mPa s]**: 0.400^0; 0.369^{25}; 0.284^{50}; 0.234^{75}
k **[W/m °C]**: 0.208^{-25}; 0.198^0; 0.188^{25}; 0.178^{50}; 0.168^{75}
MS: 41(100) 40(46) 39(13) 14(9) 38(6) 28(4) 26(4) 25(3) 42(2) 27(2)
IR [cm^{-1}]: 2940 2270 2220 1430 1410 1370 1030 920 790 750
Raman [cm^{-1}]: 3010 2950 2890 2850 2730 2290 2260 1450 1380 920 750 380
UV [nm]: 338(126) C_2Cl_4
^{13}C NMR [ppm]: 1.3 117.7 diox
^{1}H NMR [ppm]: 1.9 CCl_4
Flammability: Flash pt. = 6°C; ign. temp. = 524°C; flam. lim. = 3.0-16.0%
TLV/TWA: 40 ppm (67 mg/m^3)
Reg. Lists: CERCLA (RQ = 5000 lb.); SARA 313 (1.0%); RCRA U003 (toxic)

8. Acetophenone

Syst. Name: Ethanone, 1-phenyl-
Synonyms: Methyl phenyl ketone; 1-Phenylethanone

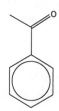

CASRN: 98-86-2   **Beil Ref.**: 4-07-00-00619
Merck No.: 65
Beil RN: 605842
MF: C_8H_8O **MP[°C]**: 20   **Den.[g/cm^3]**: 1.0281^{20}
MW: 120.15   **BP[°C]**: 202 n_D: 1.5372^{20}
Sol.: H_2O 1; EtOH 3; eth
  3; ace 3; bz 3; chl 3;
  con sulf 3
Crit. Const.: T_c = 436.4°C; V_c = 386 cm^3
Vap. Press. [kPa]: 0.049^{25}; 0.172^{100}; 0.640^{125}; $2.03^{150.}$
Therm. Prop.[kJ/mol]: $\Delta_{vap}H$ = 38.8; $\Delta_f H°$(l, 25°C) = -142.5
C_p **(liq.) [J/mol °C]**: 204.6^{25}
γ **[mN/m]**: 39.04^{25}; dγ/dT = 0.1154 mN/m °C
Elec. Prop.: μ = 3.02 D; ε = 17.44^{25}; IP = 9.29 eV
η **[mPa s]**: 1.68^{25}; 0.634^{100}
MS: 105(100) 77(83) 120(33) 51(31) 43(16) 50(13) 78(9) 106(7) 39(6) 74(5)
IR [cm^{-1}]: 3100 1700 1600 1580 1450 1360 1300 1270 1180 1080 1030 950
  920 750 690
Raman [cm^{-1}]: 3080 2940 1680 1600 1490 1450 1440 1310 1270 1180 1160
  1080 1030 1000 950 850 760 730 620 590 470 410 370 160
UV [nm]: 318(60) 279(1050) 242(12600) EtOH
^{13}C NMR [ppm]: 26.3 128.2 128.4 132.9 137.1 197.6 CDCl$_3$
^{1}H NMR [ppm]: 2.6 7.9 CDCl$_3$
Flammability: Flash pt. = 77°C; ign. temp. = 570°C
TLV/TWA: 10 ppm (49 mg/m^3)
Reg. Lists: CERCLA (RQ = 5000 lb.); RCRA U004 (toxic)

9. Acetylacetone

Syst. Name: 2,4-Pentanedione
Synonyms: Diacetylmethane

CASRN: 123-54-6
Merck No.: 75
Beil RN: 741937
DOT No.: 2310
Beil Ref.: 4-01-00-03662

MF: $C_5H_8O_2$
MW: 100.12
MP[°C]: -23
BP[°C]: 138
Den.[g/cm^3]: 0.9721^{25}
n_D: 1.4494^{20}

Sol.: H_2O 4; EtOH 5; eth 5; ace 5; chl 5

Vap. Press. [kPa]: 1.02^{25}; 4.35^{50}; 13.6^{75}; 33.9^{100}; 72.1^{125}

Therm. Prop.[kJ/mol]: $\Delta_{vap}H$ = 34.3; $\Delta_f H°$(l, 25°C) = -423.8

C_p **(liq.) [J/mol °C]**: 208.2^{25}

γ **[mN/m]**: 32.13^{11}

Elec. Prop.: μ = (2.8 D); ε = 26.52^{30}; IP = 8.85 eV

η **[mPa s]**: 0.483^{65}

MS: 43(100) 85(31) 100(20) 27(12) 42(10) 29(10) 41(7) 39(7) 31(5) 26(5)

IR [cm^{-1}]: 2920 1730 1710 1620 1420 1360 1300 1250 1160 1000 950 920 780 630 530 510 410 380

Raman [cm^{-1}]: 3100 2940 2850 2730 1730 1600 1430 1370 1300 1250 1170 1040 1000 930 790 640 550 510 410 370 230

UV [nm]: 272(6800) MeOH

^{13}C NMR [ppm]: 24.3 30.2 58.2 100.3 191.4 201.9 diox

^{1}H NMR [ppm]: 2.0 2.2 3.5 5.4 14.7 CCl_4

Flammability: Flash pt. = 34°C; ign. temp. = 340°C

10. Acrolein

Syst. Name: 2-Propenal
Synonyms: Acrylaldehyde

CASRN: 107-02-8   **DOT No.**: 1092
Merck No.: 122   **Beil Ref.**: 4-01-00-03435
Beil RN: 741856
MF: C_3H_4O **MP[°C]**: -87.7   **Den.[g/cm³]**: 0.840^{20}
MW: 56.06   **BP[°C]**: 52.6 n_D: 1.4017^{20}
Sol.: H_2O 4; EtOH 3; eth
    3; ace 3; chl 2
Vap. Press. [kPa]: 2.86^{-25}; 11.7^{0}; 36.2^{25}; 91.3^{50}; 198^{75}
Therm. Prop.[kJ/mol]: $\Delta_{vap}H = 28.3$
Elec. Prop.: $\mu = 3.12$ D; IP = 10.10 eV
MS: 27(100) 56(74) 28(65) 26(54) 55(52) 29(37) 25(8) 53(5) 38(5) 57(4)
IR [cm⁻¹]: 3030 2780 2700 2630 1700 1410 1370 1270 1240 1150 1080 980
    920 740
UV [nm]: 238(9333) sulf
¹H NMR [ppm]: 6.4 9.5 $CDCl_3$
Flammability: Flash pt. = -26°C; ign. temp. = 220°C; flam. lim. = 2.8-31%
TLV/TWA: 0.1 ppm (0.23 mg/m³)
Reg. Lists: CERCLA (RQ = 1 lb.); SARA 313 (1.0%); RCRA P003 (accutely
    hazardous)

11. Acrylic acid

Syst. Name: 2-Propenoic acid

CASRN: 79-10-7   **DOT No.:** 2218
Merck No.: 124   **Beil Ref.:** 4-02-00-01455
Beil RN: 635743
MF: $C_3H_4O_2$ **MP[°C]:** 12.3   **Den.[g/cm³]:** 1.0511^{20}
MW: 72.06   **BP[°C]:** 141 n_D: 1.4224^{20}
Sol.: H_2O 5; EtOH 5; eth
   5; ace 3; bz 3; ctc 3
Vap. Press. [kPa]: 2.45^{50}; 8.75^{75}; 25.2^{100}; 61.3^{125}; 131^{150}
Therm. Prop.[kJ/mol]: $\Delta_{fus}H = 11.16$; $\Delta_f H°(l, 25°C) = -383.8$
C_p **(liq.) [J/mol °C]:** 145.7^{25}
γ **[mN/m]:** 27.6^{25}
Elec. Prop.: IP = 10.60 eV
MS: 27(100) 72(76) 55(60) 26(54) 45(35) 28(22) 44(13) 25(13) 43(12) 29(10)
IR [cm⁻¹]: 3000 2660 1710 1640 1620 1430 1300 1240 1040 980 970 930 820
   650
Raman [cm⁻¹]: 3130 3050 3010 2950 1730 1660 1640 1440 1400 1290 1240
   1070 1050 980 930 870 830 650 630 550 520 350 310
¹H NMR [ppm]: 5.9 6.2 6.5 12.4 CCl_4
Flammability: Flash pt. = 50°C; ign. temp. = 438°C; flam. lim. = 2.4-8.0%
TLV/TWA: 2 ppm (5.9 mg/m³)
Reg. Lists: CERCLA (RQ = 5000 lb.); SARA 313 (1.0%); RCRA U008 (toxic)

12. Acrylonitrile

Syst. Name: 2-Propenenitrile
Synonyms: Vinyl cyanide; Cyanoethylene

CASRN: 107-13-1   **DOT No.:** 1093
Merck No.: 125   **Beil Ref.:** 4-02-00-01473
Beil RN: 605310
MF: C_3H_3N **MP[°C]:** -83.5   **Den.[g/cm^3]:** 0.8060^{20}
MW: 53.06   **BP[°C]:** 77.3 n_D: 1.3911^{20}
Sol.: ace 4; bz 4; eth 4; EtOH 4
Vap. Press. [kPa]: 3.97^0; 14.1^{25}; 39.7^{50}; 93.9^{75}; 194^{100}
Therm. Prop.[kJ/mol]: $\Delta_{fus}H = 6.23$; $\Delta_{vap}H = 32.6$; $\Delta_fH°(l, 25°C) = 147.1$
C_p **(liq.) [J/mol °C]:** 108.8^{25}
γ **[mN/m]:** 27.3^{24}
Elec. Prop.: $\mu = 3.87$ D; $\varepsilon = 33.0^{20}$; IP = 10.91 eV
η **[mPa s]:** 0.362^{20}
MS: 53(100) 26(85) 52(79) 51(34) 27(13) 50(8) 25(7) 38(5) 54(3) 37(3)
IR [cm^{-1}]: 3070 3040 2240 1650 1610 1410 1090 960 870 690
Raman [cm^{-1}]: 3130 3040 3000 2290 2230 1610 1410 1290 1100 970 870 800 690 570 240
UV [nm]: 206(6166) EtOH
^{13}C NMR [ppm]: 108.1 117.5 137.5 diox
^{1}H NMR [ppm]: 6.3 CDCl$_3$
Flammability: Flash pt. = 0°C; ign. temp. = 481°C; flam. lim. = 3.0-17.0%
TLV/TWA: 2 ppm (4.3 mg/m^3)
Reg. Lists: CERCLA (RQ = 100 lb.); SARA 313 (0.1%); RCRA U009 (toxic); confirmed or suspected carcinogen

15. Allyl alcohol

Syst. Name: 2-Propen-1-ol
Synonyms: Vinylcarbinol

CASRN: 107-18-6   **DOT No.**: 1098
Merck No.: 284   **Beil Ref.**: 4-01-00-02079
Beil RN: 605307
MF: C_3H_6O **MP[°C]**: -129   **Den.[g/cm^3]**: 0.8540^{20}
MW: 58.08   **BP[°C]**: 97.0 n_D: 1.4135^{20}
Sol.: H_2O 5; EtOH 5; eth
  5; chl 3
Crit. Const.: $T_c = 272.0$°C
Vap. Press. [kPa]: 3.14^{25}; 13.3^{50}; 42.8^{75}; 113^{100}; 254^{125}
Therm. Prop.[kJ/mol]: $\Delta_{vap}H = 40.0$; $\Delta_f H°(l, 25°C) = -171.8$
C_p **(liq.) [J/mol °C]**: 138.9^{25}
γ **[mN/m]**: 25.28^{25}; $d\gamma/dT = 0.0902$ mN/m °C
Elec. Prop.: $\mu = 1.60$ D; $\varepsilon = 19.7^{20}$; IP = 9.67 eV
η **[mPa s]**: 1.22^{25}; 0.759^{50}; 0.505^{75}
MS: 57(100) 31(34) 29(32) 28(31) 58(25) 39(22) 27(20) 30(16) 32(14) 26(11)
IR [cm^{-1}]: 3320 2850 1650 1420 1230 1110 1030 990 920
Raman [cm^{-1}]: 3280 3080 3010 2980 2920 2860 2710 1640 1450 1420 1400
  1340 1290 1230 1110 1020 990 940 910 880 640 600 550 440 350 280
UV [nm]: 273(4677) sulf
^{13}C NMR [ppm]: 63.4 114.9 137.5 CDCl$_3$
^{1}H NMR [ppm]: 3.6 4.1 5.1 5.3 6.0 CDCl$_3$
Flammability: Flash pt. = 21°C; ign. temp. = 378°C; flam. lim. = 2.5-18.0%
TLV/TWA: 2 ppm (4.8 mg/m^3)
Reg. Lists: CERCLA (RQ = 100 lb.); SARA 313 (1.0%); RCRA P005 (accutely
  hazardous)

16. Allylamine

Syst. Name: 2-Propen-1-amine
Synonyms: 2-Propenylamine

CASRN: 107-11-9
Merck No.: 285
Beil RN: 635703
MF: C_3H_7N
MW: 57.10
Sol.: H_2O 5; EtOH 5; eth
  5; chl 3

DOT No.: 2334
Beil Ref.: 4-04-00-01057

MP[°C]: -88.2
BP[°C]: 53.3

Den.[g/cm^3]: 0.758^{20}
n_D: 1.4205^{20}

Vap. Press. [kPa]: 9.77^0; 33.1^{25}
Therm. Prop.[kJ/mol]: $\Delta_f H°(l, 25°C) = -10.0$
γ [mN/m]: 24.2^{25}
Elec. Prop.: $\mu = 1.2$ D; IP = 8.76 eV
η [mPa s]: 0.374^{25}
MS: 30(100) 56(80) 28(76) 57(33) 39(21) 29(20) 27(18) 26(13) 41(8) 18(8)
IR [cm^{-1}]: 3330 3230 3130 3030 2940 2860 1640 1610 1450 1430 1370 1330
  1280 1140 1110 1050 1000 920 770 650
^{1}H NMR [ppm]: 1.5 3.3 5.0 5.1 5.9 $CDCl_3$
Flammability: Flash pt. = -29°C; ign. temp. = 374°C; flam. lim. = 2.2-22%

17. 2-Aminoisobutanol

Syst. Name: 1-Propanol, 2-amino-2-methyl-
Synonyms: 2-Amino-2-methyl-1-propanol; AMP

CASRN: 124-68-5   **Beil Ref.**: 4-04-00-01740
Merck No.: 461
Beil RN: 505979
MF: $C_4H_{11}NO$ **MP[°C]**: 25.5   **Den.[g/cm³]**: 0.934^{20}
MW: 89.14   **BP[°C]**: 165.5 n_D: 1.449^{20}
Sol.: H_2O 5; ctc 3
Therm. Prop.[kJ/mol]: $\ni 45_{vap}H = 50.6$
η **[mPa s]**: 102^{30}; 25^{50}
MS: 58(100) 41(18) 18(17) 42(13) 28(11) 56(10) 30(10) 29(8) 43(6) 59(5)
IR [cm⁻¹]: 3230 2940 1590 1470 1390 1350 1240 1180 1050 950 940 910 760
Raman [cm⁻¹]: 360 3290 3180 2950 2920 2860 2700 1600 1450 1440 1370
    1310 1270 1230 1180 1080 1050 970 940 900 890 760 550 530 450 410 360
    350 250
¹³C NMR [ppm]: 26.7 50.6 71.1 $CDCl_3$
¹H NMR [ppm]: 1.1 2.8 3.2 CCl_4
Flammability: Flash pt. = 67°C

18. Aniline

Syst. Name: Benzenamine
Synonyms: Phenylamine

CASRN: 62-53-3
Merck No.: 687
Beil RN: 605631

DOT No.: 1547
Beil Ref.: 4-12-00-00223

MF: C_6H_7N
MW: 93.13
MP[°C]: -6.0
BP[°C]: 184.1
Den.[g/cm^3]: 1.0217^{20}
n_D: 1.5863^{20}

Sol.: H_2O 3; EtOH 5; eth 5; ace 5; bz 5; ctc 3; lig 3

Crit. Const.: $T_c = 426°C$; $P_c = 4.89$ MPa; $V_c = 287$ cm^3

Vap. Press. [kPa]: 0.090^{25}; 6.10^{100}; 16.4^{125}; 38.2^{150}

Therm. Prop.[kJ/mol]: $\Delta_{fus}H = 10.56$; $\Delta_{vap}H = 42.4$; $\Delta_f H°(l, 25°C) = 31.3$

C_p **(liq.) [J/mol °C]:** 191.9^{25}

γ **[mN/m]:** 42.12^{25}; $d\gamma/dT = 0.1085$ mN/m °C

Elec. Prop.: $\mu = 1.13$ D; $\varepsilon = 7.06^{20}$; IP = 7.72 eV

η **[mPa s]:** 3.85^{25}; 2.03^{50}; 1.25^{75}; 0.850^{100}

MS: 93(100) 66(32) 65(16) 39(13) 92(10) 94(7) 41(5) 40(5) 67(4) 64(3)

IR [cm^{-1}]: 3450 3330 1610 1490 1470 1280 1180 1120 1060 1030 1000 880 750 690

UV [nm]: 287 235 cyhex

^{13}C NMR [ppm]: 116.3 119.2 130.0 147.9

^{1}H NMR [ppm]: 3.3 6.4 6.6 7.0 CCl$_4$

Flammability: Flash pt. = 70°C; ign. temp. = 615°C; flam. lim. = 1.3-11%

TLV/TWA: 2 ppm (7.6 mg/m^3)

Reg. Lists: CERCLA (RQ = 5000 lb.); SARA 313 (1.0%); RCRA U012 (toxic); confirmed or suspected carcinogen

19. Anisole

Syst. Name: Benzene, methoxy-
Synonyms: Methyl phenyl ether

CASRN: 100-66-3   **DOT No.**: 2222
Merck No.: 699   **Beil Ref.**: 4-06-00-00548
Beil RN: 506892
MF: C_7H_8O **MP[°C]**: -37.5   **Den.[g/cm^3]**: 0.9940^{20}
MW: 108.14   **BP[°C]**: 153.7 n_D: 1.5174^{20}
Sol.: H_2O 1; EtOH 3; eth
  3; ace 4; bz 4; chl 3
Crit. Const.: T_c = 372.5°C; P_c = 4.25 MPa
Vap. Press. [kPa]: 0.472^{25}; 18.5^{100}; 43.9^{125}; 91.9^{150}
Therm. Prop.[kJ/mol]: $\Delta_{vap}H$ = 39.0; $\Delta_f H°$(l, 25°C) = -114.8
C_p **(liq.) [J/mol °C]**: 199.0^{25}
γ **[mN/m]**: 35.10^{25}; $d\gamma/dT$ = 0.1204 mN/m °C
Elec. Prop.: μ = 1.38 D; ϵ = 4.30^{21}; IP = 8.21 eV
η **[mPa s]**: 1.06^{25}; 0.747^{50}; 0.554^{75}; 0.427^{100}
k **[W/m °C]**: 0.170^{-25}; 0.163^{0}; 0.156^{25}; 0.150^{50}; 0.143^{75}; 0.136^{100}
MS: 108(100) 65(76) 78(60) 39(44) 51(21) 77(20) 93(16) 79(14) 50(13) 63(12)
IR [cm^{-1}]: 2941 2857 1587 1471 1299 1250 1042 877 763 690
Raman [cm^{-1}]: 3060 3040 3000 2940 2830 1590 1580 1490 1450 1430 1380
  1330 1300 1290 1240 1180 1170 1150 1070 1030 1020 990 870 810 780 750
  610 550 510 440 260 200
UV [nm]: 271 MeOH
^{13}**C NMR [ppm]**: 54.7 114.1 120.7 129.5 160.2 diox
^{1}**H NMR [ppm]**: 3.8 7.1 CDCl$_3$
Flammability: Flash pt. = 52°C; ign. temp. = 475°C

20. Benzal chloride

Syst. Name: Benzene, (dichloromethyl)-
Synonyms: α, α-Dichlorotoluene; Benzylidene
chloride

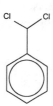

CASRN: 98-87-3   **DOT No.**: 1886
Merck No.: 1064   **Beil Ref.**: 4-05-00-00817
Beil RN: 1099407
MF: $C_7H_6Cl_2$ **MP[°C]**: -17   **Den.[g/cm^3]**: 1.26^{25}
MW: 161.03   **BP[°C]**: 205 n_D: 1.5502^{20}
Sol.: eth 4; EtOH 4
Vap. Press. [kPa]: 0.063^{25}; 0.301^{50}; 3.33^{100}; 8.54^{125}; 19.3^{150}; 39.3^{175}
Elec. Prop.: $\mu = (2.1$ D$)$; $\varepsilon = 6.9^{20}$
MS: 125(100) 127(32) 160(14) 89(13) 162(9) 63(9) 126(8) 62(7) 105(5) 39(5)
IR [cm^{-1}]: 3130 3030 1490 1470 1330 1250 1220 1190 1080 1030 1010 880
    840 810 700
UV [nm]: 288 272 265 259 247 240 220 MeOH
^{1}H NMR [ppm]: 6.6 7.4 CCl_4
Reg. Lists: CERCLA (RQ = 5000 lb.); SARA 313 (1.0%); RCRA U017 (toxic)

21. Benzaldehyde

Syst. Name: Benzaldehyde
Synonyms: Benzenecarboxaldehyde

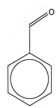

CASRN: 100-52-7   **DOT No.**: 1989
Merck No.: 1065   **Beil Ref.**: 4-07-00-00505
Beil RN: 471223
MF: C_7H_6O **MP[°C]**: -26   **Den.[g/cm^3]**: 1.0415^{10}
MW: 106.12   **BP[°C]**: 179.0 n_D: 1.5463^{20}
Sol.: H_2O 2; EtOH 5; eth
   5; ace 4; bz 4; chl 3;
   liq NH_3 3; lig 4
Crit. Const.: T_c = 422°C; P_c = 4.65 MPa
Vap. Press. [kPa]: 0.169^{25}; 0.776^{50}; 2.83^{75}; 8.37^{100}; 21.0^{125}; 46.3^{150}
Therm. Prop.[kJ/mol]: $\Delta_{fus}H$ = 9.32; $\Delta_{vap}H$ = 42.5; $\Delta_f H°$(l, 25°C) = -87.0
C_p **(liq.) [J/mol °C]**: 172.0^{25}
γ **[mN/m]**: 38.00^{25}; $d\gamma/dT$ = 0.1090 mN/m °C
Elec. Prop.: μ = (3.0 D); ϵ = 17.85^{20}; IP = 9.49 eV
η **[mPa s]**: 1.32^{25}
k **[W/m °C]**: 0.151^{25}; 0.141^{50}; 0.131^{75}; 0.121^{100}
MS: 51(100) 77(81) 50(55) 106(44) 105(43) 52(26) 78(16) 39(13) 27(10) 74(8)
IR [cm^{-1}]: 3030 2778 2703 1695 1667 1587 1449 1389 1299 1205 1163 1075
   1020 833 746 690
UV [nm]: 290(1000) 283(1259) 241(15849) hx
^{13}C NMR [ppm]: 128.9 129.5 134.2 136.4 192.0 $CDCl_3$
^{1}H NMR [ppm]: 7.7 10.0 $CDCl_3$
Flammability: Flash pt. = 63°C; ign. temp. = 192°C

22. Benzene

Syst. Name: Benzene

CASRN: 71-43-2   **DOT No.:** 1114
Merck No.: 1074   **Beil Ref.:** 4-05-00-00583
Beil RN: 969212
MF: C_6H_6 **MP[°C]:** 5.5   **Den.[g/cm^3]:** 0.8765^{20}
MW: 78.11   **BP[°C]:** 80.0 n_D: 1.5011^{20}
Sol.: H_2O 2; EtOH 5; eth
    5; ace 5; ctc 3; chl 5;
    HOAc 5
Crit. Const.: $T_c = 288.90°C$; $P_c = 4.895$ MPa; $V_c = 255$ cm^3
Vap. Press. [kPa]: 0.485^{-25}; 3.29^0; 12.7^{25}; 36.2^{50}; 86.4^{75}; 180^{100}; 338^{125}; 583^{150}
Therm. Prop.[kJ/mol]: $\Delta_{fus}H = 9.95$; $\Delta_{vap}H = 30.7$; $\Delta_fH°(l, 25°C) = 49.0$
C_p **(liq.) [J/mol °C]:** 136.1^{25}
γ **[mN/m]:** 28.22^{25}; $d\gamma/dT = 0.1291$ mN/m °C
Elec. Prop.: $\mu = 0$; $\varepsilon = 2.28^{20}$; IP = 9.25 eV
η **[mPa s]:** 0.604^{25}; 0.436^{50}; 0.335^{75}
k **[W/m °C]:** 0.1411^{25}; 0.1329^{50}; 0.1247^{75}
MS: 78(100) 77(20) 52(19) 51(17) 50(15) 39(12) 79(6) 76(5) 74(4) 38(4)
IR [cm^{-1}]: 3080 3030 1965 1815 1480 1035 670
Raman [cm^{-1}]: 3060 3040 2950 1610 1580 1180 990 610
UV [nm]: 268 260 254 248 243 cyhex
^{13}C NMR [ppm]: 128.7
^{1}H NMR [ppm]: 7.2 CCl_4
Flammability: Flash pt. = -11°C; ign. temp. = 498°C; flam. lim. = 1.2-7.8%
TLV/TWA: 10 ppm (32 mg/m^3)
Reg. Lists: CERCLA (RQ = 10 lb.); SARA 313 (0.1%); RCRA U019 (toxic);
    confirmed or suspected carcinogen

23. Benzenethiol

Syst. Name: Benzenethiol
Synonyms: Phenyl mercaptan; Thiophenol

CASRN: 108-98-5
Merck No.: 9285
Beil RN: 506523
MF: C_6H_6S **MP[°C]**: -14.9 **Den.[g/cm^3]**: 1.0775^{20}
MW: 110.18 **BP[°C]**: 169.1 n_D: 1.5893^{20}
Sol.: H_2O 1; EtOH 3; eth
   3; bz 3; ctc 2
DOT No.: 2337
Beil Ref.: 4-06-00-01463

Vap. Press. [kPa]: 2.00^{60}; 5.15^{80}; 11.7^{100}; 28.4^{125}; 60.7^{150}
Therm. Prop.[kJ/mol]: $\Delta_{fus}H = 11.48$; $\Delta_{vap}H = 39.9$; $\Delta_fH°(l, 25°C) = 63.7$
C_p (liq.) [J/mol °C]: 173.2^{25}
γ [mN/m]: 38.7^{25}
Elec. Prop.: $\mu = (1.2\ D)$; $\varepsilon = 4.26^{30}$; IP = 8.30 eV
η [mPa s]: 1.14^{25}
MS: 110(100) 66(28) 109(23) 39(15) 77(14) 51(14) 84(13) 69(11) 50(11)
   45(11)
IR [cm^{-1}]: 3030 2560 1610 1470 1450 1300 1180 1110 1090 1020 910 740 690
Raman [cm^{-1}]: 3160 3070 2960 2580 1590 1480 1440 1380 1340 1280 1180
   1160 1120 1100 1030 1010 990 920 840 740 700 620 470 420 280 190
UV [nm]: 240(7244) EtOH
^{13}C NMR [ppm]: 125.4 128.9 129.2 130.7 CDCl$_3$
^{1}H NMR [ppm]: 3.2 6.9 CCl$_4$
TLV/TWA: 0.5 ppm (2.3 mg/m^3)
Reg. Lists: CERCLA (RQ = 100 lb.); RCRA P014 (accutely hazardous)

24. Benzonitrile

Syst. Name: Benzonitrile
Synonyms: Phenyl cyanide

CASRN: 100-47-0
Merck No.: 1107
Beil RN: 506893
MF: C_7H_5N
MW: 103.12
Sol.: H_2O 2; EtOH 5; eth
  5; ace 4; bz 4; ctc 3

DOT No.: 2224
Beil Ref.: 4-09-00-00892

MP[°C]: -12.7
BP[°C]: 191.1

Den.[g/cm³]: 1.0093^{15}
n_D: 1.5289^{20}

Crit. Const.: $T_c = 426.3°C$; $P_c = 4.21$ MPa
Vap. Press. [kPa]: 0.110^{25}; 0.496^{50}; 1.80^{75}; 5.41^{100}; 14.0^{125}; 32.2^{150}
Therm. Prop.[kJ/mol]: $\Delta_{fus}H = 10.88$; $\Delta_{vap}H = 45.9$; $\Delta_f H°(l, 25°C) = 163.2$
C_p (liq.) [J/mol °C]: 165.2^{25}
γ [mN/m]: 38.79^{25}; $d\gamma/dT = 0.1159$ mN/m °C
Elec. Prop.: $\mu = 4.18$ D; $\varepsilon = 25.9^{20}$; IP = 9.62 eV
η [mPa s]: 1.27^{25}; 0.883^{50}; 0.662^{75}; 0.524^{100}
MS: 103(100) 76(34) 50(13) 104(9) 75(7) 51(7) 77(5) 52(4) 39(4) 74(3)
IR [cm⁻¹]: 3030 2240 1600 1580 1500 1450 1290 1180 1070 1030 920 760 690
Raman [cm⁻¹]: 3200 3160 3080 3000 2980 2220 1600 1450 1190 1180 1030
  1000 760 740 630 550 460 380 170 150
UV [nm]: 277(932) 263(804) 230(9100) 222(10600) MeOH
¹³C NMR [ppm]: 112.3 118.7 129.1 132.0 132.7 CDCl₃
¹H NMR [ppm]: 7.5 CCl₄
Reg. Lists: CERCLA (RQ = 5000 lb.)

25. Benzyl acetate

Syst. Name: Ethanoic acid, phenylmethyl ester
Synonyms: (Acetoxymethyl)benzene

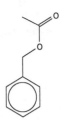

CASRN: 140-11-4   **Beil Ref.**: 4-06-00-02262
Merck No.: 1137
Beil RN: 1908121
MF: $C_9H_{10}O_2$ **MP[°C]**: -51.3   **Den.[g/cm³]**: 1.0550^{20}
MW: 150.18   **BP[°C]**: 213 n_D: 1.5232^{20}
Sol.: H_2O 2; EtOH 5; eth
    3; ace 3; chl 3
Vap. Press. [kPa]: 0.124^{50}; 0.537^{75}; 1.89^{100}; 5.65^{125}; 14.8^{150}
Therm. Prop.[kJ/mol]: $\Delta_{vap}H = 49.4$
C_p **(liq.) [J/mol °C]**: 148.5^{25}
Elec. Prop.: $\mu = (1.2\ D)$; $\varepsilon = 5.34^{30}$
η **[mPa s]**: 1.40^{45}
MS: 108(100) 43(76) 91(60) 90(48) 79(27) 107(17) 77(17) 65(15) 51(15)
    89(14)
IR [cm⁻¹]: 3030 1720 1450 1370 1350 1160 1020 960 900 830 750 700
Raman [cm⁻¹]: 3210 3170 3070 3010 2940 2900 2850 2710 1740 1610 1590
    1500 1460 1380 1360 1220 1180 1160 1080 1040 1010 940 910 840 830 760
    700 650 630 580 500 440 310 280 230 190
UV [nm]: 267(114) 263(174) 261 257(214) 252(174) MeOH
¹³C NMR [ppm]: 20.7 66.1 128.1 128.4 136.2 170.5 CDCl₃
¹H NMR [ppm]: 2.1 5.1 7.3 CDCl₃
Flammability: Flash pt. = 90°C; ign. temp. = 460°C

26. Benzyl alcohol

Syst. Name: Benzenemethanol
Synonyms: Benzenecarbinol; Phenylmethanol

CASRN: 100-51-6   **Beil Ref.**: 4-06-00-02222
Merck No.: 1138
Beil RN: 878307
MF: C_7H_8O **MP[°C]**: -15.2   **Den.[g/cm^3]**: 1.0419^{24}
MW: 108.14   **BP[°C]**: 205.3 n_D: 1.5396^{20}
Sol.: H_2O 3; EtOH 3; eth
    3; ace 3; bz 3; chl 3;
    MeOH 3
Crit. Const.: T_c = 442°C; P_c = 4.3 MPa
Vap. Press. [kPa]: 0.015^{25}; 2.27^{100}; 18.1^{150}
Therm. Prop.[kJ/mol]: $\Delta_{fus}H$ = 8.97; $\Delta_{vap}H$ = 50.5; $\Delta_f H°$(l, 25°C) = -160.7
C_p (liq.) [J/mol °C]: 217.9^{25}
γ [mN/m]: 34.10^{30}; $d\gamma/dT$ = 0.1381 mN/m °C
Elec. Prop.: μ = 1.71 D; ϵ = 11.92^{30}; IP = 8.50 eV
η [mPa s]: 5.47^{25}; 2.76^{50}; 1.62^{75}; 1.06^{100}
MS: 79(100) 108(83) 77(74) 107(66) 51(35) 105(23) 106(22) 50(17) 78(16)
    39(16)
IR [cm^{-1}]: 3570 3030 2860 1490 1450 1370 1210 1180 1080 1030 1010 900
    790 730 690
Raman [cm^{-1}]: 3060 2980 2930 2870 1600 1580 1490 1460 1390 1370 1280
    1210 1180 1150 1080 1030 1000 900 840 810 800 750 700 620 590 480 410
UV [nm]: 267(112) 263(165) 257(229) 252(233) 247 MeOH
^{13}**C NMR [ppm]**: 64.5 126.8 127.2 128.2 140.8 CDCl$_3$
^{1}**H NMR [ppm]**: 2.4 4.6 7.3 CDCl$_3$
Flammability: Flash pt. = 93°C; ign. temp. = 436°C

27. Benzyl benzoate

Syst. Name: Benzoic acid, phenylmethyl ester
Synonyms: Benzyl benzenecarboxylate

CASRN: 120-51-4   **Beil Ref.:** 4-09-00-00307
Merck No.: 1141
Beil RN: 2049280
MF: $C_{14}H_{12}O_2$ **MP[°C]:** 21   **Den.[g/cm^3]:** 1.1121^{25}
MW: 212.25   **BP[°C]:** 323.5 n_D: 1.5680^{20}
Sol.: H_2O 1; EtOH 3; eth
   3; ace 3; bz 3; chl 3;
   MeOH 3; peth 3
Therm. Prop.[kJ/mol]: $\Delta_{vap}H = 53.6$
γ **[mN/m]:** 42.82^{25}; $d\gamma/dT = 0.1102$ mN/m °C
Elec. Prop.: $\mu = (2.1$ D); $\varepsilon = 5.26^{30}$
η **[mPa s]:** 8.29^{25}; 5.24^{40}
MS: 105(100) 91(42) 77(28) 51(12) 212(10) 106(9) 65(9) 90(7) 107(6) 39(5)
IR [cm^{-1}]: 3030 1720 1590 1490 1450 1370 1280 1160 1110 1080 1030 700
Raman [cm^{-1}]: 3200 3160 3070 3000 2970 2950 2890 1710 1600 1580 1480
   1450 1370 1310 1260 1210 1170 1150 1100 1030 1000 980 950 900 880 840
   820 810 750 670 610 510 460 390 300 230 190 140
UV [nm]: 280(771) 272(968) 267(954) 263(952) 257(952) 230(1470) MeOH
^{1}H NMR [ppm]: 5.3 8.1 CDCl$_3$
Flammability: Flash pt. = 148°C; ign. temp. = 480°C

28. Benzyl chloride

Syst. Name: Benzene, (chloromethyl)-
Synonyms: (Chloromethyl)benzene; Tolyl chloride

CASRN: 100-44-7   **DOT No.**: 1738
Merck No.: 1143   **Beil Ref.**: 4-05-00-00809
Beil RN: 471308
MF: C_7H_7Cl **MP[°C]**: -45   **Den.[g/cm^3]**: 1.1004^{20}
MW: 126.59   **BP[°C]**: 179 n_D: 1.5391^{20}
Sol.: H_2O 1; EtOH 5; eth
    5; ctc 2; chl 5
Vap. Press. [kPa]: 0.164^{25}; 0.747^{50}; 2.66^{75}; 7.83^{100}; 19.8^{125}; 44.4^{150}
Therm. Prop.[kJ/mol]: $\Delta_f H°$(l, 25°C) = -32.6
C_p (liq.) [J/mol °C]: 182.4^{25}
γ [mN/m]: 37.5^{21}
Elec. Prop.: μ = (1.8 D); ε = 6.85^{20}
η [mPa s]: 1.50^{15}
MS: 91(100) 126(20) 65(14) 92(9) 39(9) 63(8) 128(6) 45(6) 89(5) 125(3)
IR [cm^{-1}]: 3060 3030 1500 1460 1270 1210 1070 1030 810 760 700 670
Raman [cm^{-1}]: 3060 3000 2960 1600 1580 1440 1260 1210 1180 1150 1030
    1000 980 810 800 760 700 670 610 550 460 330 270 120
UV [nm]: 265(211) 259(231) 254(190) 217(6970) MeOH
^{1}H NMR [ppm]: 4.5 7.3 CCl_4
Flammability: Flash pt. = 67°C; ign. temp. = 585°C; lower flam. lim. = 1.1%
TLV/TWA: 1 ppm (5.2 mg/m^3)
Reg. Lists: CERCLA (RQ = 100 lb.); SARA 313 (1.0%); RCRA P028 (accutely
    hazardous); confirmed or suspected carcinogen

29. Benzyl ethyl ether

Syst. Name: Benzene, (ethoxymethyl)-
Synonyms: (Ethoxymethyl)benzene; Ethyl benzyl
  ether

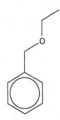

CASRN: 539-30-0   **Beil Ref.**: 4-06-00-02229
Merck No.: 1147
Beil RN: 2040908
MF: $C_9H_{12}O$ **BP[°C]**: 186   **Den.[g/cm^3]**: 0.9478^{20}
MW: 136.19 n_D: 1.4955^{20}
Sol.: H_2O 1; EtOH 5; eth
  5
Vap. Press. [kPa]: 0.136^{25}; 0.608^{50}; 2.16^{75}; 6.42^{100}; 16.5^{125}; 37.5^{150}
γ [mN/m]: 32.18^{25}
Elec. Prop.: $\varepsilon = 3.90^{25}$
MS: 91(100) 92(80) 79(30) 107(18) 65(16) 77(15) 51(10) 39(10) 29(9) 135(8)
IR [cm^{-1}]: 3030 2860 1520 1470 1370 1280 1210 1110 1080 1030 850 790 740
  710 690
Raman [cm^{-1}]: 3062 2931 2867 1607 1452 1210 1030 1003 822 619

30. Bis(2-chloroethyl) ether

Syst. Name: Ethane, 1,1'-oxybis[2-chloro-
Synonyms: Dichlorodiethyl ether

Cl⌒⌒O⌒⌒Cl

CASRN: 111-44-4   **DOT No.**: 1916
Merck No.: 3055   **Beil Ref.**: 4-01-00-01375
Beil RN: 605317
MF: $C_4H_8Cl_2O$ **MP[°C]**: -51.9   **Den.[g/cm^3]**: 1.22^{20}
MW: 143.01   **BP[°C]**: 178.5 n_D: 1.451^{20}
Sol.: H_2O 1; EtOH 3;
   eth3; ace 3; bz 5;
   MeOH 5; os 3
Vap. Press. [kPa]: 0.143^{25}; 0.681^{50}; 2.51^{75}; 7.58^{100}; 19.6^{125}; 44.8^{150}
Therm. Prop.[kJ/mol]: $\Delta_{fus}H$ = 8.66; $\Delta_{vap}H$ = 45.2
C_p **(liq.) [J/mol °C]**: 220.9^{25}
γ **[mN/m]**: 37.0^{25}
Elec. Prop.: μ = (2.6 D); ε = 21.20^{20}
η **[mPa s]**: 2.14^{25}
MS: 93(100) 63(74) 27(38) 95(32) 65(24) 31(9) 49(4) 28(4) 94(3) 62(3)
IR [cm^{-1}]: 2960 2860 2800 1450 1420 1350 1300 1200 1120 1040 740 660
Flammability: Flash pt. = 55°C; ign. temp. = 369°C; lower flam. lim. = 2.7%
TLV/TWA: 5 ppm (29 mg/m^3)
Reg. Lists: CERCLA (RQ = 10 lb.); SARA 313 (1.0%); RCRA U025 (toxic);
   confirmed or suspected carcinogen

31. Bis(2-ethylhexyl) phthalate

Syst. Name: 1,2-Benzenedicarboxylic acid, bis(2-ethylhexyl) ester

Synonyms: Di-2-ethylhexyl phthalate

CASRN: 117-81-7   **Beil Ref.**: 4-09-00-03181

Merck No.: 1262

Beil RN: 1890696

MF: $C_{24}H_{38}O_4$ **MP[°C]**: -55   **Den.[g/cm^3]**: 0.981^{25}

MW: 390.56   **BP[°C]**: 384 n_D: 1.4853^{20}

Sol.: ctc 2

Vap. Press. [kPa]: 0.001^{125}; 0.008^{150}

C_p **(liq.) [J/mol °C]**: 704.7^{25}

Elec. Prop.: $\mu = (2.8 \text{ D})$; $\varepsilon = 5.3^{20}$

η **[mPa s]**: 381^0; 81^{20}; 56.5^{25}

MS: 149(100) 57(32) 167(29) 71(21) 43(21) 70(18) 150(11) 113(10) 55(10) 41(9)

IR [cm^{-1}]: 2940 1720 1590 1470 1390 1280 1120 1080 1040 950 740 700

Raman [cm^{-1}]: 3070 3020 2930 2890 2860 2720 2580 1720 1590 1570 1480 1440 1380 1260 1150 1110 1030 950 880 850 820 750 640 390 230

^{1}H NMR [ppm]: 0.9 1.5 4.2 7.4 7.7 CCl$_4$

Flammability: Flash pt. = 218°C

TLV/TWA: 5 mg/m^3

Reg. Lists: CERCLA (RQ = 100 lb.); SARA 313 (0.1%); RCRA U028 (toxic); confirmed or suspected carcinogen

32. Bromobenzene

Syst. Name: Benzene, bromo-
Synonyms: Phenyl bromide

CASRN: 108-86-1
Merck No.: 1394
Beil RN: 1236661
MF: C$_6$H$_5$Br
MW: 157.01
Sol.: H$_2$O 1; EtOH 4; eth
  4; bz 4; ctc 3

DOT No.: 2514
Beil Ref.: 4-05-00-00670

MP[°C]: -30.6
BP[°C]: 156.0

Den.[g/cm^3]: 1.4950^{20}
n$_D$: 1.5597^{20}

Crit. Const.: T_c = 397°C; P_c = 4.52 MPa; V_c = 324 cm^3
Vap. Press. [kPa]: 0.556^{25}; 2.27^{50}; 7.20^{75}; 18.9^{100}; 43.0^{125}; 87.0^{150}
Therm. Prop.[kJ/mol]: $\Delta_{fus}H$ = 10.62; $\Delta_f H°$(l, 25°C) = 60.9
C_p (liq.) [J/mol °C]: 154.3^{25}
γ [mN/m]: 35.24^{25}; dγ/dT = 0.1160 mN/m °C
Elec. Prop.: μ = 1.70 D; ε = 5.45^{20}; IP = 8.98 eV
η [mPa s]: 1.56^0; 1.07^{25}; 0.798^{50}; 0.627^{75}; 0.512^{100}
MS: 77(100) 158(64) 156(64) 51(39) 50(17) 78(8) 76(6) 75(6) 28(5) 159(4)
IR [cm^{-1}]: 3030 1590 1470 1450 1080 1020 1000 910 740 690
Raman [cm^{-1}]: 3160 3070 1570 1170 1160 1070 1020 1000 680 610 310 250
  180
UV [nm]: 271(1190) 264(1710) 261(1730) 257(1590) 250(1390) MeOH
^{13}C NMR [ppm]: 122.4 126.7 129.8 131.4 CDCl$_3$
^{1}H NMR [ppm]: 7.1 7.4 CCl$_4$
Flammability: Flash pt. = 51°C; ign. temp. = 565°C

33. 1-Bromobutane

Syst. Name: Butane, 1-bromo-
Synonyms: Butyl bromide

CASRN: 109-65-9   **DOT No.:** 1126
Merck No.: 1553   **Beil Ref.:** 4-01-00-00258
Beil RN: 1098260
MF: C_4H_9Br **MP[°C]:** -112.4   **Den.[g/cm³]:** 1.2758^{20}
MW: 137.02   **BP[°C]:** 101.6 n_D: 1.4401^{20}
Sol.: H_2O 1; EtOH 5; eth
   5; ace 5; ctc 2; chl 3
Vap. Press. [kPa]: 0.232^{-25}; 1.41^{0}; 5.26^{25}
Therm. Prop.[kJ/mol]: $\Delta_{fus}H = 6.69$; $\Delta_{vap}H = 32.5$; $\Delta_f H°(l, 25°C) = -143.8$
C_p **(liq.) [J/mol °C]:** 109.3^{25}
γ **[mN/m]:** 25.90^{25}; $d\gamma/dT = 0.1126$ mN/m °C
Elec. Prop.: $\mu = 2.08$ D; $\varepsilon = 7.32^{10}$; IP = 10.13 eV
η **[mPa s]:** 0.815^{0}; 0.606^{25}; 0.471^{50}; 0.379^{75}
MS: 57(100) 41(56) 29(45) 27(29) 56(13) 39(12) 28(12) 55(7) 43(7) 138(6)
IR [cm⁻¹]: 2970 2930 2880 1470 1380 1260 1220 750 640 570
Raman [cm⁻¹]: 3000 2960 2940 2900 2870 2830 2730 1430 1330 1280 1250
   1210 1180 1090 1050 1010 990 960 900 890 860 790 730 640 560 450 410
   350 270 230
¹³C NMR [ppm]: 13.5 21.7 33.2 35.4
¹H NMR [ppm]: 1.0 1.4 1.8 3.4 CCl_4
Flammability: Flash pt. = 18°C; ign. temp. = 265°C; flam. lim. = 2.6-6.6%

34. 2-Bromobutane

Syst. Name: Butane, 2-bromo-, (±)-
Synonyms: *sec*-Butyl bromide

CASRN: 5787-31-5
Merck No.: 1554
Beil RN: 1718776
MF: C_4H_9Br
MW: 137.02
Sol.: ace 4; eth 4; chl 4

Rel. CASRN: 78-76-2
DOT No.: 2339
Beil Ref.: 4-01-00-00261

MP[°C]: -111.9   **Den.[g/cm^3]:** 1.2585^{20}
BP[°C]: 91.2 n_D: 1.4366^{20}

Vap. Press. [kPa]: 2.61^0; 9.32^{25}; 26.3^{50}; 62.3^{75}
Therm. Prop.[kJ/mol]: $\Delta_{fus}H = 6.89$; $\Delta_{vap}H = 30.8$; $\Delta_f H°(l, 25°C) = -154.8$
γ [mN/m]: 25.01^{20}
Elec. Prop.: $\mu = 2.23$ D; $\varepsilon = 8.64^{25}$; IP = 9.98 eV
η [mPa s]: 0.563^{25}
MS: 57(100) 41(60) 29(57) 27(34) 39(20) 28(9) 26(9) 136(1)
IR [cm^{-1}]: 2980 2930 2880 1460 1390 1280 1210 1150 1000 960 790 540
Raman [cm^{-1}]: * 1540 1454 1211 609 580 530 479 458 352 314 297
UV [nm]: 207(200) hp
^{13}C NMR [ppm]: 12.1 26.0 34.2 53.1 CDCl$_3$
^{1}H NMR [ppm]: 1.1 1.7 1.8 4.1 CDCl$_3$
Flammability: Flash pt. = 21°C

35. 1-Bromo-2-chloroethane

Syst. Name: Ethane, 1-bromo-2-chloro-
Synonyms: 2-Chloro-1-bromoethane; Ethylene
chlorobromide

CASRN: 107-04-0　　　　　　　　**Beil Ref.**: 4-01-00-00155
Beil RN: 605265
MF: C_2H_4BrCl　　　**MP[°C]**: -16.7　　**Den.[g/cm^3]**: 1.7392^{20}
MW: 143.41　　　　**BP[°C]**: 107　　　n_D: 1.4908^{20}
Sol.: H_2O 2; EtOH 3; eth
　3; chl 3
Vap. Press. [kPa]: 0.177^{-25}; 1.03^0; 4.24^{25}; 13.6^{50}; 36.2^{75}; 83.0^{100}
C_p **(liq.) [J/mol °C]**: 130.1^{27}
Elec. Prop.: $\mu = 1.09$ D; $\varepsilon = 7.41^{10}$
MS: 63(100) 27(85) 65(31) 26(23) 144(8) 81(8) 79(8) 28(7) 142(6) 93(6)
IR [cm^{-1}]: 2980 2950 1450 1420 1260 1200 860 730 640
^{13}C NMR [ppm]: 32.5 45.0
^{1}H NMR [ppm]: 3.3 4.0 CDCl$_3$

---

36. Bromochlorofluoromethane

Syst. Name: Methane, bromochlorofluoro-

CASRN: 593-98-6　　　　　　　　**Beil Ref.**: 4-01-00-00075
Beil RN: 1731043
MF: CHBrClF　　　**MP[°C]**: -115　　**Den.[g/cm^3]**: 1.9771^0
MW: 147.37　　　　**BP[°C]**: 36　　　n_D: 1.4144^{25}
Sol.: H_2O 1; eth 3; ace 3;
　chl 3
MS: 67(100) 69(32) 31(12) 113(9) 111(9) 32(9) 81(8) 79(8) 48(8) 47(8)
Raman [cm^{-1}]: 3018 1302 1205 1063 773 659 649 425 313 225

37. Bromochloromethane

Syst. Name: Methane, bromochloro-
Synonyms: Chlorobromomethane; Halon 1011

CASRN: 74-97-5
Beil RN: 1730801
MF: CH$_2$BrCl
MW: 129.38
Sol.: H$_2$O 1; EtOH 3; eth
  3; ace 3; bz 3; ctc 3;
  os 3
DOT No.: 1887
Beil Ref.: 4-01-00-00074
MP[°C]: -87.9   **Den.[g/cm^3]**: 1.9344^{20}
BP[°C]: 68.0 n_D: 1.4838^{20}

Vap. Press. [kPa]: 5.26^0; 19.5^{25}; 54.6^{50}
Therm. Prop.[kJ/mol]: $\Delta_{vap}H$ = 30.0
C_p **(liq.) [J/mol °C]**: 52.7^{25}
γ **[mN/m]**: 33.32^{20}
Elec. Prop.: μ = (1.7 D); IP = 10.77 eV
η **[mPa s]**: 0.674^{20}
MS: 49(100) 130(67) 128(52) 51(31) 93(23) 81(20) 79(20) 95(17) 132(16)
  47(8)
IR [cm^{-1}]: 3030 2940 2170 1720 1450 1410 1220 850 740
Raman [cm^{-1}]: 3060 2980 1400 1220 1120 720 600 220
^{13}C NMR [ppm]: 40.0
^{1}H NMR [ppm]: 5.2 CCl$_4$
TLV/TWA: 200 ppm (1060 mg/m^3)

38. 1-Bromodecane

Syst. Name: Decane, 1-bromo-
Synonyms: Decyl bromide

CASRN: 112-29-8   **Beil Ref.:** 4-01-00-00470
Beil RN: 1735227
MF: $C_{10}H_{21}Br$ **MP[°C]:** -29.2   **Den.[g/cm^3]:** 1.0702^{20}
MW: 221.18   **BP[°C]:** 240.6 n_D: 1.4557^{20}
Sol.: H_2O 1; eth 4; ctc 3;
   chl 4
Vap. Press. [kPa]: 2.66^{125}; 7.19^{150}
Therm. Prop.[kJ/mol]: $\Delta_f H°$(l, 25°C) = -344.7
γ [mN/m]: 29.1^{26}
Elec. Prop.: μ = (1.9 D); ε = 4.44^{25}
η [mPa s]: 2.37^{20}
MS: 43(100) 135(94) 137(91) 57(81) 41(58) 55(56) 71(38) 69(36) 85(33)
   29(27)
IR [cm^{-1}]: 2860 1470 1370 1250 720 650
Raman [cm^{-1}]: 2960 2930 2890 2870 2850 2720 1430 1370 1340 1300 1250
   1220 1150 1120 1070 1030 970 890 870 840 800 770 740 640 560 460 400
   330 220
^{13}C NMR [ppm]: 14.1 22.7 28.2 28.8 29.3 29.5 31.9 32.9 33.5 $CDCl_3$
^{1}H NMR [ppm]: 0.9 1.8 3.3 CCl_4

39. Bromoethane

Syst. Name: Ethane, bromo-
Synonyms: Ethyl bromide

CASRN: 74-96-4
Merck No.: 3730
Beil RN: 1209224
MF: C_2H_5Br
MW: 108.97
Sol.: H_2O 2; EtOH 5; eth
  5; chl 5

DOT No.: 1891
Beil Ref.: 4-01-00-00150

MP[°C]: -118.6
BP[°C]: 38.5

Den.[g/cm³]: 1.4604^{20}
n_D: 1.4239^{20}

Crit. Const.: $T_c = 230.8°C$; $P_c = 6.23$ MPa; $V_c = 215$ cm³
Vap. Press. [kPa]: 5.86^{-25}; 21.7^0; 62.5^{25}; 149^{50}; 308^{75}; 571^{100}; 970^{125}
Therm. Prop.[kJ/mol]: $\Delta_{fus}H = 5.86$; $\Delta_{vap}H = 27.0$; $\Delta_f H°(l, 25°C) = -90.1$
C_p (liq.) [J/mol °C]: 100.8^{25}
γ [mN/m]: 23.62^{25}; $d\gamma/dT = 0.1159$ mN/m °C
Elec. Prop.: $\mu = 2.03$ D; $\varepsilon = 9.01^{25}$; IP = 10.28 eV
η [mPa s]: 0.635^{-25}; 0.477^0; 0.374^{25}
MS: 108(100) 110(97) 29(62) 27(51) 28(35) 26(14) 93(6) 32(6) 95(5) 81(5)
IR [cm⁻¹]: 3030 2860 1450 1390 1250 970 950 770
¹³C NMR [ppm]: 20.3 28.3
¹H NMR [ppm]: 1.7 3.4 $CDCl_3$
Flammability: Ign. temp. = 511°C; flam. lim. = 6.8-8.0%
TLV/TWA: 5 ppm (22 mg/m³)
Reg. Lists: Confirmed or suspected carcinogen

40. Bromoethylene

Syst. Name: Ethene, bromo-
Synonyms: Vinyl bromide

CASRN: 593-60-2   **DOT No.**: 1085
Beil RN: 1361370   **Beil Ref.**: 4-01-00-00718
MF: C_2H_3Br **MP[°C]**: -137.8   **Den.[g/cm^3]**: 1.4933^{20}
MW: 106.95   **BP[°C]**: 15.8 n_D: 1.4380^{20}
Sol.: H_2O 1; EtOH 3; eth
   3; ace 3; bz 3; chl 3
Vap. Press. [kPa]: 16.9^{-25}; 54.5^0; 141^{25}; 307^{50}
Therm. Prop.[kJ/mol]: $\Delta_{fus}H$ = 5.12; $\Delta_{vap}H$ = 23.4; $\Delta_f H°$(g, 25°C) = 79.2
C_p **(liq.) [J/mol °C]**: 107.7^{15}
Elec. Prop.: μ = 1.42 D; ε = 5.63^5; IP = 9.80 eV
MS: 27(100) 106(99) 108(93) 26(15) 81(7) 79(7) 107(6) 25(5) 105(4) 104(3)
IR [cm^{-1}]: 3020 2990 1600 1390 1370 1270 1140 1000 940 900 620
^{13}C NMR [ppm]: 115.6 122.1
Flammability: Ign. temp. = 530°C; flam. lim. = 9-15%
TLV/TWA: 5 ppm (22 mg/m^3)
Reg. Lists: CERCLA (RQ = 1 lb.); SARA 313 (0.1%); confirmed or suspected
   carcinogen

41. Bromomethane

Syst. Name: Methane, bromo-
Synonyms: Methyl bromide; Halon 1001

Br

CASRN: 74-83-9   **DOT No.**: 1062
Merck No.: 5951   **Beil Ref.**: 4-01-00-00068
Beil RN: 1209223
MF: CH_3Br **MP[°C]**: -93.7   **Den.[g/cm^3]**: 1.6755^{20}
MW: 94.94   **BP[°C]**: 3.5 n_D: 1.4218^{20}
Sol.: H_2O 2; EtOH 5; eth
   5; chl 5; CS_2 5
Vap. Press. [kPa]: 28.8^{-25}; 88.0^0; 217^{25}
Therm. Prop.[kJ/mol]: $\Delta_{fus}H = 5.98$; $\Delta_{vap}H = 23.9$; $\Delta_f H°$(g, 25°C) = -35.5
C_p **(liq.) [J/mol °C]**: 78.8^7 (sat. press.)
Elec. Prop.: $\mu = 1.822$ D; $\varepsilon = 9.71^3$; IP = 10.54 eV
η **[mPa s]**: 0.409^{-10}
MS: 15(100) 94(92) 96(80) 81(14) 79(14) 95(13) 93(11) 14(10) 91(7) 13(7)
IR [cm^{-1}]: 2950 2850 1440 1310 1280 950
UV [nm]: 202(263) hp
^{13}**C NMR [ppm]**: 10.2
Flammability: Ign. temp. = 537°C; flam. lim. = 10-16.0%
TLV/TWA: 5 ppm (19 mg/m^3)
Reg. Lists: CERCLA (RQ = 1000 lb.); SARA 313 (1.0%); RCRA U029 (toxic);
   confirmed or suspected carcinogen

42. 2-Bromo-2-methylpropane

Syst. Name: Propane, 2-bromo-2-methyl-
Synonyms: *tert*-Butyl bromide

CASRN: 507-19-7 **DOT No.**: 2342
Merck No.: 1555 **Beil Ref.**: 4-01-00-00295
Beil RN: 1730892
MF: C_4H_9Br **MP[°C]**: -16.2 **Den.[g/cm^3]**: 1.4278^{20}
MW: 137.02 **BP[°C]**: 73.3 n_D: 1.4278^{20}
Sol.: H_2O 1; ctc 2
Vap. Press. [kPa]: 5.58^0; 17.7^{25}; 46.8^{50}
Therm. Prop.[kJ/mol]: $\Delta_{fus}H$ = 1.97; $\Delta_{vap}H$ = 29.2; $\Delta_fH°$(l, 25°C) = -163.8
C_p **(liq.) [J/mol °C]**: 151.0^{25}
γ **[mN/m]**: 21.23^{20}
Elec. Prop.: μ = (2.2 D); ε = 10.98^{20}; IP = 9.92 eV
η **[mPa s]**: 0.750^{25}
MS: 57(100) 41(67) 29(45) 39(30) 27(18) 28(8) 40(5) 38(5) 58(4) 55(4)
IR [cm^{-1}]: 1465 1380 1240 1200 1155 810 630
Raman [cm^{-1}]: 2980 2970 2930 2900 2780 2730 1460 1240 1140 1030 930 810
520 400 300 220
UV [nm]: 215(282) hp
^{13}C NMR [ppm]: 36.4 62.1 $CDCl_3$
^{1}H NMR [ppm]: 1.8 CCl_4

43. 1-Bromonaphthalene

Syst. Name: Naphthalene, 1-bromo-
Synonyms: 1-Naphthyl bromide

CASRN: 90-11-9   **Beil Ref.**: 4-05-00-01665
Merck No.: 1413
Beil RN: 1906414
MF: $C_{10}H_7Br$ **MP[°C]**: -0.9   **Den.[g/cm^3]**: 1.4785^{20}
MW: 207.07   **BP[°C]**: 281 n_D: 1.658^{20}
Sol.: H_2O 3; EtOH 5; eth
  5; ace 3; bz 5; ctc 2;
  chl 3
Vap. Press. [kPa]: 0.001^{25}; 0.074^{75}; 0.289^{100}; 0.930^{125}; 2.57^{150}
Therm. Prop.[kJ/mol]: $\Delta_{fus}H$ = 15.16; $\Delta_{vap}H$ = 39.3
γ **[mN/m]**: 44.19^{27}
Elec. Prop.: μ = (1.5 D); ϵ = 4.77^{25}; IP = 8.09 eV
η **[mPa s]**: 4.52^{25}
MS: 44(100) 206(39) 127(39) 208(37) 36(31) 69(29) 131(13) 29(13) 126(12)
  63(12)
IR [cm^{-1}]: 3130 1610 1560 1520 1390 1250 1210 1140 1050 1020 950 860 790
  760
UV [nm]: 319(262) 315(447) 284(8670) 274(7290) 223(96900) MeOH
^{1}H NMR [ppm]: 7.4 8.1 CCl_4

44. 1-Bromopentane

Syst. Name: Pentane, 1-bromo-
Synonyms: Amyl bromide; Pentyl bromide

CASRN: 110-53-2   **Rel. CASRN:** 29756-38-5
Merck No.: 637   **DOT No.:** 2343
Beil RN: 1730981   **Beil Ref.:** 4-01-00-00312
MF: $C_5H_{11}Br$ **MP[°C]:** -95   **Den.[g/cm^3]:** 1.2182^{20}
MW: 151.05   **BP[°C]:** 129.8 n_D: 1.4447^{20}
Sol.: H_2O 1; EtOH 3; eth
   5; bz 3; ctc 2; chl 3
Vap. Press. [kPa]: 1.68^{25}; 6.01^{50}; 17.3^{75}; 41.9^{100}; 89.3^{125}; 171^{150}
Therm. Prop.[kJ/mol]: $\Delta_{fus}H = 11.46$; $\Delta_{vap}H = 35.0$; $\Delta_f H°(l, 25°C) = -170.2$
C_p **(liq.) [J/mol °C]:** 132.2^{25}
γ **[mN/m]:** 26.80^{25}
Elec. Prop.: $\mu = 2.20$ D; $\varepsilon = 6.31^{26}$; IP = 10.09 eV
η **[mPa s]:** 0.753^{25}
MS: 43(100) 71(80) 41(56) 27(51) 42(37) 29(34) 55(33) 39(30) 28(12) 26(9)
IR [cm^{-1}]: 2940 2860 1470 1450 1430 1300 1270 1250 1210 1050 920 730
Raman [cm^{-1}]: 3000 2950 2930 2900 2860 2730 1430 1330 1290 1260 1240
   1200 1170 1100 1070 1050 1020 960 910 880 870 840 820 770 750 720 640
   560 470 430 390 350 320 260 240 230 210
^{13}C NMR [ppm]: 14.2 22.4 30.8 33.2 33.5
^{1}H NMR [ppm]: 0.9 1.4 1.9 3.3 CCl_4
Flammability: Flash pt. = 32°C

45. 1-Bromopropane
Syst. Name: Propane, 1-bromo-
Synonyms: Propyl bromide

CASRN: 106-94-5
Merck No.: 7857
Beil RN: 505936
MF: C_3H_7Br
MW: 122.99
Sol.: H_2O 2; EtOH 3; eth 3; ace 3; bz 3; ctc 3; chl 3

Rel. CASRN: 26446-77-5
DOT No.: 2344
Beil Ref.: 4-01-00-00205
MP[°C]: -110   **Den.[g/cm^3]:** 1.3537^{20}
BP[°C]: 71.1 n_D: 1.4343^{20}

Vap. Press. [kPa]: 5.56^0; 18.6^{25}; 50.1^{50}
Therm. Prop.[kJ/mol]: $\Delta_{fus}H = 6.53$; $\Delta_{vap}H = 29.8$; $\Delta_f H°(l, 25°C) = -121.8$
C_p **(liq.) [J/mol °C]:** 86.4^{25}
γ **[mN/m]:** 25.26^{25}; $d\gamma/dT = 0.1218$ mN/m °C
Elec. Prop.: $\mu = 2.18$ D; $\varepsilon = 8.09^{20}$; IP = 10.18 eV
η **[mPa s]:** 0.645^0; 0.489^{25}; 0.387^{50}
MS: 43(100) 41(77) 28(69) 27(60) 39(49) 124(42) 122(41) 42(34) 32(20) 29(15)
IR [cm^{-1}]: 2940 1470 1430 1390 1280 1220 1030 890 890 830 780 740
Raman [cm^{-1}]: 3000 2960 2930 2890 2860 2840 2730 2660 1450 1430 1320 1280 1220 1190 1090 1070 1020 880 840 770 640 560 400 310 260 210
^{13}**C NMR [ppm]:** 13.2 26.8 35.7
^{1}**H NMR [ppm]:** 1.0 1.9 3.4 CCl_4
Flammability: Flash pt. < 79°C; ign. temp. = 490°C

46. 2-Bromopropane

Syst. Name: Propane, 2-bromo-
Synonyms: Isopropyl bromide

CASRN: 75-26-3
Merck No.: 5098
Beil RN: 741852
MF: C_3H_7Br **MP[°C]**: -89
MW: 122.99   **BP[°C]**: 59.5
Sol.: H_2O 2; EtOH 5; eth
  5; ace 3; bz 3; chl 3

Rel. CASRN: 26446-77-5
DOT No.: 2344
Beil Ref.: 4-01-00-00208
Den.[g/cm^3]: 1.3140^{20}
n_D: 1.4251^{20}

Vap. Press. [kPa]: 9.23^0; 28.9^{25}; 74.1^{50}
Therm. Prop.[kJ/mol]: $\Delta_{vap}H$ = 28.3; $\Delta_f H°$(l, 25°C) = -130.5
C_p **(liq.) [J/mol °C]**: 132.2^{25}
γ **[mN/m]**: 23.25^{25}; $d\gamma/dT$ = 0.1183 mN/m °C
Elec. Prop.: μ = 2.21 D; ε = 9.46^{20}; IP = 10.07 eV
η **[mPa s]**: 0.612^0; 0.458^{25}; 0.359^{50}
MS: 43(100) 27(47) 41(43) 39(22) 124(8) 122(8) 26(7) 81(6) 79(6) 38(6)
IR [cm^{-1}]: 2990 2970 2920 2860 1470 1440 1390 1370 1230 1160 1120 1040
  930 880 870 540
Raman [cm^{-1}]: 2980 2960 2920 2860 2760 2740 1450 1440 1380 1320 1220
  1150 1120 1030 880 580 530 400 290
^{13}C NMR [ppm]: 28.2 44.2
^{1}H NMR [ppm]: 1.7 4.3 $CDCl_3$

47. 2-Bromopropene

Syst. Name: 1-Propene, 2-bromo-

CASRN: 557-93-7   **Beil Ref.**: 4-01-00-00754
Beil RN: 1731926
MF: C_3H_5Br **MP[°C]**: -126   **Den.[g/cm^3]**: 1.3965[16]
MW: 120.98   **BP[°C]**: 48.4 n_D: 1.4467[16]
Sol.: H_2O 1; eth 3; ace 3;
   chl 3
Vap. Press. [kPa]: 43.0[25]; 107[50]
Elec. Prop.: $\mu = (1.5\ D)$
MS: 41(100) 39(58) 122(37) 120(36) 38(12) 37(8) 40(6) 81(5) 79(5) 42(4)
IR [cm^{-1}]: 2940 1640 1430 1370 1300 1150 920 880 670
^{1}H NMR [ppm]: 2.3 5.3 5.5 $CDCl_3$

48. Butanal

Syst. Name: Butanal
Synonyms: Butyraldehyde

CASRN: 123-72-8
Merck No.: 1591
Beil RN: 506061
MF: C_4H_8O
MW: 72.11
DOT No.: 1129
Beil Ref.: 4-01-00-03229
MP[°C]: -99
BP[°C]: 74.8
Den.[g/cm³]: 0.8016^{20}
n_D: 1.3843^{20}
Sol.: H_2O 3; EtOH 5; eth 5; ace 4; bz 4; chl 2
Crit. Const.: T_c = 264.1°C; P_c = 4.32 MPa; V_c = 258 cm³
Vap. Press. [kPa]: 15.7^{25}; 43.2^{50}; 102^{75}
Therm. Prop.[kJ/mol]: $\Delta_{fus}H$ = 11.09; $\Delta_{vap}H$ = 31.5; $\Delta_fH°$(l, 25°C) = -239.2
C_p (liq.) [J/mol °C]: 163.7^{25}
γ [mN/m]: 29.9^{24}
Elec. Prop.: μ = 2.72 D; ε = 13.45^{25}; IP = 9.84 eV
η [mPa s]: 0.475^{18}
MS: 44(100) 43(74) 72(57) 41(56) 27(55) 29(48) 57(23) 39(22) 28(15) 42(11)
IR [cm⁻¹]: 2940 2860 2780 2700 1720 1470 1390 1140 1110 1000 960 780
Raman [cm⁻¹]: 2940 2910 2870 2810 2730 1720 1450 1390 1300 1150 1110 1040 950 930 890 870 840 780 750 680 650 520 450 380 350 280
UV [nm]: 283(13) 225(12) H_2O
¹³C NMR [ppm]: 13.8 16.0 45.9 201.9 diox
¹H NMR [ppm]: 1.0 1.7 2.4 9.7 CDCl₃
Flammability: Flash pt. = -22°C; ign. temp. = 218°C; flam. lim. = 1.9-12.5%
Reg. Lists: SARA 313 (1.0%)

49. Butane

Syst. Name: Butane

CASRN: 106-97-8   **Rel. CASRN**: 68476-85-7
Merck No.: 1507   **DOT No.**: 1011
Beil RN: 969129   **Beil Ref.**: 4-01-00-00236
MF: C_4H_{10} **MP[°C]**: -138.2   **Den.[g/cm^3]**: 0.573^{25} (sat. press.)
MW: 58.12   **BP[°C]**: -0.5 n_D: 1.3326^{20}
Sol.: H_2O 3; EtOH 4;
   eth 4; chl 4
Crit. Const.: T_c = 151.97°C; P_c = 3.784 MPa; V_c = 255 cm^3
Vap. Press. [kPa]: 1.9^{-73}; 36.9^{-25}; 103^0; 242^{25}
Therm. Prop.[kJ/mol]: $\Delta_{fus}H$ = 4.66; $\Delta_{vap}H$ = 22.4; $\Delta_f H°$(g, 25°C) = -125.6
C_p (liq.) [J/mol °C]: $134.5^{-0.5}$
γ [mN/m]: 17.28^{-20}; $d\gamma/dT$ = 0.1206 mN/m °C
Elec. Prop.: μ = 0; ϵ = 1.77^{22}; IP = 10.53 eV
η [mPa s]: 0.314^{-40}; 0.164^{20} (at saturation pressure)
MS: 43(100) 29(44) 27(43) 28(34) 41(32) 39(17) 42(13) 58(12) 15(11) 26(9)
IR [cm^{-1}]: 2860 1470 1390 1300 1140 970 740
^{13}C NMR [ppm]: 13.2 25.0
Flammability: Flash pt. = -60°C; ign. temp. = 287°C; flam. lim. = 1.9-8.5%
TLV/TWA: 800 ppm (1900 mg/m^3)

50. 1,3-Butanediol

Syst. Name: 1,3-Butanediol
Synonyms: 1,3-Butylene glycol

CASRN: 107-88-0   **Beil Ref.**: 0-01-00-00477
Merck No.: 1566
Beil RN: 1731276
MF: $C_4H_{10}O_2$ **MP[°C]**: <-50   **Den.[g/cm^3]**: 1.0053^{20}
MW: 90.12   **BP[°C]**: 207.5 n_D: 1.4401^{20}
Vap. Press. [kPa]: 0.008^{25}; 1.37^{100}; 4.59^{125}; 13.4^{150}
Therm. Prop.[kJ/mol]: $\Delta_{vap}H$ = 58.5; $\Delta_fH°$(l, 25°C) = -501.0
C_p **(liq.) [J/mol °C]**: 227.2^{30}
γ **[mN/m]**: 37.8^{25}
Elec. Prop.: $\varepsilon = 28.8^{25}$
η **[mPa s]**: 130^{20}; 98^{25}
MS: 45(100) 43(13) 29(9) 27(9) 57(8) 47(6) 44(6) 46(3) 75(2) 72(2)
Flammability: Flash pt. = 121°C; ign. temp. = 395°C

51. 1,4-Butanediol

Syst. Name: 1,4-Butanediol
Synonyms: Tetramethylene glycol; Butylene glycol

CASRN: 110-63-4   **Beil Ref.**: 4-01-00-02515
Beil RN: 1633445
MF: $C_4H_{10}O_2$ **MP[°C]**: 20.1   **Den.[g/cm³]**: 1.0171^{20}
MW: 90.12   **BP[°C]**: 235 n_D: 1.4460^{20}
Sol.: H_2O 5; EtOH 3; eth
  2; DMSO 2
Vap. Press. [kPa]: 1.61^{125}; 5.30^{150}
Therm. Prop.[kJ/mol]: $\Delta_f H°$(l, 25°C) = -503.3
C_p **(liq.) [J/mol °C]**: 200.1^{25}
γ **[mN/m]**: 44.6^{20}
Elec. Prop.: μ = (2.6 D); ϵ = 31.9^{25}
η **[mPa s]**: 71.5^{25}
MS: 42(100) 31(74) 44(66) 41(41) 43(29) 29(25) 71(24) 27(23) 57(18) 39(15)
IR [cm⁻¹]: 3400 3320 2950 2860 1460 1440 1370 1330 1160 1050 930 650
Raman [cm⁻¹]: 3300 2910 2880 2740 1470 1440 1370 1300 1220 1170 1070
  1050 1000 950 920 880 860 820 760 540 510 450 360 330
UV [nm]: 291(200) sulf
¹³C NMR [ppm]: 29.4 62.1
¹H NMR [ppm]: 1.4 3.4 4.3 DMSOd₆
Flammability: Flash pt. = 121°C

52. 2,3-Butanediol
Syst. Name: 2,3-Butanediol, (R^*,R^*)-(±)-
Synonyms: (±)-2,3-Butylene glycol

CASRN: 6982-25-8   **Beil Ref.**: 4-01-00-02525
Beil RN: 1718902
MF: $C_4H_{10}O_2$ **MP[°C]**: 7.6   **Den.[g/cm³]**: 1.0033^{20}
MW: 90.12   **BP[°C]**: 182.5 n_D: 1.4310^{25}
Sol.: H_2O 5; EtOH 5; eth
   3; ace 3; chl 3
Vap. Press. [kPa]: 3.59^{100}; 11.9^{125}; 33.6^{150}
Therm. Prop.[kJ/mol]: $\Delta_f H°$(l, 25°C) = -541.5
C_p **(liq.) [J/mol °C]**: 213.0^{25}
η **[mPa s]**: 121^{25}
MS: 45(100) 43(21) 57(13) 44(13) 55(10) 41(10) 29(10) 27(10) 30(6) 26(5)
IR [cm⁻¹]: 3330 3030 2940 1470 1410 1390 1280 1160 1120 1090 1050 1010
   990 930 890 780
Raman [cm⁻¹]: 2979 2928 2868 1450 1160 1113 928 812 759 546
UV [nm]: 285(200) sulf
¹H NMR [ppm]: 1.1 3.1 3.8 CDCl₃
Flammability: Ign. temp. = 402°C

53. Butanenitrile

Syst. Name: Butanenitrile
Synonyms: Propyl cyanide

CASRN: 109-74-0   **DOT No.:** 2411
Merck No.: 1597   **Beil Ref.:** 4-02-00-00806
Beil RN: 1361452
MF: C_4H_7N **MP[°C]:** -111.9   **Den.[g/cm³]:** 0.7936^{20}
MW: 69.11   **BP[°C]:** 117.6 n_D: 1.3842^{20}
Sol.: H_2O 2; EtOH 5; eth
   5; bz 3; ctc 2
Crit. Const.: T_c = 312.3°C; P_c = 3.88 MPa
Vap. Press. [kPa]: 2.55^{25}; 25.1^{75}; 59.5^{100}; 125^{125}
Therm. Prop.[kJ/mol]: $\Delta_{fus}H$ = 5.02; $\Delta_{vap}H$ = 33.7; $\Delta_f H°$(l, 25°C) = -5.8
C_p (liq.) [J/mol °C]: 159^{67}
γ [mN/m]: 26.92^{25}; $d\gamma/dT$ = 0.1037 mN/m °C
Elec. Prop.: μ = 4.07 D; ϵ = 24.83^{20}; IP = 11.20 eV
η [mPa s]: 0.553^{25}; 0.418^{50}; 0.330^{75}; 0.268^{100}
MS: 41(100) 29(62) 27(37) 28(10) 39(9) 26(7) 40(5) 42(4) 38(4) 15(4)
IR [cm⁻¹]: 2940 2860 2220 1470 1430 1390 1330 1100 920 840 740
Raman [cm⁻¹]: 2930 2890 2840 2750 2250 1460 1430 1330 1310 1260 1240
   1100 1050 940 920 870 840 770 560 530 380 360 180
¹H NMR [ppm]: 1.1 1.7 2.3 CCl_4
Flammability: Flash pt. = 24°C; ign. temp. = 501°C; lower flam. lim. = 1.65%

54. 1-Butanethiol

Syst. Name: 1-Butanethiol
Synonyms: Butyl mercaptan; Thiobutyl alcohol

CASRN: 109-79-5
Merck No.: 1575
Beil RN: 1730908
MF: $C_4H_{10}S$
MW: 90.19
Sol.: H_2O 2; EtOH 4; eth 4; chl 2

DOT No.: 2347
Beil Ref.: 4-01-00-01555

MP[°C]: -115.7
BP[°C]: 98.5

Den.[g/cm^3]: 0.8416^{20}
n_D: 1.4440^{20}

Crit. Const.: $T_c = 297.0°C$; $V_c = 324$ cm^3
Vap. Press. [kPa]: 1.51^0; 6.07^{25}; 18.8^{50}; 48.1^{75}; 106^{100}
Therm. Prop.[kJ/mol]: $\Delta_{fus}H = 10.46$; $\Delta_{vap}H = 32.2$; $\Delta_f H°(l, 25°C) = -124.7$
C_p **(liq.) [J/mol °C]**: 171.2^{25}
γ [mN/m]: 25.5^{22}
Elec. Prop.: $\mu = (1.5$ D); $\varepsilon = 5.20^{15}$; IP = 9.14 eV
η [mPa s]: 0.535^{20}
MS: 56(100) 41(74) 90(66) 47(43) 27(43) 28(36) 29(33) 57(17) 39(16) 61(15)
IR [cm^{-1}]: 2940 2560 1450 1370 1270 1220 1110 950 780 740 730
Raman [cm^{-1}]: 2962 2934 2905 2874 2574 1443 1301 1110 1053 654
UV [nm]: 277(158) iso
^{1}H NMR [ppm]: 0.9 1.2 1.5 2.5 CDCl$_3$
Flammability: Flash pt. = 2°C
TLV/TWA: 0.5 ppm (1.8 mg/m^3)

55. Butanoic acid

Syst. Name: Butanoic acid
Synonyms: Butyric acid

CASRN: 107-92-6   **DOT No.:** 2820
Merck No.: 1593   **Beil Ref.:** 4-02-00-00779
Beil RN: 906770
MF: $C_4H_8O_2$ **MP[°C]:** -5.7   **Den.[g/cm^3]:** 0.9577^{20}
MW: 88.11   **BP[°C]:** 163.7 n_D: 1.3980^{20}
Sol.: H_2O 5; EtOH 5; eth
  5; ctc 2
Crit. Const.: T_c = 351°C; P_c = 4.03 MPa; V_c = 290 cm^3
Vap. Press. [kPa]: 0.221^{25}; 0.895^{50}; 2.81^{75}; 9.39^{100}; 26.8^{125}; 65.5^{150}
Therm. Prop.[kJ/mol]: $\Delta_{fus}H$ = 11.08; $\Delta_f H°$(l, 25°C) = -533.8
C_p **(liq.) [J/mol °C]:** 178.6^{25}
γ **[mN/m]:** 26.05^{25}; $d\gamma/dT$ = 0.0920 mN/m °C
Elec. Prop.: μ = (1.6 D); ε = 2.98^{14}; IP = 10.17 eV
η **[mPa s]:** 2.22^0; 1.43^{25}; 0.982^{50}; 0.714^{75}; 0.542^{100}
MS: 60(100) 27(50) 73(27) 42(25) 41(24) 43(22) 29(21) 45(19) 39(15) 28(11)
IR [cm^{-1}]: 3050 2960 2880 1720 1460 1420 1280 1220 1100 940 780 490
UV [nm]: 208(50) H_2O
^{1}H NMR [ppm]: 0.9 1.7 2.3 12.0 CCl$_4$
Flammability: Flash pt. = 72°C; ign. temp. = 443°C; flam. lim. = 2.0-10.0%
Reg. Lists: CERCLA (RQ = 5000 lb.)

56. Butanoic anhydride

Syst. Name: Butanoic acid, anhydride
Synonyms: Butyric anhydride

CASRN: 106-31-0
Merck No.: 1594
Beil RN: 1099474
DOT No.: 2739
Beil Ref.: 4-02-00-00802

MF: $C_8H_{14}O_3$
MW: 158.20
MP[°C]: -75
BP[°C]: 200
Den.[g/cm^3]: 0.9668^{20}
n_D: 1.4070^{20}

Sol.: eth 3; ctc 2
Vap. Press. [kPa]: 1.25^{75}; 3.88^{100}; 10.5^{125}; 25.2^{150}
Therm. Prop.[kJ/mol]: $\Delta_{vap}H = 50.0$
C_p (liq.) [J/mol °C]: 283.7^{25}
γ [mN/m]: 28.44^{25}
Elec. Prop.: $\varepsilon = 12.8^{20}$
η [mPa s]: 1.49^{25}
MS: 71(100) 43(59) 27(26) 41(19) 42(10) 39(10) 73(9) 28(7) 72(5) 55(5)
IR [cm^{-1}]: 2940 2860 1820 1750 1470 1410 1350 1240 1110 1020 930 860 790
UV [nm]: 282(126) bz
^{1}H NMR [ppm]: 1.0 1.7 2.4 CCl$_4$
Flammability: Flash pt. = 54°C; ign. temp. = 279°C; flam. lim. = 0.9-5.8%

57. 1-Butanol

Syst. Name: 1-Butanol
Synonyms: Butyl alcohol; Propylcarbinol

CASRN: 71-36-3   **DOT No.**: 1120
Merck No.: 1540   **Beil Ref.**: 4-01-00-01506
Beil RN: 969148
MF: $C_4H_{10}O$ **MP[°C]**: -89.8   **Den.[g/cm^3]**: 0.8098^{20}
MW: 74.12   **BP[°C]**: 117.7 n_D: 1.3993^{20}
Sol.: H_2O 3; EtOH 5; eth
   5; ace 4; bz 3; ctc 2
Crit. Const.: $T_c = 289.9°C$; $P_c = 4.414$ MPa; $V_c = 275$ cm^3
Vap. Press. [kPa]: 0.103^0; 0.860^{25}; 4.52^{50}; 17.2^{75}; 51.9^{100}; 130^{125}; 283^{150}
Therm. Prop.[kJ/mol]: $\Delta_{fus}H = 9.28$; $\Delta_{vap}H = 43.3$; $\Delta_f H°(l, 25°C) = -327.3$
C_p (liq.) [J/mol °C]: 177.1^{25}
γ [mN/m]: 24.93^{25}; $d\gamma/dT = 0.0898$ mN/m °C
Elec. Prop.: $\mu = 1.66$ D; $\epsilon = 17.84^{20}$; IP = 10.06 eV
η [mPa s]: 12.2^{-25}; 5.19^0; 2.54^{25}; 1.39^{50}; 0.833^{75}; 0.533^{100}
k [W/m °C]: 0.158^0; 0.154^{25}; 0.149^{50}
MS: 31(100) 56(81) 41(62) 43(60) 27(50) 42(31) 29(31) 28(17) 39(16) 55(12)
IR [cm^{-1}]: 3340 2960 2940 2860 1460 1380 1070 1040 950 850 740 650
Raman [cm^{-1}]: 3300 2940 2910 2870 2730 1450 1290 1250 1210 1110 1070
   1050 1030 970 900 870 840 820 800 520 490 450 390 350 270 210
^{13}**C NMR [ppm]**: 13.9 19.4 35.3 61.7
^{1}**H NMR [ppm]**: 0.9 1.4 3.5 4.1 CCl_4
Flammability: Flash pt. = 37°C; ign. temp. = 343°C; flam. lim. = 1.4-11.2%
TLV/TWA: 50 ppm (152 mg/m^3)
Reg. Lists: CERCLA (RQ = 5000 lb.); SARA 313 (1.0%); RCRA U031 (toxic)

58. 2-Butanol

Syst. Name: 2-Butanol
Synonyms: *sec*-Butyl alcohol

CASRN: 78-92-2   **DOT No.:** 1120
Merck No.: 1541   **Beil Ref.:** 2-01-00-00400
Beil RN: 773649
MF: $C_4H_{10}O$ **MP[°C]:** -114.7   **Den.[g/cm^3]:** 0.8063^{20}
MW: 74.12   **BP[°C]:** 99.5 n_D: 1.3978^{20}
Sol.: H_2O 4; EtOH 5; eth
  5; ace 4; bz 3; ctc 3
Crit. Const.: T_c = 263.1°C; P_c = 4.202 MPa; V_c = 269 cm^3
Vap. Press. [kPa]: 2.32^{25}; 11.1^{50}; 37.2^{75}; 103^{100}; 240^{125}; 495^{150}
Therm. Prop.[kJ/mol]: $\Delta_{vap}H$ = 40.8; $\Delta_f H°$(l, 25°C) = -342.6
C_p **(liq.) [J/mol °C]:** 196.9^{25}
γ **[mN/m]:** 22.54^{25}; dγ/dT = 0.0795 mN/m °C
Elec. Prop.: μ = (1.8 D); ε = 17.26^{20}; IP = 9.88 eV
η **[mPa s]:** 3.10^{25}; 1.33^{50}; 0.698^{75}; 0.419^{100}
MS: 45(100) 31(22) 27(22) 59(20) 29(18) 43(13) 41(12) 44(8) 18(8) 28(5)
Flammability: Flash pt. = 24°C; ign. temp. = 405°C; flam. lim. = 1.7-9.8%
TLV/TWA: 100 ppm (303 mg/m^3)
Reg. Lists: SARA 313 (1.0%)

---

59. *cis*-2-Butene-1,4-diol

Syst. Name: 2-Butene-1,4-diol, (Z)-

CASRN: 6117-80-2   **Beil Ref.:** 4-01-00-02660
Beil RN: 1679241
MF: $C_4H_8O_2$ **MP[°C]:** 2.0   **Den.[g/cm^3]:** 1.0698^{20}
MW: 88.11   **BP[°C]:** 235 n_D: 1.4782^{20}
Sol.: H_2O 3; EtOH 4
Vap. Press. [kPa]: 0.395^{100}; 1.46^{125}; 4.64^{150}; 12.9^{175}; 32.3^{200}; 73.8^{225}
Elec. Prop.: μ = (2.5 D)
MS: 42(100) 31(64) 39(59) 41(58) 57(56) 27(48) 29(47) 70(37) 43(23) 44(19)
IR [cm^{-1}]: 3320 3010 2920 2860 1650 1460 1420 1330 1230 1080 1030 970
  680
^{1}H NMR [ppm]: 4.2 5.7 D_2O

60. *trans*-2-Butene-1,4-diol
Syst. Name: 2-Butene-1,4-diol, (*E*)-

HO ⟋⟍⟍⟋ OH

CASRN: 821-11-4   **Beil Ref.**: 4-01-00-02660
Beil RN: 1719691
MF: $C_4H_8O_2$ **MP[°C]**: 25   **Den.[g/cm³]**: 1.0700^{20}
MW: 88.11 n_D: 1.4755^{20}
Sol.: H_2O 4; EtOH 4
Vap. Press. [kPa]: 1.7^{130}
Elec. Prop.: μ = (2.5 D)

---

61. Butyl acetate
Syst. Name: Ethanoic acid, butyl ester

CASRN: 123-86-4   **DOT No.**: 1123
Merck No.: 1535   **Beil Ref.**: 4-02-00-00143
Beil RN: 1741921
MF: $C_6H_{12}O_2$ **MP[°C]**: -78   **Den.[g/cm³]**: 0.8825^{20}
MW: 116.16   **BP[°C]**: 126.1 n_D: 1.3941^{20}
Sol.: H_2O 2; EtOH 5; eth
    5; ace 3; chl 3
Crit. Const.: T_c = 306°C
Vap. Press. [kPa]: 1.66^{25}; 6.07^{50}; 19.3^{75}; 46.4^{100}; 98.4^{125}
Therm. Prop.[kJ/mol]: $\Delta_{vap}H$ = 36.3; $\Delta_f H°$(l, 25°C) = -529.2
C_p (liq.) [J/mol °C]: 227.8^{25}
γ [mN/m]: 24.88^{25}; dγ/dT = 0.1068 mN/m °C
Elec. Prop.: μ = (1.9 D); ε = 5.07^{20}; IP = 10.00 eV
η [mPa s]: 1.00^0; 0.685^{25}; 0.500^{50}; 0.383^{75}; 0.305^{100}
MS: 43(100) 56(34) 41(17) 27(16) 29(15) 73(11) 61(10) 28(7) 55(6) 39(6)
IR [cm⁻¹]: 2950 2860 1730 1450 1350 1220 1050 1020 940 830 720
Raman [cm⁻¹]: 2940 2920 2870 2740 1730 1450 1390 1370 1300 1260 1230
    1150 1120 1070 1030 1000 950 920 880 840 810 640 610 510 480 430 330
    300 260 230
UV [nm]: 214(50) 202(51) iso 208(56) MeOH
¹H NMR [ppm]: 0.9 1.4 2.0 4.1 $CDCl_3$
Flammability: Flash pt. = 22°C; ign. temp. = 425°C; flam. lim. = 1.7-7.6%
TLV/TWA: 150 ppm (713 mg/m³)
Reg. Lists: CERCLA (RQ = 5000 lb.)

62. *sec*-Butyl acetate

Syst. Name: Ethanoic acid, 1-methylpropyl ester
Synonyms: 1-Methylpropyl acetate

CASRN: 105-46-4   **Beil Ref.:** 4-02-00-00148
Merck No.: 1536
Beil RN: 1720689
MF: $C_6H_{12}O_2$ **MP[°C]:** -98.9   **Den.[g/cm^3]:** 0.8748^{20}
MW: 116.16   **BP[°C]:** 112 n_D: 1.3888^{20}
Sol.: H_2O 1; EtOH 3; eth
   3; ctc 2
γ **[mN/m]:** 23.33^{21}
Elec. Prop.: $\mu = (1.9 \text{ D})$; $\varepsilon = 5.14^{20}$; IP = 9.90 eV
η **[mPa s]:** 0.65^{25}
MS: 43(100) 56(21) 87(15) 41(14) 29(8) 57(6) 73(4) 61(4) 55(4) 27(4)
IR [cm^{-1}]: 2940 1720 1470 1370 1250 1120 1110 1100 1030 1020 940 860
Raman [cm^{-1}]: *1780 1763 1469 1395 1368 972 894 865 658 366 207
^{1}H NMR [ppm]: 0.9 1.2 1.5 2.0 4.8 CCl_4
Flammability: Flash pt. = 31°C; flam. lim. = 1.7-9.8%
TLV/TWA: 200 ppm (950 mg/m^3)
Reg. Lists: CERCLA (RQ = 5000 lb.)

63. Butylamine

Syst. Name: 1-Butanamine

CASRN: 109-73-9   **DOT No.:** 1125
Merck No.: 1543   **Beil Ref.:** 4-04-00-00540
Beil RN: 605269
MF: $C_4H_{11}N$ **MP[°C]:** -49.1   **Den.[g/cm^3]:** 0.7414^{20}
MW: 73.14   **BP[°C]:** 77.0 n_D: 1.4031^{20}
Sol.: H_2O 5; EtOH 3; eth
   3
Crit. Const.: T_c = 258.8°C; P_c = 4.25 MPa; V_c = 277 cm^3
Vap. Press. [kPa]: 3.17^0; 12.2^{25}; 36.8^{50}; 92.1^{75}; 200^{100}
Therm. Prop.[kJ/mol]: $\Delta_{vap}H$ = 31.8; $\Delta_f H°$(l, 25°C) = -127.7
C_p (liq.) [J/mol °C]: 179.2^{25}
γ [mN/m]: 23.44^{25}; $d\gamma/dT$ = 0.1122 mN/m °C
Elec. Prop.: μ = 1.0 D; ε = 4.71^{20}; IP = 8.71 eV
η [mPa s]: 0.830^0; 0.574^{25}; 0.409^{50}; 0.298^{75}
MS: 30(100) 73(10) 28(5) 41(3) 27(3) 18(3) 44(2) 42(2) 31(2) 29(2)
IR [cm^{-1}]: 3380 3300 2960 2930 2850 1610 1460 1370 1090 980 840 790
Raman [cm^{-1}]: 3370 3320 3200 2950 2940 2900 2870 2750 2740 1600 1450
   1360 1300 1250 1190 1120 1090 1050 1040 970 950 900 870 850 820 800
   750 500 480 440 400 350 280
^{13}**C NMR [ppm]:** 14.0 20.4 36.7 42.3
Flammability: Flash pt. = -12°C; ign. temp. = 312°C; flam. lim. = 1.7-9.8%
TLV/TWA: 5 ppm (15 mg/m^3)
Reg. Lists: CERCLA (RQ = 1000 lb.)

64. *sec*-Butylamine

Syst. Name: 2-Butanamine, (±)-

CASRN: 33966-50-6   **Rel. CASRN**: 13952-84-6, 513-49-5
Merck No.: 1544   **Beil Ref.**: 4-04-00-00618
Beil RN: 1361345
MF: $C_4H_{11}N$ **MP[°C]**: <-72   **Den.[g/cm³]**: 0.7246^{20}
MW: 73.14   **BP[°C]**: 63.5 n_D: 1.3932^{20}
Sol.: H_2O 3; EtOH 5; eth
     5; ace 4; chl 3
Crit. Const.: T_c = 241.2°C; P_c = 4.20 MPa; V_c = 278 cm³
Vap. Press. [kPa]: 6.61^0; 23^{65}; 64^{50}; 150^{75}
Therm. Prop.[kJ/mol]: $\Delta_{vap}H$ = 29.9; $\Delta_f H°$(l, 25°C) = -137.5
γ [mN/m]: 21.49^{21}
Elec. Prop.: μ = (1.3 D); IP = 8.70 eV
MS: 44(100) 18(15) 41(11) 58(11) 28(10) 30(10) 27(10) 73(1)
IR [cm⁻¹]: 3330 3230 2860 1590 1450 1410 1370 1330 1150 1120 1030 990
     960 930 840 800 770
Raman [cm⁻¹]: 3380 3320 2960 2940 2880 2730 2630 1520 1450 1370 1350
     1300 1270 1230 1160 1130 1050 1040 1000 920 810 770 490 420 380 270
¹H NMR [ppm]: 0.9 1.1 1.3 2.8 $CDCl_3$
Flammability: Flash pt. = -9°C
Reg. Lists: CERCLA (RQ = 1000 lb.)

65. *tert*-Butylamine

Syst. Name: 2-Propanamine, 2-methyl-
Synonyms: 2-Methyl-2-propanamine

CASRN: 75-64-9   **Beil Ref.**: 4-04-00-00657
Merck No.: 1545
Beil RN: 605267
MF: $C_4H_{11}N$ **MP[°C]**: -66.9   **Den.[g/cm³]**: 0.6958^{20}
MW: 73.14   **BP[°C]**: 44.0 n_D: 1.3784^{20}
Sol.: H_2O 5; EtOH 5; eth
  5; chl 3
Crit. Const.: T_c = 210.8°C; P_c = 3.84 MPa; V_c = 292 cm³
Vap. Press. [kPa]: 48.4^{25}; 123^{50}
Therm. Prop.[kJ/mol]: $\Delta_{fus}H$ = 0.88; $\Delta_{vap}H$ = 28.3; $\Delta_fH°$(l, 25°C) = -150.6
C_p (liq.) [J/mol °C]: 192.1^{25}
γ [mN/m]: 16.87^{25}; dγ/dT = 0.1028 mN/m °C
Elec. Prop.: μ = (1.3 D); IP = 8.64 eV
MS: 58(100) 41(21) 42(15) 18(9) 30(8) 15(8) 39(7) 57(6) 28(6) 59(4)
IR [cm⁻¹]: 3330 3230 2940 1610 1470 1390 1370 1250 1220 1040 840 740
Raman [cm⁻¹]: 3310 2960 2920 2860 2780 2720 1460 1370 1330 1250 1230
  1120 1040 1010 950 930 910 790 750 700 460 350
¹H NMR [ppm]: 1.1 1.2 CDCl₃
Flammability: Flash pt. = -9°C; ign. temp. = 380°C; flam. lim. = 1.7-8.9%
Reg. Lists: CERCLA (RQ = 1000 lb.)

66. Butylbenzene

Syst. Name: Benzene, butyl-
Synonyms: 1-Phenylbutane

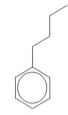

CASRN: 104-51-8
Merck No.: 1549
Beil RN: 1903395
MF: $C_{10}H_{14}$
MW: 134.22
Sol.: H_2O 1; EtOH 5;
  eth 5; ace 5; bz 5; ctc
  5; peth 5

DOT No.: 2709
Beil Ref.: 4-05-00-01033

MP[°C]: -87.9
BP[°C]: 183.3

Den.[g/cm³]: 0.8601^{20}
n_D: 1.4898^{20}

Crit. Const.: $T_c = 387.4°C$; $P_c = 2.89$ MPa; $V_c = 497$ cm³
Vap. Press. [kPa]: 0.150^{25}; 7.41^{100}; 18.7^{125}; 41.3^{150}
Therm. Prop.[kJ/mol]: $\Delta_{fus}H = 11.22$; $\Delta_{vap}H = 38.9$; $\Delta_f H°(l, 25°C) = -63.2$
C_p (liq.) **[J/mol °C]**: 243.4^{25}
γ [mN/m]: 28.71^{25}; $d\gamma/dT = 0.1025$ mN/m °C
Elec. Prop.: $\mu = 0$; $\varepsilon = 2.36^{20}$; IP = 8.69 eV
η [mPa s]: 0.950^{25}; 0.683^{50}; 0.515^{75}
MS: 91(100) 92(55) 134(20) 65(13) 27(10) 39(9) 105(8) 51(7) 78(6) 41(6)
IR [cm⁻¹]: 2940 1590 1490 1450 1390 1110 1030 740 700
Raman [cm⁻¹]: 3050 3000 2960 2940 2920 2870 2850 2740 1600 1580 1450
  1440 1330 1300 1200 1180 1150 1100 1050 1030 1000 990 970 920 780 620
  500 280 220
UV [nm]: 261 cyhex
¹³C NMR [ppm]: 13.9 22.5 33.9 35.8 125.7 128.3 142.7 diox
¹H NMR [ppm]: 0.9 1.4 2.6 7.1 CCl_4
Flammability: Flash pt. = 71°C; ign. temp. = 410°C; flam. lim. = 0.8-5.8%

67. *sec*-Butylbenzene

Syst. Name: Benzene, (1-methylpropyl)-, (±)-
Synonyms: (±)-2-Phenylbutane

CASRN: 36383-15-0   **Rel. CASRN**: 135-98-8
Merck No.: 1550   **Beil Ref.**: 4-05-00-01038
Beil RN: 3194820
MF: $C_{10}H_{14}$ **MP[°C]**: -82.7   **Den.[g/cm^3]**: 0.8621^{20}
MW: 134.22   **BP[°C]**: 173.3 n_D: 1.4902^{20}
Sol.: H_2O 1; EtOH 5;
   eth 5; ace 5; bz 5; ctc
   5; peth 5
Therm. Prop.[kJ/mol]: $\Delta_{fus}H$ = 9.83; $\Delta_f H°$(l, 25°C) = -66.4
γ **[mN/m]**: 28.03^{25}; $d\gamma/dT$ = 0.0979 mN/m °C
Elec. Prop.: μ = 0; ε = 2.36^{20}; IP = 8.68 eV
MS: 105(100) 134(18) 91(14) 77(10) 27(9) 106(9) 51(7) 79(7)
IR [cm^{-1}]: 2940 1590 1490 1450 1370 1030 900 760
Raman [cm^{-1}]: 3060 3000 2960 2940 2900 2880 2730 1600 1580 1450 1370
   1360 1320 1280 1240 1210 1180 1150 1100 1060 1030 1000 960 900 850
   840 800 770 740 730 720 640 620 590 540 510 460 330 310 230
UV [nm]: 267(151) 264(146) 261(183) 253(153) MeOH
^{1}H NMR [ppm]: 0.8 1.2 1.6 2.5 7.1 CCl_4
Flammability: Flash pt. = 52°C; ign. temp. = 418°C; flam. lim. = 0.8-6.9%

68. *tert*-Butylbenzene

Syst. Name: Benzene, (1,1-dimethylethyl)-
Synonyms: (1,1-Dimethylethyl)benzene

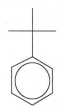

CASRN: 98-06-6   **Beil Ref.:** 4-05-00-01045
Merck No.: 1551
Beil RN: 1421537
MF: $C_{10}H_{14}$ **MP[°C]:** -57.8   **Den.[g/cm^3]:** 0.8665^{20}
MW: 134.22   **BP[°C]:** 169.1 n_D: 1.4927^{20}
Sol.: H_2O 1; EtOH 4; eth
  4; ace 5; bz 5; ctc 5;
  peth 5
Vap. Press. [kPa]: 0.280^{25}; 1.28^{50}; 4.36^{75}; 12.2^{100}; 29.0^{125}; 61.2^{150}
Therm. Prop.[kJ/mol]: $\Delta_{fus}H$ = 8.39; $\Delta_f H°$(l, 25°C) = -70.7
C_p **(liq.) [J/mol °C]:** 238.0^{25}
γ **[mN/m]:** 27.64^{25}; $d\gamma/dT$ = 0.0985 mN/m °C
Elec. Prop.: μ = 0.83 D; ε = 2.36^{20}; IP = 8.64 eV
MS: 119(100) 91(65) 41(40) 134(24) 39(15) 79(14) 77(13) 51(13) 120(11)
  65(7)
IR [cm^{-1}]: 2940 1590 1470 1370 1270 1210 1100 1030 750
Raman [cm^{-1}]: 3060 2960 2920 2900 2860 2770 2710 1600 1580 1470 1450
  1270 1210 1190 1160 1120 1030 1000 930 910 840 710 620 540 450 420 140
UV [nm]: 267 257 cyhex
^{1}H NMR [ppm]: 1.3 7.2 CCl_4
Flammability: Flash pt. = 60°C; ign. temp. = 450°C; flam. lim. = 0.7-5.7%

69. Butyl ethyl ether

Syst. Name: Butane, 1-ethoxy-
Synonyms: Ethyl butyl ether

CASRN: 628-81-9   **DOT No.**: 1179
Beil RN: 1731323   **Beil Ref.**: 4-01-00-01518
MF: $C_6H_{14}O$ **MP[°C]**: -124   **Den.[g/cm^3]**: 0.7495^{20}
MW: 102.18   **BP[°C]**: 92.3 n_D: 1.3818^{20}
Sol.: H_2O 1; EtOH 5; eth
    5; ace 4
Vap. Press. [kPa]: 7.46^{25}; 23.0^{50}; 58.3^{75}; 127^{100}
Therm. Prop.[kJ/mol]: $\Delta_{vap}H = 31.6$
C_p **(liq.) [J/mol °C]**: 159.0^{25}
γ **[mN/m]**: 20.13^{25}; $d\gamma/dT = 0.1049$ mN/m °C
Elec. Prop.: $\mu = (1.2$ D); IP = 9.36 eV
η **[mPa s]**: 0.397^{25}
MS: 59(100) 31(75) 29(35) 41(27) 57(23) 56(22) 27(11) 73(7) 45(7) 43(6)
IR [cm^{-1}]: 2940 2780 1450 1370 1350 1150 1110 1060 970 740
Raman [cm^{-1}]: * 1540 1496 1469 1313 1168 1149 1125 923 860 346 202
Flammability: Flash pt. = 4°C

---

70. Butyl formate

Syst. Name: Methanoic acid, butyl ester

CASRN: 592-84-7   **DOT No.**: 1128
Beil RN: 1742108   **Beil Ref.**: 4-02-00-00028
MF: $C_5H_{10}O_2$ **MP[°C]**: -91.5   **Den.[g/cm^3]**: 0.8885^{20}
MW: 102.13   **BP[°C]**: 106.1 n_D: 1.3912^{20}
Sol.: H_2O 2; EtOH 5; eth
    5; ace 3
Vap. Press. [kPa]: 3.53^{25}; 15.0^{50}; 41.1^{75}; 86.8^{100}
Therm. Prop.[kJ/mol]: $\Delta_{vap}H = 36.6$
C_p **(liq.) [J/mol °C]**: 200.2^{25}
γ **[mN/m]**: 24.52^{25}; $d\gamma/dT = 0.1026$ mN/m °C
Elec. Prop.: $\mu = (2.0$ D); $\varepsilon = 6.10^{30}$; IP = 10.50 eV
η **[mPa s]**: 0.937^0; 0.644^{25}; 0.472^{50}; 0.362^{75}; 0.289^{100}
MS: 56(100) 41(60) 31(58) 29(53) 27(45) 43(34) 28(21) 39(19) 55(11) 42(11)
IR [cm^{-1}]: 2940 1720 1470 1370 1180
^{1}H NMR [ppm]: 0.9 1.5 4.2 8.1 $CDCl_3$
Flammability: Flash pt. = 18°C; ign. temp. = 322°C; flam. lim. = 1.7-8.2%

71. Butyl methyl ketone

Syst. Name: 2-Hexanone
Synonyms: Methyl butyl ketone

CASRN: 591-78-6   **DOT No.:** 1224
Merck No.: 5955   **Beil Ref.:** 4-01-00-03298
Beil RN: 1737676
MF: $C_6H_{12}O$ **MP[°C]:** -55.5   **Den.[g/cm^3]:** 0.8113^{20}
MW: 100.16   **BP[°C]:** 127.6 n_D: 1.4007^{20}
Sol.: H_2O 2; EtOH 5; eth
     5; ace 3
Crit. Const.: T_c = 313.9°C; P_c = 3.32 MPa
Vap. Press. [kPa]: 1.54^{25}; 5.80^{50}; 17.3^{75}; 43.2^{100}; 94.2^{125}; 184^{150}
Therm. Prop.[kJ/mol]: $\Delta_{fus}H$ = 14.90; $\Delta_{vap}H$ = 36.4; $\Delta_f H°$(l, 25°C) = -322.0
C_p **(liq.) [J/mol °C]:** 213.3^{25}
γ **[mN/m]:** 25.45^{25}; $d\gamma/dT$ = 0.1092 mN/m °C
Elec. Prop.: μ = (2.7 D); ε = 14.56^{20}; IP = 9.35 eV
η **[mPa s]:** 1.30^{-25}; 0.840^0; 0.583^{25}; 0.429^{50}; 0.329^{75}; 0.262^{100}
k **[W/m °C]:** 0.151^{-25}; 0.145^0; 0.139^{25}; 0.133^{50}; 0.127^{75}; 0.121^{100}
MS: 43(100) 58(60) 57(17) 100(16) 29(15) 41(13) 85(8) 27(8) 71(7) 59(5)
IR [cm^{-1}]: 2940 1720 1450 1410 1350 1160 1110 730
Flammability: Flash pt. = 25°C; ign. temp. = 423°C; flam. lim. = 1-8%
TLV/TWA: 5 ppm (20 mg/m^3)

---

72. Butyl oleate

Syst. Name: 9-Octadecenoic acid (Z)-, butyl ester
Synonyms: Butyl cis-9-octadecenoate

CASRN: 142-77-8   **Beil Ref.:** 4-02-00-01653
Beil RN: 1728057
MF: $C_{22}H_{42}O_2$ **MP[°C]:** -26.4   **Den.[g/cm^3]:** 0.8704^{15}
MW: 338.57 n_D: 1.4480^{25}
Sol.: EtOH 4
Vap. Press. [kPa]: 0.001^{100}; 0.011^{125}
Therm. Prop.[kJ/mol]: $\Delta_f H°$(l, 25°C) = -816.9
Elec. Prop.: ε = 4.00^{25}
η **[mPa s]:** 7.91^{25}
Flammability: Flash pt. = 180°C

73. Butyl stearate

Syst. Name: Octadecanoic acid, butyl ester

CASRN: 123-95-5   **Beil Ref.:** 4-02-00-01219
Merck No.: 1589
Beil RN: 1792866
MF: $C_{22}H_{44}O_2$ **MP[°C]:** 27   **Den.[g/cm^3]:** 0.854^{25}
MW: 340.59   **BP[°C]:** 343 n_D: 1.4328^{50}
Sol.: H_2O 1; EtOH 3;
  ace 4
Vap. Press. [kPa]: 0.001^{100}; 0.007^{125}
Therm. Prop.[kJ/mol]: $\Delta_{fus}H = 56.90$
γ **[mN/m]:** 32.7^{30}
Elec. Prop.: $\mu = (1.9 \text{ D})$; $\varepsilon = 3.12^{25}$
η **[mPa s]:** 8.26^{25}; 4.9^{50}
MS: 56(100) 57(54) 285(44) 43(40) 73(28) 41(28) 340(26) 55(26) 60(25)
  267(24)
IR [cm^{-1}]: 2940 1720 1470 1350 1250 1180 1120 1080 790 730
Raman [cm^{-1}]: 2930 2910 2870 2850 2730 1740 1440 1370 1300 1150 1120
  1080 1070 960 930 890 870 840 740 640 600 510 400 340 280 210
Flammability: Flash pt. = 160°C; ign. temp. = 355°C

74. *p-tert*-Butyltoluene

Syst. Name: Benzene, 1-(1,1-dimethylethyl)-4-methyl-

Synonyms: 1-(1,1-Dimethylethyl)-4-methylbenzene

CASRN: 98-51-1   **Beil Ref.**: 4-05-00-01097

Beil RN: 2038670

MF: $C_{11}H_{16}$ **MP[°C]**: -52   **Den.[g/cm³]**: 0.8612^{20}

MW: 148.25   **BP[°C]**: 190 n_D: 1.4918^{20}

Sol.: H_2O 1; EtOH 2; eth 4; ace 3; bz 3; chl 4

Vap. Press. [kPa]: 0.090^{25}; 1.75^{75}; 5.41^{100}; 14.1^{125}; 32.3^{150}

Elec. Prop.: $\mu = 0$; IP = 8.28 eV

MS: 133(100) 105(38) 41(23) 148(18) 93(16) 91(14) 115(13) 134(11) 39(11) 116(10)

IR [cm⁻¹]: 3100 3060 3030 2970 2870 2740 1900 1650 1580 1510 1460 1360 1270 1199 1110 1020 810 720

Raman [cm⁻¹]: 3065 3034 3015 2983 2959 2924 2903 2877 1606 796

UV [nm]: 272(347) 266(355) 264(407) 259(282) 212(8313) iso

¹³C NMR [ppm]: 20.7 31.4 34.2 125.0 128.7 134.5 148.0 $CDCl_3$

Flammability: Flash pt. = 68°C

TLV/TWA: 1 ppm (6.1 mg/m³)

75. Butyl vinyl ether

Syst. Name: Butane, 1-(ethenyloxy)-
Synonyms: Vinyl butyl ether

CASRN: 111-34-2	**DOT No.**: 2352
Beil RN: 1560217	**Beil Ref.**: 4-01-00-02052

MF: $C_6H_{12}O$ **MP[°C]**: -92   **Den.[g/cm^3]**: 0.7888^{20}
MW: 100.16   **BP[°C]**: 94 n_D: 1.4026^{20}
Sol.: H_2O 1; EtOH 4; eth
   5; ace 4; bz 3; os 5;
   glycol 2; glycerol 2
Vap. Press. [kPa]: 1.76^0; 6.65^{25}; 20.5^{50}; 53.8^{75}; 124^{100}
Therm. Prop.[kJ/mol]: $\Delta_{vap}H = 31.6$; $\Delta_f H°(l, 25°C) = -218.8$
C_p **(liq.) [J/mol °C]**: 232.0^{25}
γ **[mN/m]**: 21.99^{20}
Elec. Prop.: $\mu = (1.3\ D)$
η **[mPa s]**: 0.5^{20}
MS: 29(100) 41(74) 56(56) 57(43) 27(42) 44(26) 15(16) 85(14) 39(14) 43(13)
IR [cm^{-1}]: 3130 2940 1640 1610 1470 1390 1320 1190 1120 1080 960 810
Raman [cm^{-1}]: 3120 3050 3020 2960 2940 2910 2880 2770 2720 1640 1620
   1470 1420 1370 1350 1330 1260 1210 1180 1170 1140 1020 970 960 940
   920 870 840 810 710 600 550 500 460 430 400 370 320 290 270 170
UV [nm]: 192(9550) hp
^{1}H NMR [ppm]: 0.9 1.4 3.6 3.8 4.0 6.3 CCl_4
Flammability: Flash pt. = -9°C; ign. temp. = 255°C

76. γ-Butyrolactone
Syst. Name: 2(3H)-Furanone, dihydro-
Synonyms: Oxolan-2-one

CASRN: 96-48-0   **Beil Ref.**: 5-17-09-00007
Merck No.: 1596
Beil RN: 105248
MF: $C_4H_6O_2$ **MP[°C]**: -43.3   **Den.[g/cm³]**: 1.1284^{16}
MW: 86.09   **BP[°C]**: 204 n_D: 1.4341^{20}
Sol.: ace 4; bz 4; eth 4;
    EtOH 4
Vap. Press. [kPa]: 0.430^{25}; 8.30^{125}; 19.7^{150}
Therm. Prop.[kJ/mol]: $\Delta_{fus}H = 9.57$; $\Delta_{vap}H = 52.2$
C_p **(liq.) [J/mol °C]**: 141.4^{25}
Elec. Prop.: $\mu = 4.27$ D; $\varepsilon = 39.0^{20}$
η **[mPa s]**: 1.7^{25}
MS: 28(100) 42(74) 29(48) 27(33) 41(27) 56(25) 86(24) 26(18) 85(10) 39(10)
IR [cm⁻¹]: 3030 2940 1750 1470 1430 1390 1280 1250 1180 1090 1040 990
    930 870 800 670
Raman [cm⁻¹]: 3010 2940 1770 1490 1470 1420 1380 1240 1200 1080 1040
    990 930 880 800 670 630 540 500 170
UV [nm]: 209(43) MeOH
¹³C NMR [ppm]: 22.2 27.7 68.6 177.9 CDCl₃
¹H NMR [ppm]: 4.4 CDCl₃
Flammability: Flash pt. = 98°C

77. Camphene

Syst. Name: Bicyclo[2.2.1]heptane, 2,2-dimethyl-3-methylene-, (±)-

CASRN: 565-00-4   **Rel. CASRN**: 79-92-5
Beil RN: 3194798   **DOT No.**: 9011
                                       Beil Ref.: 4-05-00-00462
MF: $C_{10}H_{16}$ **MP[°C]**: 51.5   **Den.[g/cm^3]**: 0.879^{20}
MW: 136.24   **BP[°C]**: 158.5 n_D: 1.4551^{54}
Sol.: H_2O 1; EtOH 4; eth
   4
MS: 93(100) 121(63) 41(59) 39(51) 27(44) 79(38) 67(34) 136(14)
IR [cm^{-1}]: 3030 2940 1670 1470 1370 1160 1110 970 890 810 750
UV [nm]: 206(5495) EtOH
^{13}C NMR [ppm]: 23.8 25.8 28.9 29.4 37.4 41.7 47.0 48.2 99.1 165.9 CDCl$_3$

78. (+)-Camphor

Syst. Name: Bicyclo[2.2.1]heptan-2-one, 1,7,7-trimethyl-, (1R)-

Synonyms: 1,7,7-Trimethylbicyclo[2.2.1]hepten-2-one

CASRN: 464-49-3   **Beil Ref.**: 4-07-00-00213
Merck No.: 1738
Beil RN: 2042745

MF: $C_{10}H_{16}O$	**MP[°C]**: 178.8	**Den.[g/cm³]**: 0.990^{25}
MW: 152.24	**BP[°C]**: 207.4	n_D: 1.5462

Sol.: H_2O 1; EtOH 4; eth 4; ace 3; bz 3; ctc 4; chl 4; MeOH 4; HOAc 4

Vap. Press. [kPa]: 0.032^{25}; 0.173^{50}; 0.735^{75}; 2.58^{100}; 7.7^{125}

Therm. Prop.[kJ/mol]: $\Delta_{fus}H$ = 6.84; $\Delta_{vap}H$ = 59.5; $\Delta_f H°$(s, 25°C) = -319.4

Elec. Prop.: μ = (3.1 D); IP = 8.76 eV

MS: 126(100) 95(97) 41(79) 81(71) 39(48) 69(43) 108(39) 152(29)

IR [cm⁻¹]: 2960 2880 1750 1470 1450 1420 1390 1370 1320 1280 1090 1050 1020 750 520

Raman [cm⁻¹]: 2960 2930 2860 2780 2730 1740 1470 1450 1420 1390 1370 1320 1300 1250 1220 1190 1160 1100 1080 1050 1020 990 950 940 920 860 830 750 720 660 610 580 560 530 480 420 400 300 290 270 250 230 130

UV [nm]: 289(36) MeOH

¹H NMR [ppm]: 0.8 0.9 1.0 0.8 0.9 1.0 CCl_4

Flammability: Flash pt. = 66°C; ign. temp. = 466°C; flam. lim. = 0.6-3.5%

TLV/TWA: 2 ppm (12 mg/m³)

79. ε-Caprolactam
Syst. Name: 2H-Azepin-2-one, hexahydro-
Synonyms: 6-Hexanelactam

CASRN: 105-60-2   **Beil Ref.**: 5-21-06-00444
Merck No.: 1762
Beil RN: 106934
MF: $C_6H_{11}NO$ **MP[°C]**: 69.3
MW: 113.16   **BP[°C]**: 270
Sol.: H_2O 4; bz 4; EtOH
    4; chl 4
Vap. Press. [kPa]: 0.143^{90}
Therm. Prop.[kJ/mol]: $\Delta_f H°$(s, 25°C) = -329.4
Elec. Prop.: μ = (3.9 D); IP = 9.07 eV
η **[mPa s]**: 19.7^{70}
MS: 55(100) 113(87) 30(81) 56(66) 84(60) 85(57) 42(51) 41(33) 28(26) 43(17)
IR [cm^{-1}]: 3130 2630 1670 1490 1430 1410 1370 1210 1120 1090 980 820
Raman [cm^{-1}]: 3200 2970 2940 2900 2860 2700 2660 1640 1490 1450 1430
    1370 1340 1320 1290 1270 1240 1200 1170 1120 1090 1030 990 970 900
    870 850 830 750 710 590 490 400 350 *
UV [nm]: 198(7762) H_2O
^{13}C NMR [ppm]: 23.2 29.7 30.6 36.8 42.6 179.5 CDCl$_3$
^{1}H NMR [ppm]: 1.7 2.4 3.2 7.8 CDCl$_3$
Flammability: Flash pt. = 125°C
TLV/TWA: 1 mg/m^3
Reg. Lists: CERCLA (RQ = 1 lb.)

80. Carbon disulfide

Syst. Name: Carbon disulfide
Synonyms: Carbon bisulfide

CASRN: 75-15-0	**DOT No.**: 1131
Merck No.: 1818	**Beil Ref.**: 4-03-00-00395

Beil RN: 1098293
MF: CS_2 **MP[°C]**: -111.5   **Den.[g/cm³]**: 1.2632^{20}
MW: 76.14   **BP[°C]**: 46 n_D: 1.6319^{20}
Sol.: H_2O 3; EtOH 5; eth
   5; chl 3
Crit. Const.: T_c = 279°C; P_c = 7.90 MPa; V_c = 173 cm³
Vap. Press. [kPa]: 4.68^{-25}; 16.9^{0}; 48.2^{25}; 114^{50}
Therm. Prop.[kJ/mol]: $\Delta_{fus}H$ = 4.40; $\Delta_{vap}H$ = 26.7; $\Delta_f H°(l, 25°C)$ = 89.0
C_p **(liq.) [J/mol °C]**: 76.4^{25}
γ **[mN/m]**: 31.58^{25}; $d\gamma/dT$ = 0.1484 mN/m °C
Elec. Prop.: μ = 0; ε = 2.63^{20}; IP = 10.07 eV
η **[mPa s]**: 0.429^{0}; 0.352^{25}
k **[W/m °C]**: 0.154^{0}; 0.149^{25}
MS: 76(100) 32(22) 44(17) 78(9) 38(6) 28(5) 77(3) 64(1) 46(1) 39(1)
IR [cm⁻¹]: 2070 1540 1480 390
Raman [cm⁻¹]: 800 660 650 400
UV [nm]: 317 cyhex
Flammability: Flash pt. = -30°C; ign. temp. = 90°C; flam. lim. = 1.3-50.0%
TLV/TWA: 10 ppm (31 mg/m³)
Reg. Lists: CERCLA (RQ = 100 lb.); SARA 313 (1.0%); RCRA P022 (accutely
   hazardous)

81. *o*-Chloroaniline

Syst. Name: Benzenamine, 2-chloro-

CASRN: 95-51-2   **Rel. CASRN:** 27134-26-5
Merck No.: 2118   **DOT No.:** 2019
Beil RN: 606077   **Beil Ref.:** 4-12-00-01115
MF: C_6H_6ClN **MP[°C]:** -14 n_D: 1.5895^{20}
MW: 127.57   **BP[°C]:** 208.8
Sol.: H_2O 1; EtOH 5; eth
   3; ace 3; os 3
Vap. Press. [kPa]: 0.034^{25}; 7.82^{125}; 19.3^{150}
Therm. Prop.[kJ/mol]: $\Delta_{fus}H = 11.88$; $\Delta_{vap}H = 44.4$
γ [mN/m]: 43.66^{20}
Elec. Prop.: $\mu = (1.8\ D)$; $\varepsilon = 13.40^{20}$; IP = 8.50 eV
η [mPa s]: 3.32^{25}; 1.91^{50}; 1.25^{75}; 0.887^{100}
MS: 127(100) 129(32) 92(17) 65(16) 128(10) 91(9) 64(9) 39(8) 63(7) 99(6)
IR [cm^{-1}]: 3450 3330 1610 1490 1450 1320 1270 1160 1090 1050 1020 940
   830 750 680
Raman [cm^{-1}]: 3380 3200 3150 3060 3030 1610 1590 1490 1450 1420 1310
   1260 1160 1150 1090 1050 1030 880 840 750 680 560 480 380 260 170
UV [nm]: 290 236 cyhex
^{1}H NMR [ppm]: 3.8 6.8 CCl_4

82. Chlorobenzene

Syst. Name: Benzene, chloro-
Synonyms: Phenyl chloride

CASRN: 108-90-7
Merck No.: 2121
Beil RN: 605632
MF: C_6H_5Cl
MW: 112.56

DOT No.: 1134
Beil Ref.: 4-05-00-00640

MP[°C]: -45.2
BP[°C]: 131.7

Den.[g/cm^3]: 1.1058^{20}
n_D: 1.5241^{20}

Sol.: H_2O 1; EtOH 5; eth 5; bz 4; ctc 4; chl 4; CS_2 4
Crit. Const.: T_c = 359.3°C; P_c = 4.52 MPa; V_c = 308 cm^3
Vap. Press. [kPa]: 1.60^{25}; 16.3^{75}; 39.5^{100}; 84.3^{125}
Therm. Prop.[kJ/mol]: $\Delta_{fus}H$ = 9.61; $\Delta_{vap}H$ = 35.2; $\Delta_f H°$(l, 25°C) = 11.0
C_p **(liq.) [J/mol °C]:** 150.1^{25}
γ **[mN/m]:** 32.99^{25}; $d\gamma/dT$ = 0.1191 mN/m °C
Elec. Prop.: μ = 1.69 D; ε = 5.69^{20}; IP = 9.06 eV
η **[mPa s]:** 1.70^{-25}; 1.06^0; 0.753^{25}; 0.575^{50}; 0.456^{75}; 0.369^{100}
k **[W/m °C]:** 0.136^{-25}; 0.131^0; 0.127^{25}; 0.122^{50}; 0.117^{75}; 0.112^{100}
MS: 112(100) 77(63) 114(33) 51(29) 50(14) 75(8) 113(7) 78(5) 76(5) 28(4)
IR [cm^{-1}]: 3030 1590 1490 1450 1140 1090 1030 890 750 690
Raman [cm^{-1}]: 3160 3060 1580 1170 1160 1080 1020 1000 700 610 420 290 190
UV [nm]: 263(191) H_2O
^{13}C NMR [ppm]: 126.5 128.6 128.8 134.3 diox
^{1}H NMR [ppm]: 7.2 CCl$_4$
Flammability: Flash pt. = 28°C; ign. temp. = 593°C; flam. lim. = 1.3-9.6%
TLV/TWA: 10 ppm (46 mg/m^3)
Reg. Lists: CERCLA (RQ = 100 lb.); SARA 313 (1.0%); RCRA U037 (toxic)

83. 1-Chlorobutane

Syst. Name: Butane, 1-chloro-
Synonyms: Butyl chloride

CASRN: 109-69-3
Merck No.: 1560
Beil RN: 1730909
MF: C_4H_9Cl
MW: 92.57
Sol.: H_2O 1; EtOH 5; eth 5; ctc 2

Rel. CASRN: 25154-42-1
DOT No.: 1127
Beil Ref.: 4-01-00-00246
MP[°C]: -123.1   **Den.[g/cm³]:** 0.8862^{20}
BP[°C]: 78.6 n_D: 1.4021^{20}

Vap. Press. [kPa]: 3.82^0; 13.7^{25}; 38.6^{50}; 91.1^{75}; 188^{100}
Therm. Prop.[kJ/mol]: $\Delta_{vap}H = 30.4$; $\Delta_f H°(l, 25°C) = -188.1$
C_p **(liq.) [J/mol °C]:** 175.0^{25}
γ **[mN/m]:** 23.18^{25}; $d\gamma/dT = 0.1117$ mN/m °C
Elec. Prop.: $\mu = 2.05$ D; $\varepsilon = 7.28^{20}$; IP = 10.67 eV
η **[mPa s]:** 0.556^0; 0.422^{25}; 0.329^{50}; 0.261^{75}
MS: 56(100) 41(65) 27(50) 43(35) 29(24) 39(18) 28(16) 26(11) 55(8) 15(8)
IR [cm⁻¹]: 2970 2940 2870 1470 1440 1380 1310 1290 750 660
Raman [cm⁻¹]: 2990 2960 2940 2900 2870 2840 2730 1450 1350 1310 1290 1230 1200 1110 1080 1060 1020 930 900 880 820 800 750 730 660 480 440 340 310 250
¹³C NMR [ppm]: 13.4 20.4 35.2 44.6
¹H NMR [ppm]: 1.0 1.5 3.5 CCl_4
Flammability: Flash pt. = -9°C; ign. temp. = 240°C; flam. lim. = 1.8-10.1%

84. 2-Chlorobutane

Syst. Name: Butane, 2-chloro-, (±)-
Synonyms: *sec*-Butyl chloride

CASRN: 53178-20-4
Merck No.: 1561
Beil RN: 1718773
MF: C_4H_9Cl
MW: 92.57
Sol.: bz 4; eth 4; EtOH 4; chl 4

Rel. CASRN: 25154-42-1
DOT No.: 1127
Beil Ref.: 4-01-00-00248
MP[°C]: -131.3   **Den.[g/cm^3]**: 0.8732^{20}
BP[°C]: 68.2 n_D: 1.3971^{20}

Vap. Press. [kPa]: 6.45^0; 21.0^{25}; 55.4^{50}
Therm. Prop.[kJ/mol]: $\Delta_{vap}H = 29.2$; $\Delta_f H°(l, 25°C) = -192.8$
Elec. Prop.: $\mu = 2.04$ D; $\varepsilon = 8.56^{20}$; IP = 10.53 eV
MS: 56(100) 57(100) 41(90) 27(77) 29(57) 63(46) 39(34) 92(1)
IR [cm^{-1}]: 3000 2950 2900 1475 1450 1380 1300 1280 1480 1160 1060 990 850 790 670
Raman [cm^{-1}]: 2970 2930 2880 2850 2730 2700 1460 1300 1240 1160 1110 1070 1060 1030 990 960 850 830 800 670 630 610 530 470 420 390 380 330 250
^{13}C NMR [ppm]: 11.1 25.0 33.7 60.1 diox
^{1}H NMR [ppm]: 1.1 1.5 1.7 3.9 CCl$_4$
Flammability: Flash pt. < 0°C

85. 1-Chloro-1,1-difluoroethane

Syst. Name: Ethane, 1-chloro-1,1-difluoro
Synonyms: R 142b

CASRN: 75-68-3   **DOT No.:** 2517
MF: $C_2H_3ClF_2$ **MP[°C]:** -130.8   **Den.[g/cm³]:** 1.107^{25}
MW: 100.50   **BP[°C]:** -9.7
Sol.: H_2O 1; bz 3
Crit. Const.: T_c = 137.1°C; P_c = 4.041 MPa; V_c = 225 cm³
Vap. Press. [kPa]: 14^{-50}; 53^{-25}
Therm. Prop.[kJ/mol]: $\Delta_{fus}H$ = 2.69
C_p **(liq.) [J/mol °C]:** 130.5^{21} (sat. press.)
Elec. Prop.: μ = 2.14 D; IP = 11.98 eV
η **[mPa s]:** 0.477^{-25}; 0.376^{0}

---

86. Chlorodifluoromethane

Syst. Name: Methane, chlorodifluoro-
Synonyms: Difluorochloromethane; R 22; CFC 22

CASRN: 75-45-6   **DOT No.:** 1018
Beil RN: 1731036   **Beil Ref.:** 4-01-00-00032
MF: $CHClF_2$ **MP[°C]:** -157.4   **Den.[g/cm³]:** 1.4909^{-69}
MW: 86.47   **BP[°C]:** -40.7
Sol.: H_2O 4; eth 3; ace 3;
    chl 3
Crit. Const.: T_c = 96.2°C; P_c = 4.99 MPa; V_c = 169 cm³
Vap. Press. [kPa]: 16.7^{-73}; 201^{-25}; 498^{0}; 1044^{25}; 1944^{50}; 3317^{75}
Therm. Prop.[kJ/mol]: $\Delta_{fus}H$ = 4.12; $\Delta_{vap}H$ = 20.2; $\Delta_f H°$(g, 25°C) = -482.6
C_p **(liq.) [J/mol °C]:** 93.0^{-41}
Elec. Prop.: μ = 1.42 D; ε = 6.11^{24}; IP = 12.20 eV
MS: 51(100) 31(17) 67(15) 35(12) 32(10) 50(6) 69(5) 37(4) 13(4) 47(3)
IR [cm⁻¹]: 2940 2170 1320 1110 920 820 810 800
TLV/TWA: 1000 ppm (3540 mg/m³)

87. Chloroethane

Syst. Name: Ethane, chloro-
Synonyms: Ethyl chloride; R 160

CASRN: 75-00-3   **DOT No.:** 1037
Merck No.: 3740   **Beil Ref.:** 4-01-00-00124
Beil RN: 1730751
MF: C_2H_5Cl **MP[°C]:** -138.7   **Den.[g/cm^3]:** 0.8902^{25} (sat. press.)
MW: 64.51   **BP[°C]:** 12.3 n_D: 1.3676^{20}
Sol.: H_2O 2; EtOH 4; eth
   5; chl 2
Crit. Const.: $T_c = 187.3°C$; $P_c = 5.3$ MPa
Vap. Press. [kPa]: 19.5^{-25}; 62.3^{0}; 160^{25}
Therm. Prop.[kJ/mol]: $\Delta_{fus}H = 4.45$; $\Delta_{vap}H = 24.7$; $\Delta_f H°(g, 25°C) = -112.2$
C_p (liq.) [J/mol °C]: 104.3^{25} (sat. press.)
γ [mN/m]: 20.64^{10}
Elec. Prop.: $\mu = 2.05$ D; $\varepsilon = 9.45^{20}$; IP = 10.97 eV
η [mPa s]: 0.416^{-25}; 0.319^{0}
k [W/m °C]: 0.145^{-25}; 0.132^{0}; 0.119^{25}; 0.106^{50}; 0.093^{75}
MS: 64(100) 28(91) 29(84) 27(75) 66(32) 26(28) 49(25) 51(8) 63(6) 65(4)
IR [cm^{-1}]: 2940 2860 1450 1410 1300 1280 990 970 790 690
^{13}C NMR [ppm]: 18.7 39.9
^{1}H NMR [ppm]: 1.5 3.6 CDCl$_3$
Flammability: Flash pt. = -50°C; ign. temp. = 519°C; flam. lim. = 3.8-15.4%
TLV/TWA: 1000 ppm (2640 mg/m^3)
Reg. Lists: CERCLA (RQ = 100 lb.); SARA 313 (1.0%)

88. Chloroethylene

Syst. Name: Ethene, chloro-
Synonyms: Vinyl chloride; R 1140

CASRN: 75-01-4   **DOT No.**: 1086
Merck No.: 9898   **Beil Ref.**: 4-01-00-00700
Beil RN: 1731576
MF: C_2H_3Cl **MP[°C]**: -153.7   **Den.[g/cm^3]**: 0.9106^{20}
MW: 62.50   **BP[°C]**: -13.3 n_D: 1.3700^{20}
Sol.: H_2O 2; EtOH 3; eth
    4
Vap. Press. [kPa]: 62.6^{-25}; 170^0; 355^{25}
Therm. Prop.[kJ/mol]: $\Delta_{fus}H = 4.75$; $\Delta_{vap}H = 20.8$; $\Delta_f H°(g, 25°C) = 37.3$
C_p **(liq.) [J/mol °C]**: 89.4^{25} (sat. press.)
γ **[mN/m]**: 22.27^{-20}
Elec. Prop.: $\mu = 1.45$ D; $\epsilon = 6.26^{17}$; IP = 9.99 eV
MS: 27(100) 62(77) 26(34) 64(24) 25(14) 35(9) 61(7) 60(5) 63(4) 47(4)
IR [cm^{-1}]: 3080 2980 1600 1500 1350 1260 1020 900 700 600
^{13}C NMR [ppm]: 117.4 126.1
Flammability: Flash pt. = -78°C; ign. temp. = 472°C; flam. lim. = 3.6-33.0%
TLV/TWA: 5 ppm (13 mg/m^3)
Reg. Lists: CERCLA (RQ = 1 lb.); SARA 313 (0.1%); RCRA U043 (toxic);
    confirmed or suspected carcinogen

89. Chloromethane

Syst. Name: Methane, chloro-
Synonyms: Methyl chloride; R 40

CASRN: 74-87-3
Merck No.: 5964
Beil RN: 1696839
MF: CH_3Cl
MW: 50.49
Sol.: H_2O 3; EtOH 3; eth
 5; ace 5; bz 5; chl 5;
 HOAc 5

DOT No.: 1063
Beil Ref.: 4-01-00-00028

MP[°C]: -97.7 **Den.[g/cm^3]**: 0.911^{25} (sat. press.)
BP[°C]: -24.0 n_D: 1.3389^{20}

Crit. Const.: T_c = 143.10°C; P_c = 6.679 MPa; V_c = 139 cm^3
Vap. Press. [kPa]: 6.7^{-73}; 97.8^{-25}; 259^0; 574^{25}; 1115^{50}; 1965^{75}; 3220^{100}
Therm. Prop.[kJ/mol]: $\Delta_{fus}H$ = 6.43; $\Delta_{vap}H$ = 21.4; $\Delta_f H°$(g, 25°C) = -81.9
C_p (liq.) [J/mol °C]: 75.6^{-24}
Elec. Prop.: μ = 1.892 D; ε = 10.0^{22}; IP = 11.22 eV
MS: 50(100) 15(72) 52(31) 49(11) 14(8) 47(7) 35(6) 13(5) 51(4) 48(3)
IR [cm^{-1}]: 2980 2950 2880 1480 1450 1420 1360 1330 1020 740 730 710
^{13}C NMR [ppm]: 25.1
Flammability: Flash pt. < 0°C; ign. temp. = 632°C; flam. lim. = 8.1-17.4%
TLV/TWA: 50 ppm (103 mg/m^3)
Reg. Lists: CERCLA (RQ = 100 lb.); SARA 313 (1.0%); RCRA U045 (toxic);
 confirmed or suspected carcinogen

90. 1-Chloro-3-methylbutane

Syst. Name: Butane, 1-chloro-3-methyl-
Synonyms: Isoamyl chloride

CASRN: 107-84-6   **Beil Ref.:** 4-01-00-00325
Merck No.: 4998
Beil RN: 1730988
MF: $C_5H_{11}Cl$ **MP[°C]:** -104.4   **Den.[g/cm³]:** 0.8750^{20}
MW: 106.60   **BP[°C]:** 98.9 n_D: 1.4084^{20}
Sol.: H_2O 2; EtOH 5; eth
  5; chl 4
Therm. Prop.[kJ/mol]: $\Delta_{vap}H$ = 32.0; $\Delta_f H°$(l, 25°C) = -216.0
C_p **(liq.) [J/mol °C]:** 175.1^{25}
γ **[mN/m]:** 22.90^{25}
Elec. Prop.: μ = (1.9 D); ϵ = 6.10^{19}
MS: 43(100) 55(56) 41(55) 27(51) 70(49) 42(37) 29(34) 57(30) 39(30) 56(15)
IR [cm⁻¹]: 2960 2870 1460 1370 1270 1160 1020 870 720 660
Raman [cm⁻¹]: 2970 2930 2880 2840 2770 2730 1470 1450 1360 1340 1310
    1290 1240 1190 1170 1130 1100 1040 970 900 870 830 800 760 730 660 490
    450 420 380 310 280
¹³C NMR [ppm]: 22.0 25.7 41.6 43.1 $CDCl_3$
¹H NMR [ppm]: 0.9 1.7 3.6 $CDCl_3$
Flammability: Flash pt. < 21°C; flam. lim. = 1.5-7.4%

91. 3-(Chloromethyl)heptane

Syst. Name: Heptane, 3-(chloromethyl)-
Synonyms: 2-Ethylhexyl chloride

CASRN: 123-04-6   **Beil Ref.**: 4-01-00-00430
Beil RN: 1697456
MF: $C_8H_{17}Cl$ **BP[°C]**: 172   **Den.[g/cm^3]**: 0.8769^{20}
MW: 148.68 n_D: 1.4319^{20}
Sol.: H_2O 1; EtOH 3; eth
   3; ace 3; bz 3; ctc 2;
   os 3
Vap. Press. [kPa]: 9.89^{100}; 24.2^{125}; 53.1^{150}
η **[mPa s]**: 1.0^{20}
MS: 57(100) 41(39) 43(30) 29(29) 27(29) 55(28) 99(16) 39(15) 56(9) 42(7)
IR [cm^{-1}]: 2940 1470 1390 1300 790 770 730 690
Raman [cm^{-1}]: 2990 2960 2940 2870 2730 1450 1350 1300 1210 1190 1150
   1100 1090 1060 1040 960 900 870 830 810 780 760 720 680 550 520 450
   400 350 300 260
^{1}H NMR [ppm]: 0.9 1.4 3.5 CCl_4
Flammability: Flash pt. = 60°C

92. 1-Chloro-2-methylpropane
Syst. Name: Propane, 1-chloro-2-methyl-
Synonyms: Isobutyl chloride

CASRN: 513-36-0   **Beil Ref.**: 4-01-00-00287
Merck No.: 5022
Beil RN: 635650
MF: C_4H_9Cl **MP[°C]**: -130.3   **Den.[g/cm^3]**: 0.8773^{20}
MW: 92.57   **BP[°C]**: 68.5 n_D: 1.3984^{20}
Sol.: eth 3; ace 3; ctc 2;
  chl 3
Vap. Press. [kPa]: 1.30^{-25}; 5.90^0; 19.9^{25}; 53.5^{50}; 122^{75}
Therm. Prop.[kJ/mol]: $\Delta_{vap}H$ = 29.2; $\Delta_f H°$(l, 25°C) = -191.1
C_p **(liq.) [J/mol °C]**: 158.6^{25}
γ **[mN/m]**: 21.99^{20}
Elec. Prop.: μ = 2.00 D; ϵ = 7.03^{20}; IP = 10.66 eV
η **[mPa s]**: 0.471^{15}
MS: 43(100) 41(67) 42(50) 27(33) 39(26) 15(11) 29(10) 56(7) 49(6) 38(5)
IR [cm^{-1}]: 2940 1470 1390 1370 1330 1270 940 880 810 800 730 690
^{1}H NMR [ppm]: 1.0 1.9 3.3 CCl$_4$
Flammability: Flash pt. < 21°C; flam. lim. = 2.0-8.8%

93. 2-Chloro-2-methylpropane

Syst. Name: Propane, 2-chloro-2-methyl-
Synonyms: *tert*-Butyl chloride

CASRN: 507-20-0   **Beil Ref.**: 4-01-00-00288
Merck No.: 1562
Beil RN: 1730872
MF: C_4H_9Cl **MP[°C]**: -26   **Den.[g/cm³]**: 0.8420^{20}
MW: 92.57   **BP[°C]**: 50.9 n_D: 1.3857^{20}
Sol.: H_2O 2; EtOH 5; eth
    5; bz 3; ctc 3; chl 3
Vap. Press. [kPa]: 42.7^{25}; 99.2^{50}
Therm. Prop.[kJ/mol]: $\Delta_{fus}H$ = 2.09; $\Delta_{vap}H$ = 27.6; $\Delta_f H°(l, 25°C)$ = -211.2
C_p **(liq.) [J/mol °C]**: 172.8^0
γ **[mN/m]**: 20.06^{15}
Elec. Prop.: μ = 2.13 D; ε = 9.66^{20}; IP = 10.61 eV
η **[mPa s]**: 0.543^{15}
MS: 57(100) 41(80) 77(44) 29(26) 39(24) 79(14) 27(14) 59(7) 56(7) 38(6)
IR [cm⁻¹]: 2970 1470 1450 1390 1370 1240 1150 800
Raman [cm⁻¹]: 2970 2920 2900 2860 2780 2720 1500 1440 1230 1150 1030
    920 800 560 400 360 300
¹³C NMR [ppm]: 34.4 66.8 CDCl₃
¹H NMR [ppm]: 1.6 CCl₄
Flammability: Flash pt. < 0°C

94. 1-Chloronaphthalene

Syst. Name: Naphthalene, 1-chloro-
Synonyms: 1-Naphthyl chloride

CASRN: 90-13-1   **Beil Ref.**: 4-05-00-01658
Merck No.: 2149
Beil RN: 970836
MF: $C_{10}H_7Cl$ **MP[°C]**: -2.5   **Den.[g/cm^3]**: 1.1938^{20}
MW: 162.62   **BP[°C]**: 259 n_D: 1.6326^{20}
Sol.: H_2O 1; EtOH 3; eth
   3; bz 3; ctc 2; CS_2 3
Vap. Press. [kPa]: 0.132^{75}; 0.534^{100}; 1.73^{125}; 4.71^{150}
Therm. Prop.[kJ/mol]: $\Delta_{fus}H$ = 12.90; $\Delta_{vap}H$ = 52.1; $\Delta_f H°$(l, 25°C) = 54.6
C_p (liq.) [J/mol °C]: 212.6^{25}
γ [mN/m]: 41.04^{20}
Elec. Prop.: μ = (1.6 D); ε = 5.04^{25}; IP = 8.13 eV
η [mPa s]: 2.94^{25}
MS: 162(100) 127(36) 164(30) 126(20) 163(10) 77(8) 101(6) 75(6) 128(5)
   28(4)
IR [cm^{-1}]: 3130 1610 1590 1520 1390 1250 1210 1150 1090 1060 1020 970
   850 830 790 770 660
Raman [cm^{-1}]: 3050 1620 1590 1560 1500 1450 1430 1370 1350 1300 1250
   1200 1160 1140 1070 1060 1020 960 940 900 850 830 780 760 730 660 530
   510 460 420 390 350 240 220 180 130
UV [nm]: 320(249) 315(272) 310(443) 291(5280) 283(7490) 273(6600) 223
   MeOH
^{1}H NMR [ppm]: 7.1 7.5 8.2 CCl_4
Flammability: Flash pt. = 121°C; ign. temp. = >558°C

95. 1-Chlorooctane

Syst. Name: Octane, 1-chloro-
Synonyms: Octyl chloride

CASRN: 111-85-3   **Beil Ref.**: 4-01-00-00419
Beil RN: 1697464
MF: $C_8H_{17}Cl$ **MP[°C]**: -57.8   **Den.[g/cm^3]**: 0.8738^{20}
MW: 148.68   **BP[°C]**: 181.5 n_D: 1.4305^{20}
Sol.: H_2O 1; EtOH 4; eth
   4; ctc 2
Vap. Press. [kPa]: 2.30^{75}; 7.05^{100}; 18.1^{125}; 40.7^{150}
Therm. Prop.[kJ/mol]: $\Delta_f H°$(l, 25°C) = -291.3
C_p **(liq.) [J/mol °C]**: 198.5^{25}
γ **[mN/m]**: 27.15^{25}
Elec. Prop.: μ = (2.0 D); ϵ = 5.05^{25}
η **[mPa s]**: 1.13^{25}
MS: 91(100) 41(83) 43(76) 27(62) 29(56) 55(55) 39(34) 93(32) 57(32) 69(29)
IR [cm^{-1}]: 2860 1450 1370 1280 1050 910 810 760 720
^{1}H NMR [ppm]: 0.9 1.3 1.8 3.5 CCl$_4$
Flammability: Flash pt. = 70°C

---

96. Chloropentafluoroethane

Syst. Name: Ethane, chloropentafluoro-
Synonyms: Pentafluorochloromethane; R 115; CFC
   115

CASRN: 76-15-3   **DOT No.**: 1020
Beil RN: 1740329   **Beil Ref.**: 4-01-00-00129
MF: C_2ClF_5 **MP[°C]**: -99.4   **Den.[g/cm^3]**: 1.5678^{-42}
MW: 154.47   **BP[°C]**: -37.9 n_D: 1.2678^{-42}
Sol.: H_2O 1; EtOH 3; eth
   3
Crit. Const.: T_c = 80.1°C; P_c = 3.229 MPa; V_c = 252 cm^3
Vap. Press. [kPa]: 2.06^{-100}; 14.4^{-75}; 60.6^{-50}
Therm. Prop.[kJ/mol]: $\Delta_{fus}H$ = 1.88; $\Delta_{vap}H$ = 19.4
C_p **(liq.) [J/mol °C]**: 184.2^{25} (sat. press.)
Elec. Prop.: μ = 0.52 D; IP = 12.60 eV
MS: 85(100) 69(61) 31(38) 87(32) 50(17) 35(8) 119(6) 66(4) 100(3) 47(3)
IR [cm^{-1}]: 1540 1410 1350 1240 1180 1120 980 880 760
TLV/TWA: 1000 ppm (6320 mg/m^3)
Reg. Lists: SARA 313 (1.0%)

97. 1-Chloropentane

Syst. Name: Pentane, 1-chloro-
Synonyms: Amyl chloride; Pentyl chloride

CASRN: 543-59-9
Merck No.: 642
Beil RN: 1696936
MF: $C_5H_{11}Cl$
MW: 106.60
Sol.: H_2O 1; EtOH 5; eth
 5; bz 3; ctc 3; chl 4

DOT No.: 1107
Beil Ref.: 4-01-00-00309

MP[°C]: -99
BP[°C]: 107.8

Den.[g/cm^3]: 0.8820^{20}
n_D: 1.4127^{20}

Vap. Press. [kPa]: 4.36^{25}; 13.8^{50}; 35.7^{75}; 79.6^{100}; 158^{125}
Therm. Prop.[kJ/mol]: $\Delta_{vap}H = 33.2$; $\Delta_fH°(l, 25°C) = -213.2$
γ [mN/m]: 24.40^{25}; $d\gamma/dT = 0.1076$ mN/m °C
Elec. Prop.: $\mu = 2.16$ D; $\varepsilon = 6.65^{20}$
η [mPa s]: 0.546^{25}
MS: 42(100) 41(90) 70(89) 55(87) 27(73) 29(55) 43(40) 39(40) 28(19) 57(18)
IR [cm^{-1}]: 2860 1450 1370 1300 1270 1030 920 730 650
Raman [cm^{-1}]: 2990 2950 2910 2870 2730 1440 1340 1300 1270 1210 1180
 1110 1090 1060 1030 970 920 870 830 780 760 720 650 480 450 390 340
 290 260 210 130
^{13}C NMR [ppm]: 13.9 22.1 29.2 32.5 44.9 CDCl$_3$
^{1}H NMR [ppm]: 0.9 1.6 3.4 CCl$_4$
Flammability: Flash pt. = 13°C; ign. temp. = 260°C; flam. lim. = 1.6-8.6%

98. 1-Chloropropane

Syst. Name: Propane, 1-chloro-
Synonyms: Propyl chloride

CASRN: 540-54-5   **DOT No.**: 1278
Merck No.: 7859   **Beil Ref.**: 4-01-00-00189
Beil RN: 1730771
MF: C_3H_7Cl **MP[°C]**: -122.8   **Den.[g/cm^3]**: 0.8899^{20}
MW: 78.54   **BP[°C]**: 46.5 n_D: 1.3879^{20}
Sol.: H_2O 2; EtOH 5; eth
   5; bz 3; ctc 2; chl 3
Crit. Const.: $T_c = 230$°C; $P_c = 4.58$ MPa
Vap. Press. [kPa]: 15.1^0; 45.8^{25}
Therm. Prop.[kJ/mol]: $\Delta_{fus}H = 5.54$; $\Delta_{vap}H = 27.2$; $\Delta_f H°(l, 25$°C$) = -160.6$
C_p **(liq.) [J/mol °C]**: 132.2^{25}
γ **[mN/m]**: 21.30^{25}; $d\gamma/dT = 0.1246$ mN/m °C
Elec. Prop.: $\mu = 2.05$ D; $\varepsilon = 8.59^{20}$; IP = 10.82 eV
η **[mPa s]**: 0.436^0; 0.334^{25}
MS: 42(100) 29(46) 27(37) 41(23) 28(15) 43(14) 39(12) 78(6) 63(6) 49(5)
IR [cm^{-1}]: 3030 2940 2860 1470 1450 1390 1330 1320 1270 890 860 790 730
   650
^{13}C NMR [ppm]: 11.5 26.5 46.7
^{1}H NMR [ppm]: 1.0 1.8 3.4 CCl_4
Flammability: Flash pt. < -18°C; ign. temp. = 520°C; flam. lim. = 2.6-11.1%

99. 2-Chloropropane

Syst. Name: Propane, 2-chloro-
Synonyms: Isopropyl chloride

CASRN: 75-29-6
Merck No.: 5099
Beil RN: 1730782
MF: C_3H_7Cl
MW: 78.54
Sol.: H_2O 2; EtOH 5; eth
5; bz 3; ctc 3; chl 3

DOT No.: 2356
Beil Ref.: 4-01-00-00191

MP[°C]: -117.2   **Den.[g/cm³]:** 0.8617^{20}
BP[°C]: 35.7 n_D: 1.3777^{20}

Vap. Press. [kPa]: 25.1^0; 68.9^{25}
Therm. Prop.[kJ/mol]: $\Delta_{fus}H$ = 7.39; $\Delta_{vap}H$ = 26.3; $\Delta_f H°$(l, 25°C) = -172.1
γ **[mN/m]:** 19.16^{25}; $d\gamma/dT$ = 0.0883 mN/m °C
Elec. Prop.: μ = 2.17 D; IP = 10.78 eV
η **[mPa s]:** 0.401^0; 0.303^{25}
MS: 43(100) 27(34) 63(26) 78(24) 41(22) 39(11) 80(8) 65(8) 42(7) 44(4)
IR [cm⁻¹]: 2940 1470 1450 1390 1330 1300 1270 1060 1040 1030 900 860 790
730 650
¹³C NMR [ppm]: 27.0 53.9
¹H NMR [ppm]: 1.5 4.1 CCl_4
Flammability: Flash pt. = -32°C; ign. temp. = 593°C; flam. lim. = 2.8-10.7%

100. 3-Chloropropene

Syst. Name: 1-Propene, 3-chloro-
Synonyms: Allyl chloride; 3-Chloropropylene

CASRN: 107-05-1   **DOT No.:** 1100
Merck No.: 287   **Beil Ref.:** 4-01-00-00738
Beil RN: 635704
MF: C_3H_5Cl **MP[°C]:** -134.5   **Den.[g/cm³]:** 0.9376^{20}
MW: 76.53   **BP[°C]:** 45.1 n_D: 1.4157^{20}
Sol.: H_2O 1; EtOH 5; eth
   5; ace 5; bz 5; ctc 2;
   lig 5
Crit. Const.: $T_c = 241°C$
Vap. Press. [kPa]: 4.24^{-25}; 16.4^0; 48.9^{25}; 120^{50}
Therm. Prop.[kJ/mol]: $\Delta_{vap}H = 29.0$
C_p **(liq.) [J/mol °C]:** 125.1^{25}
γ **[mN/m]:** 23.14^{25}; $d\gamma/dT = 0.0946$ mN/m °C
Elec. Prop.: $\mu = 1.94$ D; $\epsilon = 8.2^{20}$; IP = 9.90 eV
η **[mPa s]:** 0.408^0; 0.314^{25}
MS: 41(100) 39(73) 76(28) 38(16) 37(13) 40(12) 27(12) 26(11) 78(9) 49(5)
IR [cm⁻¹]: 3130 2940 1850 1640 1450 1410 1280 1250 1210 1180 1100 990
   930 890 740
Raman [cm⁻¹]: 3100 3040 3000 2960 2880 1680 1650 1620 1530 1450 1420
   1300 1270 1210 1110 990 940 930 740 600 520 420 300 250 120
UV [nm]: 273(4467) sulf
¹³C NMR [ppm]: 45.1 118.1 134.6 diox
¹H NMR [ppm]: 4.0 5.2 5.3 5.9 CCl_4
Flammability: Flash pt. = -32°C; ign. temp. = 485°C; flam. lim. = 2.9-11.1%
TLV/TWA: 1 ppm (3 mg/m³)
Reg. Lists: CERCLA (RQ = 1000 lb.); SARA 313 (1.0%); confirmed or
   suspected carcinogen

101. *o*-Chlorotoluene

Syst. Name: Benzene, 1-chloro-2-methyl-
Synonyms: 1-Chloro-2-methylbenzene

CASRN: 95-49-8   **Rel. CASRN:** 25168-05-2
Merck No.: 2172   **DOT No.:** 2338
Beil RN: 1904175   **Beil Ref.:** 4-05-00-00805
MF: C_7H_7Cl **MP[°C]:** -35.6   **Den.[g/cm^3]:** 1.0825^{20}
MW: 126.59   **BP[°C]:** 159.0 n_D: 1.5268^{20}
Sol.: H_2O 1; EtOH 3; eth
 5; ace 5; bz 3; ctc 5;
 chl 5; hp 5
Vap. Press. [kPa]: 0.482^{25}; 6.29^{75}; 16.8^{100}; 38.8^{125}; 79.8^{150}
Therm. Prop.[kJ/mol]: $\Delta_{fus}H$ = 8.37; $\Delta_{vap}H$ = 37.5
C_p **(liq.) [J/mol °C]:** 166.8^{25}
γ **[mN/m]:** 33.44^{20}
Elec. Prop.: μ = 1.56 D; ϵ = 4.72^{20}; IP = 8.83 eV
η **[mPa s]:** 1.39^{0}; 0.964^{25}; 0.710^{50}; 0.547^{75}; 0.437^{100}
MS: 91(100) 126(68) 89(23) 128(22) 90(18) 65(17) 125(15) 92(13) 127(10)
 63(6)
IR [cm^{-1}]: 3030 2940 1590 1490 1470 1450 1390 1280 1140 1060 1040 1020
 990 940 860 800 750 700 680
Raman [cm^{-1}]: 3150 3060 3020 2990 2920 2910 2860 2740 2570 1590 1570
 1440 1400 1380 1280 1210 1160 1130 1090 1050 1040 1020 990 800 740
 680 550 510 440 360 240 160
UV [nm]: 273(250) 265(288) 259 212(9390) MeOH
^{13}C NMR [ppm]: 19.9 126.4 127.0 129.0 130.9 134.4 135.9 $CDCl_3$
^{1}H NMR [ppm]: 2.4 7.2 $CDCl_3$
TLV/TWA: 50 ppm (259 mg/m^3)

102. *m*-Chlorotoluene

Syst. Name: Benzene, 1-chloro-3-methyl-
Synonyms: 1-Chloro-3-methylbenzene

CASRN: 108-41-8   **Rel. CASRN**: 25168-05-2
Merck No.: 2172   **DOT No.**: 2238
Beil RN: 1903632   **Beil Ref.**: 4-05-00-00806
MF: C_7H_7Cl **MP[°C]**: -47.8   **Den.[g/cm^3]**: 1.075^{20}
MW: 126.59   **BP[°C]**: 161.8 n_D: 1.5214^{19}
Sol.: H_2O 1; EtOH 3; eth
  5; bz 3; ctc 3; chl 3
Vap. Press. [kPa]: 15.1^{100}; 35.3^{125}; 73.3^{150}
Therm. Prop.[kJ/mol]: $\Delta_{fus}H = 10.46$
Elec. Prop.: μ = (1.8 D); ε = 5.76^{20}; IP = 8.83 eV
η **[mPa s]**: 1.17^0; 0.823^{25}; 0.616^{50}; 0.482^{75}; 0.391^{100}
MS: 91(100) 126(27) 63(15) 65(12) 89(11) 39(11) 128(9) 125(8) 92(8) 62(6)
IR [cm^{-1}]: 3070 2930 1610 1600 1580 1480 1220 1160 1100 1080 1000 890
  870 860 770 680
UV [nm]: 274(172) 265(200) 260(150) 213(5560) MeOH
^{13}C NMR [ppm]: 21.0 125.5 127.1 129.1 129.3 134.0 139.7 CDCl$_3$
^{1}H NMR [ppm]: 2.3 7.1 CCl$_4$

103. *p*-Chlorotoluene

Syst. Name: Benzene, 1-chloro-4-methyl-
Synonyms: 1-Chloro-4-methylbenzene

CASRN: 106-43-4
Merck No.: 2172
Beil RN: 1903635
MF: C_7H_7Cl
MW: 126.59
Sol.: H_2O 1; EtOH 3; eth 5; ctc 3; chl 3; HOAc 3

Rel. CASRN: 25168-05-2
DOT No.: 2238
Beil Ref.: 4-05-00-00806
MP[°C]: 7.5   **Den.[g/cm³]**: 1.0697^{20}
BP[°C]: 162.4 n_D: 1.5150^{20}

Vap. Press. [kPa]: 5.64^{75}; 15.3^{100}; 35.7^{125}; 74.0^{150}
Therm. Prop.[kJ/mol]: $\Delta_{vap}H = 38.7$
γ [mN/m]: 32.28^{26}
Elec. Prop.: $\mu = 2.21$ D; $\varepsilon = 6.25^{20}$; IP = 8.69 eV
η [mPa s]: 0.837^{25}; 0.621^{50}; 0.483^{75}; 0.390^{100}
MS: 91(100) 126(36) 125(16) 63(15) 39(14) 128(12) 89(10) 65(10) 127(8) 92(8)
IR [cm⁻¹]: 3040 2920 2870 1880 1630 1600 1490 1450 1400 1380 1100 1020 800 640 490 380
Raman [cm⁻¹]: 3190 3100 3060 3040 2970 2920 2870 2730 1600 1580 1380 1300 1210 1170 1090 1040 820 800 690 640 380 300 250
UV [nm]: 277(507) 269(527) 263(388) 256 244 220(11000) MeOH
¹³C NMR [ppm]: 20.7 128.3 130.4 131.2 136.2 $CDCl_3$
¹H NMR [ppm]: 2.3 7.0 7.1 CCl_4

104. Chlorotrifluoromethane

Syst. Name: Methane, chlorotrifluoro-
Synonyms: Trifluorochloromethane; R 13; CFC 13

CASRN: 75-72-9   **DOT No.:** 1022
Beil RN: 1732392   **Beil Ref.:** 4-01-00-00034
MF: CClF$_3$ **MP[°C]:** -181
MW: 104.46   **BP[°C]:** -81.4
Crit. Const.: T_c = 29°C; P_c = 3.870 MPa; V_c = 180 cm^3
Vap. Press. [kPa]: 5.3^{-123}; 155^{-73}; 1044^{-23}; 3716^{27}
Therm. Prop.[kJ/mol]: $\Delta_{vap}H$ = 15.8; $\Delta_f H°$(g, 25°C) = -707.8
Elec. Prop.: μ = 0.50 D; ϵ = 3.01^{-150}; IP = 12.39 eV
MS: 69(100) 85(28) 50(15) 35(15) 87(9) 31(9) 37(5) 47(2) 104(1) 70(1)
IR [cm^{-1}]: 1330 1280 1220 1120 1110 920 780

---

105. Cineole

Syst. Name: 2-Oxabicyclo[2.2.2]octane, 1,3,3-
  trimethyl-
Synonyms: Eucalyptole

CASRN: 470-82-6   **Beil Ref.:** 5-17-01-00273
Merck No.: 3851
Beil RN: 105109
MF: C$_{10}$H$_{18}$O   **MP[°C]:** 0.8   **Den.[g/cm^3]:** 0.9267^{20}
MW: 154.25   **BP[°C]:** 176.4 n_D: 1.4586^{20}
Sol.: H$_2$O 1; EtOH 3; eth
  3; ctc 2; chl 3
Vap. Press. [kPa]: 0.260^{25}; 1.08^{50}; 3.59^{75}; 9.90^{100}; 23.7^{125}; 50.8^{150}
γ [mN/m]: 32.1^{20}
Elec. Prop.: ϵ = 4.57^{25}
MS: 43(100) 81(37) 71(37) 84(31) 108(30) 69(29) 55(27) 41(26) 111(25)
  154(24)
IR [cm^{-1}]: 2940 1470 1450 1370 1350 1320 1270 1240 1220 1160 1090 1050
  1020 990 920 840 790 760
^{1}H NMR [ppm]: 1.0 1.2 1.8 CDCl$_4$
Flammability: Flash pt. = 48°C

106. *o*-Cresol

Syst. Name: Phenol, 2-methyl-
Synonyms: 2-Methylphenol; *o*-Cresylic acid

CASRN: 95-48-7
Merck No.: 2580
Beil RN: 506917
MF: C_7H_8O
MW: 108.14
Sol.: H_2O 3; EtOH 4; eth 4; ace 5; bz 5; ctc 5; os 3

Rel. CASRN: 1319-77-3
DOT No.: 2076
Beil Ref.: 4-06-00-01940
MP[°C]: 29.8
BP[°C]: 191.0
Den.[g/cm^3]: 1.135^{25}
n_D: 1.5361^{20}

Crit. Const.: $T_c = 424.5°C$; $P_c = 5.01$ MPa
Vap. Press. [kPa]: 0.041^{25}; 4.23^{100}; 30.3^{150}
Therm. Prop.[kJ/mol]: $\Delta_{fus}H = 13.94$; $\Delta_{vap}H = 45.2$; $\Delta_f H°(s, 25°C) = -204.6$
C_p (liq.) [J/mol °C]: 233.6^{40}
γ [mN/m]: 35.39^{40}; $d\gamma/dT = 0.1011$ mN/m °C
Elec. Prop.: $\mu = (1.4 \text{ D})$; $\varepsilon = 6.76^{25}$; IP = 8.14 eV
η [mPa s]: 3.04^{50}; 1.56^{75}; 0.961^{100}
MS: 108(100) 107(75) 77(34) 79(33) 39(28) 90(25) 51(21) 27(20) 53(14) 80(12)
IR [cm^{-1}]: 3400 3020 2920 1590 1490 1460 1330 1300 1240 1200 1170 1100 1050 840 750 710
UV [nm]: 273(1820) 214(6030) MeOH
^{1}H NMR [ppm]: 2.2 5.2 6.7 CCl_4
Flammability: Flash pt. = 81°C; ign. temp. = 599°C; lower flam. lim. = 1.4%
TLV/TWA: 5 ppm (22 mg/m^3)
Reg. Lists: CERCLA (RQ = 1000 lb.); SARA 313 (1.0%); RCRA U052 (toxic)

107. *m*-Cresol

Syst. Name: Phenol, 3-methyl-
Synonyms: 3-Methylphenol; *m*-Cresylic acid

CASRN: 108-39-4   **Rel. CASRN**: 1319-77-3
Merck No.: 2579   **DOT No.**: 2076
Beil RN: 506719   **Beil Ref.**: 4-06-00-02035
MF: C_7H_8O **MP[°C]**: 11.8   **Den.[g/cm^3]**: 1.0341^{20}
MW: 108.14   **BP[°C]**: 202.2 n_D: 1.5438^{20}
Sol.: H_2O 2; EtOH 5; eth
   5; ace 5; bz 5; ctc 5;
   os 3
Crit. Const.: T_c = 432.7°C; P_c = 4.56 MPa; V_c = 309 cm^3
Vap. Press. [kPa]: 0.019^{25}; 2.50^{100}; 20.6^{150}
Therm. Prop.[kJ/mol]: $\Delta_{fus}H$ = 9.41; $\Delta_{vap}H$ = 47.4; $\Delta_f H°$(l, 25°C) = -194.0
C_p (liq.) [J/mol °C]: 224.9^{25}
γ [mN/m]: 35.69^{25}; $d\gamma/dT$ = 0.0924 mN/m °C
Elec. Prop.: μ = (1.5 D); ε = 12.44^{25}; IP = 8.29 eV
η [mPa s]: 12.9^{25}; 4.42^{50}; 2.09^{75}; 1.21^{100}
MS: 108(100) 107(85) 79(35) 39(31) 77(29) 51(17) 27(17) 53(15) 90(11)
   78(10)
IR [cm^{-1}]: 3350 3040 2920 1610 1590 1490 1460 1310 1280 1270 1230 1150
   1080 1040 1000 930 880 850 780 740 690
Raman [cm^{-1}]: 3060 3020 2940 2870 2740 1610 1590 1500 1450 1380 1280
   1270 1220 1160 1130 1090 1000 930 890 840 770 740 560 540 520 460 440
   300 240 220
UV [nm]: 278 273 MeOH
^{13}C NMR [ppm]: 21.1 112.5 116.2 121.8 129.4 139.8 155.0 CDCl$_3$
^{1}H NMR [ppm]: 2.3 5.7 6.9 CDCl$_3$
Flammability: Flash pt. = 86°C; ign. temp. = 558°C; lower flam. lim. = 1.1%
TLV/TWA: 5 ppm (22 mg/m^3)
Reg. Lists: CERCLA (RQ = 1000 lb.); SARA 313 (1.0%); RCRA U052 (toxic)

108. *p*-Cresol

Syst. Name: Phenol, 4-methyl-
Synonyms: 4-Methylphenol; *p*-Cresylic acid

CASRN: 106-44-5
Merck No.: 2581
Beil RN: 1305151
MF: C$_7$H$_8$O
MW: 108.14
Sol.: H$_2$O 2; EtOH 5; eth
   5; ace 5; bz 5; ctc 5;
   os 3

Rel. CASRN: 1319-77-3
DOT No.: 2076
Beil Ref.: 4-06-00-02093
MP[°C]: 35.5   **Den.[g/cm^3]**: 1.0185^{40}
BP[°C]: 201.9 n_D: 1.5312^{20}

Crit. Const.: T_c = 431.5°C; P_c = 5.15 MPa
Vap. Press. [kPa]: 0.017^{25}; 2.45^{100}; 20.7^{150}
Therm. Prop.[kJ/mol]: $\Delta_{fus}H$ = 11.89; $\Delta_{vap}H$ = 47.5; $\Delta_f H°$(s, 25°C) = -199.3
C_p **(liq.) [J/mol °C]**: 221.0^{40}
γ **[mN/m]**: 19.2^{195}
Elec. Prop.: μ = (1.5 D); ε = 13.05^{25}; IP = 8.13 eV
η **[mPa s]**: 9.40^{35}
MS: 107(100) 108(91) 77(28) 79(21) 51(19) 39(17) 27(17) 53(15) 50(12)
   52(10)
IR [cm^{-1}]: 3340 3020 2920 2860 1620 1600 1520 1440 1360 1240 1170 1100
   840 810 740 700
UV [nm]: 279 MeOH
^{1}H NMR [ppm]: 2.2 6.4 6.7 6.9 CDCl$_3$
Flammability: Flash pt. = 86°C; ign. temp. = 558°C; lower flam. lim. = 1.1%
TLV/TWA: 5 ppm (22 mg/m^3)
Reg. Lists: CERCLA (RQ = 1000 lb.); SARA 313 (1.0%); RCRA U052 (toxic)

109. *trans*-Crotonaldehyde

Syst. Name: 2-Butenal, (*E*)-
Synonyms: *trans*-2-Butenal

CASRN: 123-73-9
Merck No.: 2599
Beil RN: 1209254
MF: C_4H_6O
MW: 70.09
Sol.: H_2O 3; EtOH 4; eth
 4; ace 4; bz 5; chl 3

Rel. CASRN: 4170-30-3
DOT No.: 1143
Beil Ref.: 4-01-00-03447
MP[°C]: -76
BP[°C]: 102.2

Den.[g/cm^3]: 0.8516^{20}
n_D: 1.4366^{20}

Vap. Press. [kPa]: 4.92^{25}; 15.8^{50}; 41.3^{75}; 93.2^{100}
Therm. Prop.[kJ/mol]: $\Delta_f H°$(l, 25°C) = -138.7
C_p **(liq.) [J/mol °C]**: 95.4^{25}
Elec. Prop.: μ = 3.67 D; IP = 9.73 eV
MS: 41(100) 39(97) 70(82) 69(65) 27(49) 29(39) 42(30) 38(29) 40(27) 37(18)
IR [cm^{-1}]: 2940 2860 2700 1700 1640 1450 1410 1150 1080 970 930
Raman [cm^{-1}]: 3030 3000 2920 2850 2820 2730 1680 1640 1440 1390 1380
 1310 1150 1080 1050 990 930 780 550 470 440 300
UV [nm]: 301(37) 218(14500) MeOH
^{13}C NMR [ppm]: 18.2 134.9 153.7 193.4 diox
^{1}H NMR [ppm]: 2.0 6.1 6.9 9.5 CDCl$_3$
Flammability: Flash pt. = 13°C; ign. temp. = 232°C; flam. lim. = 2.1-15.9%
TLV/TWA: 2 ppm (5.7 mg/m^3)
Reg. Lists: CERCLA (RQ = 100 lb.); RCRA U053 (toxic); confirmed or
 suspected carcinogen

110. *trans*-Crotonic acid

Syst. Name: 2-Butenoic acid, (*E*)-
Synonyms: *trans*-2-Butenoic acid

CASRN: 107-93-7   **Rel. CASRN**: 3724-65-0
Merck No.: 2600   **DOT No.**: 2823
Beil RN: 1719943   **Beil Ref.**: 4-02-00-01498
MF: $C_4H_6O_2$ **MP[°C]**: 72   **Den.[g/cm^3]**: 0.9604^{77}
MW: 86.09   **BP[°C]**: 184.7 n_D: 1.4249^{77}
Sol.: H_2O 4; EtOH 4; eth
   3; ace 3; lig 3
Vap. Press. [kPa]: 1.04^{75}; 3.88^{100}; 12.0^{125}; 31.6^{150}
Therm. Prop.[kJ/mol]: $\Delta_{fus}H$ = 12.98
Elec. Prop.: μ = (2.1 D); IP = 9.90 eV
MS: 86(100) 41(63) 39(54) 69(31) 68(26) 45(17) 40(16) 71(11) 38(11) 42(10)
IR [cm^{-1}]: 2980 2920 2600 1700 1650 1530 1440 1420 1370 1320 1300 1230
   1100 1050 990 970 920 840 700 690 540 520 400
Raman [cm^{-1}]: 3050 3040 2980 2960 2920 2870 1660 1630 1440 1430 1380
   1310 1290 1220 1110 1040 990 960 950 900 840 680 510 480 410 220

---

111. *cis*-Crotonyl alcohol

Syst. Name: 2-Buten-1-ol, (*Z*)-
Synonyms: *cis*-Crotyl alcohol

CASRN: 4088-60-2   **Beil Ref.**: 4-01-00-02107
Beil RN: 1719375
MF: C_4H_8O **BP[°C]**: 123   **Den.[g/cm^3]**: 0.8662^{20}
MW: 72.11 n_D: 1.4342^{25}
MS: 57(100) 39(35) 72(33) 29(26) 41(24) 31(22) 43(19) 53(14) 54(12) 44(8)

112. *trans*-Crotonyl alcohol

Syst. Name: 2-Buten-1-ol, (*E*)-
Synonyms: *trans*-Crotyl alcohol

CASRN: 504-61-0   **Rel. CASRN**: 6117-91-5
Beil RN: 1361395   **Beil Ref.**: 4-01-00-02107
MF: C_4H_8O **MP[°C]**: <-30   **Den.[g/cm^3]**: 0.8521^{20}
MW: 72.11   **BP[°C]**: 121.2 n_D: 1.4288^{20}
Sol.: H_2O 4; EtOH 5; eth
   5; chl 3
MS: 57(100) 39(55) 27(44) 29(40) 72(32) 41(30) 43(27) 31(25) 53(21) 54(19)
IR [cm^{-1}]: 3330 3030 2940 1640 1430 1370 1250 1190 1160 1120 1100 1040
   990 960 920 870 860 800
^{1}H NMR [ppm]: 1.7 2.1 4.1 5.7 CDCl$_3$

113. Cumene

Syst. Name: Benzene, (1-methylethyl)-
Synonyms: Isopropylbenzene

CASRN: 98-82-8
Merck No.: 2619
Beil RN: 1236613
MF: C_9H_{12}
MW: 120.19
Sol.: H_2O 1; EtOH 5; eth 5; ace 5; bz 5; ctc 5; peth 5

DOT No.: 1918
Beil Ref.: 4-05-00-00985

MP[°C]: -96.0
BP[°C]: 152.4

Den.[g/cm³]: 0.8618^{20}
n_D: 1.4915^{20}

Crit. Const.: T_c = 357.9°C; P_c = 3.209 MPa
Vap. Press. [kPa]: 0.610^{25}; 2.49^{50}; 7.86^{75}; 20.7^{100}; 46.9^{125}; 95.1^{150}
Therm. Prop.[kJ/mol]: $\Delta_{fus}H$ = 7.79; $\Delta_{vap}H$ = 37.5; $\Delta_f H°$(l, 25°C) = -41.1
C_p (liq.) [J/mol °C]: 210.7^{25}
γ [mN/m]: 27.69^{25}; $d\gamma/dT$ = 0.1054 mN/m °C
Elec. Prop.: μ = 0.79 D; ϵ = 2.38^{20}; IP = 8.73 eV
η [mPa s]: 1.08^{0}; 0.737^{25}; 0.547^{50}
k [W/m °C]: 0.128^{25}; 0.120^{50}; 0.112^{75}; 0.107^{100}
MS: 105(100) 120(25) 77(13) 51(12) 79(10) 106(9) 39(9) 27(8) 103(6) 91(5)
IR [cm⁻¹]: 2940 1590 1490 1450 1390 1370 1080 1020 910 770 690
Raman [cm⁻¹]: 3060 3010 2970 2950 2920 2880 1600 1580 1460 1440 1310 1280 1210 1180 1150 1110 1080 1030 1000 950 890 860 840 620 460 310 140
UV [nm]: 258 cyhex
¹³C NMR [ppm]: 23.9 34.2 125.7 126.3 128.2 148.7 $CDCl_3$
¹H NMR [ppm]: 1.3 2.4 2.9 7.3 $CDCl_3$
Flammability: Flash pt. = 36°C; ign. temp. = 424°C; flam. lim. = 0.9-6.5%
TLV/TWA: 50 ppm (246 mg/m³)
Reg. Lists: CERCLA (RQ = 5000 lb.); SARA 313 (1.0%); RCRA U055 (toxic)

114. Cyclohexane

Syst. Name: Cyclohexane

CASRN: 110-82-7

DOT No.: 1145

Merck No.: 2729

Beil Ref.: 4-05-00-00027

Beil RN: 1900225

MF: C_6H_{12} **MP[°C]**: 6.6   **Den.[g/cm^3]**: 0.7785^{20}

MW: 84.16   **BP[°C]**: 80.7 n_D: 1.4266^{20}

Sol.: H_2O 1; EtOH 5; eth 5; ace 5; bz 5; ctc 5; lig 5

Crit. Const.: T_c = 280.4°C; P_c = 4.07 MPa; V_c = 308 cm^3

Vap. Press. [kPa]: 13.0^{25}; 36.3^{50}; 85.0^{75}; 175^{100}; 324^{125}; 553^{150}

Therm. Prop.[kJ/mol]: $\Delta_{fus}H$ = 2.63; $\Delta_{vap}H$ = 30.0; $\Delta_f H°$(l, 25°C) = -156.4

C_p **(liq.) [J/mol °C]**: 154.9^{25}

γ **[mN/m]**: 24.65^{25}; $d\gamma/dT$ = 0.1188 mN/m °C

Elec. Prop.: μ = 0; ε = 2.02^{20}; IP = 9.86 eV

η **[mPa s]**: 0.894^{25}; 0.615^{50}; 0.447^{75}

k **[W/m °C]**: 0.123^{25}; 0.117^{50}; 0.111^{75}

MS: 56(100) 84(71) 41(70) 27(37) 55(36) 39(35) 42(30) 69(23) 28(18) 43(14)

IR [cm^{-1}]: 2940 2860 1450 900 860

Raman [cm^{-1}]: 2930 2920 2890 2840 2800 2690 2660 2630 1450 1350 1270 1160 1030 810 430 390

^{13}C NMR [ppm]: 27.8

^{1}H NMR [ppm]: 1.4 CCl_4

Flammability: Flash pt. = -20°C; ign. temp. = 245°C; flam. lim. = 1.3-8%

TLV/TWA: 300 ppm (1030 mg/m^3)

Reg. Lists: CERCLA (RQ = 1000 lb.); SARA 313 (1.0%); RCRA U056 (toxic)

115. Cyclohexanol

Syst. Name: Cyclohexanol
Synonyms: Cyclohexyl alcohol

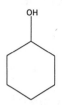

CASRN: 108-93-0
Merck No.: 2731
Beil RN: 906744
MF: $C_6H_{12}O$
MW: 100.16
Sol.: H_2O 3; EtOH 3; eth
  3; ace 3; bz 5; chl 2;
  CS_2 5

DOT No.: 1986
Beil Ref.: 4-06-00-00020

MP[°C]: 25.4
BP[°C]: 160.8

Den.[g/cm^3]: 0.9624^{20}
n_D: 1.4641^{20}

Crit. Const.: $T_c = 376.9°C$; $P_c = 4.26$ MPa
Vap. Press. [kPa]: 0.100^{25}; 10.4^{100}; 30.6^{125}; 72.7^{150}
Therm. Prop.[kJ/mol]: $\Delta_{fus}H = 1.76$
C_p (liq.) [J/mol °C]: 208.2^{26}
γ [mN/m]: 32.92^{25}; $d\gamma/dT = 0.0966$ mN/m °C
Elec. Prop.: $\varepsilon = 16.40^{20}$; IP = 9.75 eV
η [mPa s]: 57.5^{25}; 12.3^{50}; 4.27^{75}; 1.98^{100}
k [W/m °C]: 0.134^{25}; 0.131^{50}
MS: 57(100) 44(68) 41(68) 39(51) 32(40) 43(38) 31(32) 42(22) 67(18) 82(16)
IR [cm^{-1}]: 3350 2920 2850 1450 1350 1250 1130 1070 970 890 550
Raman [cm^{-1}]: 3360 2940 2850 2750 1470 1440 1360 1350 1300 1250 1170
  1140 1070 1050 1020 960 940 840 790 660 550 480 450 400 340
^{13}C NMR [ppm]: 24.3 25.7 35.5 70.0 $CDCl_3$
^{1}H NMR [ppm]: 1.6 3.5 4.2 CCl_4
Flammability: Flash pt. = 68°C; ign. temp. = 300°C; flam. lim. = 1-9%
TLV/TWA: 50 ppm (206 mg/m^3)

116. Cyclohexanone

Syst. Name: Cyclohexanone
Synonyms: Pimelic ketone

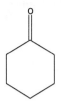

CASRN: 108-94-1   **DOT No.:** 1915
Merck No.: 2732   **Beil Ref.:** 4-07-00-00015
Beil RN: 385735
MF: $C_6H_{10}O$ **MP[°C]:** -31   **Den.[g/cm^3]:** 0.9478^{20}
MW: 98.14   **BP[°C]:** 155.4 n_D: 1.4507^{20}
Sol.: H_2O 3; EtOH 3; eth
  3; ace 3; bz 3; ctc 3;
  chl 3
Crit. Const.: T_c = 379.9°C; P_c = 4.0 MPa
Vap. Press. [kPa]: 0.530^{25}; 18.6^{100}; 42.7^{125}; 87.4^{150}
Therm. Prop.[kJ/mol]: $\Delta_{vap}H$ = 40.3; $\Delta_f H°$(l, 25°C) = -271.2
C_p **(liq.) [J/mol °C]:** 182.2^{25}
γ **[mN/m]:** 34.57^{25}; $d\gamma/dT$ = 0.1242 mN/m °C
Elec. Prop.: μ = 2.87 D; ε = 16.1^{20}; IP = 9.14 eV
η **[mPa s]:** 2.02^{25}; 1.32^{50}; 0.919^{75}; 0.671^{100}
MS: 55(100) 42(85) 41(34) 27(33) 98(31) 39(27) 69(26) 70(20) 43(14) 28(14)
IR [cm^{-1}]: 2940 2860 1710 1450 1430 1340 1310 1220 1120 1040 900 860 740
  650 490
Raman [cm^{-1}]: 2960 2900 2870 2840 2660 1710 1450 1430 1350 1320 1270
  1250 1230 1120 1070 1020 1000 910 900 850 790 760 660 490 420 320 190
UV [nm]: 280(27) 276(26) MeOH
^{13}C NMR [ppm]: 25.1 27.1 41.9 211.3 $CDCl_3$
^{1}H NMR [ppm]: 1.8 2.3 CCl_4
Flammability: Flash pt. = 44°C; ign. temp. = 420°C; flam. lim. = 1.1-9.4%
TLV/TWA: 25 ppm (100 mg/m^3)
Reg. Lists: CERCLA (RQ = 5000 lb.); RCRA U057 (toxic)

117. Cyclohexene

Syst. Name: Cyclohexene
Synonyms: Tetrahydrobenzene

CASRN: 110-83-8
Merck No.: 2733
Beil RN: 906737
MF: C_6H_{10}
MW: 82.15
Sol.: H_2O 1; EtOH 5; eth
  5; ace 5; bz 5; ctc 5;
  lig 5

DOT No.: 2256
Beil Ref.: 4-05-00-00218

MP[°C]: -103.5
BP[°C]: 82.9

Den.[g/cm³]: 0.8110^{20}
n_D: 1.4465^{20}

Crit. Const.: $T_c = 287.33$°C
Vap. Press. [kPa]: 11.8^{25}; 33.4^{50}; 79.2^{75}; 165^{100}
Therm. Prop.[kJ/mol]: $\Delta_{fus}H = 3.29$; $\Delta_{vap}H = 30.5$; $\Delta_f H°(l, 25°C) = -38.5$
C_p **(liq.) [J/mol °C]**: 148.3^{25}
γ **[mN/m]**: 26.17^{25}; $d\gamma/dT = 0.1223$ mN/m °C
Elec. Prop.: $\mu = 0.332$ D; $\varepsilon = 2.22^{20}$; IP = 8.95 eV
η **[mPa s]**: 0.882^0; 0.625^{25}; 0.467^{50}; 0.364^{75}
k **[W/m °C]**: 0.142^{-25}; 0.136^0; 0.130^{25}; 0.124^{50}; 0.118^{75}
MS: 67(100) 54(72) 82(37) 41(35) 39(33) 27(15) 53(12) 81(9) 51(8) 79(6)
IR [cm⁻¹]: 3080 2990 2930 1660 1450 1360 1140 920 720 640
Raman [cm⁻¹]: 3060 3020 2940 2910 2880 2860 2840 2700 2660 2640 1660
  1440 1350 1270 1240 1220 1140 1070 1040 970 910 880 830 790 650 500
  460 400 280 180
UV [nm]: 184(7762) cyhex
¹³C NMR [ppm]: 22.9 25.3 127.2 $CDCl_3$
¹H NMR [ppm]: 1.6 2.0 5.6 CCl_4
Flammability: Flash pt. < -7°C; ign. temp. = 244°C
TLV/TWA: 300 ppm (1010 mg/m³)

118. Cyclohexylamine

Syst. Name: Cyclohexanamine
Synonyms: Aminocyclohexane

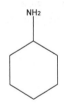

CASRN: 108-91-8
Merck No.: 2735
Beil RN: 471175
MF: $C_6H_{13}N$
MW: 99.18
Sol.: H_2O 3; EtOH 4; eth
  5; ace 5; bz 5; ctc 3;
  os 5

DOT No.: 2357
Beil Ref.: 4-12-00-00008

MP[°C]: -17.7
BP[°C]: 134

Den.[g/cm^3]: 0.8191^{20}
n_D: 1.4372^{20}

Vap. Press. [kPa]: 1.20^{25}; 14.3^{75}; 36.1^{100}; 78.9^{125}
Therm. Prop.[kJ/mol]: $\Delta_{vap}H$ = 36.1; $\Delta_f H°$(l, 25°C) = -147.7
γ [mN/m]: 31.22^{25}; $d\gamma/dT$ = 0.1188 mN/m °C
Elec. Prop.: μ = (1.3 D); ε = 4.55^{20}; IP = 8.62 eV
η [mPa s]: 1.94^{25}; 1.17^{50}; 0.782^{75}; 0.565^{100}
MS: 56(100) 43(23) 28(17) 99(10) 70(8) 57(6) 30(6) 93(5) 54(4) 41(4)
IR [cm^{-1}]: 3330 3230 2940 2860 1610 1450 1370 1250 1090 1030 930 890 830
  780
^{13}C NMR [ppm]: 25.1 25.7 36.7 50.4 $CDCl_3$
^{1}H NMR [ppm]: 1.4 1.5 2.6 CCl_4
Flammability: Flash pt. = 31°C; ign. temp. = 293°C; flam. lim. = 1-9%
TLV/TWA: 10 ppm (41 mg/m^3)

119. Cyclohexylbenzene

Syst. Name: Benzene, cyclohexyl-
Synonyms: Phenylcyclohexane

CASRN: 827-52-1   **Beil Ref.:** 4-05-00-01424
Beil RN: 1906803
MF: $C_{12}H_{16}$ **MP[°C]:** 7.3   **Den.[g/cm^3]:** 0.9427^{20}
MW: 160.26   **BP[°C]:** 240.1 n_D: 1.5329^{20}
Sol.: H_2O 1; EtOH 4; eth
    3; ctc 2
Vap. Press. [kPa]: 8.50^{150}
Therm. Prop.[kJ/mol]: $\Delta_{fus}H$ = 15.30; $\Delta_f H°$(l, 25°C) = -76.6
C_p (liq.) [J/mol °C]: 261.3^{25}
Elec. Prop.: μ = 0
MS: 104(100) 117(90) 91(74) 160(70) 92(23) 115(18) 78(17) 77(17) 118(16)
    105(16)
IR [cm^{-1}]: 3030 2941 1587 1493 1449 1075 1031 1000 735 694
Raman [cm^{-1}]: 3070 3010 2940 2930 2900 2860 1600 1580 1440 1400 1370
    1330 1300 1260 1230 1200 1180 1150 1130 1050 1030 1000 860 830 7770
    620 460 430 270 220
UV [nm]: 259 cyhex
^{1}H NMR [ppm]: 1.5 2.4 7.1 CCl_4
Flammability: Flash pt. = 99°C

120. Cyclopentane

Syst. Name: Cyclopentane
Synonyms: Pentamethylene

CASRN: 287-92-3   **DOT No.**: 1146
Merck No.: 2746   **Beil Ref.**: 4-05-00-00014
Beil RN: 1900195
MF: C_5H_{10} **MP[°C]**: -93.8   **Den.[g/cm^3]**: 0.7457^{20}
MW: 70.13   **BP[°C]**: 49.3 n_D: 1.4065^{20}
Sol.: H_2O 1; EtOH 5; eth
  5; ace 5; bz 5; ctc 5;
  peth 5
Crit. Const.: T_c = 238.6°C; P_c = 4.508 MPa; V_c = 260 cm^3
Vap. Press. [kPa]: 3.69^{-25}; 14.2^0; 42.3^{25}; 104^{50}; 220^{75}; 418^{100}; 725^{125}; 1175^{150}
Therm. Prop.[kJ/mol]: $\Delta_{fus}H$ = 0.61; $\Delta_{vap}H$ = 27.3; $\Delta_f H°$(l, 25°C) = -105.1
C_p **(liq.) [J/mol °C]**: 128.8^{25}
γ **[mN/m]**: 21.88^{25}; dγ/dT = 0.1462 mN/m °C
Elec. Prop.: μ = 0; ϵ = 1.97^{20}; IP = 10.51 eV
η **[mPa s]**: 0.555^0; 0.413^{25}; 0.321^{50}
k **[W/m °C]**: 0.140^{-25}; 0.133^0; 0.126^{25}
MS: 42(100) 70(30) 55(29) 41(29) 39(22) 27(15) 40(7) 29(5) 28(4) 43(3)
IR [cm^{-1}]: 2940 2860 1470 1320 970 890
Raman [cm^{-1}]: 2980 2960 2880 1450 1280 1220 1170 1030 890 710 570 490
  280
^{13}C NMR [ppm]: 26.5
^{1}H NMR [ppm]: 1.5 CCl$_4$
Flammability: Flash pt. < -7°C; ign. temp. = 361°C; lower flam. lim. = 1.5%
TLV/TWA: 600 ppm (1720 mg/m^3)

121. Cyclopentanone

Syst. Name: Cyclopentanone
Synonyms: Adipic ketone

CASRN: 120-92-3   **DOT No.:** 2245
Merck No.: 2748   **Beil Ref.:** 4-07-00-00005
Beil RN: 605573
MF: C_5H_8O **MP[°C]:** -51.3   **Den.[g/cm^3]:** 0.9487^{20}
MW: 84.12   **BP[°C]:** 130.5 n_D: 1.4366^{20}
Sol.: H_2O 1; EtOH 3; eth 5; ace 3; ctc 3; MeOH 3; hx 3
Crit. Const.: $T_c = 351.4°C$; $P_c = 4.60$ MPa
Vap. Press. [kPa]: 1.55^{25}; 5.82^{50}; 17.0^{75}; 41.4^{100}; 87.5^{125}
Therm. Prop.[kJ/mol]: $\Delta_{vap}H = 36.4$; $\Delta_f H°(l, 25°C) = -235.7$
C_p **(liq.) [J/mol °C]:** 154.5^{25}
γ **[mN/m]:** 32.80^{25}; $d\gamma/dT = 0.1100$ mN/m °C
Elec. Prop.: $\mu = 3.3$ D; $\varepsilon = 13.58^{25}$; IP = 9.25 eV
η **[mPa s]:** 2.25^{15}
MS: 55(100) 28(50) 84(42) 41(38) 56(29) 27(24) 39(19) 42(15) 26(9) 29(7)
IR [cm^{-1}]: 2980 2880 1700 1460 1410 1270 1150 960 840 580
Raman [cm^{-1}]: 2980 2920 2900 2810 2720 1750 1730 1480 1460 1420 1320 1280 1240 1200 1160 1030 960 900 820 710 590 480 450 240
UV [nm]: 300 cyhex
^{13}**C NMR [ppm]:** 23.5 38.0 218.2 diox
^{1}**H NMR [ppm]:** 2.0 CCl$_4$
Flammability: Flash pt. = 26°C

122. Cyclopentene

Syst. Name: Cyclopentene

CASRN: 142-29-0   **DOT No.**: 2246
Merck No.: 0   **Beil Ref.**: 4-05-00-00209
Beil RN: 635707
MF: C_5H_8 **MP[°C]**: -135.1   **Den.[g/cm^3]**: 0.7720^{20}
MW: 68.12   **BP[°C]**: 44.2 n_D: 1.4225^{20}
Sol.: H_2O 1; EtOH 3; eth
   3; bz 3; ctc 3; peth 3
Crit. Const.: T_c = 233.9°C; P_c = 4.802 MPa; V_c = 245 cm^3
Vap. Press. [kPa]: 4.59^{-25}; 17.4^0; 50.7^{25}; 122^{50}
Therm. Prop.[kJ/mol]: $\Delta_f H°$(l, 25°C) = 4.4
C_p **(liq.) [J/mol °C]**: 122.4^{25}
γ **[mN/m]**: 22.20^{25}; dγ/dT = 0.1495 mN/m °C
Elec. Prop.: μ = 0.20 D; ϵ = 2.08^{22}; IP = 9.01 eV
η **[mPa s]**: 0.21^{38}
k **[W/m °C]**: 0.143^{-25}; 0.136^0; 0.129^{25}
MS: 67(100) 68(43) 39(38) 53(23) 41(19) 40(16) 27(15) 66(10) 42(8) 38(8)
IR [cm^{-1}]: 3030 2940 2860 1640 1470 1370 1330 1220 1060 1050 1030 920
   900 720 690 680
Raman [cm^{-1}]: 3056 2953 2993 2847 1613 1442 1108 965 895 385
UV [nm]: 183(13489) gas
^{1}H NMR [ppm]: 1.9 2.3 5.7 CCl$_4$
Flammability: Flash pt. = -29°C; ign. temp. = 395°C

123. *p*-Cymene

Syst. Name: Benzene, 1-methyl-4-(1-methylethyl)-
Synonyms: 1-Methyl-4-isopropylbenzene; 4-
 Isopropyltoluene

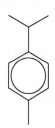

CASRN: 99-87-6
Merck No.: 2770
Beil RN: 1903377
MF: $C_{10}H_{14}$
MW: 134.22
Sol.: H_2O 1; EtOH 5; eth
 5; ace 5; bz 5; ctc 5;
 peth 5

Rel. CASRN: 25155-15-1
DOT No.: 2046
Beil Ref.: 4-05-00-01060
MP[°C]: -68.9   **Den.[g/cm^3]**: 0.8573^{20}
BP[°C]: 177.1 n_D: 1.4909^{20}

Crit. Const.: T_c = 379°C; P_c = 2.8 MPa
Vap. Press. [kPa]: 0.190^{25}; 9.20^{100}; 22.6^{125}; 49.1^{150}
Therm. Prop.[kJ/mol]: $\Delta_{fus}H$ = 9.60; $\Delta_{vap}H$ = 38.2; $\Delta_f H°$(l, 25°C) = -78.0
C_p (liq.) [J/mol °C]: 236.4^{25}
γ [mN/m]: 26.64^{25}; $d\gamma/dT$ = 0.0877 mN/m °C
Elec. Prop.: μ = 0; ε = 2.23^{25}; IP = 8.29 eV
η [mPa s]: 3.40^{20}; 1.60^{30}
k [W/m °C]: 0.132^{-25}; 0.127^0; 0.122^{25}; 0.117^{50}; 0.112^{75}; 0.107^{100}
MS: 119(100) 91(42) 134(33) 39(27) 41(20) 117(18) 65(18) 77(17) 27(16)
 120(15)
IR [cm^{-1}]: 3100 3060 3030 2970 2930 2880 1520 1460 1380 1360 1110 1060
 1020 820 720
Raman [cm^{-1}]: 3060 3020 2980 2930 2880 2760 2740 2720 1620 1580 1460
 1440 1380 1340 1300 1280 1210 1180 1140 1100 1060 950 890 820 800 720
 690 640 530 440 390 350 300 220
UV [nm]: 279 cyhex
H NMR [ppm]: 1.2 2.3 2.9 7.1 $CDCl_3$
Flammability: Flash pt. = 47°C; ign. temp. = 436°C; flam. lim. = 0.7-5.6%

124. *cis*-Decahydronaphthalene

Syst. Name: Naphthalene, decahydro-, *cis*-
Synonyms: *cis*-Decalin; *cis*-Bicyclo[4.4.0]decane

CASRN: 493-01-6   **Beil Ref.:** 4-05-00-00310
Merck No.: 2839
Beil RN: 1900822
MF: $C_{10}H_{18}$ **MP[°C]:** -42.9   **Den.[g/cm³]:** 0.8965^{20}
MW: 138.25   **BP[°C]:** 195.8 n_D: 1.4810^{20}
Sol.: H_2O 1; EtOH 5; eth
   4; ace 4; bz 5; chl 4
Crit. Const.: T_c = 429.2°C; P_c = 3.20 MPa
Vap. Press. [kPa]: 0.100^{25}; 1.87^{75}; 5.56^{100}; 14.0^{125}; 30.8^{150}
Therm. Prop.[kJ/mol]: $\Delta_{fus}H$ = 9.49; $\Delta_{vap}H$ = 41.0; $\Delta_f H°$(l, 25°C) = -219.4
C_p (liq.) [J/mol °C]: 232.0^{25}
γ [mN/m]: 32.18^{20}
Elec. Prop.: μ = 0; ε = 2.22^{20}; IP = 9.26 eV
η [mPa s]: 12.8^{-25}; 5.65^{0}; 3.04^{25}; 1.88^{50}; 1.27^{75}; 0.924^{100}
MS: 67(100) 81(87) 41(81) 138(67) 96(62) 82(62) 39(50) 55(45) 27(44) 95(42)
IR [cm⁻¹]: 2920 2870 1460 1440 1010 970 860 600
Raman [cm⁻¹]: 2940 2910 2870 2750 2670 1450 1360 1270 1250 1160 1050
   1040 1010 980 930 880 850 800 750 600 500 450 410 380 350 320 150
¹H NMR [ppm]: 1.3 CCl_4

125. *trans*-Decahydronaphthalene

Syst. Name: Naphthalene, decahydro-, *trans*-
Synonyms: *trans*-Decalin; *trans*-
  Bicyclo[4.4.0]decane

CASRN: 493-02-7   **Beil Ref.:** 4-05-00-00311
Merck No.: 2839
Beil RN: 2036251
MF: $C_{10}H_{18}$ **MP[°C]:** -30.3   **Den.[g/cm³]:** 0.8699^{20}
MW: 138.25   **BP[°C]:** 187.3 n_D: 1.4695^{20}
Sol.: H_2O 1; EtOH 4; eth
  4; ace 4; bz 5; chl;
  MeOH 2
Crit. Const.: T_c = 414.0°C
Vap. Press. [kPa]: 0.164^{25}; 2.62^{75}; 7.46^{100}; 18.1^{125}; 38.9^{150}
Therm. Prop.[kJ/mol]: $\Delta_{fus}H$ = 14.41; $\Delta_{vap}H$ = 40.2; $\Delta_f H°$(l, 25°C) = -230.6
C_p (liq.) [J/mol °C]: 228.5^{25}
γ [mN/m]: 32.15^{20}
Elec. Prop.: μ = 0; ε = 2.18^{20}; IP = 9.24 eV
η [mPa s]: 6.19^{-25}; 3.24^{0}; 1.95^{25}; 1.29^{50}; 0.917^{75}; 0.689^{100}
MS: 41(100) 68(91) 67(88) 82(67) 27(65) 96(61) 95(55) 138(51) 81(51) 29(51)
IR [cm⁻¹]: 2920 2860 1460 1440 1340 1300 1250 1140 970 920 840 820
Flammability: Flash pt. = 54°C; ign. temp. = 255°C; flam. lim. = 0.7-5.4%

126. Decane

Syst. Name: Decane

CASRN: 124-18-5   **DOT No.**: 2247
Beil RN: 1696981   **Beil Ref.**: 4-01-00-00464
MF: $C_{10}H_{22}$ **MP[°C]**: -29.7   **Den.[g/cm^3]**: 0.7300^{20}
MW: 142.28   **BP[°C]**: 174.1 n_D: 1.4102^{20}
Sol.: H_2O 1; EtOH 5; eth
   3; ctc 2
Crit. Const.: T_c = 344.6°C; P_c = 2.11 MPa; V_c = 624 cm^3
Vap. Press. [kPa]: 0.170^{25}; 3.22^{50}; 9.56^{100}; 23.9^{125}; 52.5^{150}
Therm. Prop.[kJ/mol]: $\Delta_{fus}H$ = 28.78; $\Delta_{vap}H$ = 38.8; $\Delta_f H°$(l, 25°C) = -300.9
C_p **(liq.) [J/mol °C]**: 314.4^{25}
γ **[mN/m]**: 23.37^{25}; dγ/dT = 0.0920 mN/m °C
Elec. Prop.: μ = 0; ε = 1.99^{20}; IP = 9.65 eV
η **[mPa s]**: 2.19^{-25}; 1.28^{0}; 0.838^{25}; 0.598^{50}; 0.453^{75}; 0.359^{100}
k **[W/m °C]**: 0.144^{-25}; 0.138^{0}; 0.132^{25}; 0.126^{50}; 0.119^{75}; 0.113^{100}
MS: 43(100) 57(90) 41(41) 71(33) 29(30) 85(24) 27(20) 56(17) 55(14) 42(14)
IR [cm^{-1}]: 2950 2850 1470 1380 720
Raman [cm^{-1}]: 2980 2950 2910 2890 2860 2740 2680 1460 1450 1370 1340
   1300 1160 1140 1080 1060 1020 990 920 890 850 810 770 520 400 350 250
   230
^{13}C NMR [ppm]: 14.1 23.0 32.4 29.9 30.3
^{1}H NMR [ppm]: 0.9 1.3 CCl_4
Flammability: Flash pt. = 46°C; ign. temp. = 210°C; flam. lim. = 0.8-5.4%

127. 1-Decene

Syst. Name: 1-Decene

CASRN: 872-05-9   **Beil Ref.**: 3-01-00-00858
Beil RN: 1737236
MF: $C_{10}H_{20}$ **MP[°C]**: -66.3   **Den.[g/cm^3]**: 0.7408^{20}
MW: 140.27   **BP[°C]**: 170.5 n_D: 1.4215^{20}
Sol.: H_2O 1; EtOH 5; eth
   5
Crit. Const.: $T_c = 343.3$°C; $P_c = 2.218$ MPa; $V_c = 584$ cm^3
Vap. Press. [kPa]: 0.210^{25}; 10.9^{100}; 26.8^{125}; 58.1^{150}
Therm. Prop.[kJ/mol]: $\Delta_{fus}H = 21.10$; $\Delta_{vap}H = 38.7$; $\Delta_f H°(l, 25°C) = -173.8$
C_p **(liq.) [J/mol °C]**: 300.8^{25}
γ **[mN/m]**: 23.54^{25}; $d\gamma/dT = 0.0919$ mN/m °C
Elec. Prop.: $\mu = 0$; $\varepsilon = 2.14^{20}$; IP = 9.42 eV
η **[mPa s]**: 0.756^{25}
MS: 41(100) 56(82) 55(81) 43(77) 70(69) 57(62) 29(57) 69(53) 27(50) 39(43)
Raman [cm^{-1}]: 3080 2998 2960 2899 2852 2730 1641 1439 1416 1300
Flammability: Flash pt. < 55°C; ign. temp. = 235°C

128. Diacetone alcohol

Syst. Name: 2-Pentanone, 4-hydroxy-4-methyl-
Synonyms: 4-Hydroxy-4-methyl-2-pentanone

CASRN: 123-42-2   **DOT No.**: 1148
Merck No.: 2944   **Beil Ref.**: 4-01-00-04023
Beil RN: 1740440
MF: $C_6H_{12}O_2$ **MP[°C]**: -44   **Den.[g/cm^3]**: 0.9387^{20}
MW: 116.16   **BP[°C]**: 167.9 n_D: 1.4213^{20}
Sol.: H$_2$O 5; EtOH 5; eth
    5; chl 3
Vap. Press. [kPa]: 0.224^{25}; 0.989^{50}; 3.53^{75}; 10.6^{100}
C_p **(liq.) [J/mol °C]**: 221.3^{25}
γ **[mN/m]**: 31.0^{20}
Elec. Prop.: μ = (3.2 D); ε = 18.2^{25}
η **[mPa s]**: 28.7^{-25}; 6.62^0; 2.80^{25}; 1.83^{50}; 1.65^{75}
MS: 43(100) 59(41) 58(17) 101(10) 41(9) 31(9) 83(6) 56(6) 55(6) 29(6)
IR [cm^{-1}]: 3450 2940 1700 1390 1330 1220 1180 1150 950 920
UV [nm]: 281 238 MeOH
^{1}**H NMR [ppm]**: 1.3 2.2 2.6 3.7 CDCl$_3$
Flammability: Flash pt. = 58°C; ign. temp. = 643°C; flam. lim. = 1.8-6.9%
TLV/TWA: 50 ppm (238 mg/m^3)

129. Dibenzyl ether

Syst. Name: Benzene, 1,1'-[oxybis(methylene)]bis-
Synonyms: Benzyl ether

CASRN: 103-50-4   **Beil Ref.:** 4-06-00-02240
Merck No.: 1146
Beil RN: 1911156
MF: $C_{14}H_{14}O$ **MP[°C]:** 1.8   **Den.[g/cm³]:** 1.0428^{20}
MW: 198.26   **BP[°C]:** 298 n_D: 1.5168^{20}
Sol.: H_2O 1; EtOH 5; eth
   5; ctc 3
γ **[mN/m]:** 38.2^{35}
Elec. Prop.: $\varepsilon = 3.82^{20}$
η **[mPa s]:** 5.33^{20}
MS: 92(100) 91(83) 79(20) 65(18) 77(17) 107(15) 51(12) 39(11) 93(8) 50(5)
IR [cm⁻¹]: 3060 3030 2850 1500 1460 1390 1370 1210 1100 1070 1030 910
   740 700 680
Raman [cm⁻¹]: 3060 2980 2930 2860 2780 1600 1580 1470 1450 1400 1390
   1200 1170 1150 1070 1030 1000 940 900 840 810 800 740 700 620 480 220
   200
UV [nm]: 258 cyhex
¹H NMR [ppm]: 4.5 7.3 CCl_4
Flammability: Flash pt. = 135°C

---

130. 1,2-Dibromo-1,1-difluoroethane

Syst. Name: Ethane, 1,2-dibromo-1,1-difluoro-
Synonyms: 1,1-Difluoro-1,2-dibromoethane

CASRN: 75-82-1   **Beil Ref.:** 4-01-00-00160
Beil RN: 1733225
MF: $C_2H_2Br_2F_2$ **MP[°C]:** -61.3   **Den.[g/cm³]:** 2.2238^{20}
MW: 223.84   **BP[°C]:** 92.5 n_D: 1.4456^{20}
MS: 143(100) 145(98) 64(42) 123(17) 125(15) 93(15) 81(13) 95(11) 79(11)
   63(11)

131. 1,2-Dibromoethane

Syst. Name: Ethane, 1,2-dibromo-
Synonyms: Ethylene dibromide

CASRN: 106-93-4   **DOT No.:** 1605
Merck No.: 3753   **Beil Ref.:** 4-01-00-00158
Beil RN: 605266
MF: $C_2H_4Br_2$ **MP[°C]:** 9.9   **Den.[g/cm^3]:** 2.1791^{20}
MW: 187.86   **BP[°C]:** 131.6 n_D: 1.5387^{20}
Sol.: ace 4; bz 4; eth 4;
 EtOH 4
Crit. Const.: T_c = 309.9°C; P_c = 7.2 MPa
Vap. Press. [kPa]: 1.55^{25}; 5.79^{50}; 16.9^{75}; 40.7^{100}; 85.5^{125}
Therm. Prop.[kJ/mol]: $\Delta_{fus}H$ = 10.84; $\Delta_{vap}H$ = 34.8; $\Delta_f H°$(l, 25°C) = -79.2
C_p (liq.) [J/mol °C]: 136.0^{25}
γ [mN/m]: 39.55^{25}; dγ/dT = 0.1320 mN/m °C
Elec. Prop.: μ = (1.2 D); ε = 4.96^{20}; IP = 10.37 eV
η [mPa s]: 1.60^{25}; 1.12^{50}; 0.837^{75}; 0.661^{100}
MS: 27(100) 107(77) 109(72) 26(24) 28(10) 81(5) 79(5) 25(5) 95(4) 93(4)
TLV/TWA: Carcinogen
Reg. Lists: CERCLA (RQ = 1 lb.); SARA 313 (0.1%); RCRA U067 (toxic);
 confirmed or suspected carcinogen

---

132. Dibromofluoromethane

Syst. Name: Methane, dibromofluoro-
Synonyms: Fluorodibromomethane

CASRN: 1868-53-7   **Beil Ref.:** 4-01-00-00080
Beil RN: 1697007
MF: $CHBr_2F$ **MP[°C]:** -78   **Den.[g/cm^3]:** 2.421^{20}
MW: 191.83   **BP[°C]:** 64.9 n_D: 1.4685^{20}
Sol.: H_2O 1; EtOH 3; eth
 3; ace 3; bz 3; chl 3
MS: 111(100) 113(98) 192(29) 43(16) 41(16) 190(15) 194(14) 81(9) 79(9)
 122(7)
IR [cm^{-1}]: 2940 1300 1220 1160 1060 700 650

133. Dibromomethane

Syst. Name: Methane, dibromo-
Synonyms: Methylene bromide

CASRN: 74-95-3
Merck No.: 5980
Beil RN: 969143
MF: CH_2Br_2
MW: 173.83
Sol.: H_2O 2; EtOH 5; eth
  5; ace 5; ctc 3

DOT No.: 2664
Beil Ref.: 4-01-00-00078

MP[°C]: -52.5
BP[°C]: 97

Den.[g/cm^3]: 2.4969^{20}
n_D: 1.5420^{20}

Vap. Press. [kPa]: 1.53^0; 6.12^{25}; 19.1^{50}; 49.6^{75}; 111^{100}
Therm. Prop.[kJ/mol]: $\Delta_{vap}H$ = 32.9
C_p (liq.) [J/mol °C]: 105.3^{27}
γ [mN/m]: 39.05^{25}; $d\gamma/dT$ = 0.1488 mN/m °C
Elec. Prop.: μ = 1.43 D; ε = 7.77^{10}; IP = 10.50 eV
η [mPa s]: 1.95^{-25}; 1.32^0; 0.980^{25}; 0.779^{50}; 0.652^{75}
k [W/m °C]: 0.120^{-25}; 0.114^0; 0.108^{25}; 0.103^{50}; 0.097^{75}
MS: 174(100) 93(96) 95(84) 172(53) 176(50) 91(11) 81(9) 79(9) 94(5) 65(5)
IR [cm^{-1}]: 3060 1180 800 630
Raman [cm^{-1}]: 3070 2990 1400 1280 1200 1100 820 860 650 590 410 180
UV [nm]: 218(1148) MeOH
^{13}C NMR [ppm]: 21.6
^{1}H NMR [ppm]: 4.9 CCl_4
Reg. Lists: CERCLA (RQ = 1000 lb.); SARA 313 (1.0%); RCRA U068 (toxic)

134. 1,2-Dibromopropane

Syst. Name: Propane, 1,2-dibromo-
Synonyms: Propylene dibromide

CASRN: 78-75-1   **Beil Ref.:** 4-01-00-00215
Merck No.: 7866
Beil RN: 1718884
MF: $C_3H_6Br_2$ **MP[°C]:** -55.2   **Den.[g/cm^3]:** 1.9324^{20}
MW: 201.89   **BP[°C]:** 141.9   **n_D:** 1.5201^{20}
Sol.: EtOH 3; eth 3; ctc
  2; chl 3
Vap. Press. [kPa]: 1.07^{25}; 2.96^{50}; 9.86^{75}; 27.1^{100}; 64.1^{125}; 135^{150}
Therm. Prop.[kJ/mol]: $\Delta_{fus}H = 8.94$; $\Delta_{vap}H = 35.6$
C_p (liq.) [J/mol °C]: 160.0^{25}
γ [mN/m]: 34.14^{20}
Elec. Prop.: $\mu = 1.13$ D; $\varepsilon = 4.60^{10}$; IP = 10.10 eV
η [mPa s]: 1.62^{20}
MS: 41(100) 121(66) 123(65) 39(48) 27(28) 107(11) 38(10) 26(10) 109(9)
  42(9)
IR [cm^{-1}]: 2980 2930 2860 1460 1450 1430 1380 1320 1240 1210 1160 1040
  1000 900 840 650
Raman [cm^{-1}]: 2960 2930 1440 1230 1210 1150 1120 1040 1000 900 840 640
  560 540 520 460 400 360 290 180
^{13}C NMR [ppm]: 24.1 37.6 45.7 CDCl$_3$
^{1}H NMR [ppm]: 1.8 3.5 3.8 4.2 CCl$_4$

135. 1,2-Dibromotetrafluoroethane

Syst. Name: Ethane, 1,2-dibromo-1,1,2,2-
    tetrafluoro-
Synonyms: R 114B2; Halon 2402

CASRN: 124-73-2   **Beil Ref.**: 4-01-00-00160
Beil RN: 1740342
MF: $C_2Br_2F_4$ **MP[°C]**: -110.4   **Den.[g/cm^3]**: 2.149^{25}
MW: 259.82   **BP[°C]**: 47.3 n_D: 1.361^{25}
Crit. Const.: $T_c = 214.7°C$; $P_c = 3.393$ MPa; $V_c = 341$ cm^3
Vap. Press. [kPa]: 3.79^{-25}; 14.3^0; 43.4^{25}
Therm. Prop.[kJ/mol]: $\Delta_{fus}H = 7.04$; $\Delta_{vap}H = 27.0$
C_p **(liq.) [J/mol °C]**: 180.3^{25}
γ **[mN/m]**: 18.1^{25}
Elec. Prop.: $\varepsilon = 2.34^{25}$; IP = 11.10 eV
η **[mPa s]**: 0.72^{25}
MS: 179(100) 181(97) 129(34) 131(33) 100(17) 31(13) 260(12) 50(8) 69(7)
    262(6)
IR [cm^{-1}]: 1180 1120 1010 870 820
Reg. Lists: SARA 313 (1.0%)

136. Dibutylamine

Syst. Name: 1-Butanamine, *N*-butyl-
Synonyms: *N*-Butyl-1-butanamine

CASRN: 111-92-2   **DOT No.:** 2248
Merck No.: 3019   **Beil Ref.:** 4-04-00-00550
Beil RN: 506001
MF: $C_8H_{19}N$ **MP[°C]:** -62   **Den.[g/cm^3]:** 0.7670^{20}
MW: 129.25   **BP[°C]:** 159.6 n_D: 1.4177^{20}
Sol.: H_2O 3; EtOH 4; eth
  4; ace 3; bz 3
Crit. Const.: $T_c = 334.4°C$; $P_c = 3.11$ MPa
Vap. Press. [kPa]: 0.340^{25}; 5.01^{75}; 14.5^{100}; 35.6^{125}; 77.3^{150}
Therm. Prop.[kJ/mol]: $\Delta_{vap}H = 38.4$; $\Delta_fH°(l, 25°C) = -206.0$
C_p **(liq.) [J/mol °C]:** 292.9^{25}
γ **[mN/m]:** 24.12^{25}; $d\gamma/dT = 0.0952$ mN/m °C
Elec. Prop.: $\mu = (1.0$ D); $\varepsilon = 2.77^{20}$; IP = 7.69 eV
η **[mPa s]:** 1.51^0; 0.918^{25}; 0.619^{50}; 0.449^{75}; 0.345^{100}
MS: 86(100) 72(52) 30(48) 44(40) 29(31) 57(24) 41(21) 73(15) 28(15) 43(13)
IR [cm^{-1}]: 3330 2940 2860 1470 1390 1250 1140 1060 1000 970
UV [nm]: 301(7) EtOH
^{1}H NMR [ppm]: 0.5 0.9 1.4 2.5 CCl_4
Flammability: Flash pt. = 47°C; flam. lim. = 1.1-6%

137. Dibutyl ether

Syst. Name: Butane, 1,1'-oxybis-
Synonyms: Butyl ether

CASRN: 142-96-1   **DOT No.**: 1149
Merck No.: 1568   **Beil Ref.**: 4-01-00-01520
Beil RN: 1732752
MF: $C_8H_{18}O$ **MP[°C]**: -95.2   **Den.[g/cm^3]**: 0.7684^{20}
MW: 130.23   **BP[°C]**: 140.2 n_D: 1.3992^{20}
Sol.: H_2O 1; EtOH 5; eth
   5; ace 4; ctc 2
Crit. Const.: $T_c = 311.0°C$; $P_c = 3.01$ MPa
Vap. Press. [kPa]: 0.898^{25}; 10.9^{75}; 27.9^{100}; 62.6^{125}; 127^{150}
Therm. Prop.[kJ/mol]: $\Delta_{vap}H = 36.5$; $\Delta_f H°(l, 25°C) = -377.9$
C_p **(liq.) [J/mol °C]**: 278.2^{25}
γ **[mN/m]**: 22.44^{25}; $d\gamma/dT = 0.0934$ mN/m °C
Elec. Prop.: $\mu = 1.17$ D; $\varepsilon = 3.08^{20}$; IP = 9.43 eV
η **[mPa s]**: 1.42^{-25}; 0.918^{0}; 0.637^{25}; 0.466^{50}; 0.356^{75}; 0.281^{100}
MS: 57(100) 41(34) 29(30) 56(25) 87(21) 27(9) 58(8) 55(6) 39(5) 28(5)
IR [cm^{-1}]: 2960 2940 2860 1470 1380 1240 1120 960 740
Raman [cm^{-1}]: 3020 2970 2940 2920 2880 2740 2600 1480 1450 1430 1370
   1300 1260 1220 1150 1130 1120 1060 1040 1020 980 960 930 900 880 840
   810 740 640 460 400 390 290
^{13}C NMR [ppm]: 14.6 20.3 33.1 71.2
^{1}H NMR [ppm]: 0.9 1.4 3.3 CCl_4
Flammability: Flash pt. = 25°C; ign. temp. = 194°C; flam. lim. = 1.5-7.6%

138. Dibutyl maleate

Syst. Name: 2-Butenedioic acid (*Z*)-, dibutyl ester
Synonyms: Butyl maleate; Butyl *cis*-butenedioate

CASRN: 105-76-0   **Beil Ref.:** 4-02-00-02209
Beil RN: 1726634
MF: $C_{12}H_{20}O_4$ **MP[°C]:** <-80
MW: 228.29   **BP[°C]:** 280
Vap. Press. [kPa]: 0.002^{25}; 0.010^{50}; 0.038^{75}; 0.134^{100}; 0.431^{125}; 1.28^{150}
η [mPa s]: 4.76^{25}
MS: 99(100) 57(66) 117(35) 29(35) 41(31) 56(19) 100(18) 27(15) 155(11) 173(8)
IR [cm^{-1}]: 2950 2920 2870 1730 1640 1460 1400 1370 1280 1200 1160 1060 1020 970 840 810 740 600 510 430
Raman [cm^{-1}]: 3050 2940 2910 2870 2730 1730 1640 1450 1400 1370 1300 1260 1230 1160 1120 1060 1020 960 900 880 840 810 740 670 590 530 500 470 380 330 280
^{1}H NMR [ppm]: 0.9 1.5 4.1 6.1 CCl_4
Flammability: Flash pt. = 141°C

139. Dibutyl phthalate

Syst. Name: 1,2-Benzenedicarboxylic acid, dibutyl ester

Synonyms: Butyl phthalate

CASRN: 84-74-2

Merck No.: 1586

Beil RN: 1914064

MF: $C_{16}H_{22}O_4$

MW: 278.35

Sol.: H_2O 1; EtOH 5; eth 5; bz 5; ctc 3

DOT No.: 9095

Beil Ref.: 4-09-00-03175

MP[°C]: -35

BP[°C]: 340

Den.[g/cm^3]: 1.0465^{20}

n_D: 1.4911^{20}

Therm. Prop.[kJ/mol]: $\Delta_{vap}H = 79.2$; $\Delta_f H°(1, 25°C) = -842.6$

C_p (liq.) [J/mol °C]: 498.0^{25}

γ [mN/m]: 33.40^{20}

Elec. Prop.: $\mu = (2.8\ D)$; $\varepsilon = 6.58^{20}$

η [mPa s]: 483^{-25}; 66.4^0; 16.6^{25}; 6.47^{50}; 3.50^{75}; 2.43^{100}

k [W/m °C]: 0.144^{-25}; 0.140^0; 0.136^{25}; 0.133^{50}; 0.129^{75}; 0.125^{100}

MS: 149(100) 86(18) 57(18) 223(17) 205(17) 150(17) 104(17) 56(17) 41(17) 65(16)

IR [cm^{-1}]: 2940 1720 1610 1590 1470 1390 1280 1140 1080 1040 960 940 840 790 750 700

Raman [cm^{-1}]: 3070 2960 2930 2910 2870 2730 2600 1720 1600 1570 1480 1450 1430 1380 1270 1220 1150 1110 1030 990 960 940 900 840 800 740 700 640 550 540 490 470 390 330 280

UV [nm]: 274(1230) 225(7960) MeOH

^{1}H NMR [ppm]: 1.0 1.6 4.2 7.4 7.6 CCl_4

Flammability: Flash pt. = 157°C; ign. temp. = 402°C; lower flam. lim. = 0.5%

TLV/TWA: 5 mg/m^3

Reg. Lists: CERCLA (RQ = 10 lb.); SARA 313 (1.0%); RCRA U069 (toxic)

140. Dibutyl sebacate

Syst. Name: Decanedioic acid, dibutyl ester
Synonyms: Butyl sebacate

CASRN: 109-43-3
Beil RN: 1798308
Beil Ref.: 4-02-00-02081
MF: $C_{18}H_{34}O_4$ **MP[°C]:** -10   **Den.[g/cm^3]:** 0.9405^{15}
MW: 314.47   **BP[°C]:** 344.5 n_D: 1.4433^{15}
Sol.: H_2O 1; eth 3; ctc 3
C_p **(liq.) [J/mol °C]:** 619^{39}
Elec. Prop.: $\mu = (2.5\ D)$; $\varepsilon = 4.54^{20}$
η **[mPa s]:** 9.03^{20}
MS: 241(100) 185(71) 41(37) 56(35) 55(32) 57(31) 242(23) 143(21) 98(21)
    125(20)
IR [cm^{-1}]: 2960 2930 2860 1740 1460 1350 1240 1170 1080 1020 730
Raman [cm^{-1}]: 2940 2910 2870 2730 1730 1440 1300 1260 1230 1150 1120
    1090 1060 1020 970 930 900 870 830 810 740 640 600 510 470 430 340 300
    220
^{1}H NMR [ppm]: 1.0 1.5 2.2 4.0 CCl_4
Flammability: Flash pt. = 178°C; ign. temp. = 365°C; lower flam. lim. = 0.44%

---

141. Dibutyl sulfide

Syst. Name: Butane, 1,1'-thiobis-
Synonyms: 5-Thianonane; Butyl sulfide

CASRN: 544-40-1
Merck No.: 1590
Beil RN: 1732829
Beil Ref.: 4-01-00-01559
MF: $C_8H_{18}S$ **MP[°C]:** -79.7   **Den.[g/cm^3]:** 0.8386^{20}
MW: 146.30   **BP[°C]:** 185 n_D: 1.4530^{20}
Sol.: eth 4; EtOH 4; chl
    4
Vap. Press. [kPa]: 15.3^{125}; 34.8^{150}
Therm. Prop.[kJ/mol]: $\Delta_{vap}H = 41.3$; $\Delta_f H°(l, 25°C) = -220.7$
C_p **(liq.) [J/mol °C]:** 284.3^{25}
γ **[mN/m]:** 26.8^{25}
Elec. Prop.: $\mu = (1.6\ D)$; $\varepsilon = 4.29^{25}$; IP = 8.20 eV
η **[mPa s]:** 0.98^{25}
MS: 56(100) 61(92) 29(38) 41(35) 57(30) 90(25) 146(24) 55(24) 27(18) 47(15)
Flammability: Flash pt. = 76°C

142. *o*-Dichlorobenzene

Syst. Name: Benzene, 1,2-dichloro-
Synonyms: 1,2-Dichlorobenzene

CASRN: 95-50-1   **Rel. CASRN**: 25321-22-6
Merck No.: 3044   **DOT No.**: 1591
Beil RN: 606078   **Beil Ref.**: 4-05-00-00654
MF: $C_6H_4Cl_2$ **MP[°C]**: -16.7   **Den.[g/cm^3]**: 1.3059^{20}
MW: 147.00   **BP[°C]**: 180 n_D: 1.5515^{20}
Sol.: H_2O 1; EtOH 3; eth
  3; ace 5; bz 5; ctc 5;
  lig 5
Vap. Press. [kPa]: 0.180^{25}; 8.39^{100}; 20.7^{125}; 45.0^{150}
Therm. Prop.[kJ/mol]: $\Delta_{fus}H$ = 12.93; $\Delta_{vap}H$ = 39.7; $\Delta_f H°$(l, 25°C) = -17.5
C_p (liq.) [J/mol °C]: 162.4^{25}
γ [mN/m]: 26.84^{20}
Elec. Prop.: μ = 2.50 D; ε = 10.12^{20}; IP = 9.08 eV
η [mPa s]: 1.96^0; 1.32^{25}; 0.962^{50}; 0.739^{75}; 0.593^{100}
MS: 146(100) 148(64) 111(38) 75(23) 113(12) 74(12) 50(11) 150(10) 73(9)
  147(7)
IR [cm^{-1}]: 3070 1570 1480 1460 1440 1130 1040 750 660
Raman [cm^{-1}]: 3150 3070 3030 3000 1570 1390 1280 1160 1130 1040 1020
  660 490 470 430 330 240 200 160
UV [nm]: 269 220 iso
^{13}C NMR [ppm]: 127.7 130.5 132.6 CDCl$_3$
^{1}H NMR [ppm]: 7.2 CCl$_4$
Flammability: Flash pt. = 66°C; ign. temp. = 648°C; flam. lim. = 2.2-9.2%
TLV/TWA: 25 ppm (150 mg/m^3)
Reg. Lists: CERCLA (RQ = 100 lb.); SARA 313 (0.1%); RCRA U070 (toxic)

143. *m*-Dichlorobenzene

Syst. Name: Benzene, 1,3-dichloro-
Synonyms: 1,3-Dichlorobenzene

CASRN: 541-73-1   **Rel. CASRN**: 25321-22-6
Merck No.: 3043   **DOT No.**: 9255
Beil RN: 956618   **Beil Ref.**: 4-05-00-00657
MF: $C_6H_4Cl_2$ **MP[°C]**: -24.8   **Den.[g/cm^3]**: 1.2884^{20}
MW: 147.00   **BP[°C]**: 173 n_D: 1.5459^{20}
Sol.: H_2O 1; EtOH 3; eth
  3; ace 5; bz 3; ctc 5;
  lig 5
Vap. Press. [kPa]: 0.252^{25}; 10.9^{100}; 26.1^{125}; 55.3^{150}
Therm. Prop.[kJ/mol]: $\Delta_{fus}H$ = 12.64; $\Delta_{vap}H$ = 38.6; $\Delta_f H°$(l, 25°C) = -20.7
C_p **(liq.) [J/mol °C]**: 171^{25}
γ **[mN/m]**: 35.43^{25}; $d\gamma/dT$ = 0.1147 mN/m °C
Elec. Prop.: μ = 1.72 D; ϵ = 5.02^{20}; IP = 9.11 eV
η **[mPa s]**: 1.49^0; 1.04^{25}; 0.787^{50}; 0.628^{75}; 0.525^{100}
MS: 146(100) 148(65) 111(36) 75(25) 50(19) 74(16) 150(11) 113(11) 73(11)
  147(8)
IR [cm^{-1}]: 3130 1560 1470 1410 1120 1090 1000 860 780
Raman [cm^{-1}]: 3160 3080 1580 1130 1110 1070 1000 660 430 400 370 220
  200 180
UV [nm]: 278(2800) 270(3340) 263(2330) 256(1420) 250(830) 216(12800)
  MeOH
^{13}C NMR [ppm]: 127.0 128.9 130.6 135.1 diox
^{1}H NMR [ppm]: 7.2 7.4 CCl_4
Flammability: Flash pt. = 72°C
Reg. Lists: CERCLA (RQ = 100 lb.); SARA 313 (0.1%); RCRA U071 (toxic)

144. *p*-Dichlorobenzene

Syst. Name: Benzene, 1,4-dichloro-
Synonyms: 1,4-Dichlorobenzene

CASRN: 106-46-7   **Rel. CASRN**: 25321-22-6
Merck No.: 3045   **DOT No.**: 1592
Beil RN: 1680023   **Beil Ref.**: 4-05-00-00658
MF: $C_6H_4Cl_2$ **MP[°C]**: 52.7   **Den.[g/cm^3]**: 1.2475^{55}
MW: 147.00   **BP[°C]**: 174 n_D: 1.5285^{20}
Sol.: H_2O 1; EtOH 5; eth
  3; ace 5; bz 3; ctc 3;
  chl 3; CS_2 3
Vap. Press. [kPa]: 0.235^{25}; 10.4^{100}; 25.1^{125}; 53.5^{150}
Therm. Prop.[kJ/mol]: $\Delta_{fus}H = 17.15$; $\Delta_{vap}H = 38.8$; $\Delta_f H°(s, 25°C) = -42.3$
γ **[mN/m]**: 30.69^{68}
Elec. Prop.: $\mu = 0$; $\varepsilon = 2.39^{55}$; IP = 8.89 eV
η **[mPa s]**: 0.839^{55}
MS: 146(100) 148(64) 111(35) 75(22) 74(14) 50(12) 113(11) 73(11) 150(10)
  147(7)
IR [cm^{-1}]: 3030 1890 1610 1470 1410 1390 1100 1020 810
Raman [cm^{-1}]: 3150 3080 1580 1380 1290 1170 1110 1080 1070 750 630 350
  330 300
UV [nm]: 280(301) 272(377) 264(287) 224(12800) MeOH
^{1}H NMR [ppm]: 7.2 CCl_4
Flammability: Flash pt. = 66°C
TLV/TWA: 10 ppm (60 mg/m^3)
Reg. Lists: CERCLA (RQ = 100 lb.); SARA 313 (0.1%); RCRA U072 (toxic)

145. 1,1-Dichloro-2,2-difluoroethylene

Syst. Name: Ethene, 1,1-dichloro-2,2-difluoro-
Synonyms: 2,2-Difluoro-1,1-dichloroethylene; R
  1112a; Genetron 1112a

CASRN: 79-35-6   **Rel. CASRN:** 27156-03-2
Beil RN: 1740374   **DOT No.:** 9018
                            Beil Ref.: 4-01-00-00711
MF: $C_2Cl_2F_2$ **MP[°C]:** -116   **Den.[g/cm^3]:** 1.555^{-20}
MW: 132.92   **BP[°C]:** 19 n_D: 1.383^{-20}
Elec. Prop.: $\mu = 0.50$ D
MS: 132(100) 134(63) 82(46) 47(40) 84(30) 31(17) 49(14) 35(13) 97(11)
  136(10)

---

146. Dichlorodifluoromethane

Syst. Name: Methane, dichlorodifluoro-
Synonyms: Difluorodichloromethane; R 12; CFC 12

CASRN: 75-71-8   **DOT No.:** 1028
Merck No.: 3053   **Beil Ref.:** 4-01-00-00040
Beil RN: 1732393
MF: CCl_2F_2 **MP[°C]:** -158
MW: 120.91   **BP[°C]:** -29.8
Sol.: H_2O 3; EtOH 3; eth
  3; HOAc 3
Crit. Const.: $T_c = 111.80$°C; $P_c = 4.136$ MPa; $V_c = 217$ cm^3
Vap. Press. [kPa]: 9.9^{-73}; 123^{-25}; 308^0; 651^{25}; 1216^{50}; 2076^{75}; 3332^{100}
Therm. Prop.[kJ/mol]: $\Delta_{fus}H = 4.14$; $\Delta_{vap}H = 20.1$; $\Delta_f H°(g, 25°C) = -477.4$
C_p (liq.) [J/mol °C]: 117.2^{25} (sat. press.)
Elec. Prop.: $\mu = 0.51$ D; $\varepsilon = 3.50^{-150}$; IP = 11.75 eV
MS: 85(100) 87(33) 50(12) 101(9) 35(7) 31(7) 103(6) 66(4) 47(3) 37(2)
IR [cm^{-1}]: 1160 1100 930 920 890
TLV/TWA: 1000 ppm (4950 mg/m^3)
Reg. Lists: CERCLA (RQ = 5000 lb.); SARA 313 (1.0%); RCRA U075 (toxic)

147. 1,1-Dichloroethane

Syst. Name: Ethane, 1,1-dichloro-
Synonyms: Ethylidene dichloride

CASRN: 75-34-3   **DOT No.**: 2362
Merck No.: 3766   **Beil Ref.**: 4-01-00-00130
Beil RN: 1696901
MF: $C_2H_4Cl_2$ **MP[°C]**: -96.9   **Den.[g/cm^3]**: 1.1757^{20}
MW: 98.96   **BP[°C]**: 57.4 n_D: 1.4164^{20}
Sol.: H_2O 2; EtOH 4; eth
  4; ace 3; bz 3
Crit. Const.: T_c = 250°C; P_c = 5.07 MPa; V_c = 236 cm^3
Vap. Press. [kPa]: 2.25^{-25}; 9.55^0; 30.5^{25}; 79.2^{50}
Therm. Prop.[kJ/mol]: $\Delta_{fus}H$ = 8.84; $\Delta_{vap}H$ = 28.9; $\Delta_f H°$(l, 25°C) = -158.4
C_p **(liq.) [J/mol °C]**: 126.3^{25}
γ **[mN/m]**: 24.07^{25}; dγ/dT = 0.1186 mN/m °C
Elec. Prop.: μ = 2.06 D; ϵ = 10.10^{25}; IP = 11.06 eV
η **[mPa s]**: 0.464^{25}; 0.362^{50}
MS: 63(100) 27(71) 65(31) 26(19) 83(11) 85(7) 61(7) 35(6) 98(5) 62(5)
IR [cm^{-1}]: 3030 1450 1390 1280 1240 1060 990 700
^{13}C NMR [ppm]: 32.5 69.2
Flammability: Flash pt. = -17°C; ign. temp. = 458°C; flam. lim. = 5.4-11.4%
TLV/TWA: 100 ppm (405 mg/m^3)
Reg. Lists: CERCLA (RQ = 1000 lb.); RCRA U076 (toxic)

148. 1,2-Dichloroethane

Syst. Name: Ethane, 1,2-dichloro-
Synonyms: Ethylene dichloride; R 150

CASRN: 107-06-2
Merck No.: 3754
Beil RN: 605264
MF: $C_2H_4Cl_2$
MW: 98.96
DOT No.: 1184
Beil Ref.: 4-01-00-00131

MP[°C]: -35.5
BP[°C]: 83.5
Den.[g/cm^3]: 1.2351^{20}
n_D: 1.4448^{20}

Sol.: H_2O 2; EtOH 4; eth 5; ace 3; bz 3; ctc 3; chl 3; os 3

Crit. Const.: $T_c = 288°C$; $P_c = 5.4$ MPa; $V_c = 225$ cm^3

Vap. Press. [kPa]: 0.541^{-25}; 2.84^{0}; 10.6^{25}; 31.4^{50}; 77.2^{75}; 162^{100}

Therm. Prop.[kJ/mol]: $\Delta_{fus}H = 8.83$; $\Delta_{vap}H = 32.0$; $\Delta_fH°(l, 25°C) = -167.4$

C_p (liq.) [J/mol °C]: 128.4^{25}

γ [mN/m]: 31.86^{25}; $d\gamma/dT = 0.1428$ mN/m °C

Elec. Prop.: $\mu = (1.8$ D); $\varepsilon = 10.42^{20}$; IP = 11.04 eV

η [mPa s]: 1.13^{0}; 0.779^{25}; 0.576^{50}; 0.447^{75}

MS: 62(100) 27(91) 49(40) 64(32) 26(31) 63(19) 98(14) 51(13) 61(12) 100(9)

IR [cm^{-1}]: 2960 1430 1290 1240 1010 950 880 710 670 650 420

Raman [cm^{-1}]: 3000 2960 2870 2840 1440 1430 1300 1260 1200 1140 1050 1030 940 880 750 670 650 410 300 260 120

^{13}C NMR [ppm]: 51.7

^{1}H NMR [ppm]: 3.7 CCl_4

Flammability: Flash pt. = 13°C; ign. temp. = 413°C; flam. lim. = 6.2-16%

TLV/TWA: 10 ppm (40 mg/m^3)

Reg. Lists: CERCLA (RQ = 100 lb.); SARA 313 (0.1%); RCRA U077 (toxic); confirmed or suspected carcinogen

149. 1,1-Dichloroethylene
Syst. Name: Ethene, 1,1-dichloro-
Synonyms: Vinylidene chloride

CASRN: 75-35-4   **DOT No.**: 1303
Merck No.: 9900   **Beil Ref.**: 4-01-00-00706
Beil RN: 1733365
MF: $C_2H_2Cl_2$ **MP[°C]**: -122.5   **Den.[g/cm^3]**: 1.213^{20}
MW: 96.94   **BP[°C]**: 31.6 n_D: 1.4249^{20}
Sol.: H_2O 1; EtOH 3; eth
   4; ace 3; bz 3; ctc 3;
   chl 4
Vap. Press. [kPa]: 8.22^{-25}; 28.9^0; 80.0^{25}
Therm. Prop.[kJ/mol]: $\Delta_{fus}H = 6.51$; $\Delta_{vap}H = 26.1$; $\Delta_f H°(l, 25°C) = -23.9$
C_p **(liq.) [J/mol °C]**: 111.3^{25}
Elec. Prop.: $\mu = 1.34$ D; $\varepsilon = 4.60^{20}$; IP = 9.79 eV
η **[mPa s]**: 0.358^{20}
MS: 61(100) 96(61) 98(38) 63(32) 26(16) 60(15) 62(7) 25(7) 100(6) 35(6)
IR [cm^{-1}]: 3130 3030 1610 1560 1140 1120 1100 1080 860 790 780
^{1}H NMR [ppm]: 5.5 CCl_4
Flammability: Flash pt. = -28°C; ign. temp. = 570°C; flam. lim. = 6.5-15.5%
TLV/TWA: 5 ppm (20 mg/m^3)
Reg. Lists: CERCLA (RQ = 100 lb.); SARA 313 (1.0%); RCRA U078 (toxic);
   confirmed or suspected carcinogen

150. *cis*-1,2-Dichloroethylene

Syst. Name: Ethene, 1,2-dichloro-, (*Z*)-
Synonyms: *cis*-Acetylene dichloride; R 1130

CASRN: 156-59-2
Merck No.: 86
Beil RN: 1071208
MF: $C_2H_2Cl_2$
MW: 96.94
Sol.: H_2O 2; EtOH 5; eth
5; ace 5; bz 4; chl 4
Crit. Const.: T_c = 271.1°C

Rel. CASRN: 540-59-0
DOT No.: 1150
Beil Ref.: 4-01-00-00707
MP[°C]: -80
BP[°C]: 60.1

Den.[g/cm^3]: 1.2837^{20}
n_D: 1.4490^{20}

Vap. Press. [kPa]: 26.8^{25}; 70.6^{50}; 159^{75}; 320^{100}
Therm. Prop.[kJ/mol]: $\Delta_{fus}H$ = 7.20; $\Delta_{vap}H$ = 30.2; $\Delta_f H°$(l, 25°C) = -26.4
C_p (liq.) [J/mol °C]: 116.4^{25}
γ [mN/m]: 28^{20}
Elec. Prop.: μ = 1.90 D; ϵ = 9.20^{25}; IP = 9.66 eV
η [mPa s]: 0.786^{-25}; 0.575^{0}; 0.445^{25}
MS: 61(100) 96(73) 98(47) 63(32) 26(30) 60(21) 25(13) 35(12) 62(9) 100(8)
IR [cm^{-1}]: 3030 1590 1300 860 850 710 690
^{13}C NMR [ppm]: 119.3
Flammability: Flash pt. = 2°C; ign. temp. = 460°C; flam. lim. = 3-15%
TLV/TWA: 200 ppm (793 mg/m^3)
Reg. Lists: SARA 313 (1.0%)

151. *trans*-1,2-Dichloroethylene

Syst. Name: Ethene, 1,2-dichloro-, (*E*)-
Synonyms: trans-Acetylene dichloride

CASRN: 156-60-5
Merck No.: 86
Beil RN: 1420761
MF: $C_2H_2Cl_2$
MW: 96.94
Sol.: H_2O 2; EtOH 5; eth 5; ace 5; bz 4; ctc 3; chl 4

Rel. CASRN: 540-59-0
DOT No.: 1150
Beil Ref.: 4-01-00-00709
MP[°C]: -49.8   **Den.[g/cm³]**: 1.2565^{20}
BP[°C]: 48.7 n_D: 1.4454^{20}

Crit. Const.: $T_c = 243.4°C$; $P_c = 5.51$ MPa
Vap. Press. [kPa]: 3.73^{-25}; 14.7^0; 44.2^{25}; 110^{50}; 234^{75}
Therm. Prop.[kJ/mol]: $\Delta_{fus}H = 11.98$; $\Delta_{vap}H = 28.9$; $\Delta_f H°(l, 25°C) = -23.1$
C_p **(liq.) [J/mol °C]**: 116.8^{25}
γ **[mN/m]**: 25^{20}
Elec. Prop.: $\mu = 0$; $\varepsilon = 2.14^{20}$; IP = 9.65 eV
η **[mPa s]**: 0.522^{-25}; 0.398^0; 0.317^{25}; 0.261^{50}
MS: 61(100) 96(67) 98(43) 26(34) 63(32) 60(24) 25(15) 62(10) 100(7) 47(7)
IR [cm⁻¹]: 3130 1670 1300 1210 900 820
Raman [cm⁻¹]: 3150 3070 1690 1680 1570 1270 840 760 440 350 230
¹³C NMR [ppm]: 121.1
¹H NMR [ppm]: 6.4 CCl_4
Flammability: Flash pt. = 2°C; ign. temp. = 460°C; flam. lim. = 6-13%
TLV/TWA: 200 ppm (793 mg/m³)
Reg. Lists: CERCLA (RQ = 1000 lb.); SARA 313 (1.0%); RCRA U079 (toxic)

152. Dichlorofluoromethane

Syst. Name: Methane, dichlorofluoro-
Synonyms: Fluorodichloromethane; R 21; CFC 21

CASRN: 75-43-4   **DOT No.**: 1029
Beil RN: 1731041   **Beil Ref.**: 4-01-00-00039
MF: $CHCl_2F$ **MP[°C]**: -135   **Den.[g/cm^3]**: 1.405^9
MW: 102.92   **BP[°C]**: 8.9 n_D: 1.3724^9
Sol.: H_2O 1; EtOH 3; eth
    3; ctc 3; chl 3; HOAc
    3
Crit. Const.: T_c = 178.43°C; P_c = 5.18 MPa; V_c = 196 cm^3
Vap. Press. [kPa]: 13.7^{-25}; 67.2^0; 187^{25}
Therm. Prop.[kJ/mol]: $\Delta_{vap}H$ = 25.2
C_p **(liq.) [J/mol °C]**: 112.6^{25} (sat. press.)
Elec. Prop.: μ = 1.29 D; ε = 5.34^{28}; IP = 11.50 eV
η **[mPa s]**: 0.337^{18} (at saturation pressure)
MS: 67(100) 69(32) 47(13) 35(13) 31(9) 32(8) 48(7) 83(5) 102(4) 49(4)
IR [cm^{-1}]: 3130 3030 2130 1540 1410 1320 1250 1090 1060 810 800 750
Raman [cm^{-1}]: 3020 1310 1254 1067 795 738 727 457 366 277
^{1}H NMR [ppm]: 7.5 CCl_4
TLV/TWA: 10 ppm (42 mg/m^3)

153. Dichloromethane

Syst. Name: Methane, dichloro-
Synonyms: Methylene chloride; R 30

CASRN: 75-09-2   **DOT No.**: 1593
Merck No.: 5982   **Beil Ref.**: 4-01-00-00035
Beil RN: 1730800
MF: CH_2Cl_2 **MP[°C]**: -95.1   **Den.[g/cm³]**: 1.3266^{20}
MW: 84.93   **BP[°C]**: 40 n_D: 1.4242^{20}
Sol.: H_2O 2; EtOH 5; eth
    5; ctc 3
Crit. Const.: $T_c = 237°C$; $P_c = 6.10$ MPa
Vap. Press. [kPa]: 0.1^{-73}; 4.77^{-25}; 19.2^0; 58.2^{25}; 145^{50}
Therm. Prop.[kJ/mol]: $\Delta_{fus}H = 6.00$; $\Delta_{vap}H = 28.1$; $\Delta_f H°(l, 25°C) = -124.1$
C_p **(liq.) [J/mol °C]**: 101.2^{25}
γ **[mN/m]**: 27.20^{25}; $d\gamma/dT = 0.1284$ mN/m °C
Elec. Prop.: $\mu = 1.60$ D; $\epsilon = 8.93^{25}$; IP = 11.32 eV
η **[mPa s]**: 0.727^{-25}; 0.533^0; 0.413^{25}
MS: 49(100) 84(64) 86(39) 51(31) 47(14) 48(8) 88(6) 50(3) 85(2) 83(2)
IR [cm⁻¹]: 3050 2990 2300 1420 1260 730 700
¹³C NMR [ppm]: 54.2
¹H NMR [ppm]: 5.3 CCl_4
Flammability: Ign. temp. = 556°C; flam. lim. = 13-23%
TLV/TWA: 50 ppm (174 mg/m³)
Reg. Lists: CERCLA (RQ = 1000 lb.); SARA 313 (0.1%); RCRA U080 (toxic);
    confirmed or suspected carcinogen

154. 1,2-Dichloropropane

Syst. Name: Propane, 1,2-dichloro-
Synonyms: Propylene dichloride

CASRN: 78-87-5
Merck No.: 7867
Beil RN: 1718883
MF: $C_3H_6Cl_2$
MW: 112.99

Rel. CASRN: 26198-63-0, 26638-19-7
DOT No.: 1279
Beil Ref.: 3-01-00-00225

MP[°C]: -100.4   **Den.[g/cm³]:** 1.1560^{20}
BP[°C]: 96.4 n_D: 1.4394^{20}

Sol.: H_2O 2; EtOH 3; eth 3; bz 3; chl 3
Vap. Press. [kPa]: 0.299^{-25}; 1.68^{0}; 6.62^{25}; 20.3^{50}; 51.4^{75}; 113^{100}
Therm. Prop.[kJ/mol]: $\Delta_{fus}H = 6.40$; $\Delta_f H°(l, 25°C) = -198.8$
γ [mN/m]: 28.65^{20}
Elec. Prop.: $\mu = (1.8$ D$)$
η [mPa s]: 0.857^{20}
Flammability: Flash pt. = 21°C
TLV/TWA: 75 ppm (347 mg/m³)
Reg. Lists: CERCLA (RQ = 1000 lb.); SARA 313 (1.0%); RCRA U083 (toxic); confirmed or suspected carcinogen

---

155. 1,1-Dichlorotetrafluoroethane

Syst. Name: Ethane, 1,1-dichloro-1,2,2,2-
tetrafluoro-
Synonyms: R 114a; CFC 114a

CASRN: 374-07-2
Beil RN: 1740332
MF: $C_2Cl_2F_4$
MW: 170.92

Beil Ref.: 4-01-00-00136

MP[°C]: -56.6   **Den.[g/cm³]:** 1.455^{25} (sat. press.)
BP[°C]: 4 n_D: 1.3092^{0}

Sol.: bz 4; eth 4; EtOH 4
Crit. Const.: $T_c = 145.5°C$; $P_c = 3.30$ MPa; $V_c = 294$ cm³
Vap. Press. [kPa]: 30.2^{-25}; 90.2^{0}; 218^{25}; 452^{50}; 842^{75}; 1443^{100}; 2322^{125}
MS: 135(100) 101(84) 85(57) 103(53) 31(50) 69(39) 137(31) 87(18) 66(17) 35(16)
IR [cm⁻¹]: 1610 1520 1500 1480 1420 1390 1290 1230 1100 1050 940 910 890 840 720 580 560 510

156. 1,2-Dichlorotetrafluoroethane

Syst. Name: Ethane, 1,2-dichloro-1,1,2,2-
tetrafluoro-
Synonyms: R 114; CFC 114

CASRN: 76-14-2   **DOT No.:** 1958
Merck No.: 2608   **Beil Ref.:** 4-01-00-00137
Beil RN: 1740333
MF: $C_2Cl_2F_4$ **MP[°C]:** -94   **Den.[g/cm^3]:** 1.455^{25} (sat. press.)
MW: 170.92   **BP[°C]:** 3.8 n_D: 1.3092^0
Sol.: eth 4; EtOH 4
Crit. Const.: T_c = 145.63°C; P_c = 3.252 MPa; V_c = 297 cm^3
Vap. Press. [kPa]: 29.4^{-25}; 88.3^0; 215^{25}; 447^{50}; 830^{75}; 1413^{100}; 2262^{125}
Therm. Prop.[kJ/mol]: $\Delta_{fus}H$ = 6.32; $\Delta_{vap}H$ = 23.3; $\Delta_f H°$(g, 25°C) = -916.3
C_p (liq.) [J/mol °C]: 164.2^{25} (sat. press.)
Elec. Prop.: μ = 0.5 D; ϵ = 2.48^0; IP = 12.20 eV
η **[mPa s]:** 0.38^{25}
MS: 85(100) 135(52) 87(33) 137(17) 101(9) 31(9) 103(6) 100(6) 50(5) 69(4)
TLV/TWA: 1000 ppm (6990 mg/m^3)
Reg. Lists: SARA 313 (1.0%)

157. 2,4-Dichlorotoluene

Syst. Name: Benzene, 2,4-dichloro-1-methyl-
Synonyms: 2,4-Dichloro-1-methylbenzene

CASRN: 95-73-8   **Beil Ref.:** 4-05-00-00815
Beil RN: 1931691
MF: $C_7H_6Cl_2$ **MP[°C]:** -13.5   **Den.[g/cm^3]:** 1.2476^{20}
MW: 161.03   **BP[°C]:** 201 n_D: 1.5511^{20}
Sol.: H_2O 1; ctc 3
Vap. Press. [kPa]: 0.055^{25}; 1.43^{75}; 4.61^{100}; 12.2^{125}; 27.7^{150}
Elec. Prop.: $\mu = (1.7\ D)$; $\varepsilon = 5.68^{28}$
η **[mPa s]:** 1.53^{25}
MS: 125(100) 160(61) 162(40) 127(32) 89(23) 159(16) 161(14) 63(13) 62(11) 126(10)
IR [cm^{-1}]: 3060 2920 1590 1560 1470 1440 1380 1260 1110 1050 990 860 840 800 700 690 650
Raman [cm^{-1}]: 3070 3020 2960 2930 2860 2740 2550 1600 1570 1480 1450 1390 1300 1280 1260 1210 1150 1130 1110 1060 880 840 810 750 720 690 650 470 440 410 390 320 280 210 190 130
UV [nm]: 281(588) 273(588) 266(378) 218(3740) MeOH
^{1}H NMR [ppm]: 2.3 7.0 7.3 CCl_4

158. 3,4-Dichlorotoluene

Syst. Name: Benzene, 1,2-dichloro-4-methyl-
Synonyms: 1,2-Dichloro-4-methylbenzene

CASRN: 95-75-0   **Beil Ref.**: 4-05-00-00815
Beil RN: 1931687
MF: $C_7H_6Cl_2$ **MP[°C]**: -15.2   **Den.[g/cm^3]**: 1.2564^{20}
MW: 161.03   **BP[°C]**: 208.9 n_D: 1.5471^{20}
Sol.: H_2O 1; EtOH 5; eth
   5; ace 5; bz 5; ctc 5;
   lig 5
Vap. Press. [kPa]: 3.11^{100}; 8.53^{125}; 20.2^{150}
Therm. Prop.[kJ/mol]: $\Delta_{fus}H = 10.68$
γ **[mN/m]**: 36.50^{20}
Elec. Prop.: $\mu = (3.0\ D)$; $\varepsilon = 9.39^{28}$
η **[mPa s]**: 1.57^{20}
MS: 125(100) 160(47) 127(32) 162(31) 89(15) 159(11) 161(10) 63(10) 126(8)
   62(8)
IR [cm^{-1}]: 3020 2920 1600 1560 1470 1380 1260 1210 1130 1030 870 800 680
   640
Raman [cm^{-1}]: 3060 3040 2920 2870 2740 2540 1600 1570 1470 1450 1380
   1280 1260 1220 1150 1140 1100 1040 1010 880 840 820 730 720 690 650
   570 550 500 470 440 410 380 370 320 260 210 130
UV [nm]: 281(519) 273(513) 266(328) 218(8060) MeOH
^{1}H NMR [ppm]: 2.3 7.0 CCl_4

159. Diethanolamine

Syst. Name: Ethanol, 2,2'-iminobis-
Synonyms: Bis(2-hydroxyethyl)amine

CASRN: 111-42-2   **Beil Ref.**: 4-04-00-01514
Merck No.: 3097
Beil RN: 605315
MF: $C_4H_{11}NO_2$ **MP[°C]**: 28   **Den.[g/cm³]**: 1.0966^{20}
MW: 105.14   **BP[°C]**: 268.8 n_D: 1.4776^{20}
Sol.: H_2O 4; EtOH 4; eth
   2; bz 2
Vap. Press. [kPa]: 11.1^{200}; 22.9^{220}; 44.2^{240}
Therm. Prop.[kJ/mol]: $\Delta_{fus}H = 25.10$; $\Delta_{vap}H = 65.2$
C_p **(liq.) [J/mol °C]**: 233.5^{30}
Elec. Prop.: $\mu = (2.8\ D)$; $\varepsilon = 25.75^{20}$
η **[mPa s]**: 110^{50}; 28.7^{75}; 9.10^{100}
MS: 30(100) 74(82) 28(77) 56(69) 18(50) 42(46) 29(36) 27(34) 45(30) 43(19)
IR [cm⁻¹]: 3300 2950 2850 1450 1360 1130 1070 940 860
Raman [cm⁻¹]: 3310 2930 2870 2770 1460 1360 1340 1290 1240 1200 1120
   1060 1030 930 860 820 520 440 380 340
¹H NMR [ppm]: 2.7 3.7 D_2O
Flammability: Flash pt. = 172°C; ign. temp. = 662°C; flam. lim. = 2-13%
TLV/TWA: 3 ppm (13 mg/m³)
Reg. Lists: CERCLA (RQ = 1 lb.); SARA 313 (1.0%)

160. Diethylamine
Syst. Name: Ethanamine, *N*-ethyl-
Synonyms: *N*-Ethylethanamine

CASRN: 109-89-7
Merck No.: 3100
Beil RN: 605268
MF: $C_4H_{11}N$
MW: 73.14
Sol.: H_2O 4; EtOH 5; eth 3; ctc 3

DOT No.: 1154
Beil Ref.: 4-04-00-00313

MP[°C]: -49.8
BP[°C]: 55.5

Den.[g/cm³]: 0.7056^{20}
n_D: 1.3864^{20}

Crit. Const.: $T_c = 226.84°C$; $P_c = 3.758$ MPa
Vap. Press. [kPa]: 30.1^{25}; 83.6^{50}
Therm. Prop.[kJ/mol]: $\Delta_{vap}H = 29.1$; $\Delta_f H°(l, 25°C) = -103.7$
C_p (liq.) [J/mol °C]: 169.2^{25}
γ [mN/m]: 19.85^{25}; $d\gamma/dT = 0.1143$ mN/m °C
Elec. Prop.: $\mu = 0.92$ D; $\varepsilon = 3.68^{20}$; IP = 8.01 eV
η [mPa s]: 0.319^{25}; 0.239^{50}
MS: 30(100) 58(81) 44(28) 73(18) 29(18) 28(17) 72(12) 42(11) 27(11) 59(4)
IR [cm⁻¹]: 3330 2940 2860 1450 1390 1320 1140 1040 720
UV [nm]: 222(295) 194(2951) gas
¹³C NMR [ppm]: 15.4 44.1 $CDCl_3$
¹H NMR [ppm]: 0.9 1.0 2.6 CCl_4
Flammability: Flash pt. = -23°C; ign. temp. = 312°C; flam. lim. = 1.8-10.1%
TLV/TWA: 5 ppm (15 mg/m³)
Reg. Lists: CERCLA (RQ = 1000 lb.)

161. Diethyl carbonate

Syst. Name: Carbonic acid, diethyl ester
Synonyms: Ethyl carbonate

CASRN: 105-58-8
Merck No.: 3738
Beil RN: 956591
MF: $C_5H_{10}O_3$
MW: 118.13
Sol.: H_2O 1; EtOH 3; eth 3; chl 3

DOT No.: 2366
Beil Ref.: 4-03-00-00005

MP[°C]: -43
BP[°C]: 126

Den.[g/cm^3]: 0.9752^{20}
n_D: 1.3845^{20}

Vap. Press. [kPa]: 1.63^{25}; 5.91^{50}; 17.5^{75}; 44.0^{100}; 97.3^{125}
Therm. Prop.[kJ/mol]: $\Delta_{vap}H = 36.2$; $\Delta_f H°(l, 25°C) = -681.5$
C_p (liq.) [J/mol °C]: 212.4^{25}
γ [mN/m]: 26.44^{20}
Elec. Prop.: $\mu = 1.10$ D; $\varepsilon = 2.82^{24}$
η [mPa s]: 0.748^{25}
MS: 29(100) 45(70) 31(53) 27(39) 91(24) 28(15) 63(11) 26(10) 30(6) 43(5)
IR [cm^{-1}]: 2990 2940 2910 1700 1670 1650 1610 1580 1500 1450 1020 860 790
Raman [cm^{-1}]: 2970 2940 2870 2810 2770 2730 2720 1740 1480 1450 1300 1270 1180 1120 1090 1020 980 900 850 800 700 520 350
^{1}H NMR [ppm]: 1.3 4.2 CDCl$_3$
Flammability: Flash pt. = 25°C

162. Diethylene glycol

Syst. Name: Ethanol, 2,2'-oxybis-
Synonyms: Bis(2-hydroxyethyl) ether; Diglycol

CASRN: 111-46-6   **Beil Ref.:** 4-01-00-02390
Merck No.: 3109
Beil RN: 969209
MF: $C_4H_{10}O_3$ **MP[°C]:** -10.4   **Den.[g/cm^3]:** 1.1197^{15}
MW: 106.12   **BP[°C]:** 245.8 n_D: 1.4472^{20}
Sol.: H_2O 3; EtOH 3; eth
    3; chl 3
Vap. Press. [kPa]: 0.001^{25}; 0.273^{100}; 0.935^{125}; 3.72^{150}
Therm. Prop.[kJ/mol]: $\Delta_{vap}H$ = 52.3; $\Delta_f H°$(l, 25°C) = -628.5
C_p **(liq.) [J/mol °C]:** 244.8^{25}
γ **[mN/m]:** 44.77^{25}; $d\gamma/dT$ = 0.0880 mN/m °C
Elec. Prop.: μ = (2.3 D); ϵ = 31.82^{20}
η **[mPa s]:** 30.2^{25}; 11.1^{50}; 4.92^{75}; 2.51^{100}
MS: 45(100) 75(23) 31(20) 44(16) 27(14) 76(12) 29(12) 43(11) 42(9) 41(4)
IR [cm^{-1}]: 3250 2900 1640 1540 1460 1380 1090 1040 980 760 570 530 450
Raman [cm^{-1}]: 3350 2940 2870 2700 1460 1400 1280 1240 1110 1080 1050
    900 820 530 440 390 350 300
^{1}H NMR [ppm]: 3.7 4.2 $CDCl_3$
Flammability: Flash pt. = 124°C; ign. temp. = 224°C; flam. lim. = 2-17%

---

163. Diethylene glycol dibutyl ether

Syst. Name: Butane, 1,1'-[oxybis(2,1-
    ethanediyloxy)]bis-

Synonyms: Bis(2-butoxyethyl) ether; Dibutyl
    carbitol; Butyl diglyme
CASRN: 112-73-2   **Beil Ref.:** 4-01-00-02395
Beil RN: 1750713
MF: $C_{12}H_{26}O_3$ **MP[°C]:** -60   **Den.[g/cm^3]:** 0.885^{25}
MW: 218.34   **BP[°C]:** 256 n_D: 1.4235^{20}
Vap. Press. [kPa]: 0.005^{25}; 0.029^{50}; 0.132^{75}; 0.488^{100}; 1.53^{125}; 4.21^{150}
C_p **(liq.) [J/mol °C]:** 452^{20}
γ **[mN/m]:** 27.0^{20}
η **[mPa s]:** 2.39^{20}
MS: 57(100) 75(36) 45(34) 41(34) 85(31) 56(27) 101(24) 87(21) 100(20)
    59(20)
Flammability: Flash pt. = 118°C; ign. temp. = 310°C

164. Diethylene glycol diethyl ether

Syst. Name: Ethane, 1,1'-oxybis[2-ethoxy-
Synonyms: Bis(2-ethoxyethyl) ether; Diethyl
  carbitol; Ethyl diglyme
CASRN: 112-36-7   **Beil Ref.**: 4-01-00-02394
Merck No.: 3108
Beil RN: 1699259
MF: $C_8H_{18}O_3$ **MP[°C]**: -45   **Den.[g/cm^3]**: 0.9063^{20}
MW: 162.23   **BP[°C]**: 188 n_D: 1.4115^{20}
Sol.: H_2O 4; EtOH 4; eth
  3; os 4
Vap. Press. [kPa]: 0.453^{50}; 1.65^{75}; 5.05^{100}; 13.4^{125}; 31.8^{150}
Therm. Prop.[kJ/mol]: $\Delta_{vap}H = 49.0$
C_p **(liq.) [J/mol °C]**: 341.4^{15}
γ **[mN/m]**: 27.0^{22}
Elec. Prop.: $\varepsilon = 5.70$
η **[mPa s]**: 1.40^{20}
MS: 45(100) 59(53) 72(45) 31(45) 73(43) 43(43) 29(25) 28(15) 44(12) 27(11)
IR [cm^{-1}]: 2940 2860 1450 1390 1350 1300 1250 1120 1050 1000 940 920 840
  790
Raman [cm^{-1}]: 2980 2940 2870 2820 2700 1480 1450 1270 1240 1140 1120
  1070 1050 1020 1000 910 880 840 830 800 550 500 430 270
Flammability: Flash pt. = 82°C

165. Diethylene glycol dimethyl ether

Syst. Name: Ethane, 1,1'-oxybis[2-methoxy-
Synonyms: Bis(2-methoxyethyl) ether; Diglyme

CASRN: 111-96-6   **Beil Ref.:** 4-01-00-02393
Merck No.: 3148
Beil RN: 1736101
MF: $C_6H_{14}O_3$ **MP[°C]:** -68   **Den.[g/cm^3]:** 0.9434^{20}
MW: 134.18   **BP[°C]:** 162 n_D: 1.4097^{20}
Sol.: H_2O 5; EtOH 5; eth
　5
Vap. Press. [kPa]: 0.054^0; 0.315^{25}; 1.37^{50}; 4.73^{75}; 13.6^{100}; 34.1^{125}; 76.0^{150}
Therm. Prop.[kJ/mol]: $\Delta_{fus}H$ = 13.60; $\Delta_{vap}H$ = 36.2
C_p (liq.) [J/mol °C]: 274.1^{25}
γ [mN/m]: 27.0^{20}
Elec. Prop.: μ = (2.0 D); IP = 9.8 eV
η [mPa s]: 0.989^{25}
MS: 59(100) 58(43) 31(34) 29(32) 45(28) 28(19) 89(15) 43(9) 27(5) 60(4)
IR [cm^{-1}]: 2880 1450 1350 1245 1195 1100 1020 850
Raman [cm^{-1}]: 2980 2940 2920 2890 2830 2730 1470 1450 1280 1240 1200
　1130 1100 1020 980 960 930 850 800 530 430 320
^{1}H NMR [ppm]: 3.3 3.5 CDCl$_3$
Flammability: Flash pt. = 67°C

166. Diethylene glycol monoethyl ether

Syst. Name: Ethanol, 2-(2-ethoxyethoxy)-
Synonyms: 2-(2-Ethoxyethoxy)ethanol; Carbitol

CASRN: 111-90-0 **Beil Ref.:** 4-01-00-02393
Merck No.: 1806
Beil RN: 1736441
MF: $C_6H_{14}O_3$ **BP[°C]:** 196 **Den.[g/cm^3]:** 0.9885^{20}
MW: 134.18 n_D: 1.4300^{20}
Sol.: H_2O 5; EtOH 5; eth
4; ace 5; bz 5; os 5
Vap. Press. [kPa]: 0.017^{25}; 0.189^{50}; 0.761^{75}; 2.57^{100}; 7.48^{125}; 19.3^{150}
Therm. Prop.[kJ/mol]: $\Delta_{vap}H = 47.5$
C_p **(liq.) [J/mol °C]:** 301.0^{25}
γ **[mN/m]:** 29.53^{25}
Elec. Prop.: $\mu = (1.6\ D)$
η **[mPa s]:** 3.85^{25}; 1.72^{60}
MS: 45(100) 59(56) 72(37) 73(22) 60(14) 31(13) 75(11) 44(9) 104(8) 103(7)
IR [cm^{-1}]: 3450 2860 1450 1370 1350 1110 1060 930 890
^{1}H NMR [ppm]: 1.2 3.1 3.5 3.6 CCl_4
Flammability: Flash pt. = 96°C

167. Diethylene glycol monoethyl ether acetate

Syst. Name: Ethanol, 2-(2-ethoxyethoxy)-, ethanoate
Synonyms: Carbitol acetate

CASRN: 112-15-2 **Beil Ref.:** 3-02-00-00308
Beil RN: 1764643
MF: $C_8H_{16}O_4$ **MP[°C]:** -25 **Den.[g/cm^3]:** 1.0096^{20}
MW: 176.21 **BP[°C]:** 218.5 n_D: 1.4213^{20}
Sol.: H_2O 4; ace 4; eth 4;
EtOH 4
Vap. Press. [kPa]: 0.029^{25}; 0.145^{50}; 0.575^{75}; 1.90^{100}; 5.41^{125}; 13.6^{150}
Elec. Prop.: $\mu = (1.8\ D)$
η **[mPa s]:** 2.8^{20}
MS: 43(100) 29(51) 31(42) 45(40) 59(24) 72(18) 44(10) 73(9) 42(9) 30(6)
Flammability: Flash pt. = 110°C; ign. temp. = 425°C

168. Diethylene glycol monomethyl ether

Syst. Name: Ethanol, 2-(2-methoxyethoxy)-
Synonyms: 2-(2-Methoxyethoxy)ethanol; Methyl
  Carbitol

CASRN: 111-77-3   **Beil Ref.:** 4-01-00-02392
Merck No.: 5959
Beil RN: 1697812
MF: $C_5H_{12}O_3$ **BP[°C]:** 193   **Den.[g/cm^3]:** 1.035^{20}
MW: 120.15 n_D: 1.4264^{20}
Sol.: H_2O 5; EtOH 4; eth
  4; ace 5
Vap. Press. [kPa]: 0.024^{25}; 3.65^{100}; 10.4^{125}; 26.3^{150}
Therm. Prop.[kJ/mol]: $\Delta_{vap}H = 46.6$
C_p **(liq.) [J/mol °C]:** 271.1^{25}
γ **[mN/m]:** 28.49^{25}
Elec. Prop.: $\mu = (1.6\ D)$
η **[mPa s]:** 3.48^{25}; 1.61^{60}
MS: 45(100) 31(42) 59(41) 29(38) 28(32) 58(21) 43(14) 27(13) 44(11) 32(10)
IR [cm^{-1}]: 3450 2860 1450 1350 1280 1240 1190 1100 1060 1020 980 930 890
  850
Raman [cm^{-1}]: 3400 2940 2870 2820 1470 1450 1280 1250 1200 1120 1090
  1060 1020 980 920 890 850 800 550 530 380 330
^{1}H NMR [ppm]: 3.3 3.4 3.6 CDCl$_3$
Flammability: Flash pt. = 96°C; ign. temp. = 240°C; flam. lim. = 1.4-22.7%

169. Diethylenetriamine

Syst. Name: 1,2-Ethanediamine, *N*-(2-aminoethyl)-
Synonyms: 2,2'-Diaminodiethylamine

H_2N ～～ NH ～～ NH_2

CASRN: 111-40-0   **DOT No.**: 2079
Beil RN: 605314   **Beil Ref.**: 4-04-00-01238
MF: $C_4H_{13}N_3$ **MP[°C]**: -39   **Den.[g/cm^3]**: 0.9569^{20}
MW: 103.17   **BP[°C]**: 207 n_D: 1.4810^{25}
Sol.: H_2O 5; EtOH 5; eth
   1; lig 3
Vap. Press. [kPa]: 0.030^{25}; 2.74^{100}; 8.29^{125}
C_p **(liq.) [J/mol °C]**: 254^{40}
Elec. Prop.: μ = (1.9 D); ε = 12.62^{20}
η **[mPa s]**: 7^{20}
MS: 44(100) 73(59) 30(35) 19(18) 56(16) 28(16) 27(16) 42(11) 99(8) 43(8)
IR [cm^{-1}]: 3230 2940 1590 1450 1350 1190 1060 900 840 780
Flammability: Flash pt. = 98°C; ign. temp. = 358°C; flam. lim. = 2-6.7%
TLV/TWA: 1 ppm (4.2 mg/m^3)

170. Diethyl ether

Syst. Name: Ethane, 1,1'-oxybis-
Synonyms: Ethyl ether

CASRN: 60-29-7   **DOT No.:** 1155
Merck No.: 3762   **Beil Ref.:** 4-01-00-01314
Beil RN: 1696894
MF: $C_4H_{10}O$ **MP[°C]:** -116.3   **Den.[g/cm^3]:** 0.7138^{20}
MW: 74.12   **BP[°C]:** 34.5 n_D: 1.3526^{20}
Sol.: H_2O 2; EtOH 5; eth
    5; ace 4; bz 5; chl 5;
    oils 5; lig 5
Crit. Const.: T_c = 193.59°C; P_c = 3.638 MPa; V_c = 280 cm^3
Vap. Press. [kPa]: 6.61^{-25}; 24.9^0; 71.7^{25}; 171^{50}
Therm. Prop.[kJ/mol]: $\Delta_{fus}H$ = 7.27; $\Delta_{vap}H$ = 26.5; $\Delta_f H°$(l, 25°C) = -279.3
C_p (liq.) [J/mol °C]: 172.6^{25}
γ [mN/m]: 16.65^{25}; dγ/dT = 0.0908 mN/m °C
Elec. Prop.: μ = 1.15 D; ε = 4.27^{20}; IP = 9.51 eV
η [mPa s]: 0.283^0; 0.224^{25}
***k* [W/m °C]:** 0.150^{-25}; 0.140^0; 0.130^{25}; 0.120^{50}; 0.110^{75}; 0.100^{100}
MS: 31(100) 29(63) 59(40) 27(35) 45(33) 74(23) 15(17) 43(9) 28(9) 26(9)
IR [cm^{-1}]: 2970 2930 2860 1490 1460 1440 1380 1350 1300 1120 1070 1040
    840
Raman [cm^{-1}]: 2980 2930 2900 2870 2800 2770 2730 2690 1490 1450 1270
    1150 1120 1090 1040 1020 930 910 880 840 830 790 500 430 370
UV [nm]: 188(1995) 171(3981) gas
^{13}C NMR [ppm]: 17.1 67.4
^{1}H NMR [ppm]: 1.3 3.4 CCl_4
Flammability: Flash pt. = -45°C; ign. temp. = 180°C; flam. lim. = 1.9-36.0%
TLV/TWA: 400 ppm (1210 mg/m^3)
Reg. Lists: CERCLA (RQ = 100 lb.); RCRA U117 (toxic)

171. Diethyl ketone

Syst. Name: 3-Pentanone

CASRN: 96-22-0   **DOT No.:** 1156
Merck No.: 3111   **Beil Ref.:** 4-01-00-03279
Beil RN: 635749
MF: $C_5H_{10}O$ **MP[°C]:** -39   **Den.[g/cm^3]:** 0.816^{19}
MW: 86.13   **BP[°C]:** 101.9 n_D: 1.3905^{25}
Sol.: H_2O 4; EtOH 5; eth
   5; ctc 3
Crit. Const.: $T_c = 288.31°C$; $P_c = 3.729$ MPa; $V_c = 336$ cm^3
Vap. Press. [kPa]: 1.18^0; 4.72^{25}; 15.6^{50}; 41.7^{75}; 95.5^{100}; 195^{125}; 363^{150}
Therm. Prop.[kJ/mol]: $\Delta_{fus}H = 11.59$; $\Delta_{vap}H = 33.5$; $\Delta_f H°(l, 25°C) = -296.5$
C_p (liq.) [J/mol °C]: 190.9^{25}
γ [mN/m]: 24.74^{25}; $d\gamma/dT = 0.1047$ mN/m °C
Elec. Prop.: $\mu = (2.8$ D); $\varepsilon = 17.00^{20}$; IP = 9.31 eV
η [mPa s]: 0.592^0; 0.444^{25}; 0.345^{50}; 0.276^{75}; 0.227^{100}
MS: 57(100) 29(50) 86(26) 27(13) 58(4) 56(4) 28(4) 26(3) 43(2) 42(2)
IR [cm^{-1}]: 2940 1720 1450 1370 1350 1110 1000 950 940 810
Raman [cm^{-1}]: 2990 2950 2910 2820 2750 1710 1460 1420 1380 1330 1240
   1100 1020 960 780 750 620 520 400 300 220
UV [nm]: 278 cyhex
^{13}C NMR [ppm]: 7.9 35.4 211.4 CDCl$_3$
^{1}H NMR [ppm]: 1.0 2.4 CCl$_4$
Flammability: Flash pt. = 13°C; ign. temp. = 450°C; lower flam. lim. = 1.6%
TLV/TWA: 200 ppm (705 mg/m^3)

172. Diethyl malonate

Syst. Name: Propanedioic acid, diethyl ester
Synonyms: Ethyl malonate

CASRN: 105-53-3   **Beil Ref.**: 4-02-00-01881
Merck No.: 3779
Beil RN: 774687
MF: $C_7H_{12}O_4$ **MP[°C]**: -50   **Den.[g/cm^3]**: 1.0551^{20}
MW: 160.17   **BP[°C]**: 200 n_D: 1.4139^{20}
Sol.: H_2O 2; EtOH 5; eth
   5; ace 4; bz 4; chl 3;
   HOAc 3
Vap. Press. [kPa]: 0.237^{50}; 0.932^{75}; 3.07^{100}; 8.73^{125}; 22.0^{150}
Therm. Prop.[kJ/mol]: $\Delta_{vap}H$ = 54.8
C_p **(liq.) [J/mol °C]**: 285.0^{25}
γ **[mN/m]**: 31.71^{20}
Elec. Prop.: μ = (2.5 D); ϵ = 7.55^{31}
η **[mPa s]**: 1.94^{25}
MS: 29(100) 115(53) 43(53) 133(29) 88(20) 42(19) 60(17) 27(17) 45(13) 31(9)
IR [cm^{-1}]: 2980 1730 1460 1440 1410 1360 1320 1260 1180 1140 1090 1030
   940 860
Raman [cm^{-1}]: 2970 2940 2870 2770 2730 1740 1450 1410 1360 1300 1270
   1150 1110 1090 1030 950 930 870 850 780 660 600 570 430 370 340 280
   210
UV [nm]: 235(50) MeOH
Flammability: Flash pt. = 93°C

173. Diethyl oxalate

Syst. Name: Ethanedioic acid, diethyl ester
Synonyms: Ethyl oxalate

CASRN: 95-92-1
Merck No.: 3115
Beil RN: 606350
MF: $C_6H_{10}O_4$
MW: 146.14
Sol.: H_2O 2; EtOH 5; eth
   5; ace 5; ctc 3

DOT No.: 2525
Beil Ref.: 4-02-00-01848

MP[°C]: -40.6
BP[°C]: 185.7

Den.[g/cm^3]: 1.0785[20]
n_D: 1.4101[20]

Vap. Press. [kPa]: 0.030[25]; 0.146[50]; 0.769[75]; 3.12[100]; 10.3[125]; 29.0[150]
Therm. Prop.[kJ/mol]: $\Delta_{vap}H$ = 42.0; $\Delta_f H°$(l, 25°C) = -805.5
C_p (liq.) [J/mol °C]: 260.7[25]
γ [mN/m]: 32.22[20]
Elec. Prop.: μ = (2.5 D); ε = 8.27[20]; IP = 9.80 eV
η [mPa s]: 2.31[15]
MS: 29(100) 31(16) 45(14) 27(14) 74(11) 28(9) 43(4) 30(4) 73(3) 75(2)
IR [cm^{-1}]: 3000 1770 1750 1470 1450 1390 1370 1320 1190 1160 1110 1010
   880
UV [nm]: 225(398) EtOH
^{13}C NMR [ppm]: 13.9 62.9 158.0 $CDCl_3$
^{1}H NMR [ppm]: 1.4 4.3 CCl_4
Flammability: Flash pt. = 76°C

---

174. 2,3-Diethylpentane

Syst. Name: Hexane, 3-ethyl-4-methyl-

CASRN: 3074-77-9
Beil RN: 1718784
MF: C_9H_{20}
MW: 128.26

DOT No.: 1920
Beil Ref.: 3-01-00-00514
BP[°C]: 140

Den.[g/cm^3]: 0.7420[20]
n_D: 1.4134[20]

Vap. Press. [kPa]: 4.05[50]; 12.1[75]; 30.3[100]; 66.2[125]; 130[150]
Elec. Prop.: μ = 0
MS: 57(100) 43(85) 70(64) 29(51) 41(50) 56(34) 27(33) 71(30) 55(28) 39(17)
Flammability: Flash pt. = 24°C

175. 3,3-Diethylpentane

Syst. Name: Pentane, 3,3-diethyl-
Synonyms: Tetraethylmethane

CASRN: 1067-20-5
Beil RN: 1696916
MF: C_9H_{20}
MW: 128.26
Sol.: H_2O 1; eth 3; bz 3; os 3

DOT No.: 1920
Beil Ref.: 4-01-00-00462
MP[°C]: -33.1   **Den.[g/cm³]:** 0.7536^{20}
BP[°C]: 146.3 n_D: 1.4206^{20}

Vap. Press. [kPa]: 10.5^{75}; 26.2^{100}; 57.1^{125}
Therm. Prop.[kJ/mol]: $\Delta_{fus}H = 10.09$; $\Delta_{vap}H = 34.6$; $\Delta_f H°(l, 25°C) = -275.4$
C_p **(liq.) [J/mol °C]:** 278.2^{25}
Elec. Prop.: $\mu = 0$
η **[mPa s]:** 1.20^{38}
MS: 57(100) 43(25) 41(24) 29(20) 99(19) 98(11) 27(11) 55(9) 69(5) 58(5)
Raman [cm⁻¹]: * 1540 1450 1267 1087 1048 1015 986 909 684 664 398
¹³C NMR [ppm]: 7.1 27.1 37.1 diox
Flammability: Ign. temp. = 290°C; flam. lim. = 0.7-5.7%

176. Diethyl sulfide

Syst. Name: Ethane, 1,1'-thiobis-
Synonyms: 3-Thiapentane; Ethyl sulfide

CASRN: 352-93-2
Merck No.: 3809
Beil RN: 1696909
MF: $C_4H_{10}S$
MW: 90.19
Sol.: H_2O 2; EtOH 3; eth
   3; ctc 2

DOT No.: 2375
Beil Ref.: 4-01-00-01394

MP[°C]: -103.9
BP[°C]: 92.1

Den.[g/cm³]: 0.8362^{20}
n_D: 1.4430^{20}

Crit. Const.: $T_c = 284°C$; $P_c = 3.96$ MPa; $V_c = 318$ cm³
Vap. Press. [kPa]: 7.78^{25}; 23.5^{50}; 58.9^{75}; 128^{100}
Therm. Prop.[kJ/mol]: $\Delta_{fus}H = 11.90$; $\Delta_{vap}H = 31.8$; $\Delta_f H°(l, 25°C) = -119.4$
C_p (liq.) [J/mol °C]: 171.4^{25}
γ [mN/m]: 24.57^{25}; $d\gamma/dT = 0.1106$ mN/m °C
Elec. Prop.: $\mu = 1.54$ D; $\varepsilon = 5.72^{25}$; IP = 8.43 eV
η [mPa s]: 0.558^0; 0.422^{25}; 0.331^{50}; 0.267^{75}
MS: 75(100) 47(81) 90(73) 62(60) 29(55) 61(54) 27(48) 28(23) 46(17) 45(17)
IR [cm⁻¹]: 2940 1450 1370 1250 1080 970 780 760 740 690
Raman [cm⁻¹]: 2980 2940 2900 2770 2730 1460 1400 1370 1290 1210 1150
   1100 1010 900 880 820 790 740 590 530 450 370 330 270 210
UV [nm]: 215(1585) 194(4786) hp
¹³C NMR [ppm]: 15.8 26.5
¹H NMR [ppm]: 1.2 2.5 CCl₄

177. 1,1-Difluoroethane
Syst. Name: Ethane, 1,1-difluoro-
Synonyms: Ethylidene difluoride; R 152a

CASRN: 75-37-6   **DOT No.**: 1030
Beil RN: 1696900   **Beil Ref.**: 4-01-00-00120
MF: $C_2H_4F_2$ **MP[°C]**: -117   **Den.[g/cm^3]**: 0.896^{20} (sat. press.)
MW: 66.05   **BP[°C]**: -24.9 n_D: 1.3011^{-72}
Crit. Const.: T_c = 113.6°C; P_c = 4.50 MPa; V_c = 181 cm^3
Vap. Press. [kPa]: 601^{25}; 1175^{50}; 2106^{75}; 3518^{100}
Therm. Prop.[kJ/mol]: $\Delta_{vap}H$ = 21.6; $\Delta_f H°$(g, 25°C) = -497.0
C_p **(liq.) [J/mol °C]**: 118.4^{25} (sat. press.)
Elec. Prop.: μ = 2.27 D; IP = 11.87 eV
MS: 51(100) 65(49) 47(13) 45(12) 27(5) 64(4) 46(4) 44(3) 26(3) 31(2)
IR [cm^{-1}]: 3010 2980 1420 1130 950 480

---

178. 1,1-Difluoroethylene
Syst. Name: Ethene, 1,1-difluoro-
Synonyms: Vinylidene fluoride

CASRN: 75-38-7   **DOT No.**: 1959
Beil RN: 1733321   **Beil Ref.**: 4-01-00-00696
MF: $C_2H_2F_2$ **MP[°C]**: -144
MW: 64.03   **BP[°C]**: -85.7
Sol.: eth 4; EtOH 4
Crit. Const.: T_c = 29.8°C; P_c = 4.46 MPa; V_c = 154 cm^3
Therm. Prop.[kJ/mol]: $\Delta_f H°$(g, 25°C) = -335.0
Elec. Prop.: μ = 1.368 D; IP = 10.29 eV
MS: 64(100) 45(67) 31(51) 33(42) 44(39) 14(22) 63(14) 26(6) 12(6) 25(5)
Flammability: Flam. lim. = 5.5-21.3%

179. Diiodomethane
Syst. Name: Methane, diiodo-
Synonyms: Methylene iodide

CASRN: 75-11-6 **Beil Ref.**: 4-01-00-00096
Merck No.: 5985
Beil RN: 1696892
MF: CH_2I_2 **MP[°C]**: 6.1 **Den.[g/cm^3]**: 3.3212^{20}
MW: 267.84 **BP[°C]**: 182 n_D: 1.7425^{20}
Sol.: H_2O 2; EtOH 3; eth
   3; bz 3; ctc 2; chl 3
Vap. Press. [kPa]: 0.172^{25}; 0.783^{50}; 2.75^{75}; 7.93^{100}; 19.6^{125}; 42.8^{150}
Therm. Prop.[kJ/mol]: $\Delta_{fus}H$ = 44.80; $\Delta_{vap}H$ = 42.5; $\Delta_f H°$(l, 25°C) = 66.9
C_p **(liq.) [J/mol °C]**: 134.0^{25}
γ **[mN/m]**: 50.88^{20}
Elec. Prop.: μ = (1.1 D); ε = 5.32^{25}; IP = 9.46 eV
η **[mPa s]**: 3.04^{15}
MS: 268(100) 141(68) 127(39) 254(12) 140(7) 139(6) 134(2) 269(1)
IR [cm^{-1}]: 3050 2970 1110 720
UV [nm]: 290(1318) 255(631) MeOH
^{13}C NMR [ppm]: -53.8
^{1}H NMR [ppm]: 3.9 CCl$_4$

180. Diisobutyl ketone

Syst. Name: 4-Heptanone, 2,6-dimethyl-
Synonyms: 2,6-Dimethyl-4-heptanone; Isovalerone

CASRN: 108-83-8   **DOT No.**: 1157
Beil RN: 1743163   **Beil Ref.**: 4-01-00-03360
MF: $C_9H_{18}O$ **MP[°C]**: -41.5   **Den.[g/cm³]**: 0.8062^{20}
MW: 142.24   **BP[°C]**: 169.4 n_D: 1.412^{21}
Sol.: H_2O 1; EtOH 5; eth
   5; ctc 3
Vap. Press. [kPa]: 0.230^{25}; 4.00^{75}; 11.6^{100}; 28.6^{125}; 61.8^{150}
Therm. Prop.[kJ/mol]: $\Delta_{vap}H$ = 39.9; $\Delta_f H°$(l, 25°C) = -408.5
C_p (liq.) [J/mol °C]: 297.3^{25}
γ [mN/m]: 25.54^{20}
Elec. Prop.: μ = (2.7 D); ε = 9.91^{20}; IP = 9.04 eV
η [mPa s]: 0.903^{20}
MS: 57(100) 85(82) 41(46) 43(39) 58(33) 28(30) 26(30) 39(22) 42(12) 142(11)
IR [cm⁻¹]: 2960 2860 1710 1460 1380 1360 1160 1040
Raman [cm⁻¹]: 2960 2940 2920 2880 2770 2730 1700 1620 1460 1450 1400
   1380 1360 1330 1290 1270 1220 1170 1130 1070 1050 1030 990 950 920
   870 840 810 780 620 600 570 530 440 420 350 290 260
UV [nm]: 233(287) cyhex
¹H NMR [ppm]: 0.9 2.1 CCl_4
Flammability: Flash pt. = 49°C; ign. temp. = 396°C; flam. lim. = 0.8-7.1%
TLV/TWA: 25 ppm (145 mg/m³)

181. Diisopentyl ether

Syst. Name: Butane, 1,1'-oxybis[3-methyl-
Synonyms: Isoamyl ether; Isopentyl ether

CASRN: 544-01-4   **Beil Ref.**: 4-01-00-01682
Merck No.: 5000
Beil RN: 1698014
MF: $C_{10}H_{22}O$ **BP[°C]**: 172.5   **Den.[g/cm^3]**: 0.7777^{20}
MW: 158.28 n_D: 1.4085^{20}
Sol.: ace 4; EtOH 4; chl
  4
Vap. Press. [kPa]: 0.210^{25}; 0.922^{50}; 3.21^{75}; 9.32^{100}; 23.4^{125}; 52.3^{150}
Therm. Prop.[kJ/mol]: $Э45_{vap}H = 35.1$
C_p **(liq.) [J/mol °C]**: 379^{100}
γ **[mN/m]**: 22.58^{25}; $d\gamma/dT = 0.0871$ mN/m °C
Elec. Prop.: $\mu = (1.2$ D); $\varepsilon = 2.82^{20}$
η **[mPa s]**: 1.01^{20}
MS: 71(100) 43(38) 41(30) 59(26) 29(24) 45(23) 55(20) 70(19) 27(17) 42(15)

182. Diisopropylamine
Syst. Name: 2-Propanamine, *N*-(1-methylethyl)-
Synonyms: *N*-(1-methylethyl)-2-propanamine

CASRN: 108-18-9   **DOT No.**: 1158
Merck No.: 3181   **Beil Ref.**: 4-04-00-00510
Beil RN: 605284
MF: $C_6H_{15}N$ **MP[°C]**: -61   **Den.[g/cm^3]**: 0.7153^{20}
MW: 101.19   **BP[°C]**: 83.9 n_D: 1.3924^{20}
Sol.: ace 4; bz 4; eth 4;
    EtOH 4
Crit. Const.: T_c = 250.0°C; P_c = 3.02 MPa
Vap. Press. [kPa]: 3.00^0; 10.7^{25}; 31.0^{50}; 75.3^{75}; 160^{100}; 306^{125}
Therm. Prop.[kJ/mol]: $\Delta_{vap}H$ = 30.4; $\Delta_fH°$(l, 25°C) = -178.5
γ **[mN/m]**: 19.14^{25}; dγ/dT = 0.1077 mN/m °C
Elec. Prop.: μ = (1.1 D); IP = 7.73 eV
η **[mPa s]**: 0.393^{25}; 0.300^{50}; 0.237^{75}
MS: 44(100) 86(30) 58(14) 42(13) 28(13) 41(12) 43(11) 27(11) 15(11) 39(6)
IR [cm^{-1}]: 3030 2940 1470 1390 1330 1180 1140 1090 1020 850 830 700
Raman [cm^{-1}]: 3310 3200 2960 2930 2920 2870 2850 2820 2710 1670 1450
    1370 1340 1190 1170 1130 1110 1090 1020 940 920 850 830 700 490 450
    400 320 250 190
UV [nm]: 301(7) gas
^{1}H NMR [ppm]: 0.7 1.0 2.9 CCl$_4$
Flammability: Flash pt. = -1°C; ign. temp. = 316°C; flam. lim. = 1.1-7.1%
TLV/TWA: 5 ppm (21 mg/m^3)

183. Diisopropyl ether

Syst. Name: Propane, 2,2'-oxybis-
Synonyms: Isopropyl ether

CASRN: 108-20-3   **DOT No.**: 1159
Merck No.: 5100   **Beil Ref.**: 4-01-00-01471
Beil RN: 1731256
MF: $C_6H_{14}O$ **MP[°C]**: -86.8   **Den.[g/cm^3]**: 0.7241^{20}
MW: 102.18   **BP[°C]**: 68.5 n_D: 1.3679^{20}
Sol.: H_2O 2; EtOH 5; eth
   5; ace 3; ctc 3
Crit. Const.: T_c = 227.17°C; P_c = 2.832 MPa; V_c = 386 cm^3
Vap. Press. [kPa]: 19.9^{25}; 54.0^{50}
Therm. Prop.[kJ/mol]: $\Delta_{fus}H$ = 11.03; $\Delta_{vap}H$ = 29.1; $\Delta_f H°$(l, 25°C) = -351.5
C_p (liq.) [J/mol °C]: 216.8^{25}
γ **[mN/m]**: 17.27^{25}; dγ/dT = 0.1048 mN/m °C
Elec. Prop.: μ = 1.13 D; ε = 3.81^{30}; IP = 9.20 eV
η **[mPa s]**: 0.379^{25}
MS: 45(100) 43(39) 87(15) 41(12) 59(10) 27(8) 39(4) 69(3) 42(3) 31(3)
IR [cm^{-1}]: 2940 2860 2630 1470 1390 1330 1180 1110 1020 910 790
Raman [cm^{-1}]: 2970 2940 2920 2870 2710 2620 1450 1380 1360 1330 1180
   1150 1120 930 900 850   790 760 490 390 300 250 190
^{1}H NMR [ppm]: 1.0 3.5 CCl$_4$
Flammability: Flash pt. = -28°C; ign. temp. = 443°C; flam. lim. = 1.4-7.9%
TLV/TWA: 250 ppm (1040 mg/m^3)

184. Diisopropyl ketone

Syst. Name: 3-Pentanone, 2,4-dimethyl-
Synonyms: 2,4-Dimethyl-3-pentanone

CASRN: 565-80-0   **Beil Ref.**: 4-01-00-03334
Beil RN: 773782
MF: $C_7H_{14}O$ **MP[°C]**: -69   **Den.[g/cm^3]**: 0.8108^{20}
MW: 114.19   **BP[°C]**: 125.4 n_D: 1.3999^{20}
Sol.: H_2O 1; EtOH 5; eth
   5; bz 3; ctc 2
Vap. Press. [kPa]: 6.87^{50}; 19.6^{75}; 47.4^{100}; 101^{125}
Therm. Prop.[kJ/mol]: $\Delta_{fus}H$ = 11.18; $\Delta_{vap}H$ = 34.6; $\Delta_f H°$(l, 25°C) = -352.9
C_p **(liq.) [J/mol °C]**: 233.7^{25}
Elec. Prop.: μ = (2.7 D); IP = 8.95 eV
η **[mPa s]**: 0.632^{20}
MS: 43(100) 71(31) 41(13) 27(9) 70(6) 39(6) 114(5) 42(5) 44(3) 72(2)
IR [cm^{-1}]: 2980 1710 1460 1380 1020
Raman [cm^{-1}]: 2980 2940 2920 2880 2760 2720 1710 1470 1450 1390 1330
   1280 1250 1220 1180 1110 1100 1070 970 930 900 860 830 800 790 740 720
   640 610 580 530 470 410 370 340 300 210
UV [nm]: 284(26) MeOH
^{1}H NMR [ppm]: 1.0 2.6 CCl$_4$

185. *N,N*-Dimethylacetamide
Syst. Name: Ethanamide, *N,N*-dimethyl-

CASRN: 127-19-5   **Beil Ref.**: 4-04-00-00180
Merck No.: 3216
Beil RN: 1737614
MF: C_4H_9NO **MP[°C]**: -20   **Den.[g/cm^3]**: 0.9366^{25}
MW: 87.12   **BP[°C]**: 165 n_D: 1.4380^{20}
Sol.: H_2O 5; EtOH 5; eth
    5; ace 5; bz 5; chl 5
Vap. Press. [kPa]: 0.075^{25}; 0.639^{50}; 3.16^{75}; 10.9^{100}; 29.1^{125}; 65.0^{150}
Therm. Prop.[kJ/mol]: $\Delta_{fus}H$ = 10.42; $\Delta_{vap}H$ = 43.4; $\Delta_f H°$(l, 25°C) = -278.3
C_p **(liq.) [J/mol °C]**: 175.6^{25}
γ **[mN/m]**: 32.43^{30}
Elec. Prop.: μ = (3.7 D); ε = 38.85^{21}; IP = 8.81 eV
η **[mPa s]**: 1.96^{25}; 1.28^{50}; 0.896^{75}; 0.661^{100}
MS: 44(100) 87(69) 43(46) 45(23) 42(19) 72(15) 15(11) 30(8) 28(5) 88(4)
IR [cm^{-1}]: 2940 1640 1560 1520 1390 1270 1190 1010
Raman [cm^{-1}]: 3020 2940 2880 2820 2790 1640 1460 1420 1360 1270 1190
    1010 960 740 590 470 420
UV [nm]: 218(1000) hp
^{13}**C NMR [ppm]**: 21.3 34.5 37.5 169.6 diox
^{1}**H NMR [ppm]**: 2.1 2.9 3.0 CDCl$_3$
Flammability: Flash pt. = 70°C; ign. temp. = 490°C; flam. lim. = 1.8-11.5
TLV/TWA: 10 ppm (36 mg/m^3)

186. Dimethyl adipate

Syst. Name: Hexanedioic acid, dimethyl ester
Synonyms: Methyl adipate

CASRN: 627-93-0   **Beil Ref.**: 4-02-00-01959
Beil RN: 1707443
MF: $C_8H_{14}O_4$ **MP[°C]**: 10.3   **Den.[g/cm^3]**: 1.0600^{20}
MW: 174.20 n_D: 1.4283^{20}
Sol.: H_2O 1; al 3; eth 3;
   ctc 3; HOAc 3
Vap. Press. [kPa]: 4.32^{135}; 23.5^{180}; 93.9^{225}
Elec. Prop.: $\varepsilon = 6.84^{20}$
MS: 59(100) 55(72) 114(46) 43(42) 101(39) 111(38) 74(36) 41(30) 73(25)
   27(24)
IR [cm^{-1}]: 2960 2860 1740 1430 1360 1200 1070 1000 880 750
^{1}H NMR [ppm]: 1.6 2.3 3.6 CCl_4

187. Dimethylamine

Syst. Name: Methanamine, *N*-methyl-
Synonyms: *N*-Methylmethanamine

CASRN: 124-40-3   **DOT No.**: 1032
Merck No.: 3217   **Beil Ref.**: 4-04-00-00128
Beil RN: 605257
MF: C_2H_7N **MP[°C]**: -92.2   **Den.[g/cm^3]**: 0.6804^0
MW: 45.08   **BP[°C]**: 6.8 n_D: 1.350^{17}
Sol.: H_2O 4; EtOH 3; eth
   3
Crit. Const.: T_c = 164.07°C; P_c = 5.340 MPa
Vap. Press. [kPa]: 21.1^{-25}; 75.0^0; 203^{25}
Therm. Prop.[kJ/mol]: $\Delta_{fus}H$ = 5.94; $\Delta_{vap}H$ = 26.4; $\Delta_f H°$(g, 25°C) = -18.5
C_p **(liq.) [J/mol °C]**: 137.7^{25} (sat. press.)
γ **[mN/m]**: 26.34^{25} (at saturation press.); dγ/dT = 0.1265 mN/m °C
Elec. Prop.: μ = 1.01 D; ε = 5.26^{25}; IP = 8.23 eV
η **[mPa s]**: 0.300^{-25}; 0.232^0
MS: 44(100) 45(81) 18(32) 28(30) 43(19) 42(15) 15(9) 46(5) 41(5) 27(5)
IR [cm^{-1}]: 3330 2940 2780 1470 1160 1040 1010 940 920 730
UV [nm]: 222(100) 191(3236) gas
Flammability: Flash pt. = 20°C; ign. temp. = 400°C; flam. lim. = 2.8-14.4
TLV/TWA: 5 ppm (9.2 mg/m^3)
Reg. Lists: CERCLA (RQ = 1000 lb.); RCRA U092 (toxic)

188. *N,N*-Dimethylaniline

Syst. Name: Benzenamine, *N,N*-dimethyl-
Synonyms: *N,N*-Dimethylbenzenamine

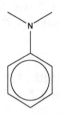

CASRN: 121-69-7   **DOT No.**: 2253
Merck No.: 3223   **Beil Ref.**: 4-12-00-00243
Beil RN: 507140
MF: $C_8H_{11}N$ **MP[°C]**: 2.4   **Den.[g/cm^3]**: 0.9557^{20}
MW: 121.18   **BP[°C]**: 194.1 n_D: 1.5582^{20}
Sol.: H$_2$O 2; EtOH 3; eth
   3; ace 3; bz 3; ctc 3;
   chl 4; os 3
Crit. Const.: T_c = 414°C; P_c = 3.63 MPa
Vap. Press. [kPa]: 1.59^{75}; 4.90^{100}; 12.9^{125}; 29.7^{150}
Therm. Prop.[kJ/mol]: $\Delta_f H°$(l, 25°C) = 47.7
C_p **(liq.) [J/mol °C]**: 214.6^{29}
γ **[mN/m]**: 35.52^{25}; dγ/dT = 0.1049 mN/m °C
Elec. Prop.: μ = 1.68 D; ε = 4.90^{25}; IP = 7.12 eV
η **[mPa s]**: 2.00^0; 1.30^{25}; 0.911^{50}; 0.675^{75}; 0.523^{100}
MS: 120(100) 121(70) 77(25) 51(16) 105(13) 104(11) 42(10) 50(7) 122(6)
   78(6)
IR [cm^{-1}]: 2860 1610 1520 1450 1350 1240 1190 1160 1140 1060 1030 990
   940 860 750 690
Raman [cm^{-1}]: 3160 3070 2920 2840 2790 1600 1510 1480 1450 1410 1340
   1190 1030 990 950 860 800 840 610 510 470 400 300 70
UV [nm]: 298(2290) 251(11600) MeOH
^{1}H NMR [ppm]: 2.9 6.5 7.1 CCl$_4$
Flammability: Flash pt. = 63°C; ign. temp. = 371°C
TLV/TWA: 5 ppm (25 mg/m^3)
Reg. Lists: CERCLA (RQ = 1 lb.); SARA 313 (1.0%)

189. 2,2-Dimethylbutane

Syst. Name: Butane, 2,2-dimethyl-
Synonyms: Neohexane

CASRN: 75-83-2   **DOT No.:** 2457
Beil RN: 1730736   **Beil Ref.:** 4-01-00-00367
MF: C_6H_{14} **MP[°C]:** -99   **Den.[g/cm^3]:** 0.6444^{25}
MW: 86.18   **BP[°C]:** 49.7 n_D: 1.3688^{20}
Sol.: H_2O 1; EtOH 3; eth 3; ace 4; bz 4; ctc 4; peth 4
Crit. Const.: $T_c = 215.7°C$; $P_c = 3.090$ MPa; $V_c = 359$ cm^3
Vap. Press. [kPa]: 3.88^{-25}; 14.6^0; 42.5^{25}; 102^{50}; 213^{75}; 397^{100}; 684^{125}; 1101^{150}
Therm. Prop.[kJ/mol]: $\Delta_{fus}H = 0.58$; $\Delta_{vap}H = 26.3$; $\Delta_f H°(l, 25°C) = -213.8$
C_p (liq.) [J/mol °C]: 191.9^{25}
γ [mN/m]: 15.81^{25}; $d\gamma/dT = 0.0990$ mN/m °C
Elec. Prop.: $\mu = 0$; $\varepsilon = 1.87^{20}$; IP = 10.06 eV
η [mPa s]: 0.351^{25}
MS: 43(100) 57(98) 71(73) 41(61) 29(51) 27(36) 56(29) 39(25) 55(15) 15(12)
IR [cm^{-1}]: 2950 2870 1470 1370 1220 1070 1010 920 780
Raman [cm^{-1}]: 2950 2920 2880 2780 2720 1470 1450 1340 1300 1250 1220 1070 1020 1000 930 870 710 670 480 410 360 340 260 100
^{13}C NMR [ppm]: 8.8 28.9 30.4 36.5 CDCl$_3$
^{1}H NMR [ppm]: 0.9 1.1 1.3 1.1 1.3 CCl$_4$
Flammability: Flash pt. = -48°C; ign. temp. = 405°C; flam. lim. = 1.2-7.0%

190. 2,3-Dimethylbutane
Syst. Name: Butane, 2,3-dimethyl-
Synonyms: Diisopropyl

CASRN: 79-29-8   **DOT No.**: 2457
Beil RN: 1730737   **Beil Ref.**: 4-01-00-00371
MF: C_6H_{14} **MP[°C]**: -128.8   **Den.[g/cm^3]**: 0.6616^{20}
MW: 86.18   **BP[°C]**: 57.9 n_D: 1.3750^{20}
Sol.: H_2O 1; EtOH 3; eth
   3; ace 4; bz 4; ctc 4;
   peth 4
Crit. Const.: T_c = 226.9°C; P_c = 3.131 MPa; V_c = 358 cm^3
Vap. Press. [kPa]: 2.56^{-25}; 10.3^0; 31.3^{25}; 78.2^{50}; 168^{75}; 323^{100}; 567^{125}; 929^{150}
Therm. Prop.[kJ/mol]: $\Delta_{fus}H$ = 0.80; $\Delta_{vap}H$ = 27.4; $\Delta_f H°$(l, 25°C) = -207.4
C_p **(liq.) [J/mol °C]**: 189.7^{25}
γ **[mN/m]**: 16.88^{25}; $d\gamma/dT$ = 0.1000 mN/m °C
Elec. Prop.: μ = 0; ε = 1.89^{20}; IP = 10.02 eV
η **[mPa s]**: 0.361^{25}
MS: 43(100) 42(97) 41(27) 71(25) 27(11) 86(10) 39(9) 29(6) 55(5) 44(4)
IR [cm^{-1}]: 2960 2880 1460 1380 1130 1040
Raman [cm^{-1}]: 2940 2910 2870 2780 2760 2720 2670 2600 1470 1450 1390
   1350 1300 1260 1200 1160 1100 1070 1030 940 870 840 820 760 740 700
   590 570 510 480 450 380 350 300
^{13}C NMR [ppm]: 19.5 34.0 CDCl$_3$
^{1}H NMR [ppm]: 0.9 1.5 CCl$_4$
Flammability: Flash pt. = -29°C; ign. temp. = 405°C; flam. lim. = 1.2-7.0%

191. 3,3-Dimethyl-1-butanol

Syst. Name: 1-Butanol, 3,3-dimethyl-
Synonyms: Dimbunol

CASRN: 624-95-3   **DOT No.:** 2282
Beil RN: 1731466   **Beil Ref.:** 4-01-00-01729
MF: $C_6H_{14}O$ **MP[°C]:** -60   **Den.[g/cm^3]:** 0.844[15]
MW: 102.18   **BP[°C]:** 143 n_D: 1.4323[15]
Sol.: H_2O 2; EtOH 3; eth
   3; ace 3
Vap. Press. [kPa]: 1.65[50]; 6.38[75]; 20.1[100]; 54.1[125]; 128[150]
MS: 57(100) 69(67) 41(56) 56(40) 43(23) 29(23) 45(12) 39(10) 31(10) 27(9)
IR [cm^{-1}]: 3450 3030 1490 1370 1250 1210 1080 1040 1000 980 930 840

192. 2,3-Dimethyl-2-butanol

Syst. Name: 2-Butanol, 2,3-dimethyl-
Synonyms: Isopropyldimethylcarbinol

CASRN: 594-60-5   **DOT No.:** 2282
Beil RN: 1731443   **Beil Ref.:** 4-01-00-01729
MF: $C_6H_{14}O$ **MP[°C]:** -14   **Den.[g/cm^3]:** 0.8236[20]
MW: 102.18   **BP[°C]:** 118.4 n_D: 1.4176[20]
Sol.: H_2O 3; EtOH 5; eth
   5
Vap. Press. [kPa]: 1.18[25]; 5.52[50]; 19.1[75]; 52.9[100]; 124[125]
MS: 59(100) 41(27) 87(26) 43(26) 69(18) 31(14) 28(13) 27(12) 39(11) 15(11)

193. 3,3-Dimethyl-2-butanol
Syst. Name: 2-Butanol, 3,3-dimethyl-, (±)-

CASRN: 20281-91-8   DOT No.: 2282
Beil RN: 1718950   Beil Ref.: 4-01-00-01727
MF: $C_6H_{14}O$ MP[°C]: 5.6   Den.[g/cm^3]: 0.8122^{25}
MW: 102.18   BP[°C]: 120.4 n_D: 1.4148^{20}
Sol.: H_2O 2; EtOH 3; eth
   5
MS: 57(100) 45(92) 41(69) 56(50) 29(42) 27(29) 43(26) 102(2)
IR [cm^{-1}]: 3330 2940 1470 1390 1370 1300 1210 1090 1010 920

---

194. *cis*-1,2-Dimethylcyclohexane
Syst. Name: Cyclohexane, 1,2-dimethyl-, *cis*-

CASRN: 2207-01-4   Beil Ref.: 4-05-00-00118
Beil RN: 1900322
MF: C_8H_{16} MP[°C]: -49.9   Den.[g/cm^3]: 0.7963^{20}
MW: 112.22   BP[°C]: 129.8 n_D: 1.4360^{20}
Sol.: H_2O 1; EtOH 3; eth
   5; ace 5; bz 3; ctc 3;
   lig 4
Vap. Press. [kPa]: 1.93^{25}; 6.59^{50}; 18.2^{75}; 43.0^{100}; 89.3^{125}
Therm. Prop.[kJ/mol]: $\Delta_{fus}H$ = 1.64; $\Delta_{vap}H$ = 33.5; $\Delta_f H°$(l, 25°C) = -211.8
C_p (liq.) [J/mol °C]: 210.2^{25}
γ [mN/m]: 25.20^{25}; dγ/dT = 0.1058 mN/m °C
Elec. Prop.: μ = 0; ε = 2.06^{25}; IP = 9.78 eV
η [mPa s]: 1.14^{20}
MS: 55(100) 97(64) 41(60) 83(39) 70(39) 56(37) 112(32) 42(32) 27(28) 69(26)
IR [cm^{-1}]: 2950 2920 2860 1450 1380 1000
^{13}C NMR [ppm]: 16.3 24.3 32.1 35.0
Flammability: Flash pt. = 16°C; ign. temp. = 304°C

195. *trans*-1,2-Dimethylcyclohexane
Syst. Name: Cyclohexane, 1,2-dimethyl-, *trans*-

CASRN: 6876-23-9   **Beil Ref.**: 4-05-00-00118
Beil RN: 3193642
MF: C_8H_{16} **MP[°C]**: -90   **Den.[g/cm^3]**: 0.7760^{20}
MW: 112.22   **BP[°C]**: 123.5 n_D: 1.4270^{20}
Sol.: H_2O 1; EtOH 3; eth
   3; ace 5; bz 5; ctc 3;
   lig 4
Vap. Press. [kPa]: 2.58^{25}; 8.46^{50}; 22.7^{75}; 52.0^{100}; 106^{125}
Therm. Prop.[kJ/mol]: $\Delta_{fus}H$ = 10.49; $\Delta_{vap}H$ = 33.0; $\Delta_f H°$(l, 25°C) = -218.2
C_p **(liq.) [J/mol °C]**: 209.4^{25}
γ **[mN/m]**: 23.57^{25}; dγ/dT = 0.0941 mN/m °C
Elec. Prop.: μ = 0; ϵ = 2.04^{25}; IP = 9.41 eV
η **[mPa s]**: 0.817^{20}
MS: 55(100) 97(91) 41(58) 56(37) 112(34) 69(30) 42(29) 70(28) 27(28) 39(25)
IR [cm^{-1}]: 2960 2900 2850 1450 1380 1250 1160 1090 1000 970 880 830 590
^{13}C NMR [ppm]: 20.8 27.4 36.6 40.1
Flammability: Flash pt. = 11°C; ign. temp. = 304°C

196. Dimethyl disulfide

Syst. Name: Disulfide, dimethyl
Synonyms: 2,3-Dithiabutane; Methyl disulfide

CASRN: 624-92-0   **DOT No.**: 2381
Beil RN: 1730824   **Beil Ref.**: 4-01-00-01281
MF: $C_2H_6S_2$ **MP[°C]**: -85   **Den.[g/cm^3]**: 1.0625^{20}
MW: 94.20   **BP[°C]**: 109.8 n_D: 1.5289^{20}
Sol.: H_2O 1; EtOH 5; eth
   5
Vap. Press. [kPa]: 0.893^0; 3.82^{25}; 12.5^{50}; 33.2^{75}; 76.0^{100}
Therm. Prop.[kJ/mol]: $\Delta_{fus}H$ = 9.19; $\Delta_{vap}H$ = 33.8; $\Delta_f H°$(l, 25°C) = -62.6
C_p (liq.) [J/mol °C]: 146.1^{25}
γ [mN/m]: 33.39^{25}; $d\gamma/dT$ = 0.1343 mN/m °C
Elec. Prop.: μ = (1.8 D); ϵ = 9.6^{25}; IP = 7.40 eV
η [mPa s]: 0.585^{25}
MS: 94(100) 45(63) 79(59) 46(38) 47(26) 15(18) 48(14) 61(12) 64(11) 96(9)
IR [cm^{-1}]: 3030 2940 2860 1430 1300 950
UV [nm]: 256(347) MeOH
Flammability: Flash pt. = 24°C

197. Dimethyl ether

Syst. Name: Methane, oxybis-
Synonyms: Methyl ether

CASRN: 115-10-6

Merck No.: 5990

Beil RN: 1730743

MF: C_2H_6O

MW: 46.07

Sol.: H_2O 3; EtOH 3; eth 3; ace 3; bz 2; chl 3

DOT No.: 1033

Beil Ref.: 4-01-00-01245

MP[°C]: -141.5

BP[°C]: -24.8

Crit. Const.: T_c = 126.9°C; P_c = 5.37 MPa; V_c = 190 cm^3

Vap. Press. [kPa]: 6.8^{-73}; 108^{-23}; 623^{27}

Therm. Prop.[kJ/mol]: $\Delta_{fus}H$ = 4.94; $\Delta_{vap}H$ = 21.5; $\Delta_f H°$(g, 25°C) = -184.1

C_p **(liq.) [J/mol °C]**: 102.4^{-33}

γ [mN/m]: 18.67^{-25}; $d\gamma/dT$ = 0.1478 mN/m °C

Elec. Prop.: μ = 1.30 D; ε = 6.18^{-15}; IP = 10.03 eV

MS: 45(100) 29(79) 15(57) 46(46) 14(13) 31(7) 13(5) 30(4) 28(3) 43(2)

IR [cm^{-1}]: 3000 2900 2840 1450 1170 1100 930

UV [nm]: 184(2512) 163(3981) gas

^{13}C NMR [ppm]: 59.4

Flammability: Flash pt. = -41°C; ign. temp. = 350°C; flam. lim. = 3.4-27.0%

198. *N,N*-Dimethylformamide

Syst. Name: Methanamide, *N,N*-dimethyl-
Synonyms: Dimethylformamide; DMF

CASRN: 68-12-2   **DOT No.**: 2265
Merck No.: 3232   **Beil Ref.**: 4-04-00-00171
Beil RN: 605365
MF: C_3H_7NO **MP[°C]**: -60.4   **Den.[g/cm³]**: 0.944^{25}
MW: 73.09   **BP[°C]**: 153 n_D: 1.4305^{20}
Sol.: H_2O 5; EtOH 5; eth
  5; ace 5; bz 5; chl 5;
  lig 2
Crit. Const.: T_c = 376.5°C; V_c = 262 cm³
Vap. Press. [kPa]: 0.439^{25}; 1.99^{50}; 6.82^{75}
Therm. Prop.[kJ/mol]: $\Delta_{fus}H$ = 16.15; $\Delta_{vap}H$ = 38.4; $\Delta_f H°$(l, 25°C) = -239.3
C_p (liq.) [J/mol °C]: 150.6^{25}
Elec. Prop.: μ = 3.82 D; ε = 38.25^{20}; IP = 9.13 eV
η [mPa s]: 1.18^{0}; 0.794^{25}; 0.624^{50}
k [W/m °C]: 0.184^{25}; 0.178^{50}; 0.171^{75}; 0.165^{100}
MS: 73(100) 44(86) 42(36) 30(22) 28(20) 29(8) 43(7) 72(6) 58(5) 74(4)
IR [cm⁻¹]: 2940 1670 1490 1430 1390 1250 1090
Raman [cm⁻¹]: 3000 2940 2860 2810 1670 1510 1440 1410 1100 1070 1010
  860 410 360 320
UV [nm]: 197(8710) 162(6918) gas
¹³C NMR [ppm]: 31.1 36.2 162.4 $CDCl_3$
¹H NMR [ppm]: 2.9 3.0 8.0 $CDCl_3$
Flammability: Flash pt. = 58°C; ign. temp. = 445°C; flam. lim. = 2.2-15.2%
TLV/TWA: 10 ppm (30 mg/m³)
Reg. Lists: CERCLA (RQ = 1 lb.)

199. Dimethyl glutarate

Syst. Name: Pentanedioic acid, dimethyl ester
Synonyms: Methyl glutarate

CASRN: 1119-40-0   **Beil Ref.:** 4-02-00-01937
Beil RN: 1771444
MF: $C_7H_{12}O_4$ **MP[°C]:** -42.5   **Den.[g/cm^3]:** 1.0876^{20}
MW: 160.17   **BP[°C]:** 214 n_D: 1.4242^{20}
Sol.: EtOH 4; eth 4; chl
   3
Vap. Press. [kPa]: 0.514^{75}; 1.83^{100}; 5.53^{125}; 14.7^{150}; 35^{175}; 76^{200}
Elec. Prop.: $\varepsilon = 7.87^{20}$
MS: 59(100) 100(51) 55(49) 42(33) 129(32) 101(32) 41(26) 43(22) 128(19)
   87(15)
IR [cm^{-1}]: 2950 1740 1430 1360 1310 1200 1060 1020 990
Raman [cm^{-1}]: 3030 2950 2930 2850 1730 1440 1430 1370 1290 1190 1160
   1070 1020 970 900 870 840 810 640 610 470 440 360 320 270 250
^{1}H NMR [ppm]: 2.0 2.4 3.7 CDCl$_3$

---

200. 2,2-Dimethylheptane

Syst. Name: Heptane, 2,2-dimethyl-

CASRN: 1071-26-7   **DOT No.:** 1920
Beil RN: 1730925   **Beil Ref.:** 4-01-00-00457
MF: C_9H_{20} **MP[°C]:** -113   **Den.[g/cm^3]:** 0.7105^{20}
MW: 128.26   **BP[°C]:** 132.7 n_D: 1.4016^{20}
Sol.: H_2O 1; eth 3; ace 4;
   bz 5; ctc 3; chl 4
Crit. Const.: $T_c = 303.7°C$; $P_c = 2.350$ MPa
Vap. Press. [kPa]: 5.30^{50}; 15.5^{75}; 38.1^{100}; 81.9^{125}; 158^{150}
Therm. Prop.[kJ/mol]: $\Delta_{fus}H = 8.90$; $\Delta_f H°(l, 25°C) = -288.2$
Elec. Prop.: $\mu = 0$
MS: 57(100) 56(46) 41(30) 29(22) 43(18) 27(17) 71(15) 39(11) 55(6) 113(5)
IR [cm^{-1}]: 2940 2860 1470 1370 1250 730
^{13}C NMR [ppm]: 13.8 22.8 24.4 29.2 30.2 33.0 44.4 diox

201. 2,2-Dimethylhexane
Syst. Name: Hexane, 2,2-dimethyl-

CASRN: 590-73-8   **DOT No.:** 1262
Beil RN: 1696861   **Beil Ref.:** 4-01-00-00432
MF: C_8H_{18} **MP[°C]:** -121.1   **Den.[g/cm³]:** 0.6953^{20}
MW: 114.23   **BP[°C]:** 106.8 n_D: 1.3935^{20}
Sol.: ace 4; bz 4; eth 4;
    EtOH 4
Crit. Const.: $T_c = 276.8°C$; $P_c = 2.529$ MPa; $V_c = 478$ cm³
Vap. Press. [kPa]: 1.10^0; 4.54^{25}; 14.4^{50}; 37.3^{75}; 83.1^{100}; 165^{125}; 298^{150}
Therm. Prop.[kJ/mol]: $\Delta_{fus}H = 6.78$; $\Delta_{vap}H = 32.1$; $\Delta_f H°(l, 25°C) = -261.9$
Elec. Prop.: $\mu = 0$; $\varepsilon = 1.95^{20}$
η [mPa s]: 0.43^{38}
MS: 57(100) 56(32) 41(26) 29(19) 43(16) 27(11) 39(7) 114(1)
IR [cm⁻¹]: 2900 1470 1380 1360 1250 1100 900 730
Raman [cm⁻¹]: 2958 2937 2906 2886 2876 2862 1448 1248 746
¹³C NMR [ppm]: 13.9 23.7 27.0 29.2 30.1 44.1 diox

---

202. 2,3-Dimethylhexane
Syst. Name: Hexane, 2,3-dimethyl-

CASRN: 584-94-1   **DOT No.:** 1262
Beil RN: 1718745   **Beil Ref.:** 4-01-00-00432
MF: C_8H_{18} **BP[°C]:** 115.6   **Den.[g/cm³]:** 0.6912^{25}
MW: 114.23 n_D: 1.4011^{20}
Sol.: ace 4; bz 4; EtOH
    4; lig 4
Crit. Const.: $T_c = 290.4°C$; $P_c = 2.628$ MPa; $V_c = 468$ cm³
Vap. Press. [kPa]: 3.13^{25}; 10.4^{50}; 28.0^{75}; 64.3^{100}; 131^{125}; 241^{150}
Therm. Prop.[kJ/mol]: $\Delta_{vap}H = 33.2$; $\Delta_f H°(l, 25°C) = -252.6$
Elec. Prop.: $\mu = 0$
η [mPa s]: 0.40^{38}
MS: 43(100) 70(58) 71(46) 41(29) 27(22) 55(20) 42(19) 29(19) 57(17) 39(11)
Flammability: Flash pt. = 7°C; ign. temp. = 438°C

203. 2,4-Dimethylhexane
Syst. Name: Hexane, 2,4-dimethyl-

CASRN: 589-43-5   **DOT No.**: 1262
MF: C_8H_{18} **BP[°C]**: 109.5   **Den.[g/cm³]**: 0.6962^{25}
MW: 114.23 n_D: 1.3929^{25}
Crit. Const.: $T_c = 280.5°C$; $P_c = 2.556$ MPa; $V_c = 472$ cm³
Elec. Prop.: $\mu = 0$
η **[mPa s]**: 0.49^{38}
Flammability: Flash pt. = 10°C

204. 2,5-Dimethylhexane
Syst. Name: Hexane, 2,5-dimethyl-
Synonyms: Biisobutyl

CASRN: 592-13-2   **DOT No.**: 1262
Beil RN: 1696877   **Beil Ref.**: 4-01-00-00434
MF: C_8H_{18} **MP[°C]**: -91   **Den.[g/cm³]**: 0.6901^{25}
MW: 114.23   **BP[°C]**: 109.1 n_D: 1.3925^{20}
Sol.: H_2O 1; EtOH 5; eth
   3; ace 5; bz 5; ctc 2;
   chl 5; lig 5
Crit. Const.: $T_c = 277.0°C$; $P_c = 2.487$ MPa; $V_c = 482$ cm³
Vap. Press. [kPa]: 4.06^{25}; 13.1^{50}; 34.4^{75}; 77.7^{100}; 156^{125}; 284^{150}
Therm. Prop.[kJ/mol]: $\Delta_{fus}H = 12.95$; $\Delta_{vap}H = 32.5$; $\Delta_fH°(l, 25°C) = -260.4$
C_p **(liq.) [J/mol °C]**: 249.2^{25}
γ **[mN/m]**: 19.40^{25}; $d\gamma/dT = 0.0912$ mN/m °C
Elec. Prop.: $\mu = 0$; $\varepsilon = 1.96^{21}$
η **[mPa s]**: 0.40^{38}
MS: 57(100) 43(93) 42(31) 41(26) 99(19) 71(19) 29(11) 70(10) 55(9) 27(9)
IR [cm⁻¹]: 2950 2860 1450 1370 1350 1330 1210 1160 910
Raman [cm⁻¹]: 2959 2870 2719 1462 1451 1337 1172 1148 960 838
¹³C NMR [ppm]: 22.4 28.4 36.9 diox
¹H NMR [ppm]: 0.9 1.4 CCl₄

205. 3,3-Dimethylhexane
Syst. Name: Hexane, 3,3-dimethyl-

CASRN: 563-16-6   **DOT No.**: 1262
Beil RN: 1696868   **Beil Ref.**: 4-01-00-00435
MF: C_8H_{18} **MP[°C]**: -126.1   **Den.[g/cm^3]**: 0.7100^{20}
MW: 114.23   **BP[°C]**: 111.9 n_D: 1.4001^{20}
Sol.: H_2O 1; EtOH 5; eth
    4; ace 4; bz 4; os 3
Crit. Const.: T_c = 289.0°C; P_c = 2.654 MPa; V_c = 443 cm^3
Vap. Press. [kPa]: 3.82^{25}; 12.2^{50}; 31.9^{75}; 71.8^{100}; 144^{125}; 261^{150}
Therm. Prop.[kJ/mol]: $\Delta_{fus}H$ = 6.98; $\Delta_{vap}H$ = 32.3; $\Delta_f H°$(l, 25°C) = -257.5
C_p **(liq.) [J/mol °C]**: 246.6^{25}
Elec. Prop.: μ = 0; ε = 1.96^{20}
η **[mPa s]**: 0.42^{38}
MS: 43(100) 71(47) 57(41) 85(36) 41(29) 29(23) 27(21) 70(17) 55(12) 39(11)
IR [cm^{-1}]: 2960 2730 1470 1390 1370 1190 1010 950 780 740
^{13}C NMR [ppm]: 8.1 14.8 17.3 26.5 32.8 34.3 44.3 diox

206. 3,4-Dimethylhexane

Syst. Name: Hexane, 3,4-dimethyl-

CASRN: 583-48-2   **DOT No.**: 1262
Beil RN: 1718744   **Beil Ref.**: 4-01-00-00436
MF: C_8H_{18} **BP[°C]**: 117.7   **Den.[g/cm^3]**: 0.7151^{25}
MW: 114.23 n_D: 1.4041^{20}
Sol.: H_2O 1; EtOH 5; eth
   3; ace 5; bz 5; chl 5;
   lig 5
Crit. Const.: T_c = 295.8°C; P_c = 2.692 MPa; V_c = 466 cm^3
Vap. Press. [kPa]: 2.89^{25}; 9.63^{50}; 26.1^{75}; 60.3^{100}; 123^{125}; 229^{150}
Therm. Prop.[kJ/mol]: $\Delta_{vap}H$ = 33.2; $\Delta_f H°$(l, 25°C) = -251.8
Elec. Prop.: μ = 0; ε = 1.98^{19}
η **[mPa s]**: 0.41^{38}
MS: 56(100) 57(83) 43(65) 41(47) 85(34) 29(15) 27(9) 39(8) 84(7) 55(7)
IR [cm^{-1}]: 2940 1470 1390 1120 1020 950 780
Raman [cm^{-1}]: 2963 2934 2909 2875 2730 1457 1162 1049 1032 981
^{13}C NMR [ppm]: 11.8 13.8 15.8 25.8 27.6 38.5 39.5 diox

207. Dimethyl maleate

Syst. Name: 2-Butenedioic acid (*Z*)-, dimethyl ester
Synonyms: Methyl maleate; Methyl *cis*-butenedioate

CASRN: 624-48-6   **Beil Ref.:** 4-02-00-02204
Beil RN: 471705
MF: $C_6H_8O_4$ **MP[°C]:** -19   **Den.[g/cm^3]:** 1.1606^{20}
MW: 144.13   **BP[°C]:** 202 n_D: 1.4416^{20}
Sol.: H_2O 1; eth 3; ctc 3;
   lig 1
Vap. Press. [kPa]: 9.14^{125}; 23.1^{150}
Therm. Prop.[kJ/mol]: $\Delta_{fus}H$ = 14.73
C_p **(liq.) [J/mol °C]:** 263.2^{25}
γ **[mN/m]:** 41.2^{25}
Elec. Prop.: μ = (2.5 D)
η **[mPa s]:** 3.21^{25}
MS: 113(100) 59(83) 26(72) 29(40) 85(37) 54(31) 114(25) 53(24) 55(13) 82(9)
IR [cm^{-1}]: 3030 2940 1720 1640 1430 1390 1300 1220 1160 1010 990 860 820
Raman [cm^{-1}]: 3060 3000 2960 2910 2850 1730 1650 1450 1390 1300 1270
   1220 1190 1170 1010 990 940 870 820 800 750 680 640 610 530 500 460
   360 340 260
^{1}H NMR [ppm]: 3.7 6.2 CCl_4
Flammability: Flash pt. = 113°C

185

208. 1,2-Dimethylnaphthalene
Syst. Name: Naphthalene, 1,2-dimethyl-

CASRN: 573-98-8   **Beil Ref.**: 4-05-00-01708
Beil RN: 2039376
MF: $C_{12}H_{12}$ **MP[°C]**: 0.8   **Den.[g/cm^3]**: 1.0179^{20}
MW: 156.23   **BP[°C]**: 266.5 n_D: 1.6166^{20}
Sol.: H_2O 1; eth 3; bz 3
Vap. Press. [kPa]: 1.10^{125}; 3.26^{150}
γ **[mN/m]**: 39.2^{20}
Elec. Prop.: $\mu = 0$; $\varepsilon = 2.61^{25}$
MS: 156(100) 141(95) 155(20) 115(16) 157(13) 142(12) 76(11) 153(10) 152(10) 77(10)
IR [cm^{-1}]: 3030 2940 1610 1520 1450 1390 1180 1020 980 950 860 810 770 740 720 680
UV [nm]: 322 313 307 292 284 275 228 cyhex

209. 1,6-Dimethylnaphthalene
Syst. Name: Naphthalene, 1,6-dimethyl-

CASRN: 575-43-9   **Beil Ref.**: 4-05-00-01711
Beil RN: 1854429
MF: $C_{12}H_{12}$ **MP[°C]**: -16.9   **Den.[g/cm^3]**: 1.0021^{20}
MW: 156.23   **BP[°C]**: 264 n_D: 1.6166^{20}
Sol.: H_2O 1; eth 3; bz 3
γ **[mN/m]**: 37.42^{20}
Elec. Prop.: $\mu = 0$; $\varepsilon = 2.73^{20}$
MS: 156(100) 141(56) 155(30) 157(13) 77(12) 153(10) 115(10) 76(10) 152(9) 128(8)
IR [cm^{-1}]: 3030 2940 1640 1610 1520 1450 1390 1270 1160 1080 1040 870 810 790 780 750
Raman [cm^{-1}]: * 1700 1591 1474 1432 1383 1329 1078 1058 716 573 450
UV [nm]: 320(1585) 314(708) 310(501) 307(1122) -170

210. 2,2-Dimethylpentane

Syst. Name: Pentane, 2,2-dimethyl-

CASRN: 590-35-2

DOT No.: 1206

Beil RN: 1730757

Beil Ref.: 4-01-00-00403

MF: C_7H_{16} **MP[°C]:** -123.8   **Den.[g/cm^3]:** 0.6739^{20}

MW: 100.20   **BP[°C]:** 79.2 n_D: 1.3822^{20}

Sol.: H_2O 1; EtOH 3; eth 3; ace 5; bz 5; chl 5; hp 5

Crit. Const.: T_c = 247.4°C; P_c = 2.773 MPa; V_c = 416 cm^3

Vap. Press. [kPa]: 4.08^0; 14.0^{25}; 38.5^{50}; 89.2^{75}; 182^{100}

Therm. Prop.[kJ/mol]: $\Delta_{fus}H$ = 5.86; $\Delta_{vap}H$ = 29.2; $\Delta_f H°$(l, 25°C) = -238.3

C_p (liq.) [J/mol °C]: 221.1^{25}

Elec. Prop.: μ = 0; ε = 1.92^{20}

η [mPa s]: 0.36^{38}

MS: 57(100) 43(73) 41(46) 56(40) 85(34) 29(31) 27(23) 39(15) 15(7) 55(6)

IR [cm^{-1}]: 2960 2890 1470 1370 1250 1210 940 740

Raman [cm^{-1}]: 2970 2950 2900 2870 2840 2780 2720 2700 1470 1450 1370 1320 1270 1250 1210 1100 1050 1020 950 930 880 750 700 500 430 350 330 200

^{13}C NMR [ppm]: 15.1 18.1 29.5 30.6 47.3 diox

^{1}H NMR [ppm]: 0.9 0.9 1.2 CDCl$_3$

211. 2,3-Dimethylpentane
Syst. Name: Pentane, 2,3-dimethyl-

CASRN: 565-59-3
Beil RN: 1718734
MF: C_7H_{16}
MW: 100.20
Sol.: H_2O 1; EtOH 3; eth 3; ace 5; bz 5; ctc 2; chl 5; hp 5

DOT No.: 1206
Beil Ref.: 3-01-00-00445
BP[°C]: 89.7
Den.[g/cm³]: 0.6951^{20}
n_D: 1.3919^{20}

Crit. Const.: $T_c = 264.3°C$; $P_c = 2.908$ MPa; $V_c = 393$ cm³
Vap. Press. [kPa]: 2.48^0; 9.18^{25}; 26.6^{50}; 64.3^{75}; 135^{100}
Therm. Prop.[kJ/mol]: $\Delta_{vap}H = 30.5$; $\Delta_f H°(l, 25°C) = -233.1$
C_p (liq.) [J/mol °C]: 218.3^{25}
γ [mN/m]: 19.47^{25}; $d\gamma/dT = 0.0994$ mN/m °C
Elec. Prop.: $\mu = 0$; $\varepsilon = 1.93^{20}$
η [mPa s]: 0.436^{15}
MS: 56(100) 43(92) 57(68) 41(60) 71(35) 29(31) 42(22) 27(16) 70(13) 55(12)
IR [cm⁻¹]: 2940 1470 1370 1120 1010 1000 960 920 780
Raman [cm⁻¹]: 2950 2940 2900 2880 2750 2710 2690 2650 1470 1450 1390 1350 1310 1280 1190 1160 1120 1040 1020 990 970 920 850 800 790 750 740 710 560 500 470 460 430 370 340 330 230
¹³C NMR [ppm]: 11.6 14.5 17.7 20.0 26.8 31.9 40.6 diox
¹H NMR [ppm]: 0.9 1.4 CCl_4
Flammability: Flash pt. < -7°C; ign. temp. = 335°C; flam. lim. = 1.1-6.7%

212. 2,4-Dimethylpentane
Syst. Name: Pentane, 2,4-dimethyl-

CASRN: 108-08-7
Beil RN: 1696855
MF: C_7H_{16}
MW: 100.20
Sol.: H_2O 1; EtOH 3; eth
  3; ace 5; bz 5; ctc 2;
  chl 5; hp 5

DOT No.: 1206
Beil Ref.: 4-01-00-00406
MP[°C]: -119.9 **Den.[g/cm^3]**: 0.6727^{20}
BP[°C]: 80.4 n_D: 1.3815^{20}

Crit. Const.: $T_c = 246.7°C$; $P_c = 2.737$ MPa; $V_c = 418$ cm^3
Vap. Press. [kPa]: 3.71^0; 13.1^{25}; 36.6^{50}; 85.7^{75}; 175^{100}
Therm. Prop.[kJ/mol]: $\Delta_{fus}H = 6.69$; $\Delta_{vap}H = 29.6$; $\Delta_f H°(l, 25°C) = -234.6$
C_p (liq.) [J/mol °C]: 224.2^{25}
γ **[mN/m]**: 17.66^{25}; $d\gamma/dT = 0.0971$ mN/m °C
Elec. Prop.: $\mu = 0$; $\varepsilon = 1.90^{20}$
η **[mPa s]**: 0.360^{20}
MS: 43(100) 57(73) 56(41) 41(35) 42(24) 85(17) 29(16) 27(12) 39(9) 58(3)
IR [cm^{-1}]: 2960 1470 1380 1160 990 920 870 810
Raman [cm^{-1}]: 2960 2940 2920 2880 2850 2760 2730 1470 1460 1350 1320
  1250 1170 1160 1080 1040 990 960 920 870 810 470 430 310 250 180
^{13}C NMR [ppm]: 22.7 25.7 49.0 diox
^{1}H NMR [ppm]: 0.9 1.1 1.6 CCl$_4$
Flammability: Flash pt. = -12°C

213. 3,3-Dimethylpentane

Syst. Name: Pentane, 3,3-dimethyl-

CASRN: 562-49-2   **DOT No.**: 1206
Beil RN: 1696854   **Beil Ref.**: 4-01-00-00409
MF: C_7H_{16} **MP[°C]**: -134.9   **Den.[g/cm^3]**: 0.6936^{20}
MW: 100.20   **BP[°C]**: 86.0 n_D: 1.3909^{20}
Sol.: H_2O 1; EtOH 3; eth
    3; ace 5; bz 5; chl 5;
    hp 5
Crit. Const.: T_c = 263.3°C; P_c = 2.946 MPa; V_c = 414 cm^3
Vap. Press. [kPa]: 3.13^0; 11.0^{25}; 30.8^{50}; 72.5^{75}; 149^{100}
Therm. Prop.[kJ/mol]: $\Delta_{fus}H$ = 7.07; $\Delta_{vap}H$ = 29.6; $\Delta_fH°$(l, 25°C) = -234.2
Elec. Prop.: μ = 0; ε = 1.94^{18}
η **[mPa s]**: 0.39^{38}
MS: 43(100) 71(62) 41(20) 70(19) 29(18) 27(18) 85(16) 55(12) 39(9) 57(5)
IR [cm^{-1}]: 2940 1470 1390 1370 1280 1190 1090 1000 910 860 790 780 770
Raman [cm^{-1}]: 2970 2940 2890 2760 2740 2710 1450 1340 1290 1240 1220
    1200 1080 1040 1010 930 910 860 700 510 490 450 410 380 350 300 250
^{13}C NMR [ppm]: 4.4 6.8 25.1 36.1

214. Dimethyl phthalate

Syst. Name: 1,2-Benzenedicarboxylic acid, dimethyl ester

Synonyms: Methyl phthalate

CASRN: 131-11-3   **Beil Ref.:** 4-09-00-03170
Merck No.: 3243
Beil RN: 1911460
MF: $C_{10}H_{10}O_4$ **MP[°C]:** 5.5   **Den.[g/cm³]:** 1.1905^{20}
MW: 194.19   **BP[°C]:** 283.7 n_D: 1.5138^{20}
Sol.: H_2O 1; EtOH 5; eth 5; bz 3; ctc 2
Vap. Press. [kPa]: 0.151^{100}; 0.548^{125}; 1.71^{150}
C_p (liq.) [J/mol °C]: 303.1^{25}
Elec. Prop.: $\varepsilon = 8.66^{20}$; IP = 9.64 eV
η [mPa s]: 63.2^0; 14.4^{25}; 5.31^{50}; 2.82^{75}; 1.98^{100}
k [W/m °C]: 0.1501^0; 0.1473^{25}; 0.1443^{50}; 0.1409^{75}; 0.1373^{100}
MS: 163(100) 28(30) 164(27) 92(19) 77(14) 194(13) 135(13) 104(13) 133(8) 105(8)
IR [cm⁻¹]: 3000 2960 2820 1740 1600 1580 1490 1440 1290 1270 1120 1080 1040 970 820 750 700
Raman [cm⁻¹]: 3200 3150 3080 3030_3000 2950 2900 2840 2730 2600 1720 1600 1570 1480 1450 1430 1280 1190 1160 1120 1070 1040 960 890 810 780 750 700 650 550 510 400 380 360 340 290 250 200 100
UV [nm]: 274 224 MeOH
¹H NMR [ppm]: 3.8 7.5 7.6 CCl_4
Flammability: Flash pt. = 146°C; ign. temp. = 490°C; lower flam. lim. = 0.9%
TLV/TWA: 5 mg/m³
Reg. Lists: CERCLA (RQ = 5000 lb.); SARA 313 (1.0%); RCRA U102 (toxic)

215. 2,2-Dimethyl-1-propanol
Syst. Name: 1-Propanol, 2,2-dimethyl-
Synonyms: Neopentyl alcohol; *tert*-Butylcarbinol

CASRN: 75-84-3   **Beil Ref.**: 4-01-00-01690
Merck No.: 6373
Beil RN: 1730984
MF: $C_5H_{12}O$ **MP[°C]**: 52.5   **Den.[g/cm³]**: 0.812^{20}
MW: 88.15   **BP[°C]**: 113.5
Sol.: H_2O 2; EtOH 4; eth
 4; ctc 3
Vap. Press. [kPa]: 5.77^{50}; 20.6^{75}; 59.8^{100}; 148^{125}
Elec. Prop.: $\varepsilon = 8.35^{60}$
MS: 57(100) 41(53) 29(49) 28(42) 56(40) 55(36) 73(24) 39(21) 27(19) 31(18)
IR [cm⁻¹]: 3350 2950 2865 1480 1400 1380 1285 1180 1050 1010 930 900 745
 650 530
Raman [cm⁻¹]: 2970 2940 2910 2720 1460 1440 1330 1290 1220 1180 1060
 1010 940 920 880 730 530 490 450 420 370 350 260
¹H NMR [ppm]: 0.9 3.2 3.8 CCl_4
Flammability: Flash pt. = 37°C

---

216. Dimethyl succinate
Syst. Name: Butanedioic acid, dimethyl ester
Synonyms: Methyl succinate

CASRN: 106-65-0   **Beil Ref.**: 4-02-00-01913
Beil RN: 956776
MF: $C_6H_{10}O_4$ **MP[°C]**: 19   **Den.[g/cm³]**: 1.1198^{20}
MW: 146.14   **BP[°C]**: 196.4 n_D: 1.4197^{20}
Sol.: H_2O 2; EtOH 3; eth
 4; ace 3; ctc 2
Elec. Prop.: $\varepsilon = 7.19^{20}$
MS: 115(100) 55(58) 59(48) 43(28) 114(25) 87(19) 27(15) 31(14) 29(14)
 45(13)
IR [cm⁻¹]: 2950 1740 1430 1350 1310 1200 1150 1020 990 840
¹H NMR [ppm]: 2.5 3.6 CCl_4

217. Dimethyl sulfide

Syst. Name: Methane, thiobis-
Synonyms: 2-Thiapropane

CASRN: 75-18-3   **DOT No.**: 1164
Merck No.: 6042   **Beil Ref.**: 4-01-00-01275
Beil RN: 1696847
MF: C_2H_6S **MP[°C]**: -98.3   **Den.[g/cm^3]**: 0.8483^{20}
MW: 62.14   **BP[°C]**: 37.3 n_D: 1.4438^{20}
Sol.: H_2O 1; EtOH 3; eth
  3
Crit. Const.: T_c = 229.9°C; P_c = 5.53 MPa; V_c = 201 cm^3
Vap. Press. [kPa]: 22.3^0; 64.4^{25}
Therm. Prop.[kJ/mol]: $\Delta_{fus}H$ = 7.99; $\Delta_{vap}H$ = 27.0; $\Delta_f H°$(l, 25°C) = -65.4
C_p (liq.) [J/mol °C]: 118.1^{25}
γ **[mN/m]**: 24.06^{25}; $d\gamma/dT$ = 0.0805 mN/m °C
Elec. Prop.: μ = 1.554 D; ϵ = 6.70^{21}; IP = 8.69 eV
η **[mPa s]**: 0.356^0; 0.284^{25}
MS: 47(100) 62(83) 45(62) 46(43) 35(35) 61(30) 27(21) 15(21) 14(10) 44(6)
IR [cm^{-1}]: 2940 1540 1430 1320 1040 1020 970
Flammability: Flash pt. < -18°C; ign. temp. = 206°C; flam. lim. = 2.2-19.7

218. Dimethyl sulfoxide

Syst. Name: Methane, sulfinylbis-
Synonyms: Methyl sulfoxide; DMSO

CASRN: 67-68-5   **Beil Ref.**: 4-01-00-01277
Merck No.: 3247
Beil RN: 506008
MF: C_2H_6OS **MP[°C]**: 18.5   **Den.[g/cm³]**: 1.1014^{20}
MW: 78.14   **BP[°C]**: 189 n_D: 1.4170^{20}
Sol.: H_2O 3; EtOH 3; eth
  3; ace 3; ctc 3; AcOEt
  3
Vap. Press. [kPa]: 0.084^{25}; 0.431^{50}; 1.67^{75}; 5.27^{100}; 14.1^{125}; 33.0^{150}
Therm. Prop.[kJ/mol]: $\Delta_{fus}H$ = 14.37; $\Delta_{vap}H$ = 43.1; $\Delta_f H°$(l, 25°C) = -204.2
C_p **(liq.) [J/mol °C]**: 153.0^{25}
γ **[mN/m]**: 42.92^{25}; $d\gamma/dT$ = 0.1145 mN/m °C
Elec. Prop.: μ = 3.96 D; ε = 47.24^{20}; IP = 9.01 eV
η **[mPa s]**: 1.99^{25}; 1.29^{50}
MS: 63(100) 78(70) 15(40) 45(35) 29(16) 61(13) 46(12) 31(11) 48(10) 47(10)
IR [cm⁻¹]: 3000 2910 1440 1410 1310 1050 950 930 700 670
Raman [cm⁻¹]: 3000 2900 2810 1420 1310 1050 960 700 670 390 340 310
UV [nm]: 203(2512) cyhex
¹³C NMR [ppm]: 43.5
¹H NMR [ppm]: 2.5 CCl_4
Flammability: Flash pt. = 95°C; ign. temp. = 215°C; flam. lim. = 2.6-42%

219. 1,4-Dioxane

Syst. Name: 1,4-Dioxane
Synonyms: 1,4-Dioxacyclohexane

CASRN: 123-91-1
Merck No.: 3294
Beil RN: 102551
DOT No.: 1165
Beil Ref.: 5-19-01-00016

MF: $C_4H_8O_2$ **MP[°C]:** 11.8   **Den.[g/cm^3]:** 1.0337^{20}
MW: 88.11   **BP[°C]:** 101.5 n_D: 1.4224^{20}
Sol.: H_2O 5; EtOH 5; eth 5; ace 5; bz 5; ctc 3; os 5; HOAc 5
Crit. Const.: $T_c = 314$°C; $P_c = 5.21$ MPa; $V_c = 238$ cm^3
Vap. Press. [kPa]: 4.95^{25}; 15.8^{50}; 42.2^{75}; 97.2^{100}; 200^{125}
Therm. Prop.[kJ/mol]: $\Delta_{fus}H = 12.85$; $\Delta_{vap}H = 34.2$; $\Delta_f H°(l, 25$°C$) = -353.9$
C_p (liq.) [J/mol °C]: 153.6^{25}
γ [mN/m]: 32.75^{25}; $d\gamma/dT = 0.1391$ mN/m °C
Elec. Prop.: $\mu = 0$; $\varepsilon = 2.22^{20}$; IP = 9.19 eV
η [mPa s]: 1.18^{25}; 0.787^{50}; 0.569^{75}
k [W/m °C]: 0.159^{25}; 0.147^{50}; 0.135^{75}; 0.123^{100}
MS: 28(100) 29(37) 88(31) 58(24) 31(17) 15(17) 27(15) 30(13) 43(11) 26(9)
IR [cm^{-1}]: 2960 2910 2890 2860 2760 2700 1450 1370 1290 1250 1130 1080 1050 880 620
Raman [cm^{-1}]: 2930 2880 2850 2770 2750 2720 2670 1450 1440 1400 1340 1310 1220 1130 1110 1020 850 830 490 440 430
UV [nm]: 180(6310) gas
^{13}C NMR [ppm]: 67.4
^{1}H NMR [ppm]: 3.6 CCl$_4$
Flammability: Flash pt. = 12°C; ign. temp. = 180°C; flam. lim. = 2.0-22%
TLV/TWA: 25 ppm (90 mg/m^3)
Reg. Lists: CERCLA (RQ = 100 lb.); SARA 313 (0.1%); RCRA U108 (toxic); confirmed or suspected carcinogen

220. 1,3-Dioxolane

Syst. Name: 1,3-Dioxolane
Synonyms: 1,3-Dioxacyclopentane

CASRN: 646-06-0   **DOT No.**: 1166
Beil RN: 102453   **Beil Ref.**: 5-19-01-00006
MF: $C_3H_6O_2$ **MP[°C]**: -95   **Den.[g/cm^3]**: 1.060^{20}
MW: 74.08   **BP[°C]**: 78 n_D: 1.3974^{20}
Sol.: H_2O 5; EtOH 3; eth
   3; ace 3
Vap. Press. [kPa]: 3.51^0; 14.6^{25}; 41.6^{50}; 99.2^{75}
Therm. Prop.[kJ/mol]: $\Delta_{fus}H$ = 27.48; $\Delta_f H°$(l, 25°C) = -333.5
C_p **(liq.) [J/mol °C]**: 118.0^{25}
Elec. Prop.: μ = 1.19 D; IP = 9.90 eV
η **[mPa s]**: 0.6^{20}
MS: 73(100) 29(56) 44(53) 45(28) 28(21) 43(20) 27(13) 31(7) 74(5) 42(3)
IR [cm^{-1}]: 2940 2860 1470 1110 1050 940
Flammability: Flash pt. = 2°C

221. Dipentene

Syst. Name: Cyclohexene, 1-methyl-4-(1-
methylethenyl)-, (±)-
Synonyms: *dl*-Limonene; *p*-Menthadiene

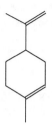

CASRN: 7705-14-8   **Beil Ref.:** 4-05-00-00440
Merck No.: 5371
Beil RN: 3195091
MF: $C_{10}H_{16}$ **MP[°C]:** -95.5   **Den.[g/cm^3]:** 0.8402^{21}
MW: 136.24   **BP[°C]:** 178 n_D: 1.4727^{20}
Vap. Press. [kPa]: 0.259^{25}; 1.07^{50}; 3.56^{75}; 9.91^{100}; 24.0^{125}; 52.0^{150}
Therm. Prop.[kJ/mol]: $\Delta_f H°(l, 25°C) = -50.8$
C_p **(liq.) [J/mol °C]:** 249.4^{20}
γ **[mN/m]:** 26.90^{24}
Elec. Prop.: $\varepsilon = 2.38^{25}$
η **[mPa s]:** 1.22^{25}
MS: 68(100) 93(50) 67(44) 94(22) 39(22) 107(18) 92(18) 53(18) 136(16)
    79(16)
IR [cm^{-1}]: 2941 1667 1449 1389 1149 917 893 800
Raman [cm^{-1}]: 3070 3020 2980 2930 2880 2850 2740 1680 1650 1610 1570
    1430 1380 1370 1330 1310 1210 1180 1160 1110 1080 1060 1020 960 920
    890 820 800 760 720 700 680 660 640 540 520 490 440 430 350 310 260 180
UV [nm]: 250(23) 220(257) iso
^{13}C NMR [ppm]: 20.7 23.4 28.0 30.7 30.9 41.2 108.4 120.7 133.5 149.9 CDCl$_3$
Flammability: Flash pt. = 45°C; ign. temp. = 237°C

222. Dipentyl ether

Syst. Name: Pentane, 1,1'-oxybis-
Synonyms: Amyl ether; Pentyl ether

CASRN: 693-65-2   **Beil Ref.:** 4-01-00-01643
Merck No.: 646
Beil RN: 1698030
MF: $C_{10}H_{22}O$ **MP[°C]:** -69   **Den.[g/cm^3]:** 0.7833^{20}
MW: 158.28   **BP[°C]:** 190 n_D: 1.4119^{20}
Sol.: H_2O 1; EtOH 5; eth
    5; chl 3
Vap. Press. [kPa]: 6.35^{100}; 16.2^{125}; 36.6^{150}
C_p **(liq.) [J/mol °C]:** 250^{25}
γ **[mN/m]:** 24.35^{25}; dγ/dT = 0.0925 mN/m °C
Elec. Prop.: μ = (1.2 D); ε = 2.80^{25}
η **[mPa s]:** 1.19^{15}; 0.922^{30}
MS: 71(100) 43(92) 29(43) 70(40) 41(36) 27(34) 42(23) 69(17) 55(16) 39(16)
IR [cm^{-1}]: 2940 1470 1390 1110 730
Raman [cm^{-1}]: 2960 2941 2921 2904 2871 2849 1483 1462 1450 1434
^{1}H NMR [ppm]: 0.9 1.4 3.4 CDCl$_3$
Flammability: Flash pt. = 57°C; ign. temp. = 170°C

223. Diphenyl ether

Syst. Name: Benzene, 1,1'-oxybis-
Synonyms: Oxybisbenzene; Phenyl ether

CASRN: 101-84-8   **Beil Ref.:** 4-06-00-00568
Merck No.: 7259
Beil RN: 1364620
MF: $C_{12}H_{10}O$ **MP[°C]:** 26.8   **Den.[g/cm^3]:** 1.0661^{30}
MW: 170.21   **BP[°C]:** 258.0 n_D: 1.5787^{25}
Sol.: H_2O 1; EtOH 3; eth
   3; bz 3; chl 2; HOAc 3
Crit. Const.: T_c = 493.7°C
Vap. Press. [kPa]: 23.5^{200}; 46.5^{225}; 84.8^{250}
Therm. Prop.[kJ/mol]: $\Delta_{fus}H$ = 17.22; $\Delta_{vap}H$ = 48.2; $\Delta_f H°$(s, 25°C) = -32.1
C_p (liq.) [J/mol °C]: 268.4^{27}
γ [mN/m]: 26.36^{30}; $d\gamma/dT$ = 0.0780 mN/m °C
Elec. Prop.: μ = 1.3 D; ε = 3.73^{10}; IP = 8.09 eV
η [mPa s]: 2.13^{50}; 1.41^{75}; 1.02^{100}
k [W/m °C]: 0.139^{50}; 0.135^{75}; 0.131^{100}
MS: 170(100) 141(39) 51(36) 77(35) 142(28) 39(17) 169(15) 171(13) 115(12)
   50(9)
IR [cm^{-1}]: 3070 3040 1585 1490 1235 1160 1070 1020 870 800 750 690 500
Raman [cm^{-1}]: 3170 3060 2970 2910 1590 1490 1460 1380 1330 1280 1230
   1200 1160 1070 1020 1000 870 820 800 750 690 620 600 500 560 470 410
   310 240 220
UV [nm]: 295(9170) 286(16100) 280(17400) 248(19500) 217(39000)
   207(39600) MeOH
^{13}C NMR [ppm]: 119.0 123.2 129.8 157.6 diox
^{1}H NMR [ppm]: 7.1 CCl$_4$
Flammability: Flash pt. = 112°C; ign. temp. = 618°C; flam. lim. = 0.8-1.5%
TLV/TWA: 1 ppm (7 mg/m^3)

224. Dipropylamine

Syst. Name: 1-Propanamine, *N*-propyl-
Synonyms: *N*-Propyl-1-propanamine

CASRN: 142-84-7   **DOT No.**: 2383
Merck No.: 3350   **Beil Ref.**: 4-04-00-00469
Beil RN: 505974
MF: $C_6H_{15}N$ **MP[°C]**: -63   **Den.[g/cm^3]**: 0.7400^{20}
MW: 101.19   **BP[°C]**: 109.3 n_D: 1.4050^{20}
Sol.: H_2O 3; EtOH 3; eth
    5; ace 4; bz 4
Crit. Const.: T_c = 282.7°C; P_c = 3.63 MPa
Vap. Press. [kPa]: 3.21^{25}; 11.2^{50}; 31.7^{75}; 76.0^{100}; 160^{125}; 306^{150}
Therm. Prop.[kJ/mol]: $\Delta_{vap}H$ = 33.5; $\Delta_f H°$(l, 25°C) = -156.1
C_p (liq.) [J/mol °C]: 253^{75}
γ [mN/m]: 22.31^{25}; $d\gamma/dT$ = 0.1022 mN/m °C
Elec. Prop.: μ = (1.0 D); ε = 3.07^{20}; IP = 7.84 eV
η [mPa s]: 0.751^{0}; 0.517^{25}; 0.377^{50}; 0.288^{75}; 0.228^{100}
MS: 30(100) 72(79) 44(40) 43(32) 27(25) 28(24) 41(22) 86(11) 58(10) 42(10)
IR [cm^{-1}]: 3330 2940 1670 1470 1390 1300 1220 1140 1090 900 760
Raman [cm^{-1}]: 3320 2960 2930 2900 2870 2810 2730 1650 1510 1450 1380
    1350 1300 1280 1230 1180 1140 1090 1060 1040 970 910 890 870 760 470
    410 380 330 300
^{1}H NMR [ppm]: 1.5 2.6 CDCl$_3$
Flammability: Flash pt. = 17°C; ign. temp. = 299°C
Reg. Lists: CERCLA (RQ = 5000 lb.); RCRA U110 (toxic)

225. Dipropyl ether

Syst. Name: Propane, 1,1'-oxybis-
Synonyms: Propyl ether

CASRN: 111-43-3   **DOT No.:** 2384
Merck No.: 7870   **Beil Ref.:** 4-01-00-01422
Beil RN: 1731312
MF: $C_6H_{14}O$ **MP[°C]:** -126.1   **Den.[g/cm^3]:** 0.7466^{20}
MW: 102.18   **BP[°C]:** 90.0 n_D: 1.3809^{20}
Sol.: eth 4; EtOH 4
Crit. Const.: $T_c = 257.5$°C; $P_c = 3.028$ MPa
Vap. Press. [kPa]: 8.35^{25}; 25.1^{50}; 62.7^{75}; 135^{100}
Therm. Prop.[kJ/mol]: $\Delta_{fus}H = 8.83$; $\Delta_{vap}H = 31.3$; $\Delta_f H°(l, 25°C) = -328.8$
C_p (liq.) [J/mol °C]: 221.6^{25}
γ [mN/m]: 19.98^{25}; dγ/d$T = 0.1047$ mN/m °C
Elec. Prop.: $\mu = 1.21$ D; $\varepsilon = 3.38^{24}$; IP = 9.27 eV
η [mPa s]: 0.542^0; 0.396^{25}; 0.304^{50}; 0.242^{75}
MS: 43(100) 73(22) 41(15) 102(12) 27(10) 39(5) 59(4) 55(4) 44(4) 42(4)
IR [cm^{-1}]: 2975 2945 2860 1470 1370 1250 1135 1010 950
Raman [cm^{-1}]: 2970 2940 2920 2880 2860 2800 2730 2680 1480 1460 1350
     1290 1250 1160 1120 1100 1050 1030 940 920 890 870 750 480 410 330 300
^{13}C NMR [ppm]: 11.1 24.0 73.2
Flammability: Flash pt. = 21°C; ign. temp. = 188°C; flam. lim. = 1.3-7.0%

226. Dodecane

Syst. Name: Dodecane

CASRN: 112-40-3   **Beil Ref.**: 4-01-00-00498
Beil RN: 1697175
MF: $C_{12}H_{26}$ **MP[°C]**: -9.6   **Den.[g/cm^3]**: 0.7487^{20}
MW: 170.34   **BP[°C]**: 216.3 n_D: 1.4216^{20}
Sol.: H_2O 1; EtOH 4; eth
   4; ace 4; ctc 4; chl 4
Crit. Const.: T_c = 385°C; P_c = 1.82 MPa; V_c = 754 cm^3
Vap. Press. [kPa]: 0.016^{25}; 2.01^{100}; 15.2^{150}
Therm. Prop.[kJ/mol]: $\Delta_{fus}H$ = 36.58; $\Delta_{vap}H$ = 44.5; $\Delta_f H°$(l, 25°C) = -350.9
C_p (liq.) [J/mol °C]: 375.8^{25}
γ [mN/m]: 24.91^{25}; dγ/dT = 0.0884 mN/m °C
Elec. Prop.: μ = 0; ε = 2.01^{20}
η [mPa s]: 2.28^{0}; 1.38^{25}; 0.930^{50}; 0.673^{75}; 0.514^{100}
k [W/m °C]: 0.157^{0}; 0.152^{25}; 0.146^{50}; 0.140^{75}; 0.135^{100}
MS: 57(100) 43(91) 71(53) 41(45) 85(31) 29(27) 55(19) 56(18) 28(16) 42(14)
IR [cm^{-1}]: 2950 2850 1470 1380 720
Raman [cm^{-1}]: 2970 2940 2880 2860 2730 1440 1370 1300 1130 1080 1060
   1030 960 890 870 840 770 400 350 240
^{13}C NMR [ppm]: 14.8 23.3 32.8 30.5
^{1}H NMR [ppm]: 0.9 1.3 CCl$_4$
Flammability: Flash pt. = 74°C; ign. temp. = 203°C; lower flam. lim. = 0.6%

227. 1-Dodecene

Syst. Name: 1-Dodecene

CASRN: 112-41-4   **Beil Ref.**: 4-01-00-00914
Beil RN: 1699848
MF: $C_{12}H_{24}$ **MP[°C]**: -35.2   **Den.[g/cm³]**: 0.7584^{20}
MW: 168.32   **BP[°C]**: 213.8 n_D: 1.4300^{20}
Sol.: H_2O 1; EtOH 3; eth
   3; ace 3; ctc 3; peth 3
Crit. Const.: T_c = 384.5°C; P_c = 1.930 MPa
Vap. Press. [kPa]: 0.019^{25}; 0.134^{50}; 0.637^{75}; 2.30^{100}; 6.73^{125}; 16.8^{150}
Therm. Prop.[kJ/mol]: $\Delta_{fus}H$ = 17.42; $\Delta_{vap}H$ = 44.0; $\Delta_f H°$(l, 25°C) = -226.2
C_p (liq.) [J/mol °C]: 360.7^{25}
γ [mN/m]: 25.15^{25}; $d\gamma/dT$ = 0.0891 mN/m °C
Elec. Prop.: μ = 0; ϵ = 2.15^{20}
η [mPa s]: 1.20^{25}
MS: 43(100) 56(83) 55(83) 41(75) 69(62) 70(58) 57(57) 83(47) 29(38) 84(33)
IR [cm⁻¹]: 3080 2940 2860 1640 1470 1380 990 910 720 640 550
Raman [cm⁻¹]: 3080 2998 2980 2960 2851 1641 1439 1416 1300 1077
¹H NMR [ppm]: 0.9 1.3 2.0 5.4 CCl_4
Flammability: Flash pt. = 79°C

228. Epichlorohydrin

Syst. Name: Oxirane, (chloromethyl)-
Synonyms: Chloropropylene oxide; 1-Chloro-2,3-
epoxypropane

CASRN: 106-89-8   **Rel. CASRN**: 13403-37-7
Merck No.: 3563   **DOT No.**: 2023
Beil RN: 79785   **Beil Ref.**: 5-17-01-00020
MF: C_3H_5ClO **MP[°C]**: -26   **Den.[g/cm^3]**: 1.1812^{20}
MW: 92.52   **BP[°C]**: 118 n_D: 1.4358^{25}
Sol.: H_2O 1; EtOH 5; eth
    5; bz 3; ctc 3
Vap. Press. [kPa]: 2.20^{25}; 8.41^{50}; 24.5^{75}; 60.8^{100}
Therm. Prop.[kJ/mol]: $\Delta_f H°$(l, 25°C) = -148.4
γ [mN/m]: 37.0^{20}
Elec. Prop.: μ = (1.8 D); ϵ = 22.6^{22}
η [mPa s]: 2.49^{-25}; 1.57^0; 1.07^{25}; 0.781^{50}; 0.597^{75}; 0.474^{100}
k [W/m °C]: 0.142^{-25}; 0.137^0; 0.131^{25}; 0.125^{50}; 0.119^{75}; 0.114^{100}
MS: 57(100) 27(39) 29(32) 49(25) 31(22) 62(18) 28(16) 92(1)
IR [cm^{-1}]: 3130 3030 1490 1450 1270 1140 1100 960 930 860 760 690
Raman [cm^{-1}]: 3060 3010 2970 2920 2870 2850 2790 1480 1450 1430 1400
    1270 1250 1210 1140 1090 960 920 900 850 760 740 720 690 510 450 440
    410 370 220
^{1}H NMR [ppm]: 2.6 2.8 3.2 3.5 3.6 CCl_4
Flammability: Flash pt. = 31°C; ign. temp. = 411°C; flam. lim. = 3.8-21%
TLV/TWA: 2 ppm (7.6 mg/m^3)
Reg. Lists: CERCLA (RQ = 100 lb.); SARA 313 (0.1%); RCRA U041 (toxic);
    confirmed or suspected carcinogen

229. 1,2-Epoxybutane

Syst. Name: Oxirane, ethyl-
Synonyms: Ethyloxirane; 1,2-Butylene oxide

CASRN: 106-88-7   **Beil Ref.**: 5-17-01-00056
Beil RN: 102411
MF: C_4H_8O **MP[°C]**: -150   **Den.[g/cm^3]**: 0.8297^{20}
MW: 72.11   **BP[°C]**: 63.3 n_D: 1.3851^{20}
Sol.: EtOH 4; eth 5; ace
   4; os 4
Vap. Press. [kPa]: 4.22^{-25}; 12.5^0; 31.7^{25}; 70.5^{50}; 142^{75}
Therm. Prop.[kJ/mol]: $\Delta_{vap}H$ = 30.3; $\Delta_f H°$(l, 25°C) = -168.9
C_p (liq.) [J/mol °C]: 147.0^{25}
Elec. Prop.: μ = (2.0 D); IP = 10.15 eV
η [mPa s]: 0.40^{25}
MS: 43(100) 44(58) 27(52) 29(50) 45(46) 26(27) 28(27) 72(5)
IR [cm^{-1}]: 3030 2940 2860 1490 1470 1410 1370 1270 1140 1100 1040 1020
   950 900 830 800 760 730
^{1}H NMR [ppm]: 1.0 1.5 2.3 2.6 2.7 CCl$_4$
Flammability: Flash pt. = -22°C; ign. temp. = 439°C; flam. lim. = 1.7-19%
Reg. Lists: CERCLA (RQ = 1 lb.); SARA 313 (1.0%)

230. Ethanol

Syst. Name: Ethanol
Synonyms: Ethyl alcohol

CASRN: 64-17-5
Merck No.: 3716
Beil RN: 1718733
MF: C_2H_6O
MW: 46.07
Sol.: H_2O 5; EtOH 5; eth
   5; ace 5; bz 3; chl 5;
   HOAc 5

DOT No.: 1170
Beil Ref.: 4-01-00-01289

MP[°C]: -114.1
BP[°C]: 78.2

Den.[g/cm^3]: 0.7893^{20}
n_D: 1.3611^{20}

Crit. Const.: T_c = 240.9°C; P_c = 6.137 MPa; V_c = 167 cm^3
Vap. Press. [kPa]: 1.50^0; 7.87^{25}; 29.5^{50}; 88.8^{75}; 224^{100}; 495^{125}; 976^{150}
Therm. Prop.[kJ/mol]: $\Delta_{fus}H$ = 5.02; $\Delta_{vap}H$ = 38.6; $\Delta_fH°$(l, 25°C) = -277.7
C_p (liq.) [J/mol °C]: 112.3^{25}
γ [mN/m]: 21.97^{25}; dγ/dT = 0.0832 mN/m °C
Elec. Prop.: μ = 1.69 D; ε = 25.3^{20}; IP = 10.47 eV
η [mPa s]: 3.26^{-25}; 1.79^0; 1.07^{25}; 0.694^{50}; 0.476^{75}
k [W/m °C]: 0.176^0; 0.169^{25}; 0.162^{50}
MS: 31(100) 45(44) 46(18) 27(18) 29(15) 43(14) 30(6) 42(3) 19(3) 14(3)
IR [cm^{-1}]: 3570 3030 2860 1390 1240 1060 890
Raman [cm^{-1}]: 3260 2980 2930 2880 2830 2750 2720 1480 1460 1280 1100
   1050 930 890 440
UV [nm]: 181(324) gas
^{13}C NMR [ppm]: 17.9 57.3
^{1}H NMR [ppm]: 1.2 2.6 3.7 CDCl$_3$
Flammability: Flash pt. = 13°C; ign. temp. = 363°C; flam. lim. = 3.3-19%
TLV/TWA: 1000 ppm (1880 mg/m^3)

231. Ethanolamine

Syst. Name: Ethanol, 2-amino-
Synonyms: 2-Aminoethanol; Glycinol

CASRN: 141-43-5

Merck No.: 3681

Beil RN: 505944

DOT No.: 2491

Beil Ref.: 4-04-00-01406

MF: C_2H_7NO	**MP[°C]**: 10.5	**Den.[g/cm³]**: 1.0180^{20}
MW: 61.08	**BP[°C]**: 171	n_D: 1.4541^{20}

Sol.: H_2O 5; EtOH 5; eth
2; bz 2; chl 3; glycerol
5; lig 2

Vap. Press. [kPa]: 0.050^{25}; 1.72^{75}; 6.56^{100}; 20.0^{125}; 51.2^{150}

Therm. Prop.[kJ/mol]: $\Delta_{fus}H = 20.50$; $\Delta_{vap}H = 49.8$

C_p **(liq.) [J/mol °C]**: 195.5^{25}

γ **[mN/m]**: 48.32^{25}; $d\gamma/dT = 0.1117$ mN/m °C

Elec. Prop.: $\mu = (2.3$ D); $\varepsilon = 31.94^{20}$; IP = 8.96 eV

η **[mPa s]**: 21.1^{25}; 8.56^{50}; 3.94^{75}; 2.00^{100}

k **[W/m °C]**: 0.299^{25}; 0.286^{50}; 0.274^{75}; 0.261^{100}

MS: 30(100) 18(30) 28(15) 42(7) 31(6) 17(6) 61(5) 15(5) 43(3) 29(3)

IR [cm⁻¹]: 3230 3130 2860 1610 1470 1350 1080 1040 960 870

Raman [cm⁻¹]: 3360 3300 3180 2930 2870 2770 2710 1600 1460 1360 1310
1260 1170 1110 1090 1040 1000 990 880 850 530 480 350

¹H NMR [ppm]: 2.7 2.8 3.5 CDCl₃

Flammability: Flash pt. = 86°C; ign. temp. = 410°C; flam. lim. = 3.0-23.5%

TLV/TWA: 3 ppm (7.5 mg/m³)

232. Ethyl acetate

Syst. Name: Ethanoic acid, ethyl ester

CASRN: 141-78-6   **DOT No.:** 1173
Merck No.: 3713   **Beil Ref.:** 4-02-00-00127
Beil RN: 506104
MF: $C_4H_8O_2$ **MP[°C]:** -83.6   **Den.[g/cm^3]:** 0.9003^{20}
MW: 88.11   **BP[°C]:** 77.1 n_D: 1.3723^{20}
Sol.: H_2O 3; EtOH 5; eth
    5; ace 4; bz 4; chl 5;
    os 4
Crit. Const.: T_c = 250.2°C; P_c = 3.882 MPa; V_c = 286 cm^3
Vap. Press. [kPa]: 3.25^0; 12.6^{25}; 37.9^{50}; 94.5^{75}; 204^{100}
Therm. Prop.[kJ/mol]: $\Delta_{fus}H$ = 10.48; $\Delta_{vap}H$ = 31.9; $\Delta_f H°$(l, 25°C) = -479.3
C_p **(liq.) [J/mol °C]:** 170.6^{25}
γ **[mN/m]:** 23.39^{25}; dγ/dT = 0.1161 mN/m °C
Elec. Prop.: μ = 1.78 D; ε = 6.08^{20}; IP = 10.01 eV
η **[mPa s]:** 0.578^0; 0.423^{25}; 0.325^{50}; 0.259^{75}
k **[W/m °C]:** 0.162^{-25}; 0.153^0; 0.144^{25}; 0.135^{50}; 0.126^{75}
MS: 43(100) 29(46) 27(33) 45(32) 61(28) 28(25) 42(18) 73(11) 88(10) 70(10)
IR [cm^{-1}]: 2980 1740 1480 1460 1440 1370 1300 1230 1100 1050 940 850 780
Raman [cm^{-1}]: 2970 2940 2880 2780 2720 1730 1450 1400 1350 1300 1270
    1240 1170 1110 1090 1040 1000 940 920 840 780 630 460 440 380 310 200
UV [nm]: 209(72) MeOH
^{1}H NMR [ppm]: 1.3 2.0 4.1 CDCl$_3$
Flammability: Flash pt. = -4°C; ign. temp. = 426°C; flam. lim. = 2.0-11.5%
TLV/TWA: 400 ppm (1440 mg/m^3)
Reg. Lists: CERCLA (RQ = 5000 lb.); RCRA U112 (toxic)

233. Ethyl acetoacetate

Syst. Name: Butanoic acid, 3-oxo-, ethyl ester
Synonyms: Ethyl 3-oxobutanoate

CASRN: 141-97-9   **Beil Ref.**: 4-03-00-01528
Merck No.: 3714
Beil RN: 385838
MF: $C_6H_{10}O_3$ **MP[°C]**: -45   **Den.[g/cm³]**: 1.0368^{10}
MW: 130.14   **BP[°C]**: 180.8 n_D: 1.4171^{20}
Sol.: H_2O 4; EtOH 5; eth
   5; bz 3; ctc 2; chl 3
Vap. Press. [kPa]: 0.095^{25}; 0.494^{50}; 1.95^{75}; 6.26^{100}; 17.0^{125}; 40.5^{150}
C_p **(liq.) [J/mol °C]**: 248.0^{25}
γ **[mN/m]**: 32.4^{15}
Elec. Prop.: $\varepsilon = 14.0^{20}$
η **[mPa s]**: 1.51^{25}
MS: 43(100) 29(24) 88(18) 28(16) 85(14) 27(12) 42(11) 60(9) 130(6) 45(6)
IR [cm⁻¹]: 2990 2960 2880 1740 1720 1640 1400 1370 1320 1260 1190 1150
   1040 550
UV [nm]: 243 cyhex
¹H NMR [ppm]: 1.3 1.9 2.2 3.3 4.1 4.9 $CDCl_3$
Flammability: Flash pt. = 57°C; ign. temp. = 295°C; flam. lim. = 1.4-9.5%

234. Ethyl acrylate

Syst. Name: 2-Propenoic acid, ethyl ester
Synonyms: Ethyl propenoate

CASRN: 140-88-5   **DOT No.:** 1917
Merck No.: 3715   **Beil Ref.:** 4-02-00-01460
Beil RN: 773866
MF: $C_5H_8O_2$ **MP[°C]:** -71.2   **Den.[g/cm^3]:** 0.9234^{20}
MW: 100.12   **BP[°C]:** 99.4 n_D: 1.4068^{20}
Sol.: H_2O 2; EtOH 5; eth 5; chl 3; DMSO-d_6 2
Vap. Press. [kPa]: 0.193^{-25}; 1.20^0; 5.14^{25}; 16.8^{50}; 44.8^{75}; 103^{100}
Therm. Prop.[kJ/mol]: $\Delta_{vap}H$ = 34.7
Elec. Prop.: μ = (2.0 D); ε = 6.05^{30}; IP = 10.30 eV
MS: 55(100) 27(32) 29(15) 56(12) 45(9) 73(8) 28(8) 26(6) 99(5) 85(5)
IR [cm^{-1}]: 3040 2980 2940 2900 1730 1640 1620 1460 1440 1410 1290 1270 1190 1060 1030 980 960 810
UV [nm]: 208(6918) EtOH
^{1}H NMR [ppm]: 1.3 4.1 5.7 6.1 6.3 CCl$_4$
Flammability: Flash pt. = 10°C; ign. temp. = 372°C; flam. lim. = 1.4-14%
TLV/TWA: 5 ppm (20 mg/m^3)
Reg. Lists: CERCLA (RQ = 1000 lb.); SARA 313 (0.1%); RCRA U113 (toxic); confirmed or suspected carcinogen

235. Ethylamine

Syst. Name: Ethanamine

CASRN: 75-04-7   **DOT No.**: 1036
Merck No.: 3718   **Beil Ref.**: 4-04-00-00307
Beil RN: 505933
MF: C_2H_7N **MP[°C]**: -80.5   **Den.[g/cm^3]**: 0.677^{25} (sat. press.)
MW: 45.08   **BP[°C]**: 16.5 n_D: 1.3663^{20}
Sol.: H_2O 5; EtOH 5; eth
  5
Crit. Const.: T_c = 183°C; P_c = 5.62 MPa; V_c = 182 cm^3
Vap. Press. [kPa]: 2.46^{-50}; 13.1^{-25}; 48.9^0
Therm. Prop.[kJ/mol]: $\Delta_f H°$(g, 25°C) = -47.5
C_p **(liq.) [J/mol °C]**: 130.0^{25} (sat. press.)
γ **[mN/m]**: 19.20^{25} (at saturation press.); $d\gamma/dT$ = 0.1372 mN/m °C
Elec. Prop.: μ = 1.22 D; ε = 8.7^0; IP = 8.86 eV
MS: 30(100) 28(32) 44(20) 45(19) 27(13) 15(10) 42(9) 29(8) 41(5) 40(5)
IR [cm^{-1}]: 3330 3200 2900 1620 1460 1400 1100 1060 780
UV [nm]: 213(794) 177(1585) gas
^{13}**C NMR [ppm]**: 19.0 36.9
^{1}**H NMR [ppm]**: 1.1 2.6 D_2O
Flammability: Flash pt. < -18°C; ign. temp. = 385°C; flam. lim. = 3.5-14%
TLV/TWA: 5 ppm (9.2 mg/m^3)
Reg. Lists: CERCLA (RQ = 100 lb.)

236. Ethylbenzene

Syst. Name: Benzene, ethyl-
Synonyms: Phenylethane

CASRN: 100-41-4   **DOT No.:** 1175
Merck No.: 3723   **Beil Ref.:** 4-05-00-00885
Beil RN: 1901871
MF: C_8H_{10} **MP[°C]:** -94.9   **Den.[g/cm³]:** 0.8670^{20}
MW: 106.17   **BP[°C]:** 136.1 n_D: 1.4959^{20}
Sol.: H_2O 1; EtOH 5; eth
   5; chl 2
Crit. Const.: T_c = 344.00°C; P_c = 3.609 MPa; V_c = 374 cm³
Vap. Press. [kPa]: 1.28^{25}; 4.70^{50}; 13.8^{75}; 34.2^{100}; 74.3^{125}
Therm. Prop.[kJ/mol]: $\Delta_{fus}H$ = 9.18; $\Delta_{vap}H$ = 35.6; $\Delta_f H°$(l, 25°C) = -12.3
C_p **(liq.) [J/mol °C]:** 183.2^{25}
γ **[mN/m]:** 28.75^{25}; $d\gamma/dT$ = 0.1094 mN/m °C
Elec. Prop.: μ = 0.59 D; ε = 2.45^{20}; IP = 8.77 eV
η **[mPa s]:** 0.872^{0}; 0.631^{25}; 0.482^{50}; 0.380^{75}; 0.304^{100}
k **[W/m °C]:** 0.130^{25}; 0.124^{50}; 0.118^{75}; 0.112^{100}
MS: 91(100) 106(31) 51(14) 39(10) 77(8) 65(8) 105(7) 92(7) 78(7) 27(6)
IR [cm⁻¹]: 2940 1590 1490 1450 1370 1040 970 910 750 710
Raman [cm⁻¹]: 3060 3010 2970 2740 2900 2880 2730 1600 1580 1450 1320
   1200 1180 1150 1060 1030 1000 960 900 840 770 620 550 480 400 300 150
UV [nm]: 270(142) 261(200) 255(168) 208(7520) MeOH
¹³C NMR [ppm]: 15.7 29.1 125.8 128.0 128.4 144.2 diox
¹H NMR [ppm]: 1.3 2.7 7.2 CDCl₃
Flammability: Flash pt. = 21°C; ign. temp. = 432°C; flam. lim. = 0.8-6.7%
TLV/TWA: 100 ppm (434 mg/m³)
Reg. Lists: CERCLA (RQ = 1000 lb.); SARA 313 (1.0%)

237. Ethyl benzoate

Syst. Name: Benzoic acid, ethyl ester
Synonyms: Ethyl benzenecarboxylate

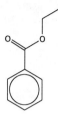

CASRN: 93-89-0   **Beil Ref.:** 4-09-00-00285
Merck No.: 3725
Beil RN: 1908172
MF: $C_9H_{10}O_2$ **MP[°C]:** -34   **Den.[g/cm³]:** 1.0511^{15}
MW: 150.18   **BP[°C]:** 212 n_D: 1.5007^{20}
Sol.: H_2O 1; EtOH 3; eth
   5; ace 3; bz 3; ctc 2;
   chl 3; peth 3
Vap. Press. [kPa]: 0.774^{75}; 2.51^{100}; 6.91^{125}; 16.7^{150}
C_p **(liq.) [J/mol °C]:** 246.0^{25}
γ **[mN/m]:** 35.4^{20}
Elec. Prop.: $\mu = 2.00$ D; $\varepsilon = 6.20^{20}$; IP = 8.90 eV
η **[mPa s]:** 2.41^{15}
MS: 105(100) 77(65) 122(34) 51(34) 27(17) 150(16) 29(14) 50(13) 106(12)
   78(6)
IR [cm⁻¹]: 3030 1750 1610 1450 1300 1190 1020 930 790 720 690
Raman [cm⁻¹]: 3070 2970 2930 2900 2870 1710 1600 1490 1450 1390 1360
   1310 1280 1170 1150 1100 1020 1000 870 850 810 770 670 610 500 400 320
   190
UV [nm]: 280(746) 273(886) 228(12600) MeOH
¹H NMR [ppm]: 1.3 4.3 7.4 8.0 CCl_4
Flammability: Flash pt. = 88°C; ign. temp. = 490°C

238. Ethyl butanoate

Syst. Name: Butanoic acid, ethyl ester
Synonyms: Ethyl butyrate

CASRN: 105-54-4 **DOT No.**: 1180
Merck No.: 3733 **Beil Ref.**: 4-02-00-00787
Beil RN: 506331
MF: $C_6H_{12}O_2$ **MP[°C]**: -98 **Den.[g/cm^3]**: 0.8844^{15}
MW: 116.16 **BP[°C]**: 121.5 n_D: 1.4000^{20}
Sol.: H_2O 2; EtOH 3; eth
3; ctc 2
Crit. Const.: $T_c = 293$°C; $P_c = 3.06$ MPa; $V_c = 421$ cm^3
Vap. Press. [kPa]: 2.01^{25}; 7.84^{50}; 22.9^{75}; 54.4^{100}; 111^{125}; 202^{150}
Therm. Prop.[kJ/mol]: $\Delta_{vap}H = 35.5$
C_p **(liq.) [J/mol °C]**: 228.0^{25}
γ **[mN/m]**: 23.94^{25}; $d\gamma/dT = 0.1045$ mN/m °C
Elec. Prop.: $\mu = (1.7$ D); $\varepsilon = 5.18^{28}$
η **[mPa s]**: 0.639^{25}; 0.453^{50}
MS: 43(100) 71(88) 29(83) 27(43) 88(40) 41(28) 60(22) 45(20) 73(17) 42(17)
IR [cm^{-1}]: 2940 1720 1470 1370 1320 1250 1190 1100 1030 930 850 790 780
750
Raman [cm^{-1}]: 2970 2940 2880 2740 1730 1450 1420 1390 1340 1300 1270
1180 1110 1090 1040 1030 920 890 870 850 800 780 630 600 470 380 320
^{1}**H NMR [ppm]**: 0.9 1.2 1.7 2.2 4.1 CCl$_4$
Flammability: Flash pt. = 24°C; ign. temp. = 463°C

239. 2-Ethyl-1-butanol

Syst. Name: 1-Butanol, 2-ethyl-
Synonyms: 2-Ethylbutyl alcohol

CASRN: 97-95-0   **DOT No.:** 2275
Beil RN: 1731254   **Beil Ref.:** 4-01-00-01725
MF: $C_6H_{14}O$ **MP[°C]:** <-15   **Den.[g/cm³]:** 0.8326[20]
MW: 102.18   **BP[°C]:** 147 n_D: 1.4220[20]
Sol.: H_2O 2; EtOH 3; eth
   3; chl 3
Vap. Press. [kPa]: 0.206[25]; 1.35[50]; 5.86[75]; 19.1[100]; 50.2[125]; 113[150]
Therm. Prop.[kJ/mol]: $\Delta_{vap}H$ = 43.2
γ **[mN/m]:** 24.32[25]
Elec. Prop.: ϵ = 6.19[89]
η **[mPa s]:** 5.89[25]
MS: 43(100) 70(44) 55(38) 71(37) 41(34) 29(32) 31(23) 27(21) 84(17) 56(16)
IR [cm⁻¹]: 3320 2950 2900 2850 1440 1370 1200 1090 1040 1010 940 880 820
   750
Raman [cm⁻¹]: 2960 2930 2870 2730 1450 1350 1300 1280 1150 1120 1040
   1030 1010 950 930 900 880 820 750 610 550 500 480 430 400 310 210
¹³C NMR [ppm]: 11.1 23.0 43.6 64.6 CDCl₃
¹H NMR [ppm]: 0.9 1.3 1.6 3.5 CDCl₃
Flammability: Flash pt. = 57°C

240. Ethyl butyl ketone

Syst. Name: 3-Heptanone
Synonyms: Butyl ethyl ketone

CASRN: 106-35-4   **DOT No.:** 1224
Beil RN: 506161   **Beil Ref.:** 4-01-00-03321
MF: $C_7H_{14}O$ **MP[°C]:** -39   **Den.[g/cm^3]:** 0.8183^{20}
MW: 114.19   **BP[°C]:** 147 n_D: 1.4057^{20}
Sol.: H_2O 1; EtOH 5; eth
    5; ctc 2
Vap. Press. [kPa]: 6.98^{75}; 19.8^{100}; 49.1^{125}; 110^{150}
γ **[mN/m]:** 26.30^{20}
Elec. Prop.: μ = (2.8 D); ε = 12.7^{20}
MS: 57(100) 29(76) 41(32) 85(29) 72(22) 43(15) 39(11) 114(10) 27(7) 55(5)
IR [cm^{-1}]: 2960 2870 1720 1460 1410 1370 1140
Raman [cm^{-1}]: 2950 2910 2890 2820 2750 1710 1450 1420 1300 1260 1210
    1110 1060 1030 970 900 880 820 790 760 740 640 530 390 340 300
UV [nm]: 282 cyhex
^{13}C NMR [ppm]: 7.8 13.8 22.5 26.2 35.8 42.1 211.2 CDCl$_3$
^{1}H NMR [ppm]: 1.0 1.4 2.3 CCl$_4$
Flammability: Flash pt. = 46°C
TLV/TWA: 50 ppm (234 mg/m^3)

241. Ethyl *trans*-cinnamate

Syst. Name: 2-Propenoic acid, 3-phenyl-, ethyl ester,
   (*E*)-
Synonyms: Ethyl trans-3-benzenepropenoate

CASRN: 4192-77-2   **Beil Ref.**: 4-09-00-02006
Merck No.: 2300
Beil RN: 775541
MF: $C_{11}H_{12}O_2$ **MP[°C]**: 10   **Den.[g/cm^3]**: 1.0491^{20}
MW: 176.22   **BP[°C]**: 271.5 n_D: 1.5598^{20}
Sol.: H_2O 1; EtOH 4; eth
   4; ace 4; bz 3; ctc 3;
   os 4
γ [mN/m]: 36.5^{25}
Elec. Prop.: μ = (1.8 D); ε = 5.63^{20}
η [mPa s]: 8.7^{20}
MS: 131(100) 103(38) 176(37) 77(24) 148(12) 51(12) 104(12) 132(12)
IR [cm^{-1}]: 3050 3010 2980 2880 1710 1640 1570 1490 1450 1360 1310 1270
   1200 1170 1040 980 860 770 710 680 570 490
Raman [cm^{-1}]: 3060 2980 2930 1700 1640 1600 1570 1490 1450 1390 1370
   1330 1310 1260 1200 1170 1110 1030 1000 880 860 840 760 730 710 610
   590 570 390 270 230 180
UV [nm]: 276(55500) 222(35800) 216(42700) MeOH
^{1}H NMR [ppm]: 1.3 4.2 6.3 7.4 7.7 CCl_4

242. Ethyl cyanoacetate

Syst. Name: Ethanoic acid, cyano-, ethyl ester

CASRN: 105-56-6   **DOT No.**: 2666
Merck No.: 3744   **Beil Ref.**: 4-02-00-01889
Beil RN: 605871
MF: $C_5H_7NO_2$ **MP[°C]**: -22.5   **Den.[g/cm^3]**: 1.0654^{20}
MW: 113.12   **BP[°C]**: 205 n_D: 1.4175^{20}
Sol.: eth 4; EtOH 4
Vap. Press. [kPa]: 0.177^{75}; 0.885^{100}; 3.51^{125}; 11.6^{150}
C_p **(liq.) [J/mol °C]**: 220.2^{25}
γ **[mN/m]**: 36.48^{20}
Elec. Prop.: $\mu = (2.2\ D)$; $\epsilon = 31.62^{-10}$
η **[mPa s]**: 5.06^{0}; 3.26^{15}; 2.50^{25}
MS: 29(100) 68(59) 27(34) 40(21) 28(16) 41(14) 15(13) 45(10) 43(9) 26(9)
IR [cm^{-1}]: 2980 2940 2260 1750 1470 1450 1400 1370 1340 1260 1200 1100
1030 930 850
Raman [cm^{-1}]: 2970 2940 2870 2770 2730 2260 1740 1450 1400 1370 1330
1300 1270 1210 1170 1110 1100 1020 970 920 850 800 780 730 680 590 490
460 400 360 300 210
^{1}H NMR [ppm]: 1.3 3.4 4.3 CCl$_4$
Flammability: Flash pt. = 110°C

243. Ethylcyclohexane
Syst. Name: Cyclohexane, ethyl-

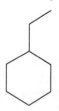

CASRN: 1678-91-7   **Beil Ref.:** 4-05-00-00115
Beil RN: 1900337
MF: C_8H_{16} **MP[°C]:** -111.3   **Den.[g/cm³]:** 0.7880^{20}
MW: 112.22   **BP[°C]:** 131.9 n_D: 1.4330^{20}
Sol.: H_2O 1; EtOH 3; eth
  3; ace 3; bz 3; ctc 5;
  lig 4
Vap. Press. [kPa]: 1.71^{25}; 5.96^{50}; 16.8^{75}; 40.1^{100}; 84.3^{125}
Therm. Prop.[kJ/mol]: $\Delta_{fus}H$ = 8.33; $\Delta_{vap}H$ = 34.0; $\Delta_f H°$(l, 25°C) = -211.9
C_p **(liq.) [J/mol °C]:** 211.8^{25}
γ [mN/m]: 25.15^{25}; dγ/dT = 0.1054 mN/m °C
Elec. Prop.: μ = 0; ε = 2.05; IP = 9.54 eV
η [mPa s]: 1.14^{0}; 0.784^{25}; 0.579^{50}
MS: 83(100) 55(65) 82(42) 41(36) 112(23) 56(11) 67(10) 39(10) 42(9) 84(8)
IR [cm⁻¹]: 2860 1450 1370 980 890 830 790 760
Raman [cm⁻¹]: 2957 2934 2915 2873 2851 2841 1442 1258 1169 1032
¹³C NMR [ppm]: 12.1 27.5 31.0 33.9 40.6
¹H NMR [ppm]: 0.9 1.9 1.4 CDCl₄
Flammability: Flash pt. = 35°C; ign. temp. = 238°C; flam. lim. = 0.9-6.6%

244. Ethylene carbonate

Syst. Name: 1,3-Dioxolan-2-one
Synonyms: Vinylene carbonate

CASRN: 96-49-1   **Beil Ref.**: 5-19-04-00006
Beil RN: 106249
MF: $C_3H_4O_3$ **MP[°C]**: 36.4   **Den.[g/cm^3]**: 1.3214^{39}
MW: 88.06   **BP[°C]**: 248 n_D: 1.4148^{50}
Sol.: H_2O 5; EtOH 5; eth
   5; bz 5; chl 5; AcOEt
   5; HOAc 5
Vap. Press. [kPa]: 0.003^{25}
Therm. Prop.[kJ/mol]: $\Delta_{fus}H$ = 13.19; $\Delta_f H°$(s, 25°C) = -581.6
C_p **(liq.) [J/mol °C]**: 133.9^{50}
Elec. Prop.: μ = (4.9 D); ε = 89.78^{40}
η **[mPa s]**: 1.93^{40}
MS: 29(100) 44(62) 43(54) 88(40) 30(16) 28(11) 45(7) 58(6) 42(6) 73(4)
IR [cm^{-1}]: 2985 2941 1786 1739 1481 1389 1163 1070 971 893 775 719
Raman [cm^{-1}]: 3050 3000 2940 1790 1760 1480 1370 1230 1160 1060 980 900
   720 690 220
^{1}H NMR [ppm]: 4.5 CDCl$_3$
Flammability: Flash pt. = 143°C

245. Ethylene chlorohydrin

Syst. Name: Ethanol, 2-chloro-

Synonyms: 2-Chloroethanol

CASRN: 107-07-3

Merck No.: 3750

Beil RN: 878139

MF: C_2H_5ClO

MW: 80.51

DOT No.: 1135

Beil Ref.: 4-01-00-01372

MP[°C]: -67.5

BP[°C]: 128.6

Den.[g/cm³]: 1.2019^{20}

n_D: 1.4419^{20}

Sol.: H_2O 5; EtOH 5; eth 2; chl 3; os 5

Vap. Press. [kPa]: 4.45^{50}; 14.2^{75}; 38.6^{100}; 92.7^{125}

Therm. Prop.[kJ/mol]: $\Delta_{vap}H = 41.4$; $\Delta_f H°(l, 25°C) = -295.4$

γ **[mN/m]**: 38.9^{20}

Elec. Prop.: $\mu = 1.78$ D; $\varepsilon = 25.80^{20}$; IP = 10.52 eV

η **[mPa s]**: 3.91^{15}

MS: 31(100) 15(13) 29(10) 28(10) 27(9) 43(8) 44(7) 26(5) 18(5) 14(5)

IR [cm⁻¹]: 3330 2940 2860 1430 1300 1160 1080 1030 940 850 750

Raman [cm⁻¹]: 3000 2960 2870 2730 1450 1430 1400 1370 1290 1240 1180 1080 1030 930 850 740   660 470 390 290

¹³C NMR [ppm]: 46.6 62.9

¹H NMR [ppm]: 2.8 3.7 3.8 CDCl₃

Flammability: Flash pt. = 41°C

TLV/TWA: 1 ppm (3.3 mg/m³)

246. Ethylenediamine

Syst. Name: 1,2-Ethanediamine

CASRN: 107-15-3   **DOT No.**: 1604
Merck No.: 3752   **Beil Ref.**: 4-04-00-01166
Beil RN: 605263
MF: $C_2H_8N_2$ **MP[°C]**: 11.1   **Den.[g/cm^3]**: 0.8979^{20}
MW: 60.10   **BP[°C]**: 117 n_D: 1.4565^{20}
Sol.: H_2O 4; EtOH 5; eth
    1; bz 1; ctc 3
Vap. Press. [kPa]: 1.62^{25}; 6.77^{50}; 21.8^{75}; 57.4^{100}
Therm. Prop.[kJ/mol]: $\Delta_{fus}H$ = 22.58; $\Delta_{vap}H$ = 38.0; $\Delta_f H°$(l, 25°C) = -63.0
C_p **(liq.) [J/mol °C]**: 172.6^{25}
γ **[mN/m]**: 40.77^{20}
Elec. Prop.: μ = 1.99 D; ε = 13.82^{20}; IP = 8.60 eV
η **[mPa s]**: 1.54^{25}
MS: 30(100) 18(13) 42(6) 43(5) 27(5) 44(4) 29(4) 17(4) 15(4) 41(3)
IR [cm^{-1}]: 3390 3300 2920 2850 1600 1450 1360 1090 1050 810 770 500
^{13}C NMR [ppm]: 45.8
^{1}H NMR [ppm]: 1.2 2.6 CCl_4
Flammability: Flash pt. = 40°C; ign. temp. = 385°C; flam. lim. = 2.5-12.0%
TLV/TWA: 10 ppm (25 mg/m^3)
Reg. Lists: CERCLA (RQ = 5000 lb.)

247. Ethylene glycol

Syst. Name: 1,2-Ethanediol

HO⌒⌒OH

CASRN: 107-21-1	**DOT No.:** 1142
Merck No.: 3755	**Beil Ref.:** 4-01-00-02369

Beil RN: 505945

MF: $C_2H_6O_2$ **MP[°C]:** -13   **Den.[g/cm^3]:** 1.1088^{20}

MW: 62.07   **BP[°C]:** 197.3 n_D: 1.4318^{20}

Sol.: H_2O 5; EtOH 5; eth 3; ace 5; bz 2; chl 3; HOAc 5

Crit. Const.: $T_c = 445°C$

Vap. Press. [kPa]: 0.010^{25}; 0.092^{50}; 0.512^{75}; 2.14^{100}; 7.18^{125}; 20.2^{150}

Therm. Prop.[kJ/mol]: $\Delta_{fus}H = 11.23$; $\Delta_{vap}H = 50.5$; $\Delta_f H°(l, 25°C) = -455.3$

C_p (liq.) [J/mol °C]: 149.3^{25}

γ [mN/m]: 47.99^{25}; $d\gamma/dT = 0.0890$ mN/m °C

Elec. Prop.: $\mu = 2.28$ D; $\varepsilon = 41.4^{20}$; IP = 10.16 eV

η [mPa s]: 16.1^{25}; 6.55^{50}; 3.34^{75}; 1.98^{100}

k [W/m °C]: 0.256^{25}

MS: 31(100) 33(35) 29(13) 32(11) 43(6) 27(5) 28(4) 62(3) 30(3) 44(2)

IR [cm^{-1}]: 3350 2940 2860 1650 1450 1400 1330 1250 1190 1080 1030 870 850 640 520

Raman [cm^{-1}]: 3300 2930 2880 2720 1450 1400 1270 1210 1090 1060 1030 860 520 480 340

^{13}C NMR [ppm]: 63.4

^{1}H NMR [ppm]: 3.7 D_2O

Flammability: Flash pt. = 111°C; ign. temp. = 398°C; flam. lim. = 3.2-22%

TLV/TWA: 50 ppm (127 mg/m^3)

Reg. Lists: CERCLA (RQ = 1 lb.); SARA 313 (1.0%)

248. Ethylene glycol diacetate

Syst. Name: 1,2-Ethanediol, diethanoate
Synonyms: 1,2-Diacetoxyethane

CASRN: 111-55-7   **Beil Ref.:** 4-02-00-00217
Merck No.: 3756
Beil RN: 1762308
MF: $C_6H_{10}O_4$ **MP[°C]:** -31   **Den.[g/cm^3]:** 1.1043^{20}
MW: 146.14   **BP[°C]:** 190 n_D: 1.4159^{20}
Sol.: H_2O 4; EtOH 5; eth
   5; ace 5; bz 5; ctc 5;
   CS_2 5; HOAc 5
Vap. Press. [kPa]: 0.030^{25}; 0.275^{50}; 1.19^{75}; 4.07^{100}; 11.7^{125}; 29.2^{150}
Therm. Prop.[kJ/mol]: $\Delta_{vap}H = 45.5$
C_p **(liq.) [J/mol °C]:** 310^{25}
Elec. Prop.: $\mu = (2.3$ D); $\varepsilon = 7.7^{17}$
η **[mPa s]:** 2.9^{20}
MS: 43(100) 86(11) 42(7) 15(7) 116(4) 73(4) 44(3) 29(3) 103(2) 45(2)
IR [cm^{-1}]: 3030 2940 1790 1450 1370 1240 1090 1050 1010 960 890
Raman [cm^{-1}]: 2940 2850 2790 2710 1730 1440 1400 1370 1280 1250 1210
   1120 1070 1050 1000 930 910 860 810 630 600 500 420 350 290 250 220
^{1}H NMR [ppm]: 2.0 4.2 CCl$_4$
Flammability: Flash pt. = 88°C; ign. temp. = 482°C; flam. lim. = 1.6-8.4%

---

249. Ethylene glycol dibutyl ether

Syst. Name: Ethane, 1,2-dibutoxy
Synonyms: 1,2-Dibutoxyethane; Dibutyl cellosolve

CASRN: 112-48-1
MF: $C_{10}H_{22}O_2$ **MP[°C]:** -69.1   **Den.[g/cm^3]:** 0.8319^{25}
MW: 174.28   **BP[°C]:** 203.3
Vap. Press. [kPa]: 0.821^{75}; 3.11^{100}; 9.22^{125}; 22.7^{150}; 48.7^{175}
C_p **(liq.) [J/mol °C]:** 350^{20}
η **[mPa s]:** 1.3^{20}
Flammability: Flash pt. = 85°C

250. Ethylene glycol diethyl ether

Syst. Name: Ethane, 1,2-diethoxy-
Synonyms: 1,2-Diethoxyethane

CASRN: 629-14-1	**DOT No.**: 1153
Beil RN: 1732917	**Beil Ref.**: 4-01-00-02379

MF: $C_6H_{14}O_2$ **MP[°C]**: -74   **Den.[g/cm^3]**: 0.8484^{20}
MW: 118.18   **BP[°C]**: 119.4 n_D: 1.3860^{20}
Sol.: ace 4; bz 4; eth 4;
   EtOH 4
Vap. Press. [kPa]: 0.226^{-25}; 1.18^0; 4.33^{25}; 12.5^{50}; 29.8^{75}; 62.0^{100}; 116^{125}
Therm. Prop.[kJ/mol]: $\Delta_{vap}H$ = 36.3; $\Delta_fH°$(l, 25°C) = -451.4
C_p **(liq.) [J/mol °C]**: 259.4^{25}
Elec. Prop.: $\varepsilon = 3.90^{20}$
η **[mPa s]**: 0.7^{20}
MS: 31(100) 59(71) 29(58) 45(43) 27(33) 74(27) 43(15) 15(14) 28(12) 44(10)
IR [cm^{-1}]: 3030 2940 1450 1390 1350 1250 1120 1050 940 910 840
Raman [cm^{-1}]: * 1580 1455 1396 1271 1139 1101 923 860 816 538 361
Flammability: Flash pt. = 35°C

251. Ethylene glycol dimethyl ether

Syst. Name: Ethane, 1,2-dimethoxy-
Synonyms: 1,2-Dimethoxyethane; Dimethyl
  cellosolve

CASRN: 110-71-4   **Beil Ref.**: 4-01-00-02376
Merck No.: 3213
Beil RN: 1209237
MF: $C_4H_{10}O_2$ **MP[°C]**: -58   **Den.[g/cm^3]**: 0.8691^{20}
MW: 90.12   **BP[°C]**: 85 n_D: 1.3796^{20}
Sol.: H_2O 3; EtOH 3; eth
  3; ace 3; bz 3; ctc 3;
  chl 3
Crit. Const.: $T_c = 263°C$; $P_c = 3.87$ MPa; $V_c = 271$ cm^3
Vap. Press. [kPa]: 2.62^0; 9.93^{25}; 29.4^{50}; 72.3^{75}; 155^{100}
Therm. Prop.[kJ/mol]: $\Delta_{fus}H = 12.60$; $\Delta_{vap}H = 32.4$; $\Delta_f H°(l, 25°C) = -376.6$
C_p **(liq.) [J/mol °C]**: 193.3^{25}
γ **[mN/m]**: 24.61^{20}; $d\gamma/dT = 0.115$ mN/m °C
Elec. Prop.: $\varepsilon = 7.30^{24}$; IP = 9.30 eV
η **[mPa s]**: 0.455^{25}
MS: 45(100) 60(13) 29(13) 90(7) 58(6) 31(5) 28(5) 43(4) 59(3) 46(2)
IR [cm^{-1}]: 3030 2940 1450 1370 1250 1190 1120 1030 990 940 860
^{13}C NMR [ppm]: 58.6 72.3 diox
^{1}H NMR [ppm]: 3.3 3.4 CCl$_4$
Flammability: Flash pt. = -2°C; ign. temp. = 202°C

252. Ethylene glycol ethyl ether acetate
Syst. Name: Ethanol, 2-ethoxy-, ethanoate
Synonyms: 2-Ethoxyethyl acetate; Cellosolve acetate

CASRN: 111-15-9   **DOT No.**: 1172
Merck No.: 3708   **Beil Ref.**: 4-02-00-00214
Beil RN: 1748677
MF: $C_6H_{12}O_3$ **MP[°C]**: -61.7   **Den.[g/cm^3]**: 0.9740^{20}
MW: 132.16   **BP[°C]**: 156.4 n_D: 1.4054^{20}
Sol.: H_2O 4; ace 4; eth 4;
   EtOH 4
Crit. Const.: $T_c = 334.2$°C; $P_c = 3.166$ MPa; $V_c = 443$ cm^3
Vap. Press. [kPa]: 0.240^{25}; 1.40^{50}; 5.50^{75}; 16.5^{100}; 40.5^{125}; 85.8^{150}
C_p **(liq.) [J/mol °C]**: 376.0^{25}
γ **[mN/m]**: 31.8^{25}
Elec. Prop.: $\mu = (2.2$ D$)$; $\varepsilon = 7.57^{30}$
η **[mPa s]**: 1.02^{25}
MS: 43(100) 31(34) 59(31) 72(28) 44(25) 29(24) 45(12) 27(11) 15(11) 87(7)
IR [cm^{-1}]: 2940 2860 1750 1450 1390 1250 1140 1060 1010 960 860
UV [nm]: 214(83) EtOH
^{1}H NMR [ppm]: 1.2 2.0 3.4 3.5 4.1 CCl_4
Flammability: Flash pt. = 56°C; ign. temp. = 379°C; flam. lim. = 2-8%
TLV/TWA: 5 ppm (27 mg/m^3)

253. Ethylene glycol momomethyl ether acetate

Syst. Name: Ethanol, 2-methoxy-, ethanoate
Synonyms: 2-Methoxyethyl acetate; Methyl
Cellosolve acetate

CASRN: 110-49-6   **DOT No.:** 1189
Merck No.: 5962   **Beil Ref.:** 4-02-00-00214
Beil RN: 1700761
MF: $C_5H_{10}O_3$ **MP[°C]:** -70   **Den.[g/cm^3]:** 1.0074^{19}
MW: 118.13   **BP[°C]:** 143 n_D: 1.4002^{20}
Sol.: H_2O 4; EtOH 4
Vap. Press. [kPa]: 0.670^{25}; 8.25^{75}; 23.0^{100}; 55.3^{125}; 119^{150}
Therm. Prop.[kJ/mol]: $\Delta_{vap}H = 43.9$
C_p **(liq.) [J/mol °C]:** 310.0^{25}
Elec. Prop.: $\mu = (2.1$ D); $\varepsilon = 8.25^{20}$
MS: 43(100) 45(48) 58(42) 29(10) 42(4) 31(4) 73(3) 27(3) 59(2) 26(2)
Flammability: Flash pt. = 49°C; ign. temp. = 392°C; flam. lim. = 1.5-12.3%
TLV/TWA: 5 ppm (24 mg/m^3)

254. Ethylene glycol monobutyl ether

Syst. Name: Ethanol, 2-butoxy-
Synonyms: Butyl cellosolve; 2-Butoxyethanol

CASRN: 111-76-2   **DOT No.:** 2369
Merck No.: 1559   **Beil Ref.:** 4-01-00-02380
Beil RN: 1732511
MF: $C_6H_{14}O_2$ **MP[°C]:** -74.8   **Den.[g/cm^3]:** 0.9015^{20}
MW: 118.18   **BP[°C]:** 168.4 n_D: 1.4198^{20}
Sol.: H_2O 5; EtOH 5; eth
     5; ctc 2
Crit. Const.: T_c = 360.8°C; V_c = 424 cm^3
Vap. Press. [kPa]: 0.150^{25}; 8.77^{100}; 23.3^{125}; 54.4^{150}
C_p **(liq.) [J/mol °C]:** 281.0^{25}
γ **[mN/m]:** 26.14^{25}; dγ/dT = 0.0816 mN/m °C
Elec. Prop.: μ = (2.1 D); ε = 9.30^{25}
η **[mPa s]:** 3.15^{25}
MS: 57(100) 45(38) 29(35) 41(31) 87(16) 27(12) 56(11) 31(9) 75(7) 28(7)
IR [cm^{-1}]: 3450 2940 2860 1450 1350 1110 1050 890
Raman [cm^{-1}]: 2940 2910 2870 2800 2790 2740 1480 1450 1360 1300 1230
     1150 1130 1050 1030 880 830 530 450 390 320 300
^{13}C NMR [ppm]: 13.9 19.3 31.8 61.6 71.1 72.2 CDCl$_3$
^{1}H NMR [ppm]: 0.9 1.3 3.3 3.7 CCl$_4$
Flammability: Flash pt. = 69°C; ign. temp. = 238°C; flam. lim. = 4-13%
TLV/TWA: 25 ppm (121 mg/m^3)

255. Ethylene glycol monoethyl ether

Syst. Name: Ethanol, 2-ethoxy-
Synonyms: Cellosolve; 2-Ethoxyethanol

CASRN: 110-80-5   **DOT No.**: 1171
Merck No.: 3707   **Beil Ref.**: 4-01-00-02377
Beil RN: 1098271
MF: $C_4H_{10}O_2$ **MP[°C]**: -70   **Den.[g/cm^3]**: 0.9297^{20}
MW: 90.12   **BP[°C]**: 135 n_D: 1.4080^{20}
Sol.: H_2O 4; ace 4; eth 4;
   EtOH 4
Vap. Press. [kPa]: 0.710^{25}; 10.7^{75}; 29.8^{100}; 72.1^{125}
Therm. Prop.[kJ/mol]: $\Delta_{vap}H = 39.2$
C_p **(liq.) [J/mol °C]**: 210.8^{25}
γ **[mN/m]**: 28.35^{25}; $d\gamma/dT = 0.0897$ mN/m °C
Elec. Prop.: $\mu = (2.1$ D); $\varepsilon = 13.38^{25}$; IP = 9.60 eV
η **[mPa s]**: 1.85^{25}
MS: 31(100) 29(52) 59(50) 27(27) 45(26) 72(14) 43(14) 15(14) 28(8) 26(6)
IR [cm^{-1}]: 3450 2940 2860 1560 1450 1370 1350 1120 1060 940 890 830
Raman [cm^{-1}]: 3450 2980 2940 2870 2800 2780 2700 1480 1450 1270 1230
   1120 1080 1040 930 920 880 830 790 520 430 400 340
^{13}C NMR [ppm]: 15.0 61.5 66.5 72.0 CDCl$_3$
Flammability: Flash pt. = 43°C; ign. temp. = 235°C; flam. lim. = 3-18%
TLV/TWA: 5 ppm (18 mg/m^3)
Reg. Lists: CERCLA (RQ = 1000 lb.); SARA 313 (1.0%); RCRA U359 (toxic)

256. Ethylene glycol monomethyl ether

Syst. Name: Ethanol, 2-methoxy-
Synonyms: 2-Methoxyethanol; Methyl cellosolve

CASRN: 109-86-4   **DOT No.:** 1188
Merck No.: 5961   **Beil Ref.:** 4-01-00-02375
Beil RN: 1731074
MF: $C_3H_8O_2$ **MP[°C]:** -85.1   **Den.[g/cm^3]:** 0.9647^{20}
MW: 76.10   **BP[°C]:** 124.1 n_D: 1.4024^{20}
Sol.: H_2O 5; EtOH 4; eth
    5; ace 3; bz 5; chl 2
Vap. Press. [kPa]: 1.31^{25}; 5.19^{50}; 16.4^{75}; 43.9^{100}; 102^{125}
Therm. Prop.[kJ/mol]: $\Delta_{vap}H = 37.5$
C_p **(liq.) [J/mol °C]:** 171.1^{25}
γ **[mN/m]:** 30.84^{25}; $d\gamma/dT = 0.0984$ mN/m °C
Elec. Prop.: $\mu = 2.36$ D; $\varepsilon = 17.2^{25}$; IP = 9.60 eV
η **[mPa s]:** 1.60^{25}
MS: 45(100) 31(15) 29(14) 28(11) 47(9) 76(6) 43(6) 58(4) 46(4) 27(4)
IR [cm^{-1}]: 3450 2920 2880 2820 1460 1400 1370 1230 1190 1120 1060 1010
    960 890 830 540
Raman [cm^{-1}]: 3440 2950 2890 2840 2770 2730 1480 1450 1410 1280 1240
    1200 1160 1130 1080 1040 1020 980 930 900 880 840 540 430 380 300
^{1}H NMR [ppm]: 2.5 3.4 3.5 3.7 3.5 3.7 CDCl$_3$
Flammability: Flash pt. = 39°C; ign. temp. = 285°C; flam. lim. = 1.8-14%
TLV/TWA: 5 ppm (16 mg/m^3)
Reg. Lists: SARA 313 (1.0%)

257. Ethyleneimine

Syst. Name: Aziridine
Synonyms: Azacyclopropane

CASRN: 151-56-4   **DOT No.**: 1185
Merck No.: 3760   **Beil Ref.**: 5-20-01-00003
Beil RN: 102380
MF: C_2H_5N **MP[°C]**: -77.9   **Den.[g/cm^3]**: 0.832^{25}
MW: 43.07   **BP[°C]**: 56
Sol.: H_2O 5; EtOH 3; eth
    4; chl 2; os 5
Vap. Press. [kPa]: 7.90^0; 28.9^{25}
Therm. Prop.[kJ/mol]: $\Delta_f H°$(l, 25°C) = 91.9
γ [mN/m]: 32.8^{25}
Elec. Prop.: μ = 1.90 D; ε = 18.3^{25}; IP = 9.20 eV
η [mPa s]: 0.418^{25}
MS: 42(100) 43(61) 28(59) 15(21) 41(20) 40(14) 27(6) 39(5) 38(4) 14(4)
IR [cm^{-1}]: 3230 3030 2940 1640 1470 1270 1220 1090 1020 930 860 780
^{1}H NMR [ppm]: 0.0 1.6 $CDCl_3$
Flammability: Flash pt. = -11°C; ign. temp. = 320°C; flam. lim. = 3.3-54.8%
TLV/TWA: 0.5 ppm (0.88 mg/m^3)
Reg. Lists: CERCLA (RQ = 1 lb.); SARA 313 (0.1%); RCRA P054 (accutely
   hazardous); confirmed or suspected carcinogen

258. Ethyl formate

Syst. Name: Methanoic acid, ethyl ester

CASRN: 109-94-4

Merck No.: 3763

Beil RN: 906769

DOT No.: 1190

Beil Ref.: 4-02-00-00023

MF: $C_3H_6O_2$ **MP[°C]:** -79.6   **Den.[g/cm^3]:** 0.9168^{20}

MW: 74.08   **BP[°C]:** 54.4 n_D: 1.3598^{10}

Sol.: H_2O 3; EtOH 5; eth 5; ace 4; ctc 2

Crit. Const.: $T_c = 235.4°C$; $P_c = 4.74$ MPa; $V_c = 229$ cm^3

Vap. Press. [kPa]: 2.06^{-25}; 9.58^0; 32.3^{25}; 87.0^{50}; 197^{75}; 393^{100}; 708^{125}; 1177^{150}

Therm. Prop.[kJ/mol]: $\Delta_{fus}H = 9.20$; $\Delta_{vap}H = 29.9$

C_p **(liq.) [J/mol °C]:** 149.3^{25}

γ **[mN/m]:** 23.18^{25}; $d\gamma/dT = 0.1315$ mN/m °C

Elec. Prop.: $\mu = (1.9$ D$)$; $\varepsilon = 8.57^{15}$; IP = 10.61 eV

η **[mPa s]:** 0.506^0; 0.380^{25}; 0.300^{50}

MS: 31(100) 28(73) 27(51) 29(38) 45(34) 26(17) 74(11) 43(9) 47(8) 56(4)

IR [cm^{-1}]: 2940 1720 1470 1450 1390 1190 1010 840

UV [nm]: 220(78) 216(81) 212(81) iso 208(52) H_2O

^{1}**H NMR [ppm]:** 1.3 4.2 7.9 CCl$_4$

Flammability: Flash pt. = -20°C; ign. temp. = 455°C; flam. lim. = 2.8-16.0%

TLV/TWA: 100 ppm (303 mg/m^3)

259. 3-Ethylhexane

Syst. Name: Hexane, 3-ethyl-

CASRN: 619-99-8
Beil RN: 1696873
MF: C_8H_{18}
MW: 114.23
Sol.: H_2O 1; EtOH 5; eth
  5; ace 5; bz 5; ctc 3;
  chl 5; lig 5

DOT No.: 1262
Beil Ref.: 4-01-00-00431
BP[°C]: 118.6   **Den.[g/cm³]:** 0.7136^{20}
       n_D: 1.4018^{20}

Crit. Const.: $T_c = 292.4°C$; $P_c = 2.608$ MPa; $V_c = 455$ cm³
Vap. Press. [kPa]: 2.68^{25}; 9.11^{50}; 25.1^{75}; 58.6^{100}; 121^{125}; 226^{150}
Therm. Prop.[kJ/mol]: $\Delta_{vap}H = 33.6$; $\Delta_f H°(l, 25°C) = -250.4$
Elec. Prop.: $\mu = 0$; $\varepsilon = 1.96^{20}$
η **[mPa s]:** 0.38^{38}
MS: 43(100) 85(29) 84(23) 41(23) 29(20) 27(19) 71(14) 70(13) 57(12) 55(12)
IR [cm⁻¹]: 2940 2860 1470 1370 890 820 780 740
Raman [cm⁻¹]: 2962 2934 2912 2874 2730 1447 1305 1295 1041 887
¹³C NMR [ppm]: 10.6 14.1 20.0 25.6 35.4 40.6 diox
¹H NMR [ppm]: 0.8 1.2 CCl_4

260. 2-Ethyl-1,3-hexanediol

Syst. Name: 1,3-Hexanediol, 2-ethyl-
Synonyms: Ethohexadiol

CASRN: 94-96-2   **Beil Ref.:** 4-01-00-02597
Merck No.: 3699
Beil RN: 1735324
MF: $C_8H_{18}O_2$ **MP[°C]:** -40   **Den.[g/cm^3]:** 0.9325^{22}
MW: 146.23   **BP[°C]:** 244 n_D: 1.4497^{20}
Sol.: H_2O 2; EtOH 3; eth
    3
γ **[mN/m]:** 29.02^{35}
Elec. Prop.: $\varepsilon = 18.73^{20}$
η **[mPa s]:** 323^{20}
MS: 56(100) 55(71) 41(60) 43(55) 29(46) 27(44) 31(34) 57(33) 73(32) 39(20)
IR [cm^{-1}]: 3330 2940 1470 1370 1110 1030 970
Raman [cm^{-1}]: 2950 2930 2910 2870 2730 1450 1370 1300 1230 1140 1130
    1110 1090 1070 1040 1020 970 890 870 810 760 730 570 520 480 440 420
    320 300
^{13}C NMR [ppm]: 11.7 12.3 14.1 18.3 19.0 19.5 21.4 35.3 37.7 46.0 46.2 63.3
    63.8 74.2
^{1}H NMR [ppm]: 1.0 1.4 3.8 CDCI$_3$
Flammability: Flash pt. = 127°C; ign. temp. = 360°C

261. 2-Ethyl-1-hexanol

Syst. Name: 1-Hexanol, 2-ethyl-

CASRN: 104-76-7   **Beil Ref.:** 4-01-00-01783
Merck No.: 3764
Beil RN: 1719280
MF: $C_8H_{18}O$ **MP[°C]:** -70   **Den.[g/cm^3]:** 0.8319^{25}
MW: 130.23   **BP[°C]:** 184.6 n_D: 1.4300^{20}
Sol.: H_2O 1; EtOH 3; eth
   3; ace 3; bz 3; chl 3
Crit. Const.: T_c = 367.5°C; P_c = 2.8 MPa
Vap. Press. [kPa]: 0.019^{25}; 4.19^{100}; 13.4^{125}; 34.7^{150}
Therm. Prop.[kJ/mol]: $\Delta_{vap}H$ = 54.2; $\Delta_f H°$(l, 25°C) = -432.8
C_p **(liq.) [J/mol °C]:** 317.5^{25}
Elec. Prop.: μ = (1.7 D); ε = 7.58^{25}
η **[mPa s]:** 20.7^0; 6.27^{25}; 2.63^{50}; 1.36^{75}; 0.810^{100}
MS: 57(100) 43(41) 41(40) 29(29) 55(28) 83(27) 56(23) 70(20) 27(17) 31(13)
IR [cm^{-1}]: 3350 2940 1450 1370 1020 720 650
Raman [cm^{-1}]: 2950 2940 2870 2730 1450 1360 1310 1280 1230 1200 1140
   1120 1050 1040 960 890 870 820 760 730 610 490 410 350 300
^{13}C NMR [ppm]: 11.1 14.1 23.2 23.5 29.3 30.3 42.1 65.1 CDCl$_3$
^{1}H NMR [ppm]: 0.9 1.3 1.8 3.5 CDCl$_3$
Flammability: Flash pt. = 73°C; ign. temp. = 231°C; flam. lim. = 0.88-9.7%

262. 2-Ethylhexyl acetate
Syst. Name: Ethanoic acid, 2-ethylhexyl ester

CASRN: 103-09-3   **Beil Ref.:** 4-02-00-00166
Merck No.: 6683
Beil RN: 1758321
MF: $C_{10}H_{20}O_2$ **MP[°C]:** -80   **Den.[g/cm^3]:** 0.8718^{20}
MW: 172.27   **BP[°C]:** 199 n_D: 1.4204^{20}
Sol.: H_2O 1; EtOH 3; eth
  3
Vap. Press. [kPa]: 0.619^{50}; 2.58^{75}; 7.68^{100}; 18.2^{125}; 36.4^{150}
Therm. Prop.[kJ/mol]: $\Delta_{vap}H = 43.5$
Elec. Prop.: $\mu = (1.8\ D)$
η **[mPa s]:** 1.5^{20}
MS: 43(100) 70(60) 57(38) 55(33) 41(29) 56(20) 83(17) 29(16) 74(12) 42(12)
Flammability: Flash pt. = 71°C; ign. temp. = 268°C; flam. lim. = 0.76-8.14%

---

263. 2-Ethylhexylamine
Syst. Name: 1-Hexanamine, 2-ethyl-

CASRN: 104-75-6   **DOT No.:** 2276
Beil RN: 1209249   **Beil Ref.:** 4-04-00-00766
MF: $C_8H_{19}N$ **BP[°C]:** 169.2
MW: 129.25
Therm. Prop.[kJ/mol]: $\ni 45_{vap}H = 40.0$
η **[mPa s]:** 1.1^{20}
MS: 30(100) 72(25) 56(16) 43(15) 44(14) 28(13) 27(13) 41(12) 29(8) 39(6)
IR [cm^{-1}]: 3370 3280 2960 2930 2860 1600 1470 1380 1300 1070 800
Raman [cm^{-1}]: 3230 2940 2900 2870 2860 2730 1450 1350 1300 1150 1060
  1040 950 890 870 820 750 720 650 590 410 300
^{1}H **NMR [ppm]:** 0.9 1.3 2.6 $CDCl_3$
Flammability: Flash pt. = 60°C

264. Ethyl isovalerate

Syst. Name: Butanoic acid, 3-methyl-, ethyl ester

Synonyms: Ethyl isopentanoate

CASRN: 108-64-5   **Beil Ref.**: 4-02-00-00898

Merck No.: 3772

Beil RN: 1744677

MF: $C_7H_{14}O_2$ **MP[°C]**: -99.3   **Den.[g/cm^3]**: 0.8656^{20}

MW: 130.19   **BP[°C]**: 135.0 n_D: 1.3962^{20}

Sol.: EtOH 4; eth 4

Crit. Const.: $T_c = 315°C$

Vap. Press. [kPa]: 1.07^{25}; 4.09^{50}; 12.6^{75}; 33.0^{100}; 75.5^{125}; 155^{150}

Therm. Prop.[kJ/mol]: $\Delta_{vap}H = 37.0$; $\Delta_f H°(l, 25°C) = -570.9$

γ [mN/m]: 23.78^{20}; $d\gamma/dT = 0.1006$ mN/m °C

Elec. Prop.: $\varepsilon = 4.71^{20}$

MS: 29(100) 41(52) 27(51) 57(43) 43(42) 60(39) 88(38) 85(38) 61(26) 45(24)

IR [cm^{-1}]: 3030 2940 1720 1470 1370 1300 1270 1190 1120 1100 1040

^{1}H NMR [ppm]: 1.0 1.3 1.9 4.1 CDCl$_3$

---

265. Ethyl lactate

Syst. Name: Propanoic acid, 2-hydroxy-, ethyl ester, (±)-

Synonyms: Ethyl 2-hydroxypropionate

CASRN: 2676-33-7

Merck No.: 3773

Beil RN: 4654400

MF: $C_5H_{10}O_3$ **MP[°C]**: -26   **Den.[g/cm^3]**: 1.0328^{20}

MW: 118.13   **BP[°C]**: 154.5 n_D: 1.4124^{20}

Sol.: H_2O 4; eth 4; EtOH 4

C_p (liq.) [J/mol °C]: 254^{25}

γ [mN/m]: 28.9^{17}

Elec. Prop.: $\mu = (2.4$ D$)$; $\varepsilon = 15.4^{30}$

η [mPa s]: 2.44^{25}

Flammability: Flash pt. = 46°C; ign. temp. = 400°C; lower flam. lim. = 1.5%

266. 3-Ethyl-2-methylpentane

Syst. Name: Pentane, 3-ethyl-2-methyl-
Synonyms: 2-Methyl-3-ethylpentane

CASRN: 609-26-7   **DOT No.:** 1262
Beil RN: 1696870   **Beil Ref.:** 4-01-00-00437
MF: C_8H_{18} **MP[°C]:** -114.9   **Den.[g/cm³]:** 0.7193^{20}
MW: 114.23   **BP[°C]:** 115.6 n_D: 1.4040^{20}
Sol.: H_2O 1; EtOH 5; eth
  3; ace 5; bz 5; chl 5;
  hp 5
Crit. Const.: $T_c = 294.0°C$; $P_c = 2.700$ MPa; $V_c = 443$ cm³
Vap. Press. [kPa]: 3.19^{25}; 10.5^{50}; 28.1^{75}; 64.3^{100}; 130^{125}; 240^{150}
Therm. Prop.[kJ/mol]: $\Delta_{fus}H = 11.34$; $\Delta_{vap}H = 32.9$; $\Delta_f H°(l, 25°C) = -249.6$
Elec. Prop.: $\mu = 0$
η **[mPa s]:** 0.38^{38}
MS: 43(100) 70(50) 41(27) 71(25) 29(19) 27(19) 85(18) 55(18) 57(15) 42(14)
IR [cm⁻¹]: 2960 2920 2870 1460 1380 1360 1180 1130 950 910 870 820 770
  550
Raman [cm⁻¹]: 2980 2970 2950 2910 2890 2770 2740 2640 1460 1450 1380
  1360 1310 1270 1190 1160 1130 1050 1040 1010 950 910 870 820 770 740
  720 560 480 450 400 370 320 220
¹³C NMR [ppm]: 11.8 19.0 22.6 29.1 47.6 diox
¹H NMR [ppm]: 0.9 0.9 1.5 CCl_4
Flammability: Flash pt. < 21°C; ign. temp. = 460°C

267. 3-Ethyl-3-methylpentane

Syst. Name: Pentane, 3-ethyl-3-methyl-
Synonyms: 3-Methyl-3-ethylpentane

CASRN: 1067-08-9
Beil RN: 1696867
MF: C_8H_{18}
MW: 114.23
Sol.: H_2O 1; EtOH 5; eth 3; ace 5; bz 5; chl 5; hp 5

DOT No.: 1262
Beil Ref.: 4-01-00-00438

MP[°C]: -90.9   **Den.[g/cm^3]**: 0.7274[20]
BP[°C]: 118.2 n_D: 1.4078[20]

Crit. Const.: $T_c = 303.5°C$; $P_c = 2.808$ MPa; $V_c = 455$ cm^3
Vap. Press. [kPa]: 3.07[25]; 9.93[50]; 26.3[75]; 60.0[100]; 121[125]; 223[150]
Therm. Prop.[kJ/mol]: $\Delta_{fus}H$ = 10.84; $\Delta_{vap}H$ = 32.8; $\Delta_f H°$(l, 25°C) = -252.8
Elec. Prop.: $\mu = 0$; $\varepsilon = 1.99$[18]
MS: 43(100) 85(64) 57(27) 41(25) 29(21) 27(18) 84(17) 55(8) 39(8) 69(6)
^{13}C NMR [ppm]: 7.5 23.2 30.6 34.8 diox

268. 3-Ethylpentane

Syst. Name: Pentane, 3-ethyl-

CASRN: 617-78-7

DOT No.: 1206

Beil RN: 1730760

Beil Ref.: 4-01-00-00402

MF: C_7H_{16}

MP[°C]: -118.6   **Den.[g/cm^3]**: 0.6982^{20}

MW: 100.20

BP[°C]: 93.5 n_D: 1.3934^{20}

Sol.: H_2O 1; EtOH 3; eth 3; ace 5; bz 5; chl 5; hp 5

Crit. Const.: T_c = 267.6°C; P_c = 2.891 MPa; V_c = 416 cm^3

Vap. Press. [kPa]: 2.03^0; 7.74^{25}; 23.0^{50}; 57.0^{75}; 122^{100}

Therm. Prop.[kJ/mol]: $\Delta_{fus}H$ = 9.55; $\Delta_{vap}H$ = 31.1; $\Delta_fH°$(l, 25°C) = -224.8

C_p **(liq.) [J/mol °C]**: 219.6^{25}

Elec. Prop.: μ = 0; ε = 1.94^{20}

η **[mPa s]**: 0.31^{38}

MS: 43(100) 70(47) 71(46) 29(24) 41(22) 27(20) 55(16) 39(10) 42(8) 57(4)

IR [cm^{-1}]: 2960 2920 2880 1460 1380 900

Raman [cm^{-1}]: 2960 2930 2870 2730 2700 1450 1380 1360 1300 1280 1270 1160 1150 1050 1040 1000 900 840 830 730 550 450 440 400 310

^{13}C NMR [ppm]: 10.5 25.2 42.4 diox

269. Ethyl propanoate

Syst. Name: Propanoic acid, ethyl ester
Synonyms: Ethyl propionate

CASRN: 105-37-3   **DOT No.**: 1195
Merck No.: 3801   **Beil Ref.**: 4-02-00-00705
Beil RN: 506287
MF: $C_5H_{10}O_2$ **MP[°C]**: -73.9   **Den.[g/cm^3]**: 0.8917^{20}
MW: 102.13   **BP[°C]**: 99.1 n_D: 1.3839^{20}
Sol.: H_2O 2; EtOH 5; eth
  5; ace 3; ctc 2
Crit. Const.: T_c = 272.9°C; P_c = 3.362 MPa; V_c = 345 cm^3
Vap. Press. [kPa]: 4.97^{25}; 16.6^{50}; 45.1^{75}; 104^{100}
Therm. Prop.[kJ/mol]: $\Delta_{vap}H$ = 33.9; $\Delta_f H°$(l, 25°C) = -502.7
C_p **(liq.) [J/mol °C]**: 196.1^{25}
γ **[mN/m]**: 23.80^{25}; $d\gamma/dT$ = 0.1168 mN/m °C
Elec. Prop.: μ = (1.7 D); ε = 5.76^{20}; IP = 10.00 eV
η **[mPa s]**: 0.691^0; 0.501^{25}; 0.380^{50}; 0.299^{75}; 0.242^{100}
MS: 57(100) 29(84) 102(17) 27(17) 75(15) 28(14) 45(13) 74(12) 73(7) 43(6)
IR [cm^{-1}]: 2980 2940 1740 1460 1430 1370 1350 1260 1180 1080 1030 860
  800
Raman [cm^{-1}]: 2980 2940 2740 1730 1450 1420 1400 1350 1300 1270 1110
  1100 1080 1030 1000 890 860 790 660 600 470 450 370 290
^{1}H NMR [ppm]: 1.1 1.2 2.2 4.0 CCl$_4$
Flammability: Flash pt. = 12°C; ign. temp. = 440°C; flam. lim. = 1.9-11%

270. Ethyl vinyl ether

Syst. Name: Ethene, ethoxy-
Synonyms: Ethoxyethylene; Vinyl ethyl ether

CASRN: 109-92-2   **DOT No.:** 1302
Beil RN: 605351   **Beil Ref.:** 4-01-00-02049
MF: C_4H_8O **MP[°C]:** -115.8   **Den.[g/cm^3]:** 0.7589^{20}
MW: 72.11   **BP[°C]:** 35.5 n_D: 1.3767^{20}
Sol.: H_2O 2; EtOH 3; eth
     5; ctc 2
Crit. Const.: $T_c = 202°C$; $P_c = 4.07$ MPa
Vap. Press. [kPa]: 6.23^{-25}; 23.1^{0}; 68.8^{25}
Therm. Prop.[kJ/mol]: $\Delta_{vap}H = 26.2$; $\Delta_f H°(l, 25°C) = -167.4$
γ **[mN/m]:** 19.0^{20}
Elec. Prop.: $\mu = (1.3$ D); IP = 8.80 eV
η **[mPa s]:** 0.2^{20}
MS: 44(100) 43(81) 72(66) 29(36) 27(29) 28(19) 45(11) 42(6) 31(6) 73(3)
IR [cm^{-1}]: 3110 2980 2900 1640 1610 1480 1440 1390 1310 1200 1110 1070
     1040 960 810
Raman [cm^{-1}]: 3130 3050 3030 2980 2940 2800 2820 2770 2730 1640 1620
     1570 1480 1450 1420 1360 1330 1280 1210 1150 1120 1090 1070 1040 970
     950 890 850 810 700 600 530 500 440 400 340 250 200
^{1}H NMR [ppm]: 1.2 3.7 3.8 4.0 6.3 CCl_4
Flammability: Flash pt. < -46°C; ign. temp. = 202°C; flam. lim. = 1.7-28%

271. Fluorobenzene

Syst. Name: Benzene, fluoro-
Synonyms: Phenyl fluoride

CASRN: 462-06-6   **DOT No.:** 2387
Merck No.: 4099   **Beil Ref.:** 4-05-00-00632
Beil RN: 1236623
MF: C_6H_5F **MP[°C]:** -42.2   **Den.[g/cm³]:** 1.0225^{20}
MW: 96.10   **BP[°C]:** 84.7 n_D: 1.4684^{30}
Sol.: bz 4; eth 4; EtOH
    4; lig 4
Crit. Const.: T_c = 286.94°C; P_c = 4.551 MPa; V_c = 269 cm³
Vap. Press. [kPa]: 2.84^0; 10.4^{25}; 30.3^{50}; 74.2^{75}; 156^{100}
Therm. Prop.[kJ/mol]: $\Delta_{fus}H$ = 11.31; $\Delta_{vap}H$ = 31.2; $\Delta_fH°$(l, 25°C) = -150.6
C_p **(liq.) [J/mol °C]:** 146.4^{25}
γ **[mN/m]:** 26.66^{25}; $d\gamma/dT$ = 0.1204 mN/m °C
Elec. Prop.: μ = 1.60 D; ϵ = 5.47^{20}; IP = 9.20 eV
η **[mPa s]:** 0.749^0; 0.550^{25}; 0.423^{50}; 0.338^{75}
MS: 96(100) 70(17) 97(7) 95(6) 75(6) 50(6) 51(4) 39(4) 69(3) 57(3)
IR [cm⁻¹]: 3030 1590 1490 1220 1150 1060 1020 900 810 750 690
Raman [cm⁻¹]: 3190 3070 3020 2980 2920 1600 1220 1160 1020 1010 990 830
    810 620 520 500 240
UV [nm]: 259(513) H_2O
¹³C NMR [ppm]: 114.6 124.3 130.3 163.8
¹H NMR [ppm]: 7.0 CCl_4
Flammability: Flash pt. = -15°C

272. *o*-Fluorotoluene

Syst. Name: Benzene, 1-fluoro-2-methyl-
Synonyms: *o*-Tolyl fluoride

CASRN: 95-52-3
Merck No.: 4108
Beil RN: 1853362
MF: C_7H_7F
MW: 110.13
Sol.: H_2O 1; EtOH 4; eth
    4

Rel. CASRN: 25496-08-6
Beil Ref.: 4-05-00-00799

MP[°C]: -62
BP[°C]: 115

Den.[g/cm³]: 1.0041^{13}
n_D: 1.4704^{20}

Vap. Press. [kPa]: 3.13^{25}; 10.5^{50}; 28.4^{75}; 66.1^{100}
Therm. Prop.[kJ/mol]: $\Delta_{vap}H = 35.4$
Elec. Prop.: $\mu = 1.37$ D; $\varepsilon = 4.23^{25}$; IP = 8.91 eV
η [mPa s]: 0.680^{20}
MS: 109(100) 110(55) 83(18) 57(11) 63(10) 39(9) 51(6) 50(6) 107(5) 62(5)
IR [cm⁻¹]: 3030 2940 1590 1490 1470 1240 1190 1110 1040 930 840 750 700
Raman [cm⁻¹]: 3069 2930 1620 1234 1038 747 577 526 272 186
UV [nm]: 268(1000) 262(1000) 257(708) 206(7943) cyhex
¹H NMR [ppm]: 2.2 6.9 CCl_4

273. *m*-Fluorotoluene

Syst. Name: Benzene, 1-fluoro-3-methyl-
Synonyms: *m*-Tolyl fluoride

CASRN: 352-70-5   **Rel. CASRN**: 25496-08-6
Merck No.: 4108   **Beil Ref.**: 4-05-00-00799
Beil RN: 1903631
MF: C_7H_7F **MP[°C]**: -87   **Den.[g/cm^3]**: 0.9974^{20}
MW: 110.13   **BP[°C]**: 115 n_D: 1.4691^{20}
Sol.: H_2O 1; EtOH 4; eth
    4
Vap. Press. [kPa]: 2.83^{25}; 9.59^{50}; 26.4^{75}; 61.8^{100}; 128^{125}
Elec. Prop.: $\mu = 1.82$ D; $\varepsilon = 5.41^{25}$; IP = 8.91 eV
η **[mPa s]**: 0.608^{20}
MS: 109(100) 110(54) 83(11) 57(5) 111(4) 107(4) 63(3) 39(3) 108(2) 89(2)
IR [cm^{-1}]: 3040 2960 2920 2860 1620 1590 1490 1460 1270 1250 1140 1080
    1000 920 880 850 770 730 680
UV [nm]: 269(955) 262(851) 257(603) 207(7586) cyhex
^{1}H NMR [ppm]: 2.3 6.9 CCl_4

274. *p*-Fluorotoluene

Syst. Name: Benzene, 1-fluoro-4-methyl-
Synonyms: *p*-Tolyl fluoride

CASRN: 352-32-9   **Rel. CASRN:** 25496-08-6
Merck No.: 4108   **Beil Ref.:** 4-05-00-00800
Beil RN: 1362373
MF: C_7H_7F **MP[°C]:** -56   **Den.[g/cm^3]:** 0.9975^{20}
MW: 110.13   **BP[°C]:** 116.6 n_D: 1.4699^{20}
Sol.: H_2O 1; EtOH 4; eth
  4
Vap. Press. [kPa]: 3.00^{25}; 26.3^{75}; 61.7^{100}; 128^{125}; 240^{150}
Therm. Prop.[kJ/mol]: $\Delta_{vap}H = 34.1$; $\Delta_f H°(l, 25°C) = -186.9$
C_p **(liq.) [J/mol °C]:** 171.2^{25}
γ **[mN/m]:** 27.3^{27}
Elec. Prop.: $\mu = 2.00$ D; $\varepsilon = 5.88^{25}$; IP = 8.79 eV
η **[mPa s]:** 0.622^{20}
MS: 109(100) 110(60) 83(13) 57(10) 108(8) 39(8) 63(7) 107(5) 51(5) 50(5)
IR [cm^{-1}]: 3040 2930 2870 1620 1600 1510 1450 1380 1230 1160 1100 1020
  920 840 820 730
Raman [cm^{-1}]: 3071 2928 1601 1382 1221 1214 1157 842 825 638 454 341
UV [nm]: 273(1660) 267(1698) 264(1096) 261(1000) 259(891) 256(631)
  253(513) 206(7244) cyhex
^{1}**H NMR [ppm]:** 2.2 6.8 7.0 CCl_4

275. Formamide

Syst. Name: Methanamide
Synonyms: Carbamaldehyde

CASRN: 75-12-7
Merck No.: 4151
Beil RN: 505995
MF: CH_3NO
MW: 45.04

DOT No.: 1993
Beil Ref.: 4-02-00-00045

MP[°C]: 2.55
BP[°C]: 220

Den.[g/cm^3]: 1.1334^{20}
n_D: 1.4472^{20}

Sol.: H_2O 5; EtOH 5; eth 2; ace 3; bz 1; chl 1; peth 1
Vap. Press. [kPa]: 4.52^{125}; 12.1^{150}
Therm. Prop.[kJ/mol]: $\Delta_{fus}H$ = 6.69; $\Delta_f H°$(l, 25°C) = -254.0
C_p **(liq.) [J/mol °C]**: 107.6^{25}
γ **[mN/m]**: 57.03^{25}; $d\gamma/dT$ = 0.0842 mN/m °C
Elec. Prop.: μ = 3.73 D; ε = 111.0^{20}; IP = 10.16 eV
η **[mPa s]**: 7.11^0; 3.34^{25}; 1.83^{50}
MS: 45(100) 29(28) 44(27) 43(12) 27(12) 28(10) 42(2) 31(1) 30(1) 15(1)
IR [cm^{-1}]: 3400 2900 1700 1600 1400 1320 1050 600
Raman [cm^{-1}]: 3300 3190 2900 2780 1670 1600 1390 1310 1100 1050 610
UV [nm]: 205(158) undiluted
^{1}H NMR [ppm]: 7.1 8.0 ace
Flammability: Flash pt. = 154°C
TLV/TWA: 10 ppm (18 mg/m^3)

276. Formic acid

Syst. Name: Methanoic acid

CASRN: 64-18-6   **DOT No.:** 1779
Merck No.: 4153   **Beil Ref.:** 4-02-00-00003
Beil RN: 1209246
MF: CH_2O_2 **MP[°C]:** 8.3   **Den.[g/cm^3]:** 1.220^{20}
MW: 46.03   **BP[°C]:** 101 n_D: 1.3714^{20}
Sol.: H_2O 5; EtOH 5; eth
   5; ace 4; bz 3; tol 3
Crit. Const.: $T_c = 315°C$
Vap. Press. [kPa]: 5.75^{25}; 17.4^{50}; 44.4^{75}; 99.6^{100}
Therm. Prop.[kJ/mol]: $\Delta_{fus}H = 12.72$; $\Delta_{vap}H = 22.7$; $\Delta_f H°(l, 25°C) = -424.7$
C_p **(liq.) [J/mol °C]:** 99.5^{25}
γ **[mN/m]:** 37.13^{25}; $d\gamma/dT = 0.1098$ mN/m °C
Elec. Prop.: $\mu = 1.41$ D; $\varepsilon = 51.1^{25}$; IP = 11.33 eV
η **[mPa s]:** 1.61^{25}; 1.03^{50}; 0.724^{75}; 0.545^{100}
MS: 29(100) 46(61) 45(49) 28(17) 17(17) 44(10) 16(5) 13(3) 12(3) 47(2)
IR [cm^{-1}]: 3100 2650 1730 1340 1160 800 660
Raman [cm^{-1}]: 2960 2780 1670 1400 1210 1060 680 190
UV [nm]: 205(45) undiluted
^{1}**H NMR [ppm]:** 8.2 D_2O
Flammability: Flash pt. = 50°C; ign. temp. = 434°C; flam. lim. = 18-57%
TLV/TWA: 5 ppm (9.4 mg/m^3)
Reg. Lists: CERCLA (RQ = 5000 lb.); RCRA U123 (toxic)

277. Furan

Syst. Name: Furan
Synonyms: Oxacyclopentadiene

CASRN: 110-00-9   **DOT No.**: 2389
Merck No.: 4206   **Beil Ref.**: 5-17-01-00291
Beil RN: 103221
MF: C_4H_4O **MP[°C]**: -85.6   **Den.[g/cm^3]**: 0.9514^{20}
MW: 68.08   **BP[°C]**: 31.5 n_D: 1.4214^{20}
Sol.: H_2O 1; EtOH 4; eth
  4; ace 3; bz 3; chl 2
Crit. Const.: T_c = 217.1°C; P_c = 5.50 MPa; V_c = 218 cm^3
Vap. Press. [kPa]: 7.37^{-25}; 27.7^0; 80.0^{25}; 191^{50}; 394^{75}; 730^{100}
Therm. Prop.[kJ/mol]: $\Delta_{fus}H$ = 3.80; $\Delta_{vap}H$ = 27.1; $\Delta_f H°$(l, 25°C) = -62.3
C_p (liq.) [J/mol °C]: 114.8^{25}
γ [mN/m]: 23.38^{25}
Elec. Prop.: μ = 0.66 D; ε = 2.88^4; IP = 8.88 eV
η [mPa s]: 0.661^{-25}; 0.475^0; 0.361^{25}
k [W/m °C]: 0.142^{-25}; 0.134^0; 0.126^{25}
MS: 68(100) 39(64) 40(9) 38(9) 42(6) 29(6) 37(5) 69(4) 34(2) 67(1)
IR [cm^{-1}]: 3130 1590 1490 1180 1000 870 750
UV [nm]: 325(33884) cyhex 208(7943) EtOH
^{13}C NMR [ppm]: 109.7 143.0 diox
^{1}H NMR [ppm]: 6.4 7.4 CDCl$_3$
Flammability: Flash pt. < 0°C; flam. lim. = 2.3-14.3%
Reg. Lists: CERCLA (RQ = 100 lb.); RCRA U124 (toxic)

278. Furfural

Syst. Name: 2-Furancarboxaldehyde
Synonyms: 2-Furaldehyde

CASRN: 98-01-1   **DOT No.**: 1199
Merck No.: 4214   **Beil Ref.**: 5-17-09-00292
Beil RN: 105755
MF: $C_5H_4O_2$ **MP[°C]**: -36.5   **Den.[g/cm³]**: 1.1594^{20}
MW: 96.09   **BP[°C]**: 161.7 n_D: 1.5261^{20}
Sol.: H_2O 3; EtOH 4; eth
  5; ace 4; bz 3; chl 3
Crit. Const.: T_c = 397°C; P_c = 5.89 MPa
Vap. Press. [kPa]: 0.290^{25}; 4.57^{75}; 13.7^{100}; 34.1^{125}; 73.4^{150}
Therm. Prop.[kJ/mol]: $\Delta_{fus}H$ = 14.35; $\Delta_{vap}H$ = 43.2; $\Delta_f H°$(l, 25°C) = -201.6
C_p (liq.) [J/mol °C]: 163.2^{25}
γ [mN/m]: 43.09^{25}; $d\gamma/dT$ = 0.1327 mN/m °C
Elec. Prop.: μ = (3.5 D); ϵ = 42.1^{20}; IP = 9.21 eV
η [mPa s]: 2.50^0; 1.59^{25}; 1.14^{50}; 0.906^{75}; 0.772^{100}
MS: 39(100) 96(55) 95(52) 38(38) 29(35) 37(29) 40(11) 97(9) 50(7) 42(7)
IR [cm⁻¹]: 3140 2840 2800 2760 2720 1690 1570 1460 1390 1360 1280 1240
  1160 1080 1020 940 930 880 830 760
UV [nm]: 271 227 220 MeOH
¹³C NMR [ppm]: 112.9 121.6 148.7 153.8 178.2 diox
¹H NMR [ppm]: 6.6 7.3 7.7 9.7 $CDCl_3$
Flammability: Flash pt. = 60°C; ign. temp. = 316°C; flam. lim. = 2.1-19.3%
TLV/TWA: 2 ppm (7.9 mg/m³)
Reg. Lists: CERCLA (RQ = 5000 lb.); RCRA U125 (toxic)

279. Furfuryl alcohol

Syst. Name: 2-Furanmethanol
Synonyms: Furfurol; 2-Furancarbinol; 2-(Hydroxymethyl)furan

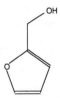

CASRN: 98-00-0
Merck No.: 4215
Beil RN: 106291
MF: $C_5H_6O_2$
MW: 98.10
Sol.: H_2O 5; EtOH 4; eth 4; chl 3

DOT No.: 2874
Beil Ref.: 5-17-03-00338

MP[°C]: -31
BP[°C]: 171

Den.[g/cm^3]: 1.1296^{20}
n_D: 1.4869^{20}

Vap. Press. [kPa]: 0.097^{25}; 0.484^{50}; 1.95^{75}; 6.60^{100}; 19.4^{125}; 50.4^{150}
Therm. Prop.[kJ/mol]: $\Delta_{fus}H$ = 13.13; $\Delta_{vap}H$ = 53.6; $\Delta_f H°$(l, 25°C) = -276.2
C_p (liq.) [J/mol °C]: 204.0^{25}
γ [mN/m]: 38^{20}
Elec. Prop.: μ = (1.9 D); ε = 16.85^{25}
η [mPa s]: 4.62^{25}
MS: 98(100) 41(65) 39(59) 81(55) 53(53) 97(51) 42(49) 69(39) 70(36) 29(28)
IR [cm^{-1}]: 3330 2910 2850 1500 1400 1210 1170 1130 1060 1000 900 870 800 730
UV [nm]: 270 MeOH
^{1}H NMR [ppm]: 2.8 4.6 6.3 7.4 CDCl$_3$
Flammability: Flash pt. = 75°C; ign. temp. = 491°C; flam. lim. = 1.8-16.3%
TLV/TWA: 10 ppm (40 mg/m^3)

280. Glycerol

Syst. Name: 1,2,3-Propanetriol
Synonyms: 1,2,3-Trihydroxypropane

CASRN: 56-81-5   **Beil Ref.**: 4-01-00-02751
Merck No.: 4379
Beil RN: 635685
MF: $C_3H_8O_3$ **MP[°C]**: 18.2   **Den.[g/cm³]**: 1.2613^{20}
MW: 92.09   **BP[°C]**: 290 n_D: 1.4746^{20}
Sol.: H_2O 5; EtOH 5; eth
   2; bz 1; ctc 1; chl 1;
   CS_2; peth 1
Vap. Press. [kPa]: 5.93^{200}; 36.2^{250}
Therm. Prop.[kJ/mol]: $\Delta_{fus}H$ = 18.28; $\Delta_{vap}H$ = 61.0; $\Delta_f H°$(l, 25°C) = -668.5
C_p **(liq.) [J/mol °C]**: 219.0^{25}
γ [mN/m]: 63.3^{20}; dγ/dT = 0.067 mN/m °C
Elec. Prop.: μ = (2.6 D); ε = 46.53^{20}
η [mPa s]: 934^{25}; 152^{50}; 39.8^{75}; 14.8^{100}
***k* [W/m °C]**: 0.292^{25}; 0.295^{50}; 0.297^{75}; 0.300^{100}
MS: 61(100) 43(90) 31(57) 44(54) 29(38) 18(32) 27(12) 42(11) 60(10) 45(10)
IR [cm⁻¹]: 3400 2930 2860 1650 1400 1200 1100 1030 600
Raman [cm⁻¹]: 3320 3100 2940 2880 2750 1460 1370 1310 1250 1210 1110
   1050 970 920 850 820 670 550 480 410 320
UV [nm]: 270
¹H NMR [ppm]: 3.6 D_2O
Flammability: Flash pt. = 199°C; ign. temp. = 370°C; flam. lim. = 3-19%
TLV/TWA: 10 mg/m³

281. Heptane

Syst. Name: Heptane

CASRN: 142-82-5   **DOT No.**: 1206
Merck No.: 4580   **Beil Ref.**: 4-01-00-00376
Beil RN: 1730763
MF: C_7H_{16} **MP[°C]**: -90.6   **Den.[g/cm^3]**: 0.6837^{20}
MW: 100.20   **BP[°C]**: 98.5 n_D: 1.3878^{20}
Sol.: H_2O 1; EtOH 4; eth
   5; ace 5; bz 5; ctc 3;
   chl 5; peth 5
Crit. Const.: T_c = 267.1°C; P_c = 2.74 MPa; V_c = 428 cm^3
Vap. Press. [kPa]: 1.52^0; 6.09^{25}; 18.9^{50}; 48.2^{75}; 106^{100}
Therm. Prop.[kJ/mol]: $\Delta_{fus}H$ = 14.16; $\Delta_{vap}H$ = 31.8; $\Delta_f H°$(l, 25°C) = -224.2
C_p **(liq.) [J/mol °C]**: 224.9^{25}
γ **[mN/m]**: 19.65^{25}; $d\gamma/dT$ = 0.0980 mN/m °C
Elec. Prop.: μ = 0; ε = 1.92^{20}; IP = 9.92 eV
η **[mPa s]**: 0.757^{-25}; 0.523^0; 0.387^{25}; 0.301^{50}; 0.243^{75}
k **[W/m °C]**: 0.1378^{-25}; 0.1303^0; 0.1228^{25}; 0.1152^{50}; 0.1077^{75}
MS: 43(100) 41(56) 29(49) 57(47) 27(46) 71(45) 56(27) 42(26) 39(23) 70(18)
IR [cm^{-1}]: 2920 2860 1460 1370 720
Raman [cm^{-1}]: 2960 2930 2910 2900 2870 2860 2730 2630 1460 1300 1240
   1210 1160 1140 1080 1060 1050 1020 950 900 850 840 780 500 460 410 400
   360 310 290
^{13}C NMR [ppm]: 14.2 23.2 29.6 32.5 diox
^{1}H NMR [ppm]: 1.0 1.3 CCl$_4$
Flammability: Flash pt. = -4°C; ign. temp. = 204°C; flam. lim. = 1.05-6.7%
TLV/TWA: 400 ppm (1640 mg/m^3)

282. 1-Heptanol

Syst. Name: 1-Heptanol

CASRN: 111-70-6   **Beil Ref.:** 4-01-00-01731
Merck No.: 4582
Beil RN: 1731686
MF: $C_7H_{16}O$ **MP[°C]:** -34   **Den.[g/cm^3]:** 0.8219^{20}
MW: 116.20   **BP[°C]:** 176.4 n_D: 1.4249^{20}
Sol.: H_2O 2; EtOH 5; eth
  5; ctc 2
Crit. Const.: T_c = 359.5°C; P_c = 3.058 MPa; V_c = 435 cm^3
Vap. Press. [kPa]: 1.37^{75}; 5.53^{100}; 17.2^{125}; 44.2^{150}
Therm. Prop.[kJ/mol]: $\Delta_f H°$(l, 25°C) = -403.3
C_p (liq.) [J/mol °C]: 272.1^{25}
Elec. Prop.: ε = 11.75^{20}; IP = 9.84 eV
η [mPa s]: 5.81^{25}; 2.60^{50}; 1.39^{75}; 0.849^{100}
k [W/m °C]: 0.166^0; 0.159^{25}; 0.153^{50}; 0.147^{75}; 0.141^{100}
MS: 41(100) 70(87) 56(86) 31(78) 43(72) 29(70) 55(67) 27(65) 42(54) 69(41)
IR [cm^{-1}]: 3330 2860 1450 1370 1050 720
Raman [cm^{-1}]: 2960 2940 2900 2870 2720 1450 1430 1300 1210 1180 1120
  1070 1050 1030 1010 960 930 920 890 870 830 790 740 520 450 400 370
  280
^{13}**C NMR [ppm]:** 14.2 23.1 26.4 29.7 32.4 33.2 62.2
^{1}**H NMR [ppm]:** 0.9 1.4 3.4 3.5 CCl_4

283. 2-Heptanol

Syst. Name: 2-Heptanol, (±)-

CASRN: 52390-72-4	**Rel. CASRN**: 543-49-7
Merck No.: 4583	**Beil Ref.**: 4-01-00-01740

Beil RN: 1719091

MF: $C_7H_{16}O$ **BP[°C]**: 159   **Den.[g/cm³]**: 0.8167^{20}

MW: 116.20 n_D: 1.4210^{20}

Sol.: H_2O 2; EtOH 3; eth 3; ctc 2

Crit. Const.: $T_c = 335.2°C$; $P_c = 3.021$ MPa

Vap. Press. [kPa]: 0.700^{50}; 3.54^{75}; 12.5^{100}; 34.1^{125}; 77.7^{150}

Elec. Prop.: $\mu = (1.7\ D)$; $\varepsilon = 9.72^{21}$; IP = 9.70 eV

η **[mPa s]**: 3.96^{25}; 1.80^{50}; 0.987^{75}; 0.615^{100}

MS: 59(100) 43(34) 41(20) 27(13) 57(12) 31(10) 29(9) 116(0)

IR [cm⁻¹]: 3380 2920 2860 1460 1380 1150 1110 1060 950 720 650

Raman [cm⁻¹]: 3340 2970 2930 2900 2880 2730 1460 1450 1370 1340 1300 1240 1200 1150 1130 1080 1060 1040 970 950 940 900 880 850 830 770 730 520 500 460 440 410 380 290 260

¹³C NMR [ppm]: 14.2 23.2 23.6 26.1 32.6 39.8 67.5

¹H NMR [ppm]: 0.9 1.1 1.3 3.4 3.6 CCl_4

Flammability: Flash pt. = 71°C

284. 3-Heptanol

Syst. Name: 3-Heptanol, (*S*)-
Synonyms: Ethylbutylcarbinol

CASRN: 26549-25-7   **Rel. CASRN**: 589-82-2
     Beil Ref.: 2-01-00-00444
MF: $C_7H_{16}O$ **MP[°C]**: -70   **Den.[g/cm^3]**: 0.8227^{20}
MW: 116.20   **BP[°C]**: 157 n_D: 1.4201^{20}
Sol.: H_2O 2; EtOH 3; eth
   3; ctc 2
Crit. Const.: T_c = 332.3°C
Vap. Press. [kPa]: 0.721^{50}; 3.72^{75}; 13.3^{100}; 36.4^{125}; 83.0^{150}
Therm. Prop.[kJ/mol]: $\Delta_{vap}H$ = 42.5
Elec. Prop.: μ = (1.7 D); ε = 7.07^{23}; IP = 9.68 eV
η **[mPa s]**: 1.96^{50}; 0.976^{75}; 0.584^{100}
MS: 59(100) 69(73) 87(37) 41(37) 29(33) 55(30) 31(24) 116(0)
IR [cm^{-1}]: 3450 2930 2870 1450 1370 1100 950 720 620
Raman [cm^{-1}]: 2940 2920 2880 1460 1450 1370 1310 1250 1210 1150 1130
   1060 1050 980 920 900 880 840 780 730 600 540 500 470 450 400 340 310
   260
^{13}C NMR [ppm]: 10.3 14.3 23.3 28.5 30.0 37.2 72.9
^{1}H NMR [ppm]: 0.9 1.4 2.3 3.4 CCl_4
Flammability: Flash pt. = 60°C

285. 1-Heptene

Syst. Name: 1-Heptene

CASRN: 592-76-7   **Rel. CASRN:** 81624-04-6
Beil RN: 1098332   **DOT No.:** 2278
                            Beil Ref.: 4-01-00-00857

MF: C_7H_{14} **MP[°C]:** -119.7   **Den.[g/cm^3]:** 0.6970^{20}
MW: 98.19   **BP[°C]:** 93.6 n_D: 1.3998^{20}
Sol.: H_2O 1; EtOH 3; eth 3; ctc 2
Crit. Const.: $T_c = 264.2$°C; $P_c = 2.921$ MPa; $V_c = 402$ cm^3
Vap. Press. [kPa]: 0.355^{-25}; 1.94^0; 7.52^{25}; 22.6^{50}; 56.4^{75}; 122^{100}; 236^{125}; 418^{150}
Therm. Prop.[kJ/mol]: $\Delta_{fus}H = 12.66$; $\Delta_f H°(l, 25°C) = -97.9$
C_p **(liq.) [J/mol °C]:** 211.8^{25}
γ **[mN/m]:** 19.80^{25}; $d\gamma/dT = 0.0991$ mN/m °C
Elec. Prop.: $\mu = 0$; $\varepsilon = 2.09^{20}$; IP = 9.44 eV
η **[mPa s]:** 0.441^0; 0.340^{25}; 0.273^{50}; 0.226^{75}
MS: 41(100) 56(88) 29(70) 55(59) 42(52) 27(45) 39(41) 70(37) 69(27) 57(27)
IR [cm^{-1}]: 3180 2920 2860 1640 1470 1380 1000 910 730
Raman [cm^{-1}]: 3080 2999 2980 2962 2903 2876 2860 1641 1440 1300 910
UV [nm]: 263(1) 237(2) hx
^{1}H NMR [ppm]: 0.9 1.4 2.0 4.9 5.7 CCl_4
Flammability: Flash pt. < 0°C; ign. temp. = 260°C

286. *cis*-2-Heptene

Syst. Name: 2-Heptene, (Z)-

CASRN: 6443-92-1

Beil RN: 1719389

Rel. CASRN: 81624-04-6

DOT No.: 2278

Beil Ref.: 3-01-00-00824

MF: C_7H_{14} **BP[°C]**: 98.4   **Den.[g/cm^3]**: 0.708^{20}

MW: 98.19 n_D: 1.406^{20}

Sol.: H_2O 1; EtOH 3; eth 3; ace 3; bz 3; ctc 2; chl 3; peth 3

Vap. Press. [kPa]: 1.66^0; 6.45^{25}; 19.5^{50}; 48.9^{75}; 106^{100}

Therm. Prop.[kJ/mol]: $\Delta_f H°$(l, 25°C) = -105.1

Elec. Prop.: $\mu = 0$

η **[mPa s]**: 0.32^{38}

MS: 41(100) 56(91) 55(82) 27(67) 39(54) 69(46) 29(40) 98(34)

IR [cm^{-1}]: 1639 1449 1408 1370 1235 1031 962 935 694

^{13}C NMR [ppm]: 12.6 13.9 17.8 22.4 26.7 32.0 32.4 123.5 124.5 130.8 131.7 CDCl$_3$

^{1}H NMR [ppm]: 0.9 1.3 1.6 2.0 5.3 CCl$_4$

287. *trans*-2-Heptene

Syst. Name: 2-Heptene, (*E*)-

CASRN: 14686-13-6   **Rel. CASRN**: 81624-04-6
Beil RN: 1719390   **DOT No.**: 2278
                              Beil Ref.: 4-01-00-00860
MF: C_7H_{14} **MP[°C]**: -109.5   **Den.[g/cm^3]**: 0.7012^{20}
MW: 98.19   **BP[°C]**: 98 n_D: 1.4045^{20}
Sol.: H_2O 1; EtOH 3; eth
   3; ace 3; bz 3; chl 3;
   peth 3
Vap. Press. [kPa]: 1.69^0; 6.56^{25}; 19.8^{50}; 49.6^{75}; 108^{100}
Therm. Prop.[kJ/mol]: $\Delta_{fus}H$ = 11.72; $\Delta_f H°$(l, 25°C) = -109.5
Elec. Prop.: $\mu = 0$
MS: 55(100) 56(94) 41(79) 27(56) 69(47) 39(44) 98(42) 29(32) 43(22) 42(20)
Raman [cm^{-1}]: 3001 2961 2937 2918 2876 2859 1672 1450 1297 1053 337
^{13}C NMR [ppm]: 12.6 13.9 17.8 22.4 26.7 32.0 32.4 123.5 124.5 130.8 131.7
   $CDCl_3$
Flammability: Flash pt. < 0°C

288. Hexafluorobenzene
Syst. Name: Benzene, hexafluoro-
Synonyms: Perfluorobenzene

CASRN: 392-56-3   **Beil Ref.**: 4-05-00-00640
Beil RN: 1683438
MF: C_6F_6 **MP[°C]**: 5.3   **Den.[g/cm^3]**: 1.6184^{20}
MW: 186.06   **BP[°C]**: 80.2 n_D: 1.3777^{20}
Crit. Const.: T_c = 243.6°C; P_c = 3.273 MPa; V_c = 335 cm^3
Vap. Press. [kPa]: 11.3^{25}; 34.1^{50}; 85.3^{75}; 184^{100}
Therm. Prop.[kJ/mol]: $\Delta_{fus}H$ = 11.58; $\Delta_{vap}H$ = 31.7; $\Delta_f H°$(l, 25°C) = -991.3
C_p (liq.) [J/mol °C]: 156.5^{25}
γ [mN/m]: 21.64^{25}
Elec. Prop.: μ = 0; ε = 2.03^{25}; IP = 9.91 eV
η [mPa s]: 2.79^{25}; 1.73^{50}; 1.15^{75}
MS: 186(100) 117(47) 93(18) 155(17) 167(13) 31(13) 136(12) 98(8) 69(8) 187(7)
IR [cm^{-1}]: 3030 2700 2440 2080 1790 1750 1540 1350 1160 1020 1000 720
UV [nm]: 230(776) EtOH

289. Hexamethylphosphoric triamide

Syst. Name: Phosphoric triamide, hexamethyl-
Synonyms: Tris(dimethylamino)phosphine oxide

CASRN: 680-31-9   **Beil Ref.**: 4-04-00-00284
Merck No.: 4568
Beil RN: 1099903
MF: $C_6H_{18}N_3OP$ **BP[°C]**: 232.5   **Den.[g/cm^3]**: 1.03^{20}
MW: 179.20 n_D: 1.4579^{20}
Sol.: EtOH 3; eth 3
Therm. Prop.[kJ/mol]: $\Delta_{fus}H = 14.28$
C_p **(liq.) [J/mol °C]**: 321^{25}
γ **[mN/m]**: 33.8^{20}
Elec. Prop.: $\mu = (5.5\ D)$; $\varepsilon = 31.3^{20}$
η **[mPa s]**: 3.10^{25}
MS: 44(100) 135(75) 45(71) 28(38) 42(35) 179(32) 92(25) 136(24) 46(20)
   32(14)
IR [cm^{-1}]: 3450 2820 1460 1290 1200 1060 980 740 680
^{13}C NMR [ppm]: 36.8 CDCl$_3$
TLV/TWA: Carcinogen
Reg. Lists: CERCLA (RQ = 1 lb.); SARA 313 (0.1%); confirmed or suspected
   carcinogen

290. Hexane

Syst. Name: Hexane

CASRN: 110-54-3   **DOT No.:** 1208
Merck No.: 4613   **Beil Ref.:** 4-01-00-00338
Beil RN: 1730733
MF: C_6H_{14} **MP[°C]:** -95.3   **Den.[g/cm^3]:** 0.6548^{25}
MW: 86.18   **BP[°C]:** 68.7 n_D: 1.3749^{20}
Sol.: H_2O 1; EtOH 4; eth
   3; chl 3
Crit. Const.: $T_c = 234.5°C$; $P_c = 3.025$ MPa; $V_c = 368$ cm^3
Vap. Press. [kPa]: 1.34^{-25}; 6.05^0; 20.2^{25}; 54.1^{50}; 123^{75}; 246^{100}; 446^{125}; 749^{150}
Therm. Prop.[kJ/mol]: $\Delta_{fus}H = 13.08$; $\Delta_{vap}H = 28.9$; $\Delta_fH°(l, 25°C) = -198.7$
C_p **(liq.) [J/mol °C]:** 195.6^{25}
γ **[mN/m]:** 17.89^{25}; $d\gamma/dT = 0.1022$ mN/m °C
Elec. Prop.: $\mu = 0$; $\varepsilon = 1.89^{20}$; IP = 10.13 eV
η **[mPa s]:** 0.405^0; 0.300^{25}; 0.240^{50}
k **[W/m °C]:** 0.137^{-25}; 0.128^0; 0.120^{25}; 0.111^{50}; 0.102^{75}; 0.093^{100}
MS: 57(100) 43(78) 41(77) 29(61) 27(57) 56(45) 42(39) 39(27) 28(16) 86(14)
IR [cm^{-1}]: 2960 2870 1460 1380 720
Raman [cm^{-1}]: 2970 2940 2920 2900 2880 2870 2730 2670 1460 1370 1340
   1310 1250 1220 1170 1140 1080 1070 1040 1010 980 960 900 870 830 820
   800 760 490 460 410 380 340 320 250
^{13}C NMR [ppm]: 13.9 22.9 32.0
Flammability: Flash pt. = -22°C; ign. temp. = 225°C; flam. lim. = 1.1-7.5%
TLV/TWA: 50 ppm (176 mg/m^3)
Reg. Lists: CERCLA (RQ = 1 lb.)

291. Hexanenitrile

Syst. Name: Hexanenitrile
Synonyms: Capronitrile; Pentyl cyanide

CASRN: 628-73-9   **Beil Ref.**: 4-02-00-00930
Beil RN: 1633601
MF: $C_6H_{11}N$ **MP[°C]**: -80.3   **Den.[g/cm^3]**: 0.8051^{20}
MW: 97.16   **BP[°C]**: 163.6 n_D: 1.4068^{20}
Sol.: H_2O 1; EtOH 3; eth
    3; chl 2
Crit. Const.: T_c = 360.7°C; P_c = 3.30 MPa
Vap. Press. [kPa]: 0.355^{25}; 13.8^{100}; 33.2^{125}; 70.5^{150}
γ **[mN/m]**: 27.37^{25}; dγ/dT = 0.0907 mN/m °C
Elec. Prop.: ε = 17.26^{25}
η **[mPa s]**: 0.912^{25}; 0.650^{50}; 0.488^{75}; 0.382^{100}
MS: 41(100) 54(68) 27(59) 55(55) 29(44) 39(35) 57(30) 43(30) 68(24) 28(22)
IR [cm^{-1}]: 2940 2220 1470 1430 1390 1110 730
^{1}H NMR [ppm]: 0.9 1.5 2.3 $CDCl_3$

---

292. 1,2,6-Hexanetriol

Syst. Name: 1,2,6-Hexanetriol
Synonyms: 1,2,6-Trihydroxyhexane

CASRN: 106-69-4   **Beil Ref.**: 4-01-00-02784
Beil RN: 1304479
MF: $C_6H_{14}O_3$ **Den.[g/cm^3]**: 1.1049^{20}
MW: 134.18 n_D: 1.58^{20}
Vap. Press. [kPa]: 0.023^{125}; 0.128^{150}
γ **[mN/m]**: 49.9^{20}
Elec. Prop.: ε = 31.5^{12}
η **[mPa s]**: 2630^{20}
MS: 85(100) 57(64) 67(38) 55(24) 56(17) 61(14) 86(8) 103(7) 68(5) 58(5)
Flammability: Flash pt. = 191°C

293. Hexanoic acid

Syst. Name: Hexanoic acid
Synonyms: Caproic acid

CASRN: 142-62-1   **DOT No.:** 1760
Merck No.: 1760   **Beil Ref.:** 4-02-00-00917
Beil RN: 773837
MF: $C_6H_{12}O_2$ **MP[°C]:** -3   **Den.[g/cm^3]:** 0.9274^{20}
MW: 116.16   **BP[°C]:** 205.2 n_D: 1.4163^{20}
Sol.: H_2O 1; EtOH 3; eth
   3; chl 3
Crit. Const.: T_c = 389°C; P_c = 3.20 MPa
Vap. Press. [kPa]: 0.005^{25}; 5.28^{125}; 15.6^{150}
Therm. Prop.[kJ/mol]: $\Delta_{fus}H$ = 15.40; $\Delta_f H°$(l, 25°C) = -583.8
C_p **(liq.) [J/mol °C]:** 225.0^{25}
γ **[mN/m]:** 27.55^{25}
Elec. Prop.: μ = (1.1 D); ε = 2.60^{25}; IP = 10.12 eV
η **[mPa s]:** 2.83^{25}
MS: 60(100) 73(42) 27(36) 41(33) 43(27) 29(26) 45(20) 39(16) 42(15) 55(14)
IR [cm^{-1}]: 2940 2630 1700 1450 1410 1280 1210 930 730
^{13}C NMR [ppm]: 13.8 22.4 24.5 31.4 34.2 180.6 CDCl$_3$
^{1}H NMR [ppm]: 0.9 1.4 2.4 11.4 CDCl$_3$
Flammability: Flash pt. = 102°C; ign. temp. = 380°C

294. 1-Hexanol

Syst. Name: 1-Hexanol
Synonyms: Caproyl alcohol; Hexyl alcohol

$\sim\!\!\sim\!\!\sim$ OH

CASRN: 111-27-3   **DOT No.**: 2282
Merck No.: 4615   **Beil Ref.**: 4-01-00-01694
Beil RN: 969167
MF: $C_6H_{14}O$ **MP[°C]**: -44.6   **Den.[g/cm^3]**: 0.8136^{20}
MW: 102.18   **BP[°C]**: 157.6 n_D: 1.4178^{20}
Sol.: H_2O 2; EtOH 3; eth
   5; ace 3; bz 5; ctc 2;
   chl 3
Crit. Const.: $T_c = 337.2$°C; $P_c = 3.417$ MPa; $V_c = 381$ cm^3
Vap. Press. [kPa]: 0.110^{25}; 0.611^{50}; 3.12^{75}; 11.4^{100}; 32.9^{125}; 79.9^{150}
Therm. Prop.[kJ/mol]: $\Delta_{fus}H = 15.40$; $\Delta_{vap}H = 44.5$; $\Delta_f H°(l, 25°C) = -377.5$
C_p **(liq.) [J/mol °C]**: 240.4^{25}
γ **[mN/m]**: 25.81^{25}; $d\gamma/dT = 0.0801$ mN/m °C
Elec. Prop.: $\epsilon = 13.03^{20}$; IP = 9.89 eV
η **[mPa s]**: 4.58^{25}; 2.27^{50}; 1.27^{75}; 0.781^{100}
k **[W/m °C]**: 0.159^{-25}; 0.154^{0}; 0.150^{25}; 0.145^{50}; 0.141^{75}; 0.137^{100}
MS: 56(100) 43(78) 31(74) 41(71) 27(64) 29(59) 55(58) 42(53) 39(37) 69(27)
IR [cm^{-1}]: 3340 2930 2860 1460 1380 1060 920 720
Raman [cm^{-1}]: 2940 2900 2870 2730 1450 1440 1300 1220 1190 1150 1140
   1120 1070 1060 1020 1000 950 920 880 860 820 750 520 500 440 400 350
   310
^{1}H NMR [ppm]: 0.9 1.3 3.5 3.7 CCl$_4$
Flammability: Flash pt. = 63°C

295. 2-Hexanol

Syst. Name: 2-Hexanol, (±)-

CASRN: 20281-86-1
Beil RN: 1718996

Rel. CASRN: 626-93-7
DOT No.: 2282
Beil Ref.: 4-01-00-01708

MF: $C_6H_{14}O$ **BP[°C]:** 140   **Den.[g/cm^3]:** 0.8159^{20}
MW: 102.18 n_D: 1.4144^{20}
Sol.: H_2O 2; EtOH 3; eth
   3; ctc 2
Crit. Const.: T_c = 312.8°C; P_c = 3.310 MPa
Vap. Press. [kPa]: 0.346^{25}; 1.84^{50}; 7.28^{75}; 22.9^{100}; 60.7^{125}
Therm. Prop.[kJ/mol]: $\Delta_{vap}H$ = 41.0; $\Delta_f H°$(l, 25°C) = -392.0
Elec. Prop.: ε = 11.06^{25}; IP = 9.80 eV
MS: 59(100) 31(61) 29(25) 57(23) 41(21) 56(20) 27(11) 102(4)
IR [cm^{-1}]: 3350 2960 2860 1460 1380 1140 1120 1000 960 650
^{13}C NMR [ppm]: 14.2 23.2 23.6 28.6 39.5 67.5
^{1}H NMR [ppm]: 0.9 1.0 1.4 2.8 3.4 CCl$_4$

---

296. 3-Hexanol

Syst. Name: 3-Hexanol, (±)-

CASRN: 17015-11-1
Beil RN: 1718964

Rel. CASRN: 623-37-0
DOT No.: 2282
Beil Ref.: 4-01-00-01711

MF: $C_6H_{14}O$ **BP[°C]:** 135   **Den.[g/cm^3]:** 0.8182^{20}
MW: 102.18 n_D: 1.4167^{20}
Sol.: H_2O 2; EtOH 3; eth
   5; ace 3
Crit. Const.: T_c = 309.3°C; P_c = 3.36 MPa
Vap. Press. [kPa]: 0.684^{25}; 2.87^{50}; 9.79^{75}; 28.26^{100}; 71.32^{125}
Therm. Prop.[kJ/mol]: $\Delta_f H°$(l, 25°C) = -392.4
C_p (liq.) [J/mol °C]: 286.2^{25}
Elec. Prop.: ε = 9.66^{25}; IP = 9.63 eV
MS: 59(100) 55(74) 31(55) 43(40) 73(39) 27(38) 29(30) 102(1)
IR [cm^{-1}]: 3330 2940 1470 1370 1330 1120 1060 1030 1000 960 840
^{13}C NMR [ppm]: 10.2 14.3 19.7 30.6 39.7 72.6

297. 1-Hexene

Syst. Name: 1-Hexene

CASRN: 592-41-6 **DOT No.**: 2370
Beil RN: 1209240 **Beil Ref.**: 4-01-00-00828
MF: C_6H_{12} **MP[°C]**: -139.7 **Den.[g/cm^3]**: 0.6731^{20}
MW: 84.16 **BP[°C]**: 63.4 n_D: 1.3837^{20}
Sol.: bz 4; eth 4; EtOH
  4; peth 4
Crit. Const.: $T_c = 231.0°C$; $P_c = 3.206$ MPa; $V_c = 348$ cm^3
Vap. Press. [kPa]: 7.68^0; 24.8^{25}; 64.7^{50}; 284^{100}
Therm. Prop.[kJ/mol]: $\Delta_{fus}H = 9.35$; $\Delta_{vap}H = 28.3$; $\Delta_fH°(l, 25°C) = -74.2$
C_p **(liq.) [J/mol °C]**: 183.3^{25}
γ **[mN/m]**: 17.90^{25}; $d\gamma/dT = 0.1027$ mN/m °C
Elec. Prop.: $\mu = 0$; $\epsilon = 2.08^{21}$; IP = 9.44 eV
η **[mPa s]**: 0.441^{-25}; 0.326^0; 0.252^{25}; 0.202^{50}
k **[W/m °C]**: 0.137^{-25}; 0.129^0; 0.121^{25}; 0.113^{50}
MS: 41(100) 56(79) 27(75) 42(68) 55(58) 39(56) 43(55) 29(30) 84(25) 69(19)
Flammability: Flash pt. < -7°C; ign. temp. = 253°C

298. *cis*-2-Hexene

Syst. Name: 2-Hexene, (Z)-

CASRN: 7688-21-3　　　　　　**Beil Ref.**: 4-01-00-00833
Beil RN: 1719019
MF: C_6H_{12}　　　　**MP[°C]**: -141.1　　**Den.[g/cm^3]**: 0.6869^{20}
MW: 84.16　　　　　**BP[°C]**: 68.8　　　n_D: 1.3979^{20}
Sol.: H_2O 1; EtOH 3; eth
　3; bz 3; chl 3; lig 3
Vap. Press. [kPa]: 1.37^{-25}; 6.04^0; 20.0^{25}; 53.6^{50}; 122^{75}
Therm. Prop.[kJ/mol]: $\Delta_{fus}H$ = 8.86; $\Delta_f H°$(l, 25°C) = -83.9
Elec. Prop.: μ = 0; IP = 8.97 eV
η **[mPa s]**: 0.25^{38}
MS: 55(100) 42(57) 41(50) 27(50) 39(38) 29(34) 84(30) 56(25) 69(21) 43(14)
IR [cm^{-1}]: 3030 2950 1660 1460 1410 1380 690
Raman [cm^{-1}]: 3014 2962 2937 2918 2874 2868 1658 1449 1255 890
UV [nm]: 183(12589) hp
^{13}C NMR [ppm]: 12.7 13.7 22.9 29.1 123.8 130.7 CDCl$_3$
Flammability: Flash pt. < -20°C

---

299. *trans*-2-Hexene

Syst. Name: 2-Hexene, (E)-

CASRN: 4050-45-7　　　　　　**Rel. CASRN**: 592-43-8, 25264-93-1
Beil RN: 1719020　　　　　　　**Beil Ref.**: 4-01-00-00834
MF: C_6H_{12}　　　　**MP[°C]**: -133　　**Den.[g/cm^3]**: 0.6732^{25}
MW: 84.16　　　　　**BP[°C]**: 67.9　　　n_D: 1.3936^{20}
Sol.: H_2O 1; EtOH 3; eth
　3; bz 3; chl 3; lig 3
Vap. Press. [kPa]: 1.37^{-25}; 6.20^0; 20.7^{25}; 55.5^{50}; 126^{75}
Therm. Prop.[kJ/mol]: $\Delta_{fus}H$ = 8.26; $\Delta_f H°$(l, 25°C) = -85.5
Elec. Prop.: μ = 0; ε = 1.98^{22}; IP = 8.97 eV
η **[mPa s]**: 0.25^{38}
MS: 55(100) 42(59) 41(50) 27(45) 29(38) 39(31) 56(27) 84(26) 69(19) 43(17)
IR [cm^{-1}]: 3020 2950 1660 1460 1380 970
Raman [cm^{-1}]: 3000 2963 2938 2919 2877 2865 2844 1673 1453 1307 389
UV [nm]: 184(10000) hp

300. *cis*-3-Hexene

Syst. Name: 3-Hexene, (*Z*)-

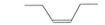

CASRN: 7642-09-3   **Beil Ref.**: 4-01-00-00837
Beil RN: 1718858
MF: C_6H_{12} **MP[°C]**: -137.8   **Den.[g/cm^3]**: 0.6796^{20}
MW: 84.16   **BP[°C]**: 66.4 n_D: 1.3947^{20}
Sol.: H_2O 1; EtOH 3; eth
  3; bz 3; chl 3; lig 3
Vap. Press. [kPa]: 1.50^{-25}; 6.66^0; 22.0^{25}; 58.4^{50}; 132^{75}
Therm. Prop.[kJ/mol]: $\Delta_{fus}H$ = 8.25; $\Delta_fH°$(l, 25°C) = -79.0
Elec. Prop.: μ = 0; ϵ = 2.07^{23}
MS: 55(100) 41(77) 42(68) 39(39) 27(39) 84(35) 56(27) 69(26) 29(25) 43(16)
Raman [cm^{-1}]: 3008 2969 2936 2908 2896 2878 2861 1654 1270 842
UV [nm]: 179(14791) gas

---

301. *trans*-3-Hexene

Syst. Name: 3-Hexene, (*E*)-

CASRN: 13269-52-8   **Beil Ref.**: 4-01-00-00837
Beil RN: 1718859
MF: C_6H_{12} **MP[°C]**: -115.4   **Den.[g/cm^3]**: 0.6772^{20}
MW: 84.16   **BP[°C]**: 67.1 n_D: 1.3943^{20}
Sol.: H_2O 1; EtOH 3; eth
  3; bz 3; chl 3; lig 3
Vap. Press. [kPa]: 1.41^{-25}; 6.37^0; 21.2^{25}; 56.9^{50}; 129^{75}
Therm. Prop.[kJ/mol]: $\Delta_{fus}H$ = 11.08; $\Delta_fH°$(l, 25°C) = -86.1
Elec. Prop.: μ = 0; ϵ = 1.95^{20}
MS: 55(100) 41(81) 42(69) 27(39) 39(38) 84(37) 69(30) 56(28) 29(25) 43(17)
IR [cm^{-1}]: 3030 2960 2940 2870 2850 1660 1460 1440 1380 960 900
Raman [cm^{-1}]: 3080 3000 2940 2920 2870 2740 1640 1610 1440 1410 1350
  1300 1210 1160 1130 1100 1050 990 910 890 870 820 780 740 620 550 450
  390 350 300 250

302. Hexyl acetate

Syst. Name: Ethanoic acid, hexyl ester

CASRN: 142-92-7   **Beil Ref.**: 4-02-00-00159
Beil RN: 1747138
MF: $C_8H_{16}O_2$ **MP[°C]**: -80.9   **Den.[g/cm^3]**: 0.8779^{15}
MW: 144.21   **BP[°C]**: 171.5 n_D: 1.4092^{20}
Sol.: eth 4; EtOH 4
Vap. Press. [kPa]: 0.185^{25}; 0.854^{50}; 3.16^{75}; 9.81^{100}
C_p **(liq.) [J/mol °C]**: 282.8^{27}
γ **[mN/m]**: 26.0^{25}
Elec. Prop.: $\varepsilon = 4.42^{20}$
η **[mPa s]**: 1.075^{25}
MS: 43(100) 56(66) 41(38) 55(37) 61(35) 42(33) 84(31) 69(19) 73(17) 57(7)
IR [cm^{-1}]: 2950 2860 1740 1470 1370 1240 1120 1040 980 950 900 730
Raman [cm^{-1}]: 2940 2900 2870 2730 1730 1440 1360 1300 1230 1200 1150
    1120 1080 1070 1030 980 950 890 870 850 820 750 630 600 510 450 400
    330 270 240
^{1}H NMR [ppm]: 0.9 1.4 2.0 4.0 CCl$_4$
Flammability: Flash pt. = 45°C

303. *sec*-Hexyl acetate

Syst. Name: 2-Pentanol, 4-methyl-, ethanoate
Synonyms: 4-Methyl-2-pentyl acetate

CASRN: 108-84-9
Beil RN: 1749848
MF: $C_8H_{16}O_2$
MW: 144.21
Sol.: eth 4; EtOH 4
Elec. Prop.: μ = (1.9 D)
η **[mPa s]**: 0.93^{20}

DOT No.: 1233
Beil Ref.: 4-02-00-00161
BP[°C]: 147.5   **Den.[g/cm³]**: 0.8805^{25}
     n_D: 1.3980^{20}

IR [cm⁻¹]: 2950 1740 1470 1370 1240 1130 1040 1020 950
Raman [cm⁻¹]: 2940 2900 2880 2770 2730 1740 1470 1450 1440 1380 1350
   1340 1280 1250 1170 1150 1120 1070 1050 1030 950 930 900 890 870 840
   830 780 640 610 570 510 490 450 420 400 310 280 180
¹H NMR [ppm]: 0.9 1.1 1.5 1.9 4.9 CCl₄
Flammability: Flash pt. = 45°C
TLV/TWA: 50 ppm (295 mg/m³)

304. Hexylene glycol

Syst. Name: 2,4-Pentanediol, 2-methyl-
Synonyms: 2-Methyl-2,4-pentanediol

CASRN: 107-41-5   **Beil Ref.**: 4-01-00-02565
Merck No.: 4631
Beil RN: 1098298
MF: $C_6H_{14}O_2$ **MP[°C]**: -50   **Den.[g/cm^3]**: 0.923^{15}
MW: 118.18   **BP[°C]**: 197.1 n_D: 1.4276^{20}
Sol.: H_2O 3; EtOH 3; eth
    3; ctc 2
Vap. Press. [kPa]: 2.06^{100}; 6.68^{125}; 18.9^{150}
Therm. Prop.[kJ/mol]: $\Delta_{vap}H = 57.3$
C_p **(liq.) [J/mol °C]**: 336.0^{25}
γ **[mN/m]**: 33.1^{20}
Elec. Prop.: $\mu = (2.9$ D$)$; $\varepsilon = 23.4^{30}$
η **[mPa s]**: 34.4^{20}
MS: 59(100) 43(61) 56(25) 45(17) 41(16) 57(13) 42(13) 85(11) 61(10) 31(10)
IR [cm^{-1}]: 3350 2980 2940 1450 1410 1380 1320 1260 1220 1200 1150 1120
    1070 1040 970 950 900 880 840 770
Raman [cm^{-1}]: 3340 2980 2930 2920 2880 2750 2720 1460 1450 1420 1350
    1330 1300 1260 1220 1200 1160 1120 1080 1050 990 970 950 930 900 880
    840 800 770 560 500 460 450 410 370 320 220 150
UV [nm]: 302(3631) sulf
^{1}H NMR [ppm]: 1.1 1.2 1.3 1.4 1.6 4.2 4.7 4.9 CCl_4
Flammability: Flash pt. = 102°C; ign. temp. = 306°C; flam. lim. = 1-9%
TLV/TWA: 25 ppm (121 mg/m^3)

305. Hexyl methyl ketone

Syst. Name: 2-Octanone
Synonyms: Methyl hexyl ketone

CASRN: 111-13-7   **Beil Ref.**: 4-01-00-03339
Merck No.: 4632
Beil RN: 635843
MF: $C_8H_{16}O$ **MP[°C]**: -16   **Den.[g/cm^3]**: 0.820^{20}
MW: 128.21   **BP[°C]**: 172.5 n_D: 1.4151^{20}
Sol.: H_2O 2; EtOH 5; eth
  5
Vap. Press. [kPa]: 0.528^{50}; 2.20^{75}; 7.34^{100}; 20.5^{125}; 50.0^{150}
Therm. Prop.[kJ/mol]: $\Delta_{fus}H$ = 24.42
C_p **(liq.) [J/mol °C]**: 273.3^{25}
γ **[mN/m]**: 25.73^{20}
Elec. Prop.: μ = (2.7 D); ε = 9.51^{20}
η **[mPa s]**: 1.19^{15}
MS: 43(100) 58(79) 41(56) 59(52) 71(49) 27(46) 29(36) 39(27) 57(18) 55(17)
IR [cm^{-1}]: 2960 2930 2860 1720 1465 1410 1355 1220 1160 1115 945 720 590
Raman [cm^{-1}]: 2930 2880 2740 1710 1440 1410 1370 1300 1160 1120 1080
   1070 1030 1010 960 890 870 850 820 770 740 720 600 470 410 350 250
^{1}H NMR [ppm]: 0.9 1.4 2.1 2.4 CCl_4
Flammability: Flash pt. = 52°C

306. Hydracrylonitrile

Syst. Name: Propanenitrile, 3-hydroxy-
Synonyms: 3-Hydroxypropanenitrile; Ethylene
cyanohydrin; 2-Cyanoethanol

CASRN: 109-78-4   **Beil Ref.**: 4-03-00-00708
Merck No.: 3751
Beil RN: 635773

MF: C_3H_5NO	**MP[°C]**: -46	**Den.[g/cm^3]**: 1.0404^{25}
MW: 71.08	**BP[°C]**: 221	n_D: 1.4248^{20}

Sol.: H_2O 5; EtOH 5; eth
2; chl 3; CS_2 1
Vap. Press. [kPa]: 0.010^{25}; 0.083^{50}; 0.345^{75}; 1.21^{100}; 3.66^{125}; 9.87^{150}
MS: 41(100) 31(97) 29(20) 42(17) 52(15) 40(13) 53(10) 51(9) 39(7) 26(7)
IR [cm^{-1}]: 3450 2900 2260 1415 1325 1245 1185 1060 850
Raman [cm^{-1}]: 3500 2980 2950 2910 2260 1480 1420 1330 1250 1220 1070
1020 1000 950 860 840 590 530 380 230
^{1}H NMR [ppm]: 2.6 3.4 3.9 CDCl$_3$
Flammability: Flash pt. = 129°C

307. Iodobenzene

Syst. Name: Benzene, iodo-
Synonyms: Phenyl iodide

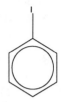

CASRN: 591-50-4   **Beil Ref.:** 4-05-00-00688
Merck No.: 4922
Beil RN: 1446140
MF: C_6H_5I **MP[°C]:** -31.3   **Den.[g/cm^3]:** 1.8308^{20}
MW: 204.01   **BP[°C]:** 188.4 n_D: 1.6200^{20}
Sol.: H_2O 1; EtOH 3; eth
   5; ace 5; bz 5; ctc 5;
   lig 5
Crit. Const.: T_c = 448°C; P_c = 4.52 MPa; V_c = 351 cm^3
Vap. Press. [kPa]: 0.133^{25}; 0.632^{50}; 2.28^{75}; 6.67^{100}; 16.7^{125}; 36.7^{150}
Therm. Prop.[kJ/mol]: $\Delta_{fus}H$ = 9.76; $\Delta_{vap}H$ = 39.5; $\Delta_f H°$(l, 25°C) = 117.2
C_p **(liq.) [J/mol °C]:** 158.7^{25}
γ **[mN/m]:** 38.71^{25}; dγ/dT = 0.1123 mN/m °C
Elec. Prop.: μ = 1.70 D; ε = 4.59^{20}; IP = 8.69 eV
η **[mPa s]:** 2.35^0; 1.55^{25}; 1.12^{50}; 0.854^{75}; 0.683^{100}
MS: 204(100) 77(82) 51(32) 50(20) 127(8) 205(6) 78(6) 74(6) 102(5) 76(5)
IR [cm^{-1}]: 3030 1590 1470 1450 1060 1010 1000 730 690
Raman [cm^{-1}]: 3150 3070 1670 1580 1470 1440 1380 1330 1270 1180 1160
   1100 1060 1040 1020 1010 990 970 910 840 730 700 660 620 540 450 410
   330 310 280 230 170
UV [nm]: 250(732) 226(11700) MeOH
^{13}C NMR [ppm]: 94.4 127.1 129.9 137.2 diox
^{1}H NMR [ppm]: 6.8 7.5 7.7 CCl$_4$

308. 1-Iodobutane

Syst. Name: Butane, 1-iodo-
Synonyms: Butyl iodide

CASRN: 542-69-8   **Beil Ref.**: 4-01-00-00271
Merck No.: 1572
Beil RN: 1420755
MF: C_4H_9I **MP[°C]**: -103   **Den.[g/cm^3]**: 1.6154^{20}
MW: 184.02   **BP[°C]**: 130.6 n_D: 1.5001^{20}
Sol.: H_2O 1; EtOH 5; eth
   5; chl 4
Vap. Press. [kPa]: 1.85^{25}; 6.37^{50}; 17.7^{75}; 42.0^{100}; 87.4^{125}; 165^{150}
Therm. Prop.[kJ/mol]: $\Delta_{vap}H$ = 34.7
γ **[mN/m]**: 28.24^{25}; $d\gamma/dT$ = 0.1031 mN/m °C
Elec. Prop.: μ = (1.9 D); ϵ = 6.27^{20}; IP = 9.23 eV
η **[mPa s]**: 0.826^{25}
MS: 57(100) 29(78) 41(56) 27(39) 184(36) 39(17) 28(16) 26(7) 127(6) 55(6)
IR [cm^{-1}]: 2950 2900 2850 1450 1400 1370 1270 1230 1170 850 720
UV [nm]: 254(501) MeOH
^{13}C NMR [ppm]: 1.7 13.7 24.2 36.1
^{1}H NMR [ppm]: 1.0 1.7 1.9 4.2 CDCl$_3$

309. 2-Iodobutane

Syst. Name: Butane, 2-iodo-, (±)-
Synonyms: *sec*-Butyl iodide

CASRN: 52152-71-3   **Rel. CASRN**: 513-48-4
Merck No.: 1573   **DOT No.**: 2390
Beil RN: 1718780   **Beil Ref.**: 4-01-00-00272
MF: C_4H_9I **MP[°C]**: -104.2   **Den.[g/cm³]**: 1.5920^{20}
MW: 184.02   **BP[°C]**: 120 n_D: 1.4991^{20}
Sol.: H_2O 1; EtOH 5; eth
    5; chl 4
Vap. Press. [kPa]: 57.8^{100}; 77.2^{110}
Therm. Prop.[kJ/mol]: $\Delta_{vap}H = 33.3$
Elec. Prop.: $\mu = 2.12$ D; $\varepsilon = 7.87^{20}$; IP = 9.09 eV
MS: 57(100) 29(36) 184(36) 41(36) 27(10) 39(9) 127(6) 58(4)
IR [cm⁻¹]: 2960 2900 2860 1440 1370 1250 1170 1130 980 940 820 770
UV [nm]: 261(550) peth
¹H NMR [ppm]: 1.0 1.7 1.9 4.2 $CDCl_3$

310. Iodoethane

Syst. Name: Ethane, iodo-
Synonyms: Ethyl iodide

CASRN: 75-03-6   **Beil Ref.:** 4-01-00-00163
Merck No.: 3769
Beil RN: 505934
MF: C_2H_5I **MP[°C]:** -111.1   **Den.[g/cm^3]:** 1.9358^{20}
MW: 155.97   **BP[°C]:** 72.5 n_D: 1.5133^{20}
Sol.: H_2O 2; EtOH 5; eth
  3; chl 3; os 5
Vap. Press. [kPa]: 1.22^{-25}; 5.46^0; 18.2^{25}; 48.5^{50}; 110^{75}
Therm. Prop.[kJ/mol]: $\Delta_{vap}H$ = 29.4; $\Delta_f H°$(l, 25°C) = -40.2
C_p **(liq.) [J/mol °C]:** 115.1^{25}
γ **[mN/m]:** 28.46^{25}; dγ/dT = 0.1286 mN/m °C
Elec. Prop.: μ = 1.91 D; ϵ = 7.82^{20}; IP = 9.35 eV
η **[mPa s]:** 0.723^0; 0.556^{25}; 0.444^{50}; 0.365^{75}
MS: 156(100) 29(75) 27(63) 127(31) 26(14) 28(9) 128(8) 141(2) 25(2) 140(1)
IR [cm^{-1}]: 2940 2860 1450 1370 1210 950 740
Raman [cm^{-1}]: 2960 2910 2850 1430 1370 1200 1050 950 500 450 260
UV [nm]: 255(468) MeOH
^{1}H NMR [ppm]: 1.2 2.6 3.7 $CDCl_3$

311. Iodomethane

Syst. Name: Methane, iodo-
Synonyms: Methyl iodide

CASRN: 74-88-4   **DOT No.:** 2644
Merck No.: 6002   **Beil Ref.:** 4-01-00-00087
Beil RN: 969135
MF: CH_3I **MP[°C]:** -66.4   **Den.[g/cm^3]:** 2.279^{20}
MW: 141.94   **BP[°C]:** 42.5 n_D: 1.5380^{20}
Sol.: H_2O 2; EtOH 5; eth
  5; ace 3; bz 3; chl 3
Crit. Const.: T_c = 255°C
Vap. Press. [kPa]: 4.97^{-25}; 18.6^0; 53.9^{25}; 130^{50}
Therm. Prop.[kJ/mol]: $\Delta_{vap}H$ = 27.3; $\Delta_f H°$(l, 25°C) = -12.3
C_p **(liq.) [J/mol °C]:** 126.0^{25}
γ **[mN/m]:** 30.34^{25}; $d\gamma/dT$ = 0.1234 mN/m °C
Elec. Prop.: μ = 1.62 D; ε = 6.97^{20}; IP = 9.54 eV
η **[mPa s]:** 0.594^0; 0.469^{25}
MS: 142(100) 127(38) 141(14) 15(13) 139(5) 140(4) 128(3) 14(1) 13(1) 71(0)
IR [cm^{-1}]: 2940 1430 1250 890
Raman [cm^{-1}]: 2940 1240 520
^{13}**C NMR [ppm]:** -20.5
^{1}**H NMR [ppm]:** 2.2 $CDCl_3$
TLV/TWA: 2 ppm (12 mg/m^3)
Reg. Lists: CERCLA (RQ = 100 lb.); SARA 313 (0.1%); RCRA U138 (toxic);
  confirmed or suspected carcinogen

312. 1-Iodo-2-methylpropane

Syst. Name: Propane, 1-iodo-2-methyl-
Synonyms: Isobutyl iodide

CASRN: 513-38-2   **DOT No.**: 2391
Merck No.: 5027   **Beil Ref.**: 4-01-00-00299
Beil RN: 1730927
MF: C_4H_9I **BP[°C]**: 121.1   **Den.[g/cm^3]**: 1.6035^{20}
MW: 184.02 n_D: 1.4959^{20}
Vap. Press. [kPa]: 0.075^{-25}; 0.468^0; 2.10^{25}; 7.36^{50}; 21.3^{75}; 52.8^{100}; 116^{125}
Therm. Prop.[kJ/mol]: $\Delta_{vap}H = 33.5$
C_p **(liq.) [J/mol °C]**: 162.3^{25}
γ **[mN/m]**: 27.97^{20}
Elec. Prop.: $\mu = (1.9\ D)$; $\varepsilon = 6.47^{20}$; IP = 9.20 eV
η **[mPa s]**: 0.777^{30}
MS: 57(100) 41(45) 184(40) 29(40) 39(15) 27(10) 127(5) 58(5) 55(4) 43(4)
IR [cm^{-1}]: 2940 1470 1390 1370 1320 1190 1100 1040 940 930 830 790
^{1}H NMR [ppm]: 1.0 1.7 3.1 CCl_4

313. 1-Iodopropane

Syst. Name: Propane, 1-iodo-
Synonyms: Propyl iodide

CASRN: 107-08-4
Merck No.: 7875
Beil RN: 505937
MF: C_3H_7I
MW: 169.99
Sol.: H_2O 2; EtOH 5; eth
  5; ctc 2

Rel. CASRN: 26914-02-3
DOT No.: 2392
Beil Ref.: 4-01-00-00222
MP[°C]: -101.3   **Den.[g/cm^3]:** 1.7489^{20}
BP[°C]: 102.6 n_D: 1.5058^{20}

Vap. Press. [kPa]: 1.48^0; 5.75^{25}; 17.4^{50}; 43.6^{75}; 94.5^{100}; 183^{125}
Therm. Prop.[kJ/mol]: $\Delta_{vap}H$ = 32.1; $\Delta_f H°$(l, 25°C) = -66.0
C_p **(liq.) [J/mol °C]:** 126.8^{25}
γ **[mN/m]:** 30.11^{15}
Elec. Prop.: μ = 2.04 D; ϵ = 7.07^{20}; IP = 9.27 eV
η **[mPa s]:** 0.970^0; 0.703^{25}; 0.541^{50}; 0.436^{75}; 0.363^{100}
MS: 43(100) 170(68) 41(35) 27(26) 127(14) 39(11) 44(4) 141(3) 128(3) 42(3)
IR [cm^{-1}]: 2940 2860 1470 1430 1390 1280 1190 1080 1010 890 880 810 760
  730
Raman [cm^{-1}]: 3000 2950 2920 2880 2860 2830 2730 1450 1420 1370 1320
  1270 1190 1180 1090 1070 1010 870 810 760 590 500 380 280 250 200
UV [nm]: 254(501) MeOH
^{13}C NMR [ppm]: 9.2 15.3 26.8 $CDCl_3$
^{1}H NMR [ppm]: 1.0 1.8 3.2 CCl_4

314. 2-Iodopropane

Syst. Name: Propane, 2-iodo-
Synonyms: Isopropyl iodide

CASRN: 75-30-9
Merck No.: 5102
Beil RN: 1098244
MF: C_3H_7I
MW: 169.99
Sol.: H_2O 2; EtOH 5; eth
　5; bz 5; chl 5

Rel. CASRN: 26914-02-3
DOT No.: 2392
Beil Ref.: 4-01-00-00223
MP[°C]: -90
BP[°C]: 89.5
Den.[g/cm^3]: 1.7042^{20}
n_D: 1.5028^{20}

Vap. Press. [kPa]: 2.65^0; 9.36^{25}; 26.7^{50}; 64.4^{75}; 137^{100}
Therm. Prop.[kJ/mol]: $\Delta_{vap}H$ = 30.7; $\Delta_fH°$(l, 25°C) = -74.8
C_p (liq.) [J/mol °C]: 91.0^{25}
γ [mN/m]: 27.42^{20}
Elec. Prop.: μ = (1.9 D); ε = 8.19^{25}; IP = 9.18 eV
η [mPa s]: 0.883^0; 0.653^{25}; 0.506^{50}; 0.407^{75}
MS: 43(100) 170(46) 41(45) 27(44) 39(19) 127(17) 42(4) 38(4) 128(3) 44(3)
IR [cm^{-1}]: 2940 1450 1370 1210 1150 1020 930 870
UV [nm]: 259(562) MeOH
^{13}C NMR [ppm]: 21.1 32.0
^{1}H NMR [ppm]: 1.9 4.3 $CDCl_3$

315. Isobutanal

Syst. Name: Propanal, 2-methyl-
Synonyms: 2-Methylpropanal; Isobutyraldehyde

CASRN: 78-84-2
Merck No.: 5038
Beil RN: 605330
MF: C_4H_8O
MW: 72.11
Sol.: H_2O 3; eth 3; ace 3; ctc 2; chl 3

DOT No.: 2045
Beil Ref.: 4-01-00-03262

MP[°C]: -65.9
BP[°C]: 64.5

Den.[g/cm^3]: 0.7891^{20}
n_D: 1.3730^{20}

Vap. Press. [kPa]: 6.66^0; 23.0^{25}; 62.5^{50}
Therm. Prop.[kJ/mol]: $\Delta_f H°(l, 25°C) = -247.4$
Elec. Prop.: $\mu = (2.6\ D)$; IP = 9.71 eV
η [mPa s]: 0.538^{28}
MS: 43(100) 41(84) 27(47) 72(46) 39(30) 29(25) 42(10) 70(5) 38(5) 28(5)
IR [cm^{-1}]: 2940 2700 1720 1470 1390 1110 910 840 790
Raman [cm^{-1}]: * 1780 1716 1698 1464 1450 1399 1276 918 801 402 341
UV [nm]: 282(8) H_2O
^{1}H NMR [ppm]: 1.1 2.4 9.6 CCl$_4$
Flammability: Flash pt. = -18°C; ign. temp. = 196°C; flam. lim. = 1.6-10.6%
Reg. Lists: SARA 313 (1.0%)

316. Isobutane

Syst. Name: Propane, 2-methyl-
Synonyms: 2-Methylpropane

CASRN: 75-28-5   **DOT No.:** 1969
Beil RN: 1730720   **Beil Ref.:** 4-01-00-00282
MF: C_4H_{10} **MP[°C]:** -138.3   **Den.[g/cm^3]:** 0.5510^{25} (sat. press.)
MW: 58.12   **BP[°C]:** -11.7 n_D: 1.3518^{-25}
Sol.: H_2O 2; EtOH 3; eth
    3; chl 3
Crit. Const.: T_c = 134.70°C; P_c = 3.630 MPa; V_c = 257 cm^3
Vap. Press. [kPa]: 3.7^{-73}; 58.7^{-25}; 156^0; 348^{25}; 677^{50}; 1190^{75}; 1928^{100}; 2929^{125}
Therm. Prop.[kJ/mol]: $\Delta_{fus}H$ = 4.66; $\Delta_{vap}H$ = 21.3; $\Delta_fH°$(g, 25°C) = -134.2
C_p **(liq.) [J/mol °C]:** 130.5^{-12}
γ **[mN/m]:** 15.30^{-20}; dγ/dT = 0.1236 mN/m °C
Elec. Prop.: μ = 0.132 D; ε = 1.75^{22}; IP = 10.57 eV
MS: 43(100) 42(37) 41(35) 27(17) 39(11) 29(5) 58(4) 57(4) 44(3) 40(2)
IR [cm^{-1}]: 2940 1470 1370 1330 1180 920
Raman [cm^{-1}]: 2960 2916 2886 2867 2713 1450 1328 1172 965 797
^{13}**C NMR [ppm]:** 24.3 25.2
Flammability: Flash pt. = -87°C; ign. temp. = 460°C; flam. lim. = 1.8-8.4%

317. Isobutyl acetate

Syst. Name: Ethanoic acid, 2-methylpropyl ester
Synonyms: 2-Methylpropyl acetate

CASRN: 110-19-0
DOT No.: 1213
Merck No.: 5014
Beil Ref.: 4-02-00-00149
Beil RN: 1741909
MF: $C_6H_{12}O_2$
MP[°C]: -98.8
Den.[g/cm^3]: 0.8712^{20}
MW: 116.16
BP[°C]: 116.5
n_D: 1.3902^{20}
Sol.: H_2O 2; EtOH 5; eth
5; ace 3; ctc 2
Crit. Const.: $T_c = 288°C$; $P_c = 3.16$ MPa
Vap. Press. [kPa]: 2.39^{25}; 8.56^{50}; 24.7^{75}; 60.3^{100}; 129^{125}
Therm. Prop.[kJ/mol]: $\Delta_{vap}H = 35.9$
C_p (liq.) [J/mol °C]: 233.8^{25}
γ [mN/m]: 23.06^{25}; $d\gamma/dT = 0.1013$ mN/m °C
Elec. Prop.: $\mu = (1.9$ D); $\varepsilon = 5.07^{20}$
η [mPa s]: 0.676^{25}; 0.493^{50}; 0.370^{75}; 0.286^{100}
MS: 43(100) 56(26) 73(15) 41(10) 29(5) 71(3) 57(3) 39(3) 27(3) 86(2)
IR [cm^{-1}]: 2950 2860 1740 1470 1390 1380 1360 1230 1180 1030 990 960 940
920 900 820 630
Raman [cm^{-1}]: 2940 2920 2880 2790 2760 2730 1730 1470 1460 1400 1370
1350 1300 1260 1180 1140 1040 990 970 950 930 910 830 810 640 610 580
510 410 390 330 280
UV [nm]: 214(56) 208(50) iso 208(58) MeOH
^{1}H NMR [ppm]: 0.9 1.9 2.0 3.8 CCl_4
Flammability: Flash pt. = 18°C; ign. temp. = 421°C; flam. lim. = 1.3-10.5%
TLV/TWA: 150 ppm (713 mg/m^3)
Reg. Lists: CERCLA (RQ = 5000 lb.)

318. Isobutylamine

Syst. Name: 1-Propanamine, 2-methyl-
Synonyms: 2-Methyl-1-propanamine

CASRN: 78-81-9   **DOT No.**: 1214
Merck No.: 5016   **Beil Ref.**: 4-04-00-00625
Beil RN: 385626
MF: $C_4H_{11}N$ **MP[°C]**: -86.7   **Den.[g/cm³]**: 0.724^{25}
MW: 73.14   **BP[°C]**: 67.7 n_D: 1.3988^{19}
Crit. Const.: $T_c = 246°C$; $P_c = 4.07$ MPa; $V_c = 278$ cm³
Vap. Press. [kPa]: 0.961^{-25}; 5.14^0; 19.0^{25}; 54.2^{50}; 128^{75}
Therm. Prop.[kJ/mol]: $\Delta_{vap}H = 30.6$; $\Delta_f H°(l, 25°C) = -132.6$
C_p **(liq.) [J/mol °C]**: 183.2^{25}
γ **[mN/m]**: 21.75^{25}; $d\gamma/dT = 0.1092$ mN/m °C
Elec. Prop.: $\mu = (1.3$ D); $\varepsilon = 4.43^{21}$; IP = 8.70 eV
η **[mPa s]**: 0.770^0; 0.571^{25}; 0.367^{50}
MS: 30(100) 28(9) 41(6) 73(5) 27(5) 39(4) 29(3) 15(3) 58(2) 56(2)
Flammability: Flash pt. = -9°C; ign. temp. = 378°C; flam. lim. = 2-12%
Reg. Lists: CERCLA (RQ = 1000 lb.)

319. Isobutylbenzene
Syst. Name: Benzene, (2-methylpropyl)-
Synonyms: (2-Methylpropyl)benzene

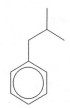

CASRN: 538-93-2   **Beil Ref.**: 4-05-00-01042
Merck No.: 5018
Beil RN: 1852218
MF: $C_{10}H_{14}$ **MP[°C]**: -51.4   **Den.[g/cm^3]**: 0.8532^{20}
MW: 134.22   **BP[°C]**: 172.7 n_D: 1.4866^{20}
Sol.: H_2O 1; EtOH 5;
  eth 5; ace 5; bz 5; ctc
  5; peth 5
Crit. Const.: T_c = 377°C; P_c = 3.05 MPa
Vap. Press. [kPa]: 0.257^{25}; 3.87^{75}; 10.9^{100}; 26.2^{125}; 55.6^{150}
Therm. Prop.[kJ/mol]: $\Delta_{fus}H$ = 12.51; $\Delta_{vap}H$ = 37.8; $\Delta_f H°$(l, 25°C) = -69.8
C_p (liq.) [J/mol °C]: 240.6^{25}
γ [mN/m]: 26.99^{25}; $d\gamma/dT$ = 0.0961 mN/m °C
Elec. Prop.: μ = 0; ϵ = 2.32^{20}; IP = 8.68 eV
MS: 91(100) 92(58) 43(22) 134(20) 65(12) 41(12) 39(11) 27(8) 51(7) 93(5)
IR [cm^{-1}]: 3030 2941 1587 1471 1370 1149 1064 1020 730 704
Raman [cm^{-1}]: 3050 3033 2971 2955 2933 2883 2865 1605 1204 1031 1003
UV [nm]: 268(174) 264(170) 261(216) 259(221) MeOH
^{1}H NMR [ppm]: 0.9 1.9 2.4 7.1 CCl_4
Flammability: Flash pt. = 55°C; ign. temp. = 427°C; flam. lim. = 0.8-6.0%

320. Isobutyl formate

Syst. Name: Methanoic acid, 2-methylpropyl ester
Synonyms: 2-Methylpropyl formate

CASRN: 542-55-2   **DOT No.**: 2393
Merck No.: 5026   **Beil Ref.**: 4-02-00-00029
Beil RN: 1738888
MF: $C_5H_{10}O_2$ **MP[°C]**: -95.8   **Den.[g/cm^3]**: 0.8776^{20}
MW: 102.13   **BP[°C]**: 98.2 n_D: 1.3857^{20}
Sol.: H_2O 2; EtOH 5; eth
  5; ace 4; chl 2
Crit. Const.: $T_c = 278°C$; $P_c = 3.88$ MPa; $V_c = 352$ cm^3
Vap. Press. [kPa]: 5.34^{25}; 17.7^{50}; 47.3^{75}; 107^{100}
Therm. Prop.[kJ/mol]: $\Delta_{vap}H = 33.6$
γ **[mN/m]**: 23.90^{20}; $d\gamma/dT = 0.1122$ mN/m °C
Elec. Prop.: $\mu = (1.9$ D$)$; $\varepsilon = 6.41^{20}$
η **[mPa s]**: 0.680^{20}
MS: 43(100) 56(82) 41(78) 31(65) 29(64) 27(53) 60(39) 39(36) 42(26) 15(14)
IR [cm^{-1}]: 2940 1720 1470 1390 1370 1220 940
Raman [cm^{-1}]: * 1780 1703 1455 1377 1341 1182 1134 967 828 762 331
^{1}H NMR [ppm]: 1.0 2.0 3.9 8.0 CDCl$_3$
Flammability: Flash pt. < 21°C; ign. temp. = 320°C; flam. lim. = 1.7-8%

321. Isobutyl isobutanoate

Syst. Name: Propanoic acid, 2-methyl-, 2-
   methylpropyl ester
Synonyms: Isobutyl isobutyrate

CASRN: 97-85-8   **DOT No.**: 2528
Merck No.: 5028   **Beil Ref.**: 4-02-00-00847
Beil RN: 1701355
MF: $C_8H_{16}O_2$ **MP[°C]**: -80.7   **Den.[g/cm^3]**: 0.8542^{20}
MW: 144.21   **BP[°C]**: 148.6 n_D: 1.3999^{20}
Sol.: H_2O 2; EtOH 3; eth
   5; ace 3; ctc 2
Crit. Const.: $T_c = 329°C$
Vap. Press. [kPa]: 0.097^0; 0.552^{25}; 2.31^{50}; 7.67^{75}; 21.2^{100}; 50.9^{125}; 109^{150}
Therm. Prop.[kJ/mol]: $\Delta_{vap}H = 38.2$
γ **[mN/m]**: 33.8^{-76}
Elec. Prop.: $\mu = (1.9 D)$
η **[mPa s]**: 0.83^{25}
MS: 43(100) 41(94) 56(76) 71(70) 57(68) 27(62) 29(59) 89(37) 39(32) 42(17)
IR [cm^{-1}]: 2940 1720 1450 1390 1330 1250 1190 1150 1080 1000 960 940 920
   860 810 750
^{1}H NMR [ppm]: 0.9 1.2 1.9 2.5 3.8 CCl_4
Flammability: Flash pt. = 38°C; ign. temp. = 432°C; flam. lim. = 0.96-7.59%

322. Isopentane

Syst. Name: Butane, 2-methyl-
Synonyms: 2-Methylbutane

CASRN: 78-78-4   **DOT No.**: 1265
Beil RN: 1730723   **Beil Ref.**: 4-01-00-00320
MF: C_5H_{12} **MP[°C]**: -159.9   **Den.[g/cm^3]**: 0.6201^{20}
MW: 72.15   **BP[°C]**: 27.8 n_D: 1.3537^{20}
Sol.: H_2O 1; EtOH 5; eth
   5
Crit. Const.: T_c = 187.28°C; P_c = 3.381 MPa; V_c = 306 cm^3
Vap. Press. [kPa]: 10.3^{-25}; 34.6^0; 91.7^{25}; 205^{50}; 404^{75}
Therm. Prop.[kJ/mol]: $\Delta_{fus}H$ = 5.15; $\Delta_{vap}H$ = 24.7; $\Delta_fH°$(l, 25°C) = -178.5
C_p (liq.) [J/mol °C]: 164.8^{25}
γ [mN/m]: 14.44^{25}; $d\gamma/dT$ = 0.1103 mN/m °C
Elec. Prop.: μ = 0.13 D; ϵ = 1.85^{20}; IP = 10.22 eV
η [mPa s]: 0.376^{-25}; 0.277^0; 0.214^{25}
MS: 43(100) 42(84) 41(60) 29(47) 27(45) 57(36) 39(17) 56(12) 72(3) 55(3)
IR [cm^{-1}]: 2960 2860 1470 1380 1280 1150 1020 970 910 760
^{13}C NMR [ppm]: 11.5 22.0 29.9 31.8
Flammability: Flash pt. < -51°C; ign. temp. = 420°C; flam. lim. = 1.4-7.6%

323. Isopentyl acetate

Syst. Name: 1-Butanol, 3-methyl-, ethanoate
Synonyms: Isoamyl acetate

CASRN: 123-92-2   **Beil Ref.:** 4-02-00-00157
Merck No.: 4993
Beil RN: 1744750
MF: $C_7H_{14}O_2$ **MP[°C]:** -78.5   **Den.[g/cm³]:** 0.876^{15}
MW: 130.19   **BP[°C]:** 142.5 n_D: 1.4000^{20}
Sol.: H_2O 2; EtOH 5; eth
  5; ace 3; chl 3; AmOH
  3
Crit. Const.: T_c = 326°C
Vap. Press. [kPa]: 0.728^{25}; 2.94^{50}; 9.49^{75}; 25.7^{100}; 60.6^{125}; 128^{150}
Therm. Prop.[kJ/mol]: $\Delta_{vap}H$ = 37.5
C_p **(liq.) [J/mol °C]:** 248.5^{25}
γ **[mN/m]:** 24.77^{20}; $d\gamma/dT$ = 0.0989 mN/m °C
Elec. Prop.: μ = (1.9 D); ε = 4.72^{20}
η **[mPa s]:** 0.790^{25}
MS: 43(100) 70(49) 55(38) 61(15) 42(15) 41(14) 27(12) 87(11) 29(10) 73(9)
IR [cm⁻¹]: 2940 1720 1470 1390 1240 1160 1050 960
Raman [cm⁻¹]: 2940 2900 2880 2730 1730 1460 1450 1360 1300 1270 1240
  1150 1120 1070 1030 970 940 900 870 840 800 770 630 610 500 440 410
  370 330 280
¹H NMR [ppm]: 0.9 1.5 2.0 4.0 CDCl₃
Flammability: Flash pt. = 25°C; ign. temp. = 360°C; flam. lim. = 1.0-7.5%
TLV/TWA: 100 ppm (532 mg/m³)
Reg. Lists: CERCLA (RQ = 5000 lb.)

324. Isopentyl isopentanoate

Syst. Name: Butanoic acid, 3-methyl-, 3-methylbutyl ester

Synonyms: 3-Methylbutyl 3-methylbutanoate; Isoamyl valerate

CASRN: 659-70-1   **Beil Ref.**: 4-02-00-00899

Merck No.: 5003

Beil RN: 1753884

MF: $C_{10}H_{20}O_2$ **BP[°C]**: 190.4   **Den.[g/cm^3]**: 0.8583[19]

MW: 172.27 n_D: 1.4130[19]

Vap. Press. [kPa]: 1.80[75]; 5.25[100]; 13.3[125]; 30.1[150]

Therm. Prop.[kJ/mol]: $\Delta_{vap}H$ = 45.9

Elec. Prop.: ε = 4.39[15]

MS: 70(100) 43(48) 85(40) 57(38) 41(33) 55(26) 71(24) 29(22) 42(14) 32(14)

---

325. Isophorone

Syst. Name: 2-Cyclohexen-1-one, 3,5,5-trimethyl-

Synonyms: 3,5,5-Trimethyl-2-cyclohexen-1-one

CASRN: 78-59-1   **DOT No.**: 1224

Beil RN: 1280721   **Beil Ref.**: 4-07-00-00165

MF: $C_9H_{14}O$ **MP[°C]**: -8.1   **Den.[g/cm^3]**: 0.9255[20]

MW: 138.21   **BP[°C]**: 215.2 n_D: 1.4766[18]

Vap. Press. [kPa]: 0.060[25]; 0.274[50]; 0.995[75]; 2.99[100]; 7.73[125]; 17.7[150]

C_p **(liq.) [J/mol °C]**: 253.5[25]

Elec. Prop.: IP = 9.07 eV

η **[mPa s]**: 4.20[0]; 2.33[25]; 1.42[50]; 0.923[75]; 0.638[100]

MS: 82(100) 39(20) 138(17) 54(13) 27(12) 41(10) 53(8) 83(7) 29(7) 55(6)

UV [nm]: 337(32) 225(14125) iso

Flammability: Flash pt. = 84°C; ign. temp. = 460°C; flam. lim. = 0.8-3.8%

TLV/TWA: 5 ppm (28 mg/m^3)

Reg. Lists: CERCLA (RQ = 5000 lb.)

326. Isopropyl acetate

Syst. Name: Ethanoic acid, 1-methylethyl ester
Synonyms: 1-Methylethyl acetate

CASRN: 108-21-4   **DOT No.**: 1220
Merck No.: 5093   **Beil Ref.**: 4-02-00-00141
Beil RN: 1740761
MF: $C_5H_{10}O_2$ **MP[°C]**: -73.4   **Den.[g/cm^3]**: 0.8718^{20}
MW: 102.13   **BP[°C]**: 88.6 n_D: 1.3773^{20}
Sol.: H_2O 3; EtOH 3; eth 5; ace 3; chl 3
Crit. Const.: T_c = 258°C
Therm. Prop.[kJ/mol]: $\Delta_{vap}H$ = 32.9; $\Delta_fH°$(l, 25°C) = -518.9
C_p **(liq.) [J/mol °C]**: 199.4^{25}
γ **[mN/m]**: 21.76^{25}; $d\gamma/dT$ = 0.1072 mN/m °C
Elec. Prop.: IP = 9.99 eV
η **[mPa s]**: 0.569^{20}
MS: 43(100) 61(17) 41(14) 87(9) 59(8) 27(8) 42(7) 39(4) 45(3) 44(2)
IR [cm^{-1}]: 2950 2900 1700 1430 1330 1200 1130 1090 1060 960 900 830 760
Raman [cm^{-1}]: * 1780 1739 1469 1359 1168 1134 981 840 658 458 336
UV [nm]: 204(62) H_2O 209(59) MeOH
^{13}C NMR [ppm]: 20.8 21.8 67.4 169.6 diox
^{1}H NMR [ppm]: 0.9 1.4 1.6 2.4 CDCl$_3$
Flammability: Flash pt. = 2°C; ign. temp. = 460°C; flam. lim. = 1.8-8%
TLV/TWA: 250 ppm (1040 mg/m^3)

327. Isopropylamine

Syst. Name: 2-Propanamine

CASRN: 75-31-0   **DOT No.**: 1221
Merck No.: 5097   **Beil Ref.**: 4-04-00-00504
Beil RN: 605259
MF: C_3H_9N **MP[°C]**: -95.1   **Den.[g/cm^3]**: 0.6891^{20}
MW: 59.11   **BP[°C]**: 31.7 n_D: 1.3742^{20}
Sol.: H_2O 5; EtOH 5; eth
   5; ace 4; bz 3; chl 3
Crit. Const.: T_c = 198.7°C; P_c = 4.54 MPa; V_c = 221 cm^3
Vap. Press. [kPa]: 78.0^{25}; 192^{50}
Therm. Prop.[kJ/mol]: $\Delta_{fus}H$ = 7.33; $\Delta_{vap}H$ = 27.8; $\Delta_fH°$(l, 25°C) = -112.3
C_p (liq.) [J/mol °C]: 163.8^{25}
γ [mN/m]: 19.53^{20}
Elec. Prop.: μ = 1.19 D; ε = 5.63^{20}; IP = 8.72 eV
η [mPa s]: 0.454^0; 0.325^{25}
MS: 44(100) 18(31) 42(12) 41(12) 28(11) 58(9) 43(9) 45(7) 27(7) 17(6)
IR [cm^{-1}]: 3330 3230 2940 1640 1470 1390 1350 1250 1180 1040 980 960 780
Raman [cm^{-1}]: 3370 3310 3200 2960 2930 2880 2740 2710 2650 1600 1460
   1390 1340 1250 1170 1130 1040 980 950 820 470 420 370 270
^{1}H NMR [ppm]: 1.0 3.0 D_2O
Flammability: Flash pt. = -37°C; ign. temp. = 402°C
TLV/TWA: 5 ppm (12 mg/m^3)

328. Isoquinoline

Syst. Name: Isoquinoline
Synonyms: 2-Azanaphthalene; Leucoline

CASRN: 119-65-3   **Beil Ref.:** 5-20-07-00333
Merck No.: 5110
Beil RN: 107549
MF: C_9H_7N **MP[°C]:** 26.47   **Den.[g/cm³]:** 1.0910^{30}
MW: 129.16   **BP[°C]:** 243.2 n_D: 1.6148^{20}
Sol.: H_2O 1; EtOH 4; eth
  5; ace 4; bz 5; chl 4
Crit. Const.: $T_c = 530°C$; $P_c = 5.10$ MPa; $V_c = 374$ cm³
Vap. Press. [kPa]: 17.4^{175}; 35.6^{200}; 67^{225}
Therm. Prop.[kJ/mol]: $\Delta_{fus}H = 7.45$; $\Delta_{vap}H = 49.0$
C_p **(liq.) [J/mol °C]:** 196.8^{27}
γ **[mN/m]:** 46.2^{27}
Elec. Prop.: $\mu = 2.73$ D; $\varepsilon = 11.0^{25}$; IP = 8.53 eV
η **[mPa s]:** 3.25^{30}
MS: 129(100) 102(26) 51(20) 128(18) 50(11) 130(10) 75(10) 76(9) 103(8)
  74(7)
IR [cm⁻¹]: 3030 1610 1590 1490 1390 1270 1250 1220 1140 1040 1020 940
  860 830 810 750
UV [nm]: 293(1259) 290(1259) 263(2512) hx
¹³C NMR [ppm]: 120.2 126.2 127.0 127.3 128.5 130.1 135.5 142.7 152.2
  $CDCl_3$
¹H NMR [ppm]: 8.5 9.3 $CDCl_3$

329. *d*-Limonene

Syst. Name: Cyclohexene, 1-methyl-4-(1-
   methylethenyl)-, (*R*)-
Synonyms: *d-p*-Mentha-1,8-diene; Citrene; Carvene

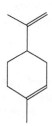

CASRN: 5989-27-5   **Beil Ref.**: 4-05-00-00438
Merck No.: 5371
Beil RN: 2204754
MF: $C_{10}H_{16}$ **MP[°C]**: -97   **Den.[g/cm^3]**: 0.8411^{20}
MW: 136.24   **BP[°C]**: 178 n_D: 1.4730^{20}
Sol.: H_2O 1; EtOH 5; eth
   5; ctc 3
Vap. Press. [kPa]: 0.277^{25}; 1.11^{50}; 3.59^{75}; 9.86^{100}; 23.7^{125}; 51.3^{150}
Therm. Prop.[kJ/mol]: $\Delta_f H°$(l, 25°C) = -54.5
C_p (liq.) [J/mol °C]: 249^{20}
γ [mN/m]: 27.17^{25}; $d\gamma/dT$ = 0.0929 mN/m °C
Elec. Prop.: $\varepsilon = 2.37^{25}$
η [mPa s]: 1.47^{25}
MS: 68(100) 93(50) 67(49) 41(22) 94(21) 79(21) 39(21) 136(20) 53(19)
   121(16)
IR [cm^{-1}]: 3080 2920 1650 1440 1370 920 890 800
UV [nm]: 250(23) 220(257) iso
^{13}C NMR [ppm]: 20.7 23.4 28.0 30.7 30.9 41.2 108.4 120.7 133.5 149.9 CDCl$_3$
^{1}H NMR [ppm]: 1.4 1.7 1.9 4.6 5.3 CCl$_4$
Flammability: Flash pt. = 49°C

330. *l*-Limonene

Syst. Name: Cyclohexene, 1-methyl-4-(1-methylethenyl)-, (*S*)-

Synonyms: *l*-*p*-Mentha-1,8-diene

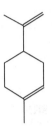

CASRN: 5989-54-8   **Beil Ref.**: 4-05-00-00440

Merck No.: 5371

Beil RN: 2323991

MF: $C_{10}H_{16}$ **BP[°C]**: 178   **Den.[g/cm³]**: 0.843^{20}

MW: 136.24 n_D: 1.4746^{20}

Sol.: eth 4; EtOH 4

Vap. Press. [kPa]: 0.254^{25}; 1.06^{50}; 3.50^{75}

γ **[mN/m]**: 26.71^{25}; $d\gamma/dT = 0.0897$ mN/m °C

Elec. Prop.: $\varepsilon = 2.37^{25}$

MS: 68(100) 93(46) 67(41) 39(31) 41(28) 27(25) 53(23) 79(20) 136(16) 121(16)

IR [cm⁻¹]: 2940 1670 1450 1390 1150 920 895 800

UV [nm]: 220(251) iso

331. 2,4-Lutidine

Syst. Name: Pyridine, 2,4-dimethyl-
Synonyms: 2,4-Dimethylpyridine

CASRN: 108-47-4   **Beil Ref.:** 5-20-06-00019
Beil RN: 1506
MF: C_7H_9N **MP[°C]:** -64   **Den.[g/cm³]:** 0.9309^{20}
MW: 107.16   **BP[°C]:** 158.5 n_D: 1.5010^{20}
Sol.: H_2O 4; EtOH 4; eth
    4; ace 3
Crit. Const.: $T_c = 374°C$
Vap. Press. [kPa]: 136^{150}
Therm. Prop.[kJ/mol]: $\Delta_{vap}H = 38.5$; $\Delta_f H°(l, 25°C) = 16.2$
C_p **(liq.) [J/mol °C]:** 184.8^{25}
γ **[mN/m]:** 33.17^{20}
Elec. Prop.: $\mu = (2.3$ D); $\varepsilon = 9.60^{20}$; IP = 8.85 eV
η **[mPa s]:** 0.887^{20}
MS: 107(100) 106(63) 79(44) 92(22) 65(22) 51(19) 77(17) 80(12) 52(11)
    50(11)
IR [cm⁻¹]: 3030 1610 1560 1450 1410 1300 1160 1040 1000 910 810 760
Raman [cm⁻¹]: 3050 2930 2880 2740 2470 1610 1570 1480 1460 1440 1380
    1300 1280 1250 1220 1180 1120 1080 1040 1010 980 880 850 820 800 760
    740 710 650 590 570 530 490 450 420 320 290 280 230 200
UV [nm]: 259(6026) hx
¹³C NMR [ppm]: 20.7 24.1 121.6 123.9 146.9 148.8 158.0 CDCl₃
¹H NMR [ppm]: 2.3 2.5 7.0 7.0 8.4 CDCl₃

332. 2,6-Lutidine

Syst. Name: Pyridine, 2,6-dimethyl-
Synonyms: 2,6-Dimethylpyridine

CASRN: 108-48-5   **Beil Ref.**: 5-20-06-00032
Merck No.: 5485
Beil RN: 105690
MF: C_7H_9N **MP[°C]**: -6.1   **Den.[g/cm^3]**: 0.9226^{20}
MW: 107.16   **BP[°C]**: 144.1 n_D: 1.4953^{20}
Sol.: H_2O 5; EtOH 2; eth
   3; ace 3; chl 3
Crit. Const.: T_c = 350.7°C
Vap. Press. [kPa]: 0.746^{25}; 3.05^{50}; 9.71^{75}; 25.6^{100}; 58.5^{125}
Therm. Prop.[kJ/mol]: $\Delta_{fus}H$ = 10.04; $\Delta_{vap}H$ = 37.5; $\Delta_f H°$(l, 25°C) = 12.7
C_p **(liq.) [J/mol °C]**: 185.2^{25}
γ [mN/m]: 31.65^{20}
Elec. Prop.: μ = (1.7 D); ε = 7.33^{20}; IP = 8.86 eV
η [mPa s]: 0.869^{20}
MS: 107(100) 39(39) 106(29) 66(22) 92(18) 65(18) 38(12) 27(11) 79(9) 63(9)
IR [cm^{-1}]: 2940 1610 1590 1470 1450 1390 1220 1160 1100 1030 1000 780
UV [nm]: 268(12589) sulf
^{13}C NMR [ppm]: 24.4 120.0 136.3 157.5 $CDCl_3$
^{1}H NMR [ppm]: 2.5 7.0 7.4 $CDCl_3$

333. Mesitylene

Syst. Name: Benzene, 1,3,5-trimethyl-
Synonyms: 1,3,5-Trimethylbenzene

CASRN: 108-67-8   **Rel. CASRN**: 25551-13-7, 15551-13-7
Merck No.: 5810   **DOT No.**: 2325
Beil RN: 906806   **Beil Ref.**: 4-05-00-01016
MF: C_9H_{12} **MP[°C]**: -44.7   **Den.[g/cm^3]**: 0.8652^{20}
MW: 120.19   **BP[°C]**: 164.7 n_D: 1.4994^{20}
Sol.: H_2O 1; EtOH 5; eth
  5; ace 5; bz 5; ctc 5;
  peth 5
Crit. Const.: T_c = 364.2°C; P_c = 3.127 MPa
Vap. Press. [kPa]: 0.330^{25}; 4.83^{75}; 13.5^{100}; 32.2^{125}; 68.2^{150}
Therm. Prop.[kJ/mol]: $\Delta_{fus}H$ = 9.51; $\Delta_{vap}H$ = 39.0; $\Delta_f H°$(l, 25°C) = -63.4
C_p (liq.) [J/mol °C]: 209.3^{25}
γ [mN/m]: 27.55^{25}; dγ/dT = 0.0897 mN/m °C
Elec. Prop.: μ = 0; ϵ = 2.28^{20}; IP = 8.41 eV
η [mPa s]: 1.154^{20}
k [W/m °C]: 0.147^{-25}; 0.141^{0}; 0.136^{25}; 0.130^{50}; 0.124^{75}; 0.118^{100}
MS: 105(100) 120(64) 119(15) 77(13) 39(11) 106(9) 91(9) 51(8) 27(7) 121(6)
IR [cm^{-1}]: 3000 2900 1770 1710 1610 1470 1370 1040 840 690
Raman [cm^{-1}]: 3000 2930 2860 2730 1610 1380 1300 1170 1040 1000 930 880
  680 610 280 230
UV [nm]: 282(17400) 217(12700) 213(12100) 207(11900) MeOH
^{13}C NMR [ppm]: 21.0 127.1 137.4 diox
^{1}H NMR [ppm]: 2.3 6.8 CDCl$_3$
Flammability: Flash pt. = 50°C; ign. temp. = 559°C; flam. lim. = 1-5%
TLV/TWA: 25 ppm (123 mg/m^3)

334. Mesityl oxide

Syst. Name: 3-Penten-2-one, 4-methyl-
Synonyms: 4-Methyl-3-penten-2-one; Isobutenyl
methyl ketone

CASRN: 141-79-7   **DOT No.**: 1229
Merck No.: 5811   **Beil Ref.**: 4-01-00-03471
Beil RN: 1361550
MF: $C_6H_{10}O$ **MP[°C]**: -59   **Den.[g/cm^3]**: 0.8653^{20}
MW: 98.14   **BP[°C]**: 130 n_D: 1.4440^{20}
Sol.: H_2O 3; EtOH 5; eth
    5; ace 3
Vap. Press. [kPa]: 1.47^{25}; 5.40^{50}; 16.2^{75}; 40.5^{100}; 88.5^{125}
Therm. Prop.[kJ/mol]: $\Delta_{vap}H = 36.1$
C_p **(liq.) [J/mol °C]**: 212.5^{25}
γ **[mN/m]**: 22.9^{20}
Elec. Prop.: $\mu = (2.8 D)$; $\varepsilon = 15.6^0$; IP = 9.08 eV
η **[mPa s]**: 1.29^{-25}; 0.838^0; 0.602^{25}; 0.465^{50}; 0.381^{75}; 0.326^{100}
k **[W/m °C]**: 0.170^{-25}; 0.163^0; 0.156^{25}; 0.149^{50}; 0.142^{75}; 0.134^{100}
MS: 55(100) 83(89) 43(73) 29(42) 98(36) 39(32) 27(28) 53(11) 41(10) 56(5)
IR [cm^{-1}]: 2980 2920 1690 1620 1440 1370 1350 1210 1160 1060 1010 960
    820
UV [nm]: 237 MeOH
Flammability: Flash pt. = 31°C; ign. temp. = 344°C; flam. lim. = 1.4-7.2%
TLV/TWA: 15 ppm (60 mg/m^3)

335. Methacrylic acid

Syst. Name: 2-Propenoic acid, 2-methyl-
Synonyms: 2-Methylpropenoic acid; α-
Methylacrylic acid

CASRN: 79-41-4   **DOT No.**: 2531
Merck No.: 5849   **Beil Ref.**: 4-02-00-01518
Beil RN: 1719937
MF: $C_4H_6O_2$ **MP[°C]**: 16   **Den.[g/cm³]**: 1.0153^{20}
MW: 86.09   **BP[°C]**: 162.5 n_D: 1.4314^{20}
Sol.: H_2O 3; EtOH 5; eth
   5; chl 3
Vap. Press. [kPa]: 0.703^{50}; 2.98^{75}; 10.1^{100}; 28.3^{125}; 69.1^{150}
C_p **(liq.) [J/mol °C]**: 161.1^{27}
γ **[mN/m]**: 26.5^{25}
Elec. Prop.: $\mu = (1.6 \text{ D})$; IP = 10.15 eV
MS: 41(100) 86(71) 39(65) 40(21) 28(17) 69(14) 45(14) 38(14) 18(10) 42(8)
IR [cm⁻¹]: 2940 2700 2600 2500 1700 1640 1450 1430 1380 1300 1220 1200
   1000 950 810
¹H NMR [ppm]: 2.0 5.7 6.3 11.6 $CDCl_3$
Flammability: Flash pt. = 77°C; ign. temp. = 68°C; flam. lim. = 1.6-8.8%
TLV/TWA: 20 ppm (70 mg/m³)

336. Methanol

Syst. Name: Methanol
Synonyms: Methyl alcohol; Carbinol

CASRN: 67-56-1   **DOT No.**: 1230
Merck No.: 5868   **Beil Ref.**: 4-01-00-01227
Beil RN: 1098229
MF: CH$_4$O   **MP[°C]**: -97.6   **Den.[g/cm^3]**: 0.7914^{20}
MW: 32.04   **BP[°C]**: 64.6 n_D: 1.3288^{20}
Sol.: H$_2$O 5; EtOH 5; eth
  5; ace 5; bz 4; chl 3
Crit. Const.: T_c = 239.4°C; P_c = 8.084 MPa; V_c = 118 cm^3
Vap. Press. [kPa]: 4.03^0; 16.9^{25}; 55.5^{50}; 151^{75}; 353^{100}; 735^{125}; 1391^{150}
Therm. Prop.[kJ/mol]: $\Delta_{fus}H$ = 3.18; $\Delta_{vap}H$ = 35.2; $\Delta_f H°$(l, 25°C) = -239.1
C_p (liq.) [J/mol °C]: 81.1^{25}
γ [mN/m]: 22.07^{25}; dγ/dT = 0.0773 mN/m °C
Elec. Prop.: μ = 1.70 D; ε = 33.0^{20}; IP = 10.85 eV
η [mPa s]: 1.26^{-25}; 0.793^0; 0.544^{25}
k [W/m °C]: 0.214^{-25}; 0.207^0; 0.200^{25}; 0.193^{50}
MS: 31(100) 29(72) 32(67) 15(42) 28(12) 14(10) 30(9) 13(6) 12(3) 16(2)
IR [cm^{-1}]: 3340 2940 2830 1450 1110 1030 640
Raman [cm^{-1}]: 3350 2940 2830 1450 1100 1030
^{13}C NMR [ppm]: 49.3
^{1}H NMR [ppm]: 1.4 3.5 CDCl$_3$
Flammability: Flash pt. = 11°C; ign. temp. = 464°C; flam. lim. = 6.0-36%
TLV/TWA: 200 ppm (262 mg/m^3)
Reg. Lists: CERCLA (RQ = 5000 lb.); SARA 313 (1.0%); RCRA U154 (toxic)

337. *N*-Methylacetamide
Syst. Name: Ethanamide, *N*-methyl-

CASRN: 79-16-3   **Beil Ref.**: 4-04-00-00176
Beil RN: 1071255
MF: C_3H_7NO **MP[°C]**: 28   **Den.[g/cm³]**: 0.9371^{25}
MW: 73.09   **BP[°C]**: 205 n_D: 1.4301^{20}
Sol.: ace 4; bz 4; eth 4;
    EtOH 4
Vap. Press. [kPa]: 0.485^{70}
Therm. Prop.[kJ/mol]: $\Delta_{fus}H = 9.72$
γ **[mN/m]**: 33.67^{30}
Elec. Prop.: $\mu = (4.3\ D)$; $\varepsilon = 179.0^{30}$
η **[mPa s]**: 3.23^{35}
MS: 73(100) 43(83) 30(67) 58(40) 15(23) 31(17) 28(17) 42(10) 45(9) 74(4)
IR [cm⁻¹]: 3280 3080 2940 1650 1560 1410 1370 1290 1150 1030 980 720
Raman [cm⁻¹]: 3300 3100 2940 2810 2720 1660 1450 1420 1380 1310 1160
    1100 990 880 630 440 290
¹H NMR [ppm]: 2.0 2.7 8.2 CCl_4

338. Methyl acetate

Syst. Name: Ethanoic acid, methyl ester

CASRN: 79-20-9
Merck No.: 5932
Beil RN: 1736662
MF: $C_3H_6O_2$
MW: 74.08
Sol.: H_2O 4; eth 4; EtOH
  4

DOT No.: 1231
Beil Ref.: 4-02-00-00122

MP[°C]: -98
BP[°C]: 56.8

Den.[g/cm³]: 0.9342^{20}
n_D: 1.3614^{20}

Crit. Const.: T_c = 233.40°C; P_c = 4.75 MPa; V_c = 228 cm³
Vap. Press. [kPa]: 8.37^0; 28.8^{25}; 79.2^{50}; 184^{75}; 371^{100}
Therm. Prop.[kJ/mol]: $\Delta_{vap}H$ = 30.3; $\Delta_fH°$(l, 25°C) = -445.8
C_p (liq.) [J/mol °C]: 141.9^{25}
γ [mN/m]: 24.73^{25}; dγ/dT = 0.1289 mN/m °C
Elec. Prop.: μ = 1.72 D; ε = 7.07^{15}; IP = 10.27 eV
η [mPa s]: 0.477^0; 0.364^{25}; 0.284^{50}
k [W/m °C]: 0.174^{-25}; 0.164^0; 0.153^{25}; 0.143^{50}; 0.133^{75}; 0.122^{100}
MS: 43(100) 74(52) 28(38) 42(19) 59(17) 44(8) 32(8) 29(6) 31(4) 75(2)
IR [cm⁻¹]: 2940 1750 1450 1370 1250 1050 980 840
Raman [cm⁻¹]: 3020 2940 2850 2720 1740 1450 1370 1190 1160 1050 980 840
  640 610 440 300
UV [nm]: 209(58) iso 203(62) H_2O
¹H NMR [ppm]: 2.0 3.7 CCl_4
Flammability: Flash pt. = -10°C; ign. temp. = 454°C; flam. lim. = 3.1-16%
TLV/TWA: 200 ppm (606 mg/m³)

339. Methyl acetoacetate

Syst. Name: Butanoic acid, 3-oxo-, methyl ester
Synonyms: Methyl 3-oxobutanoate

CASRN: 105-45-3   **Beil Ref.**: 4-03-00-01527
Merck No.: 5933
Beil RN: 506727
MF: $C_5H_8O_3$ **MP[°C]**: 27.5   **Den.[g/cm³]**: 1.0762^{20}
MW: 116.12   **BP[°C]**: 171.7 n_D: 1.4184^{20}
Sol.: H_2O 4; EtOH 5; eth
  5; ctc 3
Vap. Press. [kPa]: 0.241^{25}; 0.995^{50}; 3.35^{75}; 9.57^{100}; 24.0^{125}; 53.9^{150}
η [mPa s]: 1.70^{20}
MS: 43(100) 42(10) 59(9) 31(8) 29(7) 85(6) 74(6) 32(5) 69(4) 44(4)
IR [cm⁻¹]: 3010 2880 1760 1730 1660 1640 1440 1410 1360 1320 1270 1140
  1040 970 810 630 540
UV [nm]: 240(1514) EtOH
¹H NMR [ppm]: 2.2 3.4 3.7 $CDCl_3$
Flammability: Flash pt. = 77°C; ign. temp. = 280°C

340. Methyl acrylate

Syst. Name: 2-Propenoic acid, methyl ester
Synonyms: Methyl propenoate

CASRN: 96-33-3   **DOT No.:** 1919
Merck No.: 5935   **Beil Ref.:** 4-02-00-01457
Beil RN: 605396
MF: $C_4H_6O_2$ **MP[°C]:** <-75   **Den.[g/cm^3]:** 0.9535^{20}
MW: 86.09   **BP[°C]:** 80.7 n_D: 1.4040^{20}
Sol.: H_2O 2; EtOH 3; eth
    3; ace 3; bz 3; chl 3
Vap. Press. [kPa]: 11.0^{25}; 34.2^{50}; 85.1^{75}
Therm. Prop.[kJ/mol]: $\Delta_{vap}H$ = 33.1; $\Delta_f H°$(l, 25°C) = -362.2
C_p **(liq.) [J/mol °C]:** 158.8^{25}
Elec. Prop.: μ = (1.8 D); ε = 7.03^{30}; IP = 9.90 eV
η **[mPa s]:** 1.40^{20}
MS: 55(100) 27(41) 26(18) 85(15) 15(11) 28(10) 42(8) 58(7) 56(7) 29(7)
IR [cm^{-1}]: 3000 2960 1730 1640 1440 1400 1270 1200 1180 1070 980 960 850
    810 660
Raman [cm^{-1}]: 3120 3050 3000 2960 2920 2860 2840 1730 1640 1450 1410
    1290 1210 1190 1070 1000 980 860 820 670 630 530 470 360 240
^{1}H NMR [ppm]: 3.8 5.8 6.2 6.4 CDCl$_3$
Flammability: Flash pt. = -3°C; ign. temp. = 468°C; flam. lim. = 2.8-25%
TLV/TWA: 10 ppm (35 mg/m^3)
Reg. Lists: SARA 313 (1.0%)

341. Methylacrylonitrile

Syst. Name: 2-Propenenitrile, 2-methyl-
Synonyms: 2-Methylpropenenitrile; 2-Cyano-1-
  propene

CASRN: 126-98-7   **Beil Ref.**: 4-02-00-01539
Merck No.: 5850
Beil RN: 773708
MF: C_4H_5N **MP[°C]**: -35.8   **Den.[g/cm³]**: 0.8001^{20}
MW: 67.09   **BP[°C]**: 90.3 n_D: 1.4003^{20}
Sol.: H_2O 1; EtOH 5; eth
  5; ace 5; chl 2; peth 5;
  tol 5
Vap. Press. [kPa]: 2.13^0; 8.26^{25}; 24.9^{50}; 62.2^{75}; 135^{100}
Therm. Prop.[kJ/mol]: $\Delta_{vap}H = 31.8$
γ **[mN/m]**: 24.4^{20}
Elec. Prop.: $\mu = 3.69$ D; IP = 10.34 eV
η **[mPa s]**: 0.392^{20}
MS: 41(100) 39(54) 67(44) 40(26) 38(24) 52(23) 27(23) 37(20) 66(17) 51(12)
IR [cm⁻¹]: 3000 2960 2930 2230 1870 1620 1450 1380 1260 1020 930 750
UV [nm]: 215(676) EtOH
¹H NMR [ppm]: 2.0 5.7 5.8 $CDCl_3$
Flammability: Flash pt. = 1.1°C; flam. lim. = 2-6.8%
TLV/TWA: 1 ppm (2.7 mg/m³)
Reg. Lists: CERCLA (RQ = 1000 lb.); RCRA U152 (toxic)

342. Methylal

Syst. Name: Methane, dimethoxy-
Synonyms: Dimethoxymethane; Formal

CASRN: 109-87-5 **DOT No.**: 1234
Merck No.: 5936 **Beil Ref.**: 4-01-00-03026
Beil RN: 1697025
MF: $C_3H_8O_2$ **MP[°C]**: -104.8 **Den.[g/cm^3]**: 0.8593^{20}
MW: 76.10 **BP[°C]**: 42 n_D: 1.3513^{20}
Sol.: ace 4; bz 4; eth 4;
  EtOH 4
Vap. Press. [kPa]: 16.7^0; 53.1^{25}; 133^{50}
Therm. Prop.[kJ/mol]: $\Delta_{fus}H = 8.33$; $\Delta_f H°(l, 25°C) = -377.7$
C_p **(liq.) [J/mol °C]**: 161.3^{25}
γ **[mN/m]**: 21.12^{20}
Elec. Prop.: $\mu = (0.7$ D); $\varepsilon = 2.64^{20}$; IP = 9.50 eV
η **[mPa s]**: 0.340^{15}
MS: 45(100) 75(61) 29(59) 31(13) 30(6) 15(6) 47(5) 76(2) 46(2) 44(2)
IR [cm^{-1}]: 2980 2930 2810 2660 1450 1400 1230 1190 1130 1110 1040 930
  600 450
Raman [cm^{-1}]: 2990 2960 2930 2880 2840 2820 2770 1450 1400 1300 1180
  1150 1100 1040 910 600 550 320 100
^{1}H NMR [ppm]: 3.2 4.4 CCl$_4$
Flammability: Flash pt. = -32°C; ign. temp. = 237°C; flam. lim. = 2.2-13.8%
TLV/TWA: 1000 ppm (3110 mg/m^3)

343. Methylamine

Syst. Name: Methanamine

CASRN: 74-89-5   **DOT No.**: 1061
Merck No.: 5938   **Beil Ref.**: 4-04-00-00118
Beil RN: 741851
MF: CH_5N **MP[°C]**: -93.4   **Den.[g/cm^3]**: 0.656^{25} (sat. press.)
MW: 31.06   **BP[°C]**: -6.3
Sol.: H_2O 4; EtOH 3; eth
  5; ace 3; bz 3
Crit. Const.: T_c = 157.6°C; P_c = 7.614 MPa
Vap. Press. [kPa]: 40.4^{-25}; 353^{25}
Therm. Prop.[kJ/mol]: $\Delta_{fus}H$ = 6.13; $\Delta_{vap}H$ = 25.6; $\Delta_f H°$(g, 25°C) = -22.5
C_p **(liq.) [J/mol °C]**: 102.1^{25} (sat. press.)
γ [mN/m]: 19.15^{25} (at saturation press.); dγ/dT = 0.1488 mN/m °C
Elec. Prop.: μ = 1.31 D; ε = 16.7^{-58}; IP = 8.97 eV
η [mPa s]: 0.319^{-25}; 0.231^0
MS: 30(100) 31(87) 28(56) 29(19) 32(15) 15(12) 27(9)
IR [cm^{-1}]: 2950 1620 1440 1140 1120 1060 1040 790 770 750
UV [nm]: 215(589) 191(3236) gas
^{13}C NMR [ppm]: 28.3
Flammability: Flash pt. = 0°C; ign. temp. = 430°C; flam. lim. = 4.9-20.7%
TLV/TWA: 5 ppm (6.4 mg/m^3)
Reg. Lists: CERCLA (RQ = 100 lb.)

344. *N*-Methylaniline

Syst. Name: Benzenamine, *N*-methyl-
Synonyms: *N*-Methylbenzenamine

CASRN: 100-61-8   **DOT No.**: 2294
Merck No.: 5941   **Beil Ref.**: 4-12-00-00241
Beil RN: 741982
MF: C_7H_9N **MP[°C]**: -57   **Den.[g/cm^3]**: 0.9891^{20}
MW: 107.16   **BP[°C]**: 196.2 n_D: 1.5684^{20}
Sol.: H_2O 1; EtOH 3; eth
  3; ctc 3; chl 3
Crit. Const.: T_c = 428°C; P_c = 5.20 MPa
Vap. Press. [kPa]: 0.306^{50}; 1.28^{75}; 4.23^{100}; 11.6^{125}; 27.4^{150}
C_p **(liq.) [J/mol °C]**: 207.1^{25}
γ **[mN/m]**: 36.90^{25}; $d\gamma/dT$ = 0.0970 mN/m °C
Elec. Prop.: ε = 5.96^{20}; IP = 7.33 eV
η **[mPa s]**: 4.12^{0}; 2.04^{25}; 1.22^{50}; 0.825^{75}; 0.606^{100}
MS: 106(100) 107(79) 77(23) 51(12) 79(11) 65(9) 39(9) 78(8) 108(7) 50(6)
IR [cm^{-1}]: 3450 2940 1610 1520 1430 1320 1270 1180 1150 1080 990 870 750
  690
UV [nm]: 293 245 MeOH
^{13}C NMR [ppm]: 30.2 112.3 116.7 129.2 150.2 diox
^{1}H NMR [ppm]: 2.7 3.3 6.4 6.6 7.1 CCl_4
TLV/TWA: 0.5 ppm (2.2 mg/m^3)

345. Methyl benzoate

Syst. Name: Benzoic acid, methyl ester
Synonyms: Methyl benzenecarboxylate

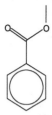

CASRN: 93-58-3
Merck No.: 5947
Beil RN: 1072099
MF: $C_8H_8O_2$ **MP[°C]:** -15
MW: 136.15   **BP[°C]:** 199

DOT No.: 2938
Beil Ref.: 4-09-00-00283

Den.[g/cm^3]: 1.0933^{15}
n_D: 1.5164^{20}

Sol.: H_2O 1; EtOH 3; eth 5; ctc 3; MeOH 3
Vap. Press. [kPa]: 0.052^{25}; 11.5^{125}; 26.3^{150}
Therm. Prop.[kJ/mol]: $\Delta_{fus}H$ = 9.74; $\Delta_{vap}H$ = 43.2; $\Delta_fH°$(l, 25°C) = -343.5
C_p **(liq.) [J/mol °C]:** 221.3^{25}
γ **[mN/m]:** 37.17^{25}; $d\gamma/dT$ = 0.1171 mN/m °C
Elec. Prop.: μ = (1.9 D); ε = 6.64^{30}; IP = 9.32 eV
η **[mPa s]:** 1.86^{25}
MS: 105(100) 77(81) 51(45) 136(24) 50(18) 106(8) 78(6) 28(6) 39(5) 27(5)
IR [cm^{-1}]: 3030 2940 1720 1610 1590 1540 1450 1430 1320 1280 1190 1180 1110 1060 1030 970 820 710 690 670
Raman [cm^{-1}]: 3070 3000 2950 2900 2840 1720 1600 1450 1310 1280 1180 1160 1110 1030 1000 970 820 680 620 360 220
UV [nm]: 280(686) 272(830) 228(11000) MeOH
^{13}C NMR [ppm]: 51.8 128.3 129.5 130.3 132.8 166.8 CDCl$_3$
^{1}H NMR [ppm]: 3.8 7.4 8.0 CCl$_4$
Flammability: Flash pt. = 83°C

346. 3-Methylbutanoic acid

Syst. Name: Butanoic acid, 3-methyl-
Synonyms: Isovaleric acid; Delphinic acid

CASRN: 503-74-2   **DOT No.**: 1760
Merck No.: 5120   **Beil Ref.**: 4-02-00-00895
Beil RN: 1098522
MF: $C_5H_{10}O_2$ **MP[°C]**: -29.3   **Den.[g/cm^3]**: 0.931^{20}
MW: 102.13   **BP[°C]**: 176.5 n_D: 1.4033^{20}
Sol.: H_2O 3; EtOH 5; eth
    5; chl 5
Crit. Const.: T_c = 356°C; P_c = 3.40 MPa
Vap. Press. [kPa]: 0.006^0; 0.067^{25}; 0.414^{50}; 1.77^{75}; 6.23^{100}; 18.0^{125}; 44.4^{150}
Therm. Prop.[kJ/mol]: $\Delta_{fus}H$ = 7.32; $\Delta_f H°$(l, 25°C) = -561.6
C_p **(liq.) [J/mol °C]**: 197.1^{25}
γ **[mN/m]**: 25.51^{20}; $d\gamma/dT$ = 0.0886 mN/m °C
Elec. Prop.: μ = (0.6 D); IP = 10.51 eV
η **[mPa s]**: 2.73^{15}
MS: 60(100) 43(61) 41(54) 27(33) 45(31) 29(27) 74(24) 39(24) 87(21) 57(20)
IR [cm^{-1}]: 2940 2630 1700 1470 1410 1300 1220 1160 1100 940 780
Raman [cm^{-1}]: 2960 2940 2920 2910 2870 2760 2730 1660 1460 1450 1410
    1340 1300 1210 1170 1120 960 920 900 880 830 800 710 680 610 550 480
    420 340 240 200
^{1}H NMR [ppm]: 1.0 2.2 11.0 CCl_4
Flammability: Ign. temp. = 416°C

347. 2-Methyl-1-butanol

Syst. Name: 1-Butanol, 2-methyl-
Synonyms: *sec*-Butylcarbinol

CASRN: 137-32-6   **Beil Ref.:** 4-01-00-01666
Merck No.: 5952
Beil RN: 1718810
MF: $C_5H_{12}O$ **BP[°C]:** 128   **Den.[g/cm^3]:** 0.8150^{25}
MW: 88.15 n_D: 1.4092^{20}
Sol.: H_2O 2; EtOH 5; eth
  5; ace 4
Crit. Const.: $T_c = 302.3°C$; $P_c = 3.94$ MPa
Vap. Press. [kPa]: 0.416^{25}; 2.59^{50}; 10.9^{75}; 34.5^{100}; 89.4^{125}; 198^{150}
Therm. Prop.[kJ/mol]: $\Delta_{vap}H = 45.2$; $\Delta_f H°(l, 25°C) = -356.6$
C_p **(liq.) [J/mol °C]:** 220.1^{25}
γ **[mN/m]:** 25.1^{25}
Elec. Prop.: $\mu = (1.9$ D$)$; $\varepsilon = 15.63^{25}$; IP $= 9.86$ eV
η **[mPa s]:** 4.45^{25}; 1.96^{50}; 1.03^{75}; 0.612^{100}
MS: 41(100) 57(99) 29(91) 56(84) 31(53) 70(35) 55(35) 27(29) 42(21) 39(20)
Raman [cm^{-1}]: * 1540 1463 1135 1037 1016 977 901 823 801 765 444
^{1}H NMR [ppm]: 0.8 0.9 1.4 2.5 3.5 $CDCl_3$
Flammability: Flash pt. = 50°C; ign. temp. = 385°C

348. 3-Methyl-1-butanol

Syst. Name: 1-Butanol, 3-methyl-
Synonyms: Isoamyl alcohol; Isopentyl alcohol

CASRN: 123-51-3
Merck No.: 5081
Beil RN: 1718835
MF: $C_5H_{12}O$
MW: 88.15
Sol.: ace 4; eth 4; EtOH 4

DOT No.: 1105
Beil Ref.: 4-01-00-01677

MP[°C]: -117.2
BP[°C]: 131.1

Den.[g/cm^3]: 0.8104^{20}
n_D: 1.4053^{20}

Crit. Const.: T_c = 304.1°C; P_c = 3.93 MPa
Vap. Press. [kPa]: 0.315^{25}; 2.17^{50}; 9.65^{75}; 31.7^{100}; 83.7^{125}; 188^{150}
Therm. Prop.[kJ/mol]: $\Delta_{vap}H$ = 44.1; $\Delta_f H°$(l, 25°C) = -356.4
C_p (liq.) [J/mol °C]: 210.0^{25}
γ [mN/m]: 23.71^{25}; $d\gamma/dT$ = 0.0820 mN/m °C
Elec. Prop.: ε = 15.63^{20}
η [mPa s]: 8.63^0; 3.69^{25}; 1.84^{50}; 1.03^{75}; 0.631^{100}
MS: 55(100) 42(90) 43(82) 41(81) 70(71) 31(61) 29(59) 27(59) 39(44) 57(31)
IR [cm^{-1}]: 3310 2950 2860 1460 1430 1380 1360 1210 1170 1120 1050 1010 960 830 650 470 390
Raman [cm^{-1}]: 2960 2930 2910 2870 2760 2720 1460 1450 1330 1300 1250 1120 1050 1010 960 880 810 520 440 330 250
^{1}H NMR [ppm]: 0.9 1.5 3.5 4.1 CCl$_4$
Flammability: Flash pt. = 43°C; ign. temp. = 350°C; flam. lim. = 1.2-9.0%
TLV/TWA: 100 ppm (361 mg/m^3)

349. 2-Methyl-2-butanol

Syst. Name: 2-Butanol, 2-methyl-
Synonyms: *tert*-Amyl alcohol; *tert*-Pentyl alcohol

CASRN: 75-85-4   **Beil Ref.:** 4-01-00-01668
Merck No.: 7096
Beil RN: 1361351
MF: $C_5H_{12}O$ **MP[°C]:** -8.8   **Den.[g/cm^3]:** 0.8096^{20}
MW: 88.15   **BP[°C]:** 102.4 n_D: 1.4052^{20}
Sol.: H_2O 3; EtOH 5; eth
  5; ace 4; bz 3; chl 3
Crit. Const.: T_c = 270.6°C; P_c = 3.71 MPa
Vap. Press. [kPa]: 2.19^{25}; 10.1^{50}; 34.3^{75}; 94.0^{100}
Therm. Prop.[kJ/mol]: $\Delta_{fus}H$ = 4.45; $\Delta_{vap}H$ = 39.0; $\Delta_f H°$(l, 25°C) = -379.5
C_p (liq.) [J/mol °C]: 247.1^{25}
γ [mN/m]: 22.77^{20}
Elec. Prop.: μ = (1.7 D); ε = 5.78^{25}; IP = 9.80 eV
η [mPa s]: 3.55^{25}
MS: 59(100) 55(60) 73(47) 43(34) 31(32) 41(24) 27(21) 39(18) 45(11) 42(9)
IR [cm^{-1}]: 3030 1695 1665 1615 1495 1470 1450 1390 1335 1280 1220 1165
  1135 1085 1065 1030 1010 990 915 900 885 840 775 760 735 690
Raman [cm^{-1}]: 3400 2980 2940 2920 2880 2730 1470 1450 1390 1340 1290
  1230 1190 1060 1020 1010 990 950 930 890 740 530 490 470 460 430 380
  350 270
UV [nm]: 299(741) 238(186) sulf
Flammability: Flash pt. = 19°C; ign. temp. = 437°C; flam. lim. = 1.2-9.0%

350. 3-Methyl-2-butanol
Syst. Name: 2-Butanol, 3-methyl-, (±)-
Synonyms: Isopropylethanol

CASRN: 70116-68-6   **Beil Ref.**: 4-01-00-01675
Merck No.: 5953
Beil RN: 1718800
MF: $C_5H_{12}O$ **BP[°C]**: 112.9   **Den.[g/cm³]**: 0.8180^{20}
MW: 88.15 n_D: 1.4089^{20}
Sol.: H_2O 2; EtOH 5; eth
   5; ace 4; bz 3; ctc 3
Crit. Const.: T_c = 283.0°C; P_c = 3.87 MPa
Therm. Prop.[kJ/mol]: $\Delta_{vap}H$ = 41.8; $\Delta_f H°$(l, 25°C) = -366.6
γ **[mN/m]**: 23.0^{25}
Elec. Prop.: ε = 12.1^{25}; IP = 10.01 eV
η **[mPa s]**: 3.51^{25}
MS: 59(100) 55(86) 45(57) 73(52) 43(43) 41(33) 29(25) 88(1)
IR [cm⁻¹]: 3450 3030 2940 1470 1370 1320 1270 1180 1150 1100 1050 1020
   1010 940 920 880
Flammability: Flash pt. = 38°C

351. Methyl cyanoacetate

Syst. Name: Ethanoic acid, cyano-, methyl ester

CASRN: 105-34-0   **Beil Ref.**: 4-02-00-01889
Beil RN: 773945
MF: $C_4H_5NO_2$ **MP[°C]**: -22.5   **Den.[g/cm³]**: 1.1225^{25}
MW: 99.09   **BP[°C]**: 200.5 n_D: 1.4176^{20}
Sol.: eth 4; EtOH 4
Vap. Press. [kPa]: 0.019^{25}; 2.20^{100}; 6.94^{125}; 18.3^{150}
Therm. Prop.[kJ/mol]: $\Delta_{vap}H = 48.2$
γ **[mN/m]**: 42.32^{20}
Elec. Prop.: $\varepsilon = 29.3^{20}$
η **[mPa s]**: 2.79^{20}
MS: 59(100) 15(65) 68(60) 40(38) 28(17) 29(16) 55(11) 54(10) 39(10) 67(8)
IR [cm⁻¹]: 2960 2930 2260 1750 1440 1390 1340 1270 1210 1180 1010 930
   890 840 710
¹H NMR [ppm]: 3.5 3.8 CDCl₃

352. Methylcyclohexane
Syst. Name: Cyclohexane, methyl-

CASRN: 108-87-2
Beil RN: 505972
MF: C_7H_{14}
MW: 98.19
Sol.: H_2O 1; EtOH 3; eth 3; ace 5; bz 5; ctc 5; lig 5

DOT No.: 2296
Beil Ref.: 4-05-00-00094
MP[°C]: -126.6   **Den.[g/cm³]**: 0.7694^{20}
BP[°C]: 100.9 n_D: 1.4231^{20}

Crit. Const.: T_c = 299.1°C; P_c = 3.471 MPa; V_c = 368 cm³
Vap. Press. [kPa]: 6.18^{25}; 18.4^{50}; 45.8^{75}; 98.7^{100}
Therm. Prop.[kJ/mol]: $\Delta_{fus}H$ = 6.75; $\Delta_{vap}H$ = 31.3; $\Delta_f H°$(l, 25°C) = -190.1
C_p (liq.) [J/mol °C]: 184.8^{25}
γ [mN/m]: 23.29^{25}; $d\gamma/dT$ = 0.1130 mN/m °C
Elec. Prop.: μ = 0; ε = 2.02^{20}; IP = 9.64 eV
η [mPa s]: 0.991^{0}; 0.679^{25}; 0.501^{50}; 0.390^{75}; 0.316^{100}
MS: 83(100) 55(82) 41(60) 98(44) 42(35) 56(30) 27(29) 39(27) 69(23) 70(22)
IR [cm⁻¹]: 2920 2860 1450 1370 1250 960 900 840
Raman [cm⁻¹]: 2940 2900 2860 2720 2630 2610 1460 1450 1370 1350 1310 1260 1210 1160 1090 1060 1030 1010 970 850 770 550 450 400 330 310 160
¹³C NMR [ppm]: 22.9 26.6 33.0 35.6 CDCl₃
¹H NMR [ppm]: 0.9 1.4 CCl₄
Flammability: Flash pt. = -4°C; ign. temp. = 250°C; flam. lim. = 1.2-6.7%
TLV/TWA: 400 ppm (1610 mg/m³)

353. 1-Methylcyclohexanol
Syst. Name: Cyclohexanol, 1-methyl-

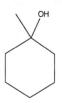

CASRN: 590-67-0

Beil RN: 1900934

MF: $C_7H_{14}O$

MW: 114.19

Sol.: H_2O 1; EtOH 3; bz 3; chl 3

Rel. CASRN: 25639-42-3

Beil Ref.: 4-06-00-00095

MP[°C]: 25   **Den.[g/cm^3]**: 0.9194^{20}

BP[°C]: 155 n_D: 1.4595^{20}

Vap. Press. [kPa]: 3.3^{70}

Therm. Prop.[kJ/mol]: $\Delta_{vap}H = 79.0$

Elec. Prop.: IP = 9.80 eV

MS: 57(100) 68(66) 81(64) 96(58) 71(49) 55(39) 41(34) 114(14)

IR [cm^{-1}]: 3330 2940 2860 1450 1370 1270 1180 1120 1040 970 910 860 830

^{13}C NMR [ppm]: 23.1 26.3 29.8 40.0 69.3

TLV/TWA: 50 ppm (234 mg/m^3)

---

354. *cis*-2-Methylcyclohexanol
Syst. Name: Cyclohexanol, 2-methyl-, *cis*-(±)-

CASRN: 615-38-3

Beil RN: 3193727

MF: $C_7H_{14}O$

MW: 114.19

Sol.: EtOH 4

Rel. CASRN: 25639-42-3

Beil Ref.: 4-06-00-00100

MP[°C]: 7   **Den.[g/cm^3]**: 0.9360^{20}

BP[°C]: 165 n_D: 1.4640^{20}

Therm. Prop.[kJ/mol]: $\Delta_{vap}H = 48.5$; $\Delta_f H°(l, 25°C) = -390.2$

C_p (liq.) [J/mol °C]: 200^{17}

η [mPa s]: 18.1^{25}

Flammability: Flash pt. = 65°C; ign. temp. = 296°C

TLV/TWA: 50 ppm (234 mg/m^3)

355. *trans*-2-Methylcyclohexanol
Syst. Name: Cyclohexanol, 2-methyl-, *trans*-(±)-

CASRN: 615-39-4   **Rel. CASRN**: 25639-42-3
Beil RN: 3193726   **Beil Ref.**: 4-06-00-00100
MF: $C_7H_{14}O$ **MP[°C]**: -2.0   **Den.[g/cm³]**: 0.9247[20]
MW: 114.19   **BP[°C]**: 167.5 n_D: 1.4616[20]
Sol.: eth 4; EtOH 4
Therm. Prop.[kJ/mol]: $\Delta_{vap}H$ = 53.0; $\Delta_fH°$(l, 25°C) = -415.7
C_p **(liq.) [J/mol °C]**: 200[17]
η **[mPa s]**: 37.1[25]
Flammability: Flash pt. = 65°C; ign. temp. = 296°C
TLV/TWA: 50 ppm (234 mg/m³)

---

356. *cis*-3-Methylcyclohexanol
Syst. Name: Cyclohexanol, 3-methyl-, *cis*-(±)-

CASRN: 5454-79-5
MF: $C_7H_{14}O$ **MP[°C]**: -5.5   **Den.[g/cm³]**: 0.9155[20]
MW: 114.19   **BP[°C]**: 168 n_D: 1.4752[20]
Sol.: eth 4; EtOH 4
Vap. Press. [kPa]: 1.6[94]
Therm. Prop.[kJ/mol]: $\Delta_fH°$(l, 25°C) = -416.1
C_p **(liq.) [J/mol °C]**: 202[17]
γ **[mN/m]**: 29.2[30]
Elec. Prop.: μ = (1.9 D); ε = 16.05[20]
η **[mPa s]**: 19.7[25]
TLV/TWA: 50 ppm (234 mg/m³)

357. *trans*-3-Methylcyclohexanol

Syst. Name: Cyclohexanol, 3-methyl-, *trans*-(±)-

CASRN: 7443-55-2
MF: $C_7H_{14}O$ **MP[°C]**: -0.5   **Den.[g/cm^3]**: 0.9214[30]
MW: 114.19   **BP[°C]**: 167 n_D: 1.4580[20]
Sol.: eth 4; EtOH 4
Vap. Press. [kPa]: 1.73[84]
Therm. Prop.[kJ/mol]: $\Delta_f H°$(l, 25°C) = -394.4
C_p **(liq.) [J/mol °C]**: 202[17]
γ **[mN/m]**: 28.8[30]
Elec. Prop.: μ = (1.7 D); ε = 8.05[20]
η **[mPa s]**: 15.6[30]
TLV/TWA: 50 ppm (234 mg/m^3)

---

358. *cis*-4-Methylcyclohexanol

Syst. Name: Cyclohexanol, 4-methyl-, *cis*-

CASRN: 7731-28-4   **Rel. CASRN**: 25639-42-3
Beil RN: 2036377   **Beil Ref.**: 4-06-00-00105
MF: $C_7H_{14}O$ **MP[°C]**: -9.2   **Den.[g/cm^3]**: 0.9170[20]
MW: 114.19   **BP[°C]**: 173 n_D: 1.4614[20]
Sol.: eth 4; EtOH 4
Therm. Prop.[kJ/mol]: $\Delta_f H°$(l, 25°C) = -413.2
C_p **(liq.) [J/mol °C]**: 202[17]
η **[mPa s]**: 0.25[25]
MS: 57(100) 58(54) 70(38) 81(36) 96(33) 55(25) 41(23) 114(17) 71(15) 56(12)
IR [cm^{-1}]: 3330 2940 2860 1450 1370 1330 1270 1150 1080 1030 990 970 930
880 820 790
^{1}H NMR [ppm]: 0.9 1.5 2.9 3.9 CDCl$_3$
Flammability: Flash pt. = 70°C; ign. temp. = 295°C
TLV/TWA: 50 ppm (234 mg/m^3)

359. *trans*-4-Methylcyclohexanol

Syst. Name: Cyclohexanol, 4-methyl-, *trans*-

CASRN: 7731-29-5   **Rel. CASRN**: 25639-42-3
Beil RN: 2036376   **Beil Ref.**: 4-06-00-00105
MF: $C_7H_{14}O$ **BP[°C]**: 174   **Den.[g/cm³]**: 0.9118^{21}
MW: 114.19 n_D: 1.4561^{20}
Sol.: H_2O 2; EtOH 5; eth
  3
Therm. Prop.[kJ/mol]: $\Delta_f H°$(l, 25°C) = -433.3
C_p **(liq.) [J/mol °C]**: 202^{17}
η **[mPa s]**: 0.38^{25}
MS: 57(100) 96(54) 81(52) 58(50) 70(31) 55(23) 41(23) 44(13) 71(12) 29(12)
IR [cm⁻¹]: 3330 2940 2860 1450 1370 1300 1240 1150 1090 1050 1010 960
  950 900 840 780
¹H NMR [ppm]: 0.9 1.5 2.6 3.5 $CDCl_3$
Flammability: Flash pt. = 70°C; ign. temp. = 295°C
TLV/TWA: 50 ppm (234 mg/m³)

360. Methylcyclopentane

Syst. Name: Cyclopentane, methyl-

CASRN: 96-37-7
Beil RN: 1900214
MF: C_6H_{12}
MW: 84.16
Sol.: H_2O 1; EtOH 5;
    eth 5; ace 5; bz 5; ctc
    5; lig 5

DOT No.: 2298
Beil Ref.: 4-05-00-00084
MP[°C]: -142.5   **Den.[g/cm³]**: 0.7486^{20}
BP[°C]: 71.8 n_D: 1.4097^{20}

Crit. Const.: T_c = 259.58°C; P_c = 3.784 MPa; V_c = 319 cm³
Vap. Press. [kPa]: 5.50^0; 18.3^{25}; 49.1^{50}
Therm. Prop.[kJ/mol]: $\Delta_{fus}H$ = 6.93; $\Delta_{vap}H$ = 29.1; $\Delta_f H°$(l, 25°C) = -137.9
C_p **(liq.) [J/mol °C]**: 158.7^{25}
γ **[mN/m]**: 21.72^{25}; $d\gamma/dT$ = 0.1163 mN/m °C
Elec. Prop.: μ = 0; ϵ = 1.99^{20}; IP = 9.85 eV
η **[mPa s]**: 0.927^{-25}; 0.653^0; 0.479^{25}; 0.364^{50}
MS: 56(100) 41(67) 69(34) 39(31) 42(29) 27(29) 55(27) 84(15) 43(13) 28(13)
IR [cm⁻¹]: 2960 2860 1450 1370 1140 980 900
Raman [cm⁻¹]: 3010 2970 2940 2880 2740 1450 1350 1320 1280 1200 1140
    1090 1010 890 850 800 530 430 310
¹³C NMR [ppm]: 20.7 25.5 34.8 34.9 CDCl₃
Flammability: Flash pt. < -7°C; ign. temp. = 258°C; flam. lim. = 1.0-8.35%

361. Methyl ethyl ketone

Syst. Name: 2-Butanone
Synonyms: Ethyl methyl ketone

CASRN: 78-93-3   **DOT No.:** 1193
Merck No.: 5991   **Beil Ref.:** 4-01-00-03243
Beil RN: 741880
MF: C_4H_8O **MP[°C]:** -86.6   **Den.[g/cm^3]:** 0.8054^{20}
MW: 72.11   **BP[°C]:** 79.5 n_D: 1.3788^{20}
Sol.: H_2O 4; EtOH 5; eth
  5; ace 5; bz 5; chl 3
Crit. Const.: T_c = 263.63°C; P_c = 4.207 MPa; V_c = 267 cm^3
Vap. Press. [kPa]: 3.51^0; 12.6^{25}; 35.6^{50}; 87.4^{75}; 184^{100}
Therm. Prop.[kJ/mol]: $\Delta_{fus}H$ = 8.44; $\Delta_{vap}H$ = 31.3; $\Delta_f H°$(l, 25°C) = -273.3
C_p (liq.) [J/mol °C]: 158.9^{25}
γ [mN/m]: 23.97^{25}; dγ/dT = 0.1122 mN/m °C
Elec. Prop.: μ = 2.78 D; ε = 18.56^{20}; IP = 9.51 eV
η [mPa s]: 0.720^{-25}; 0.533^0; 0.405^{25}; 0.315^{50}; 0.249^{75}
k [W/m °C]: 0.158^{-25}; 0.151^0; 0.145^{25}; 0.139^{50}; 0.133^{75}
MS: 43(100) 72(24) 29(19) 27(12) 57(7) 42(5) 26(4) 28(3) 44(2) 39(2)
IR [cm^{-1}]: 2980 2940 2900 1730 1460 1420 1360 1160 1090 940 750 590
Raman [cm^{-1}]: 2950 2930 2740 1710 1450 1420 1350 1250 1170 1090 990 950
  760 590 510 400 260
UV [nm]: 219(16) hx 274(17) EtOH 267(20) H_2O
^{13}C NMR [ppm]: 8.0 29.0 36.5 207.6 diox
^{1}H NMR [ppm]: 1.1 2.1 2.5 CDCl$_3$
Flammability: Flash pt. = -9°C; ign. temp. = 404°C; flam. lim. = 1.4-11.4%
TLV/TWA: 200 ppm (590 mg/m^3)
Reg. Lists: CERCLA (RQ = 5000 lb.); SARA 313 (1.0%); RCRA U159 (toxic)

362. *N*-Methylformamide

Syst. Name: Methanamide, *N*-methyl-

CASRN: 123-39-7   **Beil Ref.**: 4-04-00-00170
Beil RN: 1098352
MF: C_2H_5NO **MP[°C]**: -3.8   **Den.[g/cm^3]**: 1.011^{19}
MW: 59.07   **BP[°C]**: 199.5 n_D: 1.4319^{20}
Sol.: H_2O 4; ace 4;
     EtOH 4
Vap. Press. [kPa]: 3.04^{100}; 8.94^{125}; 22.6^{150}
C_p **(liq.) [J/mol °C]**: 123.8^{25}
γ **[mN/m]**: 39.46^{25}
Elec. Prop.: μ = 3.83 D; ε = 189.0^{20}; IP = 9.79 eV
η **[mPa s]**: 2.55^0; 1.68^{25}; 1.16^{50}; 0.824^{75}; 0.606^{100}
k **[W/m °C]**: 0.203^{25}; 0.201^{50}; 0.199^{75}; 0.196^{100}
MS: 59(100) 30(54) 28(34) 29(13) 58(8) 15(7) 60(3) 41(3) 27(3) 31(2)

363. Methyl formate

Syst. Name: Methanoic acid, methyl ester

CASRN: 107-31-3   **DOT No.:** 1243
Merck No.: 5994   **Beil Ref.:** 4-02-00-00020
Beil RN: 1734623
MF: $C_2H_4O_2$ **MP[°C]:** -99   **Den.[g/cm^3]:** 0.9742^{20}
MW: 60.05   **BP[°C]:** 31.7 n_D: 1.3433^{20}
Sol.: H_2O 4; EtOH 5; eth
    3; chl 3; MeOH 3
Crit. Const.: T_c = 214.1°C; P_c = 5.998 MPa; V_c = 172 cm^3
Vap. Press. [kPa]: 6.63^{-25}; 26.0^0; 78.1^{25}; 193^{50}; 410^{75}; 779^{100}; 1351^{125}
Therm. Prop.[kJ/mol]: $\Delta_{fus}H$ = 7.45; $\Delta_{vap}H$ = 27.9; $\Delta_f H°$(l, 25°C) = -386.1
C_p (liq.) [J/mol °C]: 119.1^{25}
γ [mN/m]: 24.36^{25}; $d\gamma/dT$ = 0.1572 mN/m °C
Elec. Prop.: μ = 1.77 D; ε = 9.20^{15}; IP = 10.82 eV
η [mPa s]: 0.424^0; 0.325^{25}
MS: 31(100) 29(63) 32(34) 60(28) 30(7) 28(7) 44(2) 18(2) 61(1) 59(1)
IR [cm^{-1}]: 2940 1720 1450 1390 1210 1160 910 770
Raman [cm^{-1}]: 3110 3040 2970 2850 2750 1720 1460 1430 1380 1210 1160
    1040 910 770 640 340 320
UV [nm]: 215(71) iso 207(63) H_2O
^{1}H NMR [ppm]: 3.8 8.1 CDCl$_3$
Flammability: Flash pt. = -19°C; ign. temp. = 449°C; flam. lim. = 4.5-23%
TLV/TWA: 100 ppm (246 mg/m^3)

364. 2-Methylheptane

Syst. Name: Heptane, 2-methyl-

CASRN: 592-27-8   **DOT No.**: 1262
Beil RN: 1696862   **Beil Ref.**: 4-01-00-00428
MF: C_8H_{18} **MP[°C]**: -108.9   **Den.[g/cm³]**: 0.6980^{20}
MW: 114.23   **BP[°C]**: 117.6 n_D: 1.3949^{20}
Sol.: H_2O 1; EtOH 5;
    eth 3; ace 5; bz 5; ctc
    3; chl 5; peth 5
Crit. Const.: T_c = 286.6°C; P_c = 2.484 MPa; V_c = 488 cm³
Vap. Press. [kPa]: 9.32^{50}; 25.7^{75}; 60.1^{100}
Therm. Prop.[kJ/mol]: $\Delta_{fus}H$ = 11.88; $\Delta_{vap}H$ = 33.3; $\Delta_f H°$(l, 25°C) = -255.0
C_p (liq.) [J/mol °C]: 252.0^{25}
Elec. Prop.: μ = 0; ε = 1.95^{20}; IP = 9.84 eV
η [mPa s]: 0.39^{38}
MS: 43(100) 57(91) 42(39) 41(27) 29(16) 70(15) 27(13) 71(12) 99(10) 55(9)
IR [cm⁻¹]: 2960 2920 2860 1460 1380 1360 1170 720
Raman [cm⁻¹]: 2964 2937 2914 2875 2848 1458 1442 1302 1044 314
¹³C NMR [ppm]: 13.8 22.4 22.8 27.2 28.1 32.4 39.3 diox
¹H NMR [ppm]: 0.9 1.3 CCl₄

---

365. 3-Methylheptane

Syst. Name: Heptane, 3-methyl-, (S)-

CASRN: 6131-25-5   **DOT No.**: 1262
Beil RN: 1718753   **Beil Ref.**: 4-01-00-00429
MF: C_8H_{18} **MP[°C]**: -120   **Den.[g/cm³]**: 0.7075^{16}
MW: 114.23   **BP[°C]**: 116.5 n_D: 1.4002^{18}
Sol.: H_2O 1; EtOH 3;
    eth 3; ace 5; bz 5; chl
    5
Crit. Const.: T_c = 290.6°C; P_c = 2.546 MPa; V_c = 464 cm³
Therm. Prop.[kJ/mol]: $\Delta_{fus}H$ = 11.38; $\Delta_{vap}H$ = 33.7; $\Delta_f H°$(l, 25°C) = -252.3
C_p (liq.) [J/mol °C]: 250.2^{25}
Elec. Prop.: μ = 0
η [mPa s]: 0.39^{38}
MS: 43(100) 57(68) 85(49) 41(46) 29(41) 56(38) 27(28) 114(0)

366. 4-Methylheptane

Syst. Name: Heptane, 4-methyl-

CASRN: 589-53-7   **DOT No.**: 1262
Beil RN: 1696874   **Beil Ref.**: 4-01-00-00431
MF: C_8H_{18} **MP[°C]**: -121   **Den.[g/cm³]**: 0.7046^{20}
MW: 114.23   **BP[°C]**: 117.7 n_D: 1.3979^{20}
Sol.: H_2O 1; EtOH 5;
   eth 3; ace 5; bz 5; chl
   5; peth 5
Crit. Const.: T_c = 288.7°C; P_c = 2.542 MPa; V_c = 476 cm³
Vap. Press. [kPa]: 2.74^{25}; 9.31^{50}; 25.6^{75}; 60.0^{100}; 124^{125}; 231^{150}
Therm. Prop.[kJ/mol]: $\Delta_{fus}H$ = 10.84; $\Delta_{vap}H$ = 33.4; $\Delta_f H°$(l, 25°C) = -251.6
C_p (liq.) [J/mol °C]: 251.1^{25}
Elec. Prop.: μ = 0
η [mPa s]: 0.38^{38}
MS: 43(100) 71(53) 70(46) 41(27) 29(23) 27(23) 55(15) 57(14) 42(14) 39(10)
IR [cm⁻¹]: 2950 2860 1450 1370 730
¹³C NMR [ppm]: 14.1 19.3 20.2 32.3 39.5 diox
¹H NMR [ppm]: 0.8 0.9 1.4 CDCl₃

367. 2-Methylhexane

Syst. Name: Hexane, 2-methyl-

CASRN: 591-76-4   **DOT No.**: 1206
Beil RN: 1696856   **Beil Ref.**: 4-01-00-00397
MF: C_7H_{16} **MP[°C]**: -118.2   **Den.[g/cm³]**: 0.6787^{20}
MW: 100.20   **BP[°C]**: 90.0 n_D: 1.3848^{20}
Sol.: H_2O 1; EtOH 3;
   eth 5; ace 5; bz 5; chl
   5; lig 5
Crit. Const.: T_c = 257.3°C; P_c = 2.734 MPa; V_c = 421 cm³
Vap. Press. [kPa]: 2.34^0; 8.78^{25}; 25.8^{50}; 63.3^{75}; 135^{100}
Therm. Prop.[kJ/mol]: $\Delta_{fus}H$ = 8.87; $\Delta_{vap}H$ = 30.6; $\Delta_f H°$(l, 25°C) = -229.5
C_p (liq.) [J/mol °C]: 222.9^{25}
γ [mN/m]: 18.81^{25}; dγ/dT = 0.0963 mN/m °C
Elec. Prop.: μ = 0; ε = 1.92^{20}
η [mPa s]: 0.378^{20}
MS: 43(100) 42(38) 41(35) 85(32) 57(26) 29(22) 27(22) 56(20) 39(11) 55(5)
IR [cm⁻¹]: 2950 2880 1470 1390 1370 1170 1070 920 730
Raman [cm⁻¹]: 2960 2940 2920 2870 2760 2720 1460 1440 1340 1300 1260
   1220 1170 1150 1080 1060 1030 1000 960 940 900 870 820 780 720 450
   430 410 370 310 290
¹³C NMR [ppm]: 13.6 22.4 23.0 28.1 29.7 38.9 diox
¹H NMR [ppm]: 0.9 0.9 1.4 CDCl₃
Flammability: Flash pt. < -18°C; ign. temp. = 280°C; flam. lim. = 1.0-6.0%

368. 3-Methylhexane

Syst. Name: Hexane, 3-methyl-, (R)-

CASRN: 78918-91-9
Beil RN: 1718738
MF: C_7H_{16}
MW: 100.20
Sol.: H_2O 1; EtOH 3;
   eth 5; ace 5; bz 5; chl
   5; lig 5

DOT No.: 1206
Beil Ref.: 3-01-00-00440
MP[°C]: -119.4   **Den.[g/cm^3]**: 0.687^{21}
BP[°C]: 92 n_D: 1.3854^{25}

Crit. Const.: $T_c = 262.2°C$; $P_c = 2.814$ MPa; $V_c = 404$ cm^3
Therm. Prop.[kJ/mol]: $\Delta_{vap}H = 30.9$; $\Delta_f H°(l, 25°C) = -226.4$
C_p **(liq.) [J/mol °C]**: 214.2^{25}
γ **[mN/m]**: 19.31^{25}; $d\gamma/dT = 0.0970$ mN/m °C
Elec. Prop.: $\mu = 0$; $\varepsilon = 1.92^{20}$
η **[mPa s]**: 0.350^{25}
MS: 43(100) 71(47) 57(47) 41(47) 29(43) 56(34) 27(32) 100(4)
Raman [cm^{-1}]: 2960 2940 2910 2870 2730 1460 1450 1380 1360 1340 1300
   1280 1230 1160 1070 1050 1030 980 930 880 870 850 820 800 770 740 450
   430 380 330
^{13}C NMR [ppm]: 10.9 13.9 18.8 20.2 29.5 34.3 39.0 diox
Flammability: Flash pt. = -4°C; ign. temp. = 280°C

369. Methyl isobutyl ketone

Syst. Name: 2-Pentanone, 4-methyl-
Synonyms: Isobutyl methyl ketone

CASRN: 108-10-1   **DOT No.:** 1245
Merck No.: 5095   **Beil Ref.:** 4-01-00-03305
Beil RN: 605399
MF: $C_6H_{12}O$ **MP[°C]:** -84   **Den.[g/cm³]:** 0.7978^{20}
MW: 100.16   **BP[°C]:** 116.5 n_D: 1.3962^{20}
Sol.: H_2O 2; EtOH 5; eth
   5; ace 5; bz 5; chl 3
Crit. Const.: $T_c = 298°C$; $P_c = 3.27$ MPa
Vap. Press. [kPa]: 0.539^0; 2.64^{25}; 9.37^{50}; 26.3^{75}; 62.0^{100}; 128^{125}
Therm. Prop.[kJ/mol]: $\Delta_{vap}H = 34.5$
C_p **(liq.) [J/mol °C]:** 213.3^{25}
γ **[mN/m]:** 23.64^{20}
Elec. Prop.: $\varepsilon = 13.11^{20}$; IP = 9.30 eV
η **[mPa s]:** 0.545^{25}; 0.406^{50}
MS: 43(100) 58(84) 29(65) 41(56) 57(44) 27(42) 39(31) 85(19) 100(14) 42(14)
IR [cm⁻¹]: 2940 1720 1470 1410 1350 1280 1240 1160 1110 940 830
Raman [cm⁻¹]: 2970 2930 2880 2770 2730 1710 1460 1450 1430 1410 1390
   1370 1340 1300 1230 1200 1170 1120 990 950 840 830 790 630 600 530 480
   440 420 350 320 250 180
UV [nm]: 283 236 cyhex
¹³C NMR [ppm]: 22.5 24.5 30.1 52.7 208.0 CDCl₃
¹H NMR [ppm]: 0.9 2.1 2.3 CDCl₃
Flammability: Flash pt. = 18°C; ign. temp. = 448°C; flam. lim. = 1.2-8.0%
TLV/TWA: 50 ppm (205 mg/m³)
Reg. Lists: CERCLA (RQ = 5000 lb.); SARA 313 (1.0%); RCRA U161 (toxic)

370. Methyl isopentyl ketone

Syst. Name: 2-Hexanone, 5-methyl-
Synonyms: Methyl isoamyl ketone; MIAK

CASRN: 110-12-3   **DOT No.**: 2302
Beil RN: 506163   **Beil Ref.**: 4-01-00-03329
MF: $C_7H_{14}O$ **BP[°C]**: 144   **Den.[g/cm^3]**: 0.888^{20}
MW: 114.19 n_D: 1.4062^{20}
Sol.: H_2O 2; EtOH 5; eth
   5; ace 4; bz 4; ctc 3
Vap. Press. [kPa]: 0.691^{25}; 2.89^{50}; 9.33^{75}; 24.8^{100}; 57.0^{125}; 116^{150}
Elec. Prop.: $\varepsilon = 13.53^{20}$; IP = 9.28 eV
MS: 43(100) 58(34) 27(14) 41(13) 15(13) 57(11) 39(9) 71(8) 59(8) 29(8)
IR [cm^{-1}]: 2940 2860 1710 1460 1400 1380 1350 1150 1000
UV [nm]: 196 192 188 176 gas
^{1}H NMR [ppm]: 0.9 1.4 2.0 2.3 CCl_4
Flammability: Flash pt. = 36°C; ign. temp. = 191°C; flam. lim. = 1.0-8.2%
TLV/TWA: 50 ppm (234 mg/m^3)

371. Methyl methacrylate

Syst. Name: 2-Propenoic acid, 2-methyl-, methyl ester

Synonyms: Methyl 2-methylpropenoate

CASRN: 80-62-6

Merck No.: 5849

Beil RN: 605459

DOT No.: 1247

Beil Ref.: 4-02-00-01519

MF: $C_5H_8O_2$

MW: 100.12

MP[°C]: -48

BP[°C]: 100.5

Den.[g/cm^3]: 0.9440^{20}

n_D: 1.4142^{20}

Sol.: H_2O 2; EtOH 5; eth 5; ace 5; chl 3; glycol 2

Vap. Press. [kPa]: 5.10^{25}; 16.0^{50}; 43.4^{75}; 100^{100}

Therm. Prop.[kJ/mol]: $\Delta_{vap}H = 36.0$

C_p (liq.) [J/mol °C]: 191.2^{25}

γ [mN/m]: 28^{20}

Elec. Prop.: $\mu = (1.7 \text{ D})$; $\varepsilon = 6.32^{30}$; IP = 9.70 eV

η [mPa s]: 0.632^{20}

MS: 41(100) 69(66) 39(40) 100(34) 15(20) 40(10) 59(8) 99(6) 38(6) 55(5)

IR [cm^{-1}]: 2940 1720 1640 1450 1320 1300 1210 1160 1020 940 820

Raman [cm^{-1}]: 3120 3060 3040 3010 2960 2940 2910 2850 2750 1720 1640 1440 1410 1380 1330 1300 1200 1160 1020 1000 930 840 660 600 510 380 220

UV [nm]: 231(100) 340 hx

^{13}C NMR [ppm]: 18.3 51.5 124.7 136.9 167.3 diox

^{1}H NMR [ppm]: 2.0 3.8 5.6 6.1 CDCl$_3$

Flammability: Flash pt. = 10°C; flam. lim. = 1.7-8.2%

TLV/TWA: 100 ppm (410 mg/m^3)

Reg. Lists: CERCLA (RQ = 1000 lb.); SARA 313 (1.0%); RCRA U162 (toxic)

372. 1-Methylnaphthalene
Syst. Name: Naphthalene, 1-methyl-

CASRN: 90-12-0   **Beil Ref.**: 4-05-00-01687
Beil RN: 506793
MF: $C_{11}H_{10}$ **MP[°C]**: -30.4   **Den.[g/cm³]**: 1.0202^{20}
MW: 142.20   **BP[°C]**: 244.7 n_D: 1.6170^{20}
Sol.: H_2O 1; EtOH 4;
    eth 4; bz 3
Crit. Const.: T_c = 499°C; P_c = 3.60 MPa
Vap. Press. [kPa]: 0.009^{25}; 7.33^{150}
Therm. Prop.[kJ/mol]: $\Delta_{fus}H$ = 6.94; $\Delta_{vap}H$ = 45.5; $\Delta_f H°$(l, 25°C) = 56.3
C_p **(liq.) [J/mol °C]**: 224.4^{25}
γ [mN/m]: 39.80^{20}
Elec. Prop.: μ = 0; ε = 2.92^{20}; IP = 7.85 eV
η [mPa s]: 3.10^{20}; 1.97^{40}
MS: 142(100) 141(81) 115(24) 139(12) 140(11) 143(10) 71(9) 63(6) 89(4)
    70(4)
IR [cm⁻¹]: 3030 2940 1610 1520 1470 1430 1390 1270 1220 1160 1090 1020
    960 860 790 770 740 700
Flammability: Ign. temp. = 529°C

373. 2-Methyloctane
Syst. Name: Octane, 2-methyl-

CASRN: 3221-61-2   **DOT No.**: 1920
Beil RN: 1696918   **Beil Ref.**: 4-01-00-00454
MF: C_9H_{20} **MP[°C]**: -80.3   **Den.[g/cm^3]**: 0.7095^{25}
MW: 128.26   **BP[°C]**: 143.2 n_D: 1.4031^{20}
Sol.: H_2O 1; EtOH 3;
    eth 3; ctc 2; peth 4;
    lig 4
Crit. Const.: T_c = 313.9°C; P_c = 2.310 MPa
Vap. Press. [kPa]: 84.9^{50}; 198^{75}; 400^{100}; 729^{125}; 1220^{150}
Therm. Prop.[kJ/mol]: $\Delta_{fus}H$ = 18.00
Elec. Prop.: μ = 0; ε = 1.97^{20}
η **[mPa s]**: 0.47^{38}
MS: 43(100) 57(51) 41(35) 71(28) 42(28) 29(23) 27(21) 85(17) 56(15) 84(13)
IR [cm^{-1}]: 2960 2860 1460 1380 1350 1150 720
^{1}H NMR [ppm]: 0.9 1.0 1.3 CCl_4

---

374. 3-Methyloctane
Syst. Name: Octane, 3-methyl-

CASRN: 2216-33-3   **DOT No.**: 1920
MF: C_9H_{20} **MP[°C]**: -107.6   **Den.[g/cm^3]**: 0.717^{25}
MW: 128.26   **BP[°C]**: 144.2 n_D: 1.4040^{25}
Elec. Prop.: μ = 0
η **[mPa s]**: 0.47^{38}

375. 4-Methyloctane
Syst. Name: Octane, 4-methyl-

CASRN: 2216-34-4   **DOT No.**: 1920
MF: C_9H_{20} **MP[°C]**: -113.3   **Den.[g/cm^3]**: 0.716^{25}
MW: 128.26   **BP[°C]**: 142.4 n_D: 1.4039^{25}
Vap. Press. [kPa]: 3.56^{50}; 11.0^{75}; 28.0^{100}; 62.3^{125}; 124^{150}
Elec. Prop.: $\mu = 0$
η **[mPa s]**: 0.47^{38}

376. Methyl oleate
Syst. Name: 9-Octadecenoic acid (Z)-, methyl ester
Synonyms: Methyl cis-9-octadecenoate

CASRN: 112-62-9   **Beil Ref.**: 4-02-00-01649
Merck No.: 6788
Beil RN: 1727037
MF: $C_{19}H_{36}O_2$ **MP[°C]**: -19.9   **Den.[g/cm^3]**: 0.8739^{20}
MW: 296.49 n_D: 1.4522^{20}
Sol.: H_2O 1; EtOH 5; eth 5; chl 3
Vap. Press. [kPa]: 0.102^{150}
Therm. Prop.[kJ/mol]: $\Delta_f H°(l, 25°C) = -734.5$
γ **[mN/m]**: 31.3^{25}
Elec. Prop.: $\varepsilon = 3.21^{20}$
η **[mPa s]**: 4.88^{30}
MS: 55(100) 41(72) 74(63) 69(59) 43(52) 67(45) 83(40) 87(39) 81(35) 97(32)
IR [cm^{-1}]: 2930 2860 1740 1470 1440 1360 1250 1200 1170 720
UV [nm]: 230(3162) EtOH
^{1}**H NMR [ppm]**: 0.9 1.3 2.0 2.3 3.7 5.4 $CDCl_3$

377. 2-Methylpentane

Syst. Name: Pentane, 2-methyl-
Synonyms: Isohexane

CASRN: 107-83-5
Beil RN: 1730735

Rel. CASRN: 43133-95-5
DOT No.: 2462
Beil Ref.: 4-01-00-00358

MF: C_6H_{14}
MW: 86.18
Sol.: H_2O 1; EtOH 3;
eth 3; ace 5; bz 5; ctc
3; chl 5; hp 5

MP[°C]: -153.7
BP[°C]: 60.2

Den.[g/cm³]: 0.650^{25}
n_D: 1.3715^{20}

Crit. Const.: $T_c = 224.6°C$; $P_c = 3.031$ MPa; $V_c = 367$ cm³
Vap. Press. [kPa]: 2.16^{-25}; 8.99^0; 28.2^{25}; 72.2^{50}; 158^{75}; 308^{100}; 548^{125}; 906^{150}
Therm. Prop.[kJ/mol]: $\Delta_{fus}H = 6.27$; $\Delta_{vap}H = 27.8$; $\Delta_f H°(l, 25°C) = -204.6$
C_p (liq.) [J/mol °C]: 193.7^{25}
γ [mN/m]: 16.88^{25}; $d\gamma/dT = 0.0997$ mN/m °C
Elec. Prop.: $\mu = 0$; $\varepsilon = 1.89^{20}$; IP = 10.12 eV
η [mPa s]: 0.372^0; 0.286^{25}; 0.226^{50}
MS: 43(100) 42(53) 41(35) 27(31) 71(29) 39(20) 29(18) 57(11) 15(10) 70(7)
IR [cm⁻¹]: 2960 2860 1460 1380 1360 1160 1140 740
Raman [cm⁻¹]: 2960 2930 2910 2870 2760 2710 2630 1450 1340 1300 1240
1180 1150 1070 1030 1010 950 940 890 860 820 780 730 550 480 450 390
320
¹³C NMR [ppm]: 14.3 20.6 22.6 27.9 41.6 CDCl₃
¹H NMR [ppm]: 0.8 0.9 1.5 CCl₄
Flammability: Flash pt. < -29°C; ign. temp. = 264°C; flam. lim. = 1.0-7.0%

378. 3-Methylpentane

Syst. Name: Pentane, 3-methyl-

CASRN: 96-14-0   **Rel. CASRN**: 43133-95-5
Beil RN: 1730734   **DOT No.**: 2462
                            Beil Ref.: 4-01-00-00363
MF: C_6H_{14} **MP[°C]**: -162.9   **Den.[g/cm^3]**: 0.6598^{25}
MW: 86.18   **BP[°C]**: 63.2 n_D: 1.3765^{20}
Sol.: H_2O 1; EtOH 3;
  eth 5; ace 5; bz 5; ctc
  3; chl 5; hp 5
Crit. Const.: T_c = 231.4°C; P_c = 3.126 MPa; V_c = 367 cm^3
Vap. Press. [kPa]: 1.86^{-25}; 7.93^0; 25.3^{25}; 65.3^{50}; 144^{75}; 282^{100}; 503^{125}; 834^{150}
Therm. Prop.[kJ/mol]: $\Delta_{fus}H$ = 5.30; $\Delta_{vap}H$ = 28.1; $\Delta_f H°$(l, 25°C) = -202.4
C_p (liq.) [J/mol °C]: 190.7^{25}
γ **[mN/m]**: 17.61^{25}; $d\gamma/dT$ = 0.1060 mN/m °C
Elec. Prop.: μ = 0; ϵ = 1.89^{20}; IP = 10.08 eV
η **[mPa s]**: 0.395^0; 0.306^{25}
MS: 57(100) 56(76) 41(68) 29(60) 27(40) 43(29) 39(22) 55(9) 15(9) 28(8)
IR [cm^{-1}]: 2940 2880 1460 1380 980 950 810 770
Raman [cm^{-1}]: 2980 2940 2890 2870 2740 1450 1380 1350 1280 1150 1050
  1040 1020 990 960 880 820 770 750 730 550 470 440 390 360 310 220
^{13}C NMR [ppm]: 11.4 18.8 29.3 36.4 CDCl$_3$
^{1}H NMR [ppm]: 0.8 1.5 CCl$_4$
Flammability: Flash pt. < -7°C; ign. temp. = 278°C; flam. lim. = 1.2-7.0%

379. 2-Methyl-1-pentanol

Syst. Name: 1-Pentanol, 2-methyl-

CASRN: 105-30-6   **DOT No.**: 2053
Beil RN: 1718974   **Beil Ref.**: 3-01-00-01665
MF: $C_6H_{14}O$ **BP[°C]**: 149   **Den.[g/cm^3]**: 0.8263^{20}
MW: 102.18 n_D: 1.4182^{20}
Sol.: EtOH 3; eth 3; ace
    3; ctc 3
Crit. Const.: T_c = 331.3°C; P_c = 3.45 MPa
Vap. Press. [kPa]: 0.236^{25}; 1.29^{50}; 5.24^{75}; 16.9^{100}; 45.8^{125}; 108^{150}
Therm. Prop.[kJ/mol]: $\Delta_{vap}H$ = 50.2
C_p **(liq.) [J/mol °C]**: 248.0^{25}
γ **[mN/m]**: 24.94^{25}
η **[mPa s]**: 5.55^{25}
MS: 43(100) 71(48) 70(36) 41(34) 69(32) 55(27) 57(25) 56(24) 29(18) 42(17)
IR [cm^{-1}]: 3440 2970 2860 1460 1410 1370 1330 1110 1030 1020 970 930 730
^{1}H NMR [ppm]: 0.9 1.0 1.4 3.3 3.6 CCl$_4$
Flammability: Flash pt. = 54°C; ign. temp. = 310°C; flam. lim. = 1.1-9.65%

---

380. 3-Methyl-1-pentanol

Syst. Name: 1-Pentanol, 3-methyl-, (±)-

CASRN: 20281-83-8   **DOT No.**: 2282
Beil RN: 1718982   **Beil Ref.**: 4-01-00-01722
MF: $C_6H_{14}O$ **BP[°C]**: 153   **Den.[g/cm^3]**: 0.8242^{20}
MW: 102.18 n_D: 1.4112^{23}
Sol.: H_2O 1; EtOH 3; eth
    3
MS: 56(100) 69(67) 55(66) 41(53) 43(52) 29(27) 57(26) 102(0)
IR [cm^{-1}]: 3330 2940 1470 1370 1210 1120 1060 1030 970 850

381. 2-Methyl-2-pentanol

Syst. Name: 2-Pentanol, 2-methyl-

CASRN: 590-36-3		**DOT No.:** 2560
Beil RN: 1718951		**Beil Ref.:** 4-01-00-01714
MF: $C_6H_{14}O$	**MP[°C]:** -103	**Den.[g/cm^3]:** 0.8350^{16}
MW: 102.18	**BP[°C]:** 121.1	n_D: 1.4100^{20}

Sol.: H_2O 2; EtOH 3; eth
   3
Crit. Const.: T_c = 286.4°C
Vap. Press. [kPa]: 0.845^{25}; 4.83^{50}; 17.9^{75}; 49.8^{100}; 113^{125}
Therm. Prop.[kJ/mol]: $\Delta_{vap}H$ = 39.6
γ **[mN/m]:** 22.90^{25}
η **[mPa s]:** 1.41^{25}
MS: 59(100) 45(34) 43(30) 87(22) 41(20) 27(18) 31(16) 29(12) 69(7) 28(4)
IR [cm^{-1}]: 3410 2980 2890 1460 1370 1150 920 890

---

382. 3-Methyl-2-pentanol

Syst. Name: 2-Pentanol, 3-methyl-

CASRN: 565-60-6		**DOT No.:** 2282
Beil RN: 1697047		**Beil Ref.:** 3-01-00-01672
MF: $C_6H_{14}O$	**BP[°C]:** 134.3	**Den.[g/cm^3]:** 0.8307^{20}
MW: 102.18		n_D: 1.4182^{20}

Sol.: H_2O 2; EtOH 3; eth
   3
Vap. Press. [kPa]: 0.431^{25}; 2.40^{50}; 9.47^{75}; 29.2^{100}; 74.5^{125}; 165^{150}
C_p **(liq.) [J/mol °C]:** 275.9
MS: 45(100) 29(15) 41(14) 27(13) 56(12) 43(11) 44(9) 39(7) 57(6) 69(4)
IR [cm^{-1}]: 3330 2940 1470 1370 1100 960 920 860

383. 4-Methyl-2-pentanol

Syst. Name: 2-Pentanol, 4-methyl-

CASRN: 108-11-2

Beil RN: 1098268

MF: $C_6H_{14}O$

MW: 102.18

Sol.: H_2O 2; EtOH 3; eth 3; ctc 2

DOT No.: 2053

Beil Ref.: 4-01-00-01717

MP[°C]: -90

BP[°C]: 131.6

Den.[g/cm^3]: 0.8075^{20}

n_D: 1.4100^{20}

Crit. Const.: T_c = 301.3°C

Vap. Press. [kPa]: 0.698^{25}; 3.26^{50}; 11.5^{75}; 33.2^{100}; 81.5^{125}

Therm. Prop.[kJ/mol]: $\Delta_{vap}H$ = 44.2; $\Delta_f H°$(l, 25°C) = -394.7

C_p (liq.) [J/mol °C]: 273.0^{25}

γ [mN/m]: 22.63^{25}

η [mPa s]: 4.07^{25}

MS: 45(100) 43(47) 69(30) 41(27) 27(19) 39(13) 29(12) 87(11) 84(10) 57(10)

IR [cm^{-1}]: 3350 2960 1460 1360 1150 1110 1050 1020 920 820

Raman [cm^{-1}]: 3340 2970 2930 2890 2870 2750 2710 1460 1450 1360 1340 1320 1240 1170 1150 1130 1110 1050 1020 970 950 920 880 830 810 770 730 670 500 460 430 410 350 300 250 170

^{13}C NMR [ppm]: 22.4 23.1 23.9 24.8 48.7 65.8 CDCl$_3$

^{1}H NMR [ppm]: 0.9 1.1 1.3 1.8 3.5 3.7 CCl$_4$

Flammability: Flash pt. = 41°C; flam. lim. = 1.0-5.5%

TLV/TWA: 25 ppm (104 mg/m^3)

384. 2-Methyl-3-pentanol

Syst. Name: 3-Pentanol, 2-methyl-

CASRN: 565-67-3 **DOT No.:** 2282
Beil RN: 1718952 **Beil Ref.:** 4-01-00-01716
MF: $C_6H_{14}O$ **BP[°C]:** 126.5 **Den.[g/cm³]:** 0.8243^{20}
MW: 102.18 n_D: 1.4175^{20}
Sol.: H_2O 2; EtOH 5; eth
  5
Crit. Const.: T_c = 302.9°C; P_c = 3.46 MPa
Vap. Press. [kPa]: 0.709^{25}; 3.64^{50}; 13.5^{75}; 39.5^{100}; 96.7^{125}
Therm. Prop.[kJ/mol]: $\Delta_f H°$(l, 25°C) = -396.4
MS: 59(100) 31(47) 73(44) 28(43) 41(31) 27(30) 43(26) 55(24) 29(24) 58(19)
IR [cm⁻¹]: 3380 2970 2870 1460 1380 1370 1300 1240 1180 1150 1100 1070
  1040 1030 970 950 880 830 800 700

---

385. 3-Methyl-3-pentanol

Syst. Name: 3-Pentanol, 3-methyl-

CASRN: 77-74-7 **DOT No.:** 2282
Beil RN: 1731456 **Beil Ref.:** 4-01-00-01723
MF: $C_6H_{14}O$ **MP[°C]:** -23.6 **Den.[g/cm³]:** 0.8286^{20}
MW: 102.18 **BP[°C]:** 122.4 n_D: 1.4186^{20}
Sol.: H_2O 2; EtOH 5; eth
  5; ctc 2
Crit. Const.: T_c = 302.5°C; P_c = 3.52 MPa
Vap. Press. [kPa]: 1.16^{25}; 5.60^{50}; 19.0^{75}; 50.5^{100}; 112^{125}
C_p (liq.) [J/mol °C]: 293.4^{25}
Elec. Prop.: ϵ = 4.32^{20}
MS: 73(100) 55(38) 43(35) 45(28) 27(25) 29(21) 41(12) 87(11) 31(11) 15(9)
IR [cm⁻¹]: 3400 2970 1460 1370 1150 1060 970 900
¹H NMR [ppm]: 0.9 1.1 1.4 1.8 CCl_4

386. 4-Methyl-4-penten-2-one

Syst. Name: 4-Penten-2-one, 4-methyl-
Synonyms: Isomesityl oxide

CASRN: 3744-02-3   **Beil Ref.**: 4-01-00-03470
Beil RN: 1740929
MF: $C_6H_{10}O$ **MP[°C]**: -72.6   **Den.[g/cm^3]**: 0.8411^{20}
MW: 98.14   **BP[°C]**: 124.2
Vap. Press. [kPa]: 1.97^{25}; 7.26^{50}; 21.2^{75}; 52.2^{100}; 112^{125}
γ **[mN/m]**: 23.0^{20}
η **[mPa s]**: 0.634^{20}
MS: 43(100) 83(24) 28(19) 39(17) 29(14) 98(13) 41(13) 55(12) 32(12) 27(11)

387. Methyl pentyl ketone

Syst. Name: 2-Heptanone
Synonyms: Amyl methyl ketone

CASRN: 110-43-0   **DOT No.**: 1110
Merck No.: 4584   **Beil Ref.**: 4-01-00-03318
Beil RN: 1699063
MF: $C_7H_{14}O$ **MP[°C]**: -35   **Den.[g/cm^3]**: 0.8111^{20}
MW: 114.19   **BP[°C]**: 151.0 n_D: 1.4088^{20}
Sol.: H_2O 4; EtOH 3; eth
   3
Crit. Const.: T_c = 338.4°C; P_c = 3.436 MPa
Vap. Press. [kPa]: 0.490^{25}; 2.17^{50}; 7.37^{75}; 20.2^{100}; 47.4^{125}; 98.5^{150}
Therm. Prop.[kJ/mol]: $\Delta_{vap}H$ = 38.3
C_p (liq.) [J/mol °C]: 232.6^{25}
γ [mN/m]: 26.12^{25}; $d\gamma/dT$ = 0.1056 mN/m °C
Elec. Prop.: μ = (2.6 D); ε = 11.95^{20}; IP = 9.30 eV
η [mPa s]: 0.714^{25}; 0.407^{50}; 0.297^{75}
MS: 43(100) 58(60) 71(14) 41(11) 27(11) 59(9) 39(8) 29(8) 42(5) 114(4)
IR [cm^{-1}]: 2950 2860 1710 1450 1400 1350 1150 720
Raman [cm^{-1}]: 3010 2970 2930 2890 2740 1720 1450 1420 1350 1300 1270
   1230 1170 1120 1070 1030 990 940 900 880 840 820 790 750 720 600 530
   480 470 400 300 230
UV [nm]: 274(22) MeOH
^{13}C NMR [ppm]: 13.9 22.6 23.6 29.6 31.5 43.7 208.4 CDCl$_3$
^{1}H NMR [ppm]: 0.9 1.3 2.0 2.3 CCl$_4$
Flammability: Flash pt. = 39°C; ign. temp. = 393°C; flam. lim. = 1.1-7.9%
TLV/TWA: 50 ppm (233 mg/m^3)

388. *N*-Methylpropanamide

Syst. Name: Propanamide, *N*-methyl-
Synonyms: *N*-Methylpropionamide

CASRN: 1187-58-2   **Beil Ref.:** 4-04-00-00183
Beil RN: 1737640
MF: C_4H_9NO **MP[°C]:** -30.9   **Den.[g/cm³]:** 0.9305^{25}
MW: 87.12   **BP[°C]:** 148 n_D: 1.4345^{25}
Vap. Press. [kPa]: 0.271^{50}; 0.708^{75}
C_p (liq.) [J/mol °C]: 179^{25}
γ [mN/m]: 31.20^{30}
Elec. Prop.: μ = 3.6 D; ε = 170.0^{20}
η [mPa s]: 5.21^{25}; 3.53^{40}
MS: 58(100) 87(68) 57(34) 29(15) 56(10) 28(7) 59(6) 55(6) 30(6) 27(6)

---

389. 2-Methylpropanenitrile

Syst. Name: Propanenitrile, 2-methyl-
Synonyms: Isobutyronitrile; Isopropyl cyanide

CASRN: 78-82-0   **DOT No.:** 2284
Beil RN: 1340512   **Beil Ref.:** 4-02-00-00853
MF: C_4H_7N **MP[°C]:** -71.5   **Den.[g/cm³]:** 0.7704^{20}
MW: 69.11   **BP[°C]:** 103.9 n_D: 1.3720^{20}
Sol.: H_2O 2; EtOH 4; eth
   4; ace 3; chl 3
Vap. Press. [kPa]: 90.9^{100}
Therm. Prop.[kJ/mol]: $\Delta_{vap}H$ = 32.4; $\Delta_f H°$(l, 25°C) = -13.8
γ [mN/m]: 24.93^{20}
Elec. Prop.: μ = 4.29 D; ε = 24.42^{20}; IP = 11.30 eV
η [mPa s]: 0.551^{15}
MS: 42(100) 68(45) 28(45) 54(26) 41(26) 27(26) 29(25) 26(15) 39(13) 15(13)
IR [cm⁻¹]: 2940 2220 1470 1390 1370 1320 1160 1100 940 750
¹³C NMR [ppm]: 19.8 19.9 123.7 $CDCl_3$
¹H NMR [ppm]: 1.3 2.7 $CDCl_3$
Flammability: Flash pt. = 8°C; ign. temp. = 482°C

390. 2-Methylpropanoic acid

Syst. Name: Propanoic acid, 2-methyl-
Synonyms: Isobutyric acid

CASRN: 79-31-2   **DOT No.:** 2529
Merck No.: 5039   **Beil Ref.:** 4-02-00-00843
Beil RN: 635770
MF: $C_4H_8O_2$ **MP[°C]:** -46   **Den.[g/cm³]:** 0.9681^{20}
MW: 88.11   **BP[°C]:** 154.4 n_D: 1.3930^{20}
Sol.: H_2O 4; EtOH 5; eth
  5; ctc 2
Crit. Const.: $T_c = 332°C$; $P_c = 3.7$ MPa; $V_c = 292$ cm³
Vap. Press. [kPa]: 4.04^{75}; 16.5^{100}; 44.0^{125}; 90.3^{150}
Therm. Prop.[kJ/mol]: $\Delta_{fus}H = 5.02$
C_p **(liq.) [J/mol °C]:** 173^{25}
γ **[mN/m]:** 25.55^{20}
Elec. Prop.: $\mu = (1.1$ D); $\varepsilon = 2.58^{20}$; IP = 10.33 eV
η **[mPa s]:** 1.86^{0}; 1.23^{25}; 0.863^{50}; 0.639^{75}; 0.492^{100}
MS: 43(100) 41(42) 27(40) 73(22) 39(15) 45(14) 42(11) 29(9) 88(7) 28(6)
IR [cm⁻¹]: 2980 2650 2560 1710 1470 1420 1280 1230 1160 1080 930 810
Raman [cm⁻¹]: 2980 2930 2880 2760 2720 1650 1450 1330 1280 1230 1170
  1100 960 910 800 740 610 540 510 340 260
¹H NMR [ppm]: 1.2 2.6 CCl_4
Flammability: Flash pt. = 56°C; ign. temp. = 481°C; flam. lim. = 2.0-9.2%
Reg. Lists: CERCLA (RQ = 5000 lb.)

391. 2-Methyl-1-propanol

Syst. Name: 1-Propanol, 2-methyl-
Synonyms: Isobutyl alcohol; Isopropylcarbinol

CASRN: 78-83-1   **DOT No.**: 1212
Merck No.: 5015   **Beil Ref.**: 4-01-00-01588
Beil RN: 1730878
MF: $C_4H_{10}O$ **MP[°C]**: -108   **Den.[g/cm^3]**: 0.8018^{20}
MW: 74.12   **BP[°C]**: 107.8 n_D: 1.3955^{20}
Sol.: H_2O 2; EtOH 3; eth
   3; ace 3; ctc 3
Crit. Const.: T_c = 274.6°C; P_c = 4.295 MPa; V_c = 273 cm^3
Vap. Press. [kPa]: 1.39^{25}; 7.10^{50}; 26.1^{75}; 75.4^{100}; 184^{125}
Therm. Prop.[kJ/mol]: $\Delta_{fus}H$ = 6.32; $\Delta_{vap}H$ = 41.8; $\Delta_f H°$(l, 25°C) = -334.7
C_p (liq.) [J/mol °C]: 181.2^{25}
γ [mN/m]: 22.98^{20}
Elec. Prop.: μ = 1.64 D; ε = 17.93^{20}; IP = 10.12 eV
η [mPa s]: 3.33^{25}
MS: 43(100) 33(73) 31(72) 41(66) 42(60) 27(43) 29(18) 39(17) 28(8) 74(6)
IR [cm^{-1}]: 3350 2950 2870 1650 1460 1380 1360 1040 1000 940 820 650
^{13}C NMR [ppm]: 18.9 30.8 69.4 CDCl$_3$
^{1}H NMR [ppm]: 0.9 1.7 3.3 4.0 CCl$_4$
Flammability: Flash pt. = 28°C; ign. temp. = 415°C; flam. lim. = 1.7-10.6%
TLV/TWA: 50 ppm (152 mg/m^3)
Reg. Lists: CERCLA (RQ = 5000 lb.); RCRA U140 (toxic)

392. 2-Methyl-2-propanol

Syst. Name: 2-Propanol, 2-methyl-
Synonyms: *tert*-Butyl alcohol; Trimethylcarbinol

CASRN: 75-65-0   **DOT No.**: 1120
Merck No.: 1542   **Beil Ref.**: 4-01-00-01609
Beil RN: 906698
MF: $C_4H_{10}O$ **MP[°C]**: 25.8   **Den.[g/cm³]**: 0.7887^{20}
MW: 74.12   **BP[°C]**: 82.4 n_D: 1.3878^{20}
Sol.: H_2O 5; EtOH 5; eth
    5; chl 3
Crit. Const.: T_c = 233.1°C; P_c = 3.972 MPa; V_c = 275 cm³
Vap. Press. [kPa]: 5.52^{25}; 23.3^{50}; 75.2^{75}; 191^{100}
Therm. Prop.[kJ/mol]: $\Delta_{fus}H$ = 6.79; $\Delta_{vap}H$ = 39.1
C_p (liq.) [J/mol °C]: 219.8^{26}
γ [mN/m]: 19.96^{25}; $d\gamma/dT$ = 0.0900 mN/m °C
Elec. Prop.: μ = (1.7 D); ε = 12.47^{25}; IP = 9.97 eV
η [mPa s]: 4.31^{25}; 1.42^{50}; 0.678^{75}
MS: 59(100) 31(33) 41(22) 43(18) 29(13) 27(11) 57(10) 42(4) 60(3) 28(3)
IR [cm⁻¹]: 3360 2970 1470 1380 1360 1240 1200 1020 910 750 630
Raman [cm⁻¹]: 3300 2970 2920 2710 1450 1240 1210 1020 910 750 470 350
UV [nm]: 300(1349) 237(302) sulf
¹³C NMR [ppm]: 31.2 68.9 $CDCl_3$
¹H NMR [ppm]: 1.3 1.4 $CDCl_3$
Flammability: Flash pt. = 11°C; ign. temp. = 478°C; flam. lim. = 2.4-8.0%
TLV/TWA: 100 ppm (303 mg/m³)
Reg. Lists: SARA 313 (1.0%)

393. Methyl propyl ketone

Syst. Name: 2-Pentanone
Synonyms: Propyl methyl ketone

CASRN: 107-87-9
Merck No.: 6032
Beil RN: 506058
DOT No.: 1249
Beil Ref.: 4-01-00-03271

MF: $C_5H_{10}O$ **MP[°C]:** -76.9   **Den.[g/cm³]:** 0.809^{20}
MW: 86.13   **BP[°C]:** 102.2 n_D: 1.3895^{20}
Sol.: H_2O 2; EtOH 5; eth 5; ctc 2

Crit. Const.: T_c = 287.93°C; P_c = 3.694 MPa; V_c = 301 cm³

Vap. Press. [kPa]: 1.25^0; 4.97^{25}; 15.7^{50}; 41.4^{75}; 94.6^{100}; 193^{125}; 360^{150}

Therm. Prop.[kJ/mol]: $\Delta_{fus}H$ = 10.63; $\Delta_{vap}H$ = 33.4; $\Delta_f H°$(l, 25°C) = -297.3

C_p (liq.) [J/mol °C]: 184.1^{25}

γ [mN/m]: 23.25^{25}; $d\gamma/dT$ = 0.0655 mN/m °C

Elec. Prop.: μ = (2.7 D); ε = 15.45^{20}; IP = 9.38 eV

η [mPa s]: 0.641^0; 0.470^{25}; 0.362^{50}; 0.289^{75}; 0.238^{100}

MS: 43(100) 41(17) 86(12) 42(12) 27(11) 39(8) 71(7) 58(7) 45(7) 44(3)

IR [cm⁻¹]: 2940 1700 1350 1160 730

Raman [cm⁻¹]: 2930 2880 2750 1720 1460 1440 1430 1410 1350 1290 1230 1170 1120 1040 970 890 820 790 730 590 530 480 420 350

UV [nm]: 280 cyhex

¹H NMR [ppm]: 0.9 1.6 2.0 2.3 CCl_4

Flammability: Flash pt. = 7°C; ign. temp. = 452°C; flam. lim. = 1.5-8.2%

TLV/TWA: 200 ppm (705 mg/m³)

394. *N*-Methyl-2-pyrrolidone

Syst. Name: 2-Pyrrolidinone, 1-methyl-
Synonyms: 1-Methyl-2-pyrrolidinone

CASRN: 872-50-4   **Beil Ref.**: 5-21-06-00321
Beil RN: 106420
MF: C_5H_9NO **MP[°C]**: -24   **Den.[g/cm^3]**: 1.0230^{25}
MW: 99.13   **BP[°C]**: 202 n_D: 1.4684^{20}
Sol.: H_2O 4; eth 3; ace 3;
   chl 3; os 3
Crit. Const.: $T_c = 448.7$°C; $V_c = 311$ cm^3
Vap. Press. [kPa]: 0.040^{25}; 3.28^{100}; 8.88^{125}; 21.4^{150}
Therm. Prop.[kJ/mol]: $\Delta_f H°$(l, 25°C) = -262.2
C_p **(liq.) [J/mol °C]**: 307.8^{25}
γ **[mN/m]**: 40.7^{25}
Elec. Prop.: μ = (4.1 D); $\varepsilon = 32.2^{25}$; IP = 9.17 eV
η **[mPa s]**: 1.67^{25}
MS: 99(100) 44(89) 98(80) 42(60) 41(38) 43(17) 28(17) 71(13) 39(11) 70(10)
IR [cm^{-1}]: 2860 1670 1490 1450 1410 1390 1280 1250 1160 1110 980 920 850
UV [nm]: 205(2884) MeOH
^{1}H NMR [ppm]: 2.1 2.4 2.8 3.4 CDCl$_3$
Flammability: Flash pt. = 96°C; ign. temp. = 346°C; flam. lim. = 1-10%

395. Methyl salicylate

Syst. Name: Benzoic acid, 2-hydroxy-, methyl ester
Synonyms: Methyl 2-hydroxybenzoate

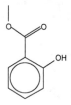

CASRN: 119-36-8   **Beil Ref.**: 4-10-00-00143
Merck No.: 6038
Beil RN: 971516
MF: $C_8H_8O_3$ **MP[°C]**: -8   **Den.[g/cm^3]**: 1.181^{25}
MW: 152.15   **BP[°C]**: 222.9 n_D: 1.535^{20}
Sol.: eth 4; EtOH 4; chl
  4
Crit. Const.: T_c = 436°C
Vap. Press. [kPa]: 0.015^{25}; 1.74^{100}; 5.24^{125}; 13.4^{150}
Therm. Prop.[kJ/mol]: $\Delta_{vap}H$ = 46.7
C_p **(liq.) [J/mol °C]**: 249.0^{25}
γ **[mN/m]**: 39.22^{25}; $d\gamma/dT$ = 0.1174 mN/m °C
Elec. Prop.: μ = (2.5 D); ε = 8.80^{41}
η **[mPa s]**: 0.815^{100}
MS: 120(100) 92(59) 152(47) 121(32) 65(22) 39(22) 93(15) 64(14) 18(14)
  63(13)
IR [cm^{-1}]: 3200 3000 2880 1680 1620 1590 1490 1440 1340 1310 1240 1210
  1160 1130 1090 1040 970 870 850 800 760 710 670
Raman [cm^{-1}]: 3070 2960 2850 1730 1670 1610 1580 1460 1440 1330 1320
  1250 1190 1150 1130 1090 1060 1030 960 860 840 810 750 660 550 530 510
  430 350 330 260 180 130
UV [nm]: 305(5300) 237(11200) MeOH
^{13}C NMR [ppm]: 52.1 112.4 117.5 119.0 129.9 135.5 161.7 170.5 CDCl$_3$
^{1}H NMR [ppm]: 3.9 6.7 6.9 7.3 7.7 10.6 CCl$_4$
Flammability: Flash pt. = 96°C; ign. temp. = 454°C

396. 2-Methyltetrahydrofuran
Syst. Name: Furan, tetrahydro-2-methyl-
Synonyms: 2-Methyloxolane

CASRN: 96-47-9   **Rel. CASRN:** 25265-68-3
Beil RN: 102448   **DOT No.:** 2536
                                       Beil Ref.: 5-17-01-00078
MF: $C_5H_{10}O$ **BP[°C]:** 78   **Den.[g/cm^3]:** 0.8552^{20}
MW: 86.13 n_D: 1.4059^{21}
Sol.: H_2O 3; EtOH 4; eth
   4; ace 4; bz 4; ctc 2;
   chl 4; os 4
Crit. Const.: T_c = 264°C; P_c = 3.76 MPa; V_c = 267 cm^3
Vap. Press. [kPa]: 3.53^0; 12.6^{25}; 36.0^{50}; 86.3^{75}
Elec. Prop.: ε = 6.97^{25}
MS: 71(100) 43(85) 41(64) 42(52) 27(43) 29(31) 39(30) 45(23) 56(22) 28(15)
IR [cm^{-1}]: 2960 2860 1450 1370 1350 1080 1020 880
Raman [cm^{-1}]: 3060 2970 2930 2870 2750 2720 2700 2630 2580 1610 1580
   1510 1490 1450 1380 1360 1280 1230 1190 1150 1090 1020 1000 920 900
   810 650 610 580 470 440 340 250
UV [nm]: 288(380) sulf
^{13}C NMR [ppm]: 21.0 26.2 33.5 67.2 75.0 diox
^{1}H NMR [ppm]: 1.3 1.7 3.6 CCl_4
Flammability: Flash pt. = -11°C

397. 2-Methylthiophene

Syst. Name: Thiophene, 2-methyl-

CASRN: 554-14-3   **Beil Ref.**: 5-17-01-00324
Beil RN: 103734
MF: C_5H_6S **MP[°C]**: -63.4   **Den.[g/cm^3]**: 1.0193^{20}
MW: 98.17   **BP[°C]**: 112.6 n_D: 1.5203^{20}
Sol.: H_2O 1; EtOH 5; eth
  5; ace 5; bz 5; ctc 5;
  hp 5;
Vap. Press. [kPa]: 3.45^{25}; 11.0^{50}; 29.4^{75}; 69.0^{100}
Therm. Prop.[kJ/mol]: $\Delta_{fus}H$ = 9.20; $\Delta_{vap}H$ = 33.9; $\Delta_f H°$(l, 25°C) = 44.6
C_p **(liq.) [J/mol °C]**: 149.8^{25}
γ **[mN/m]**: 30.95^{20}
Elec. Prop.: μ = 0.674 D; IP = 8.61 eV
η **[mPa s]**: 0.660^{25}
MS: 97(100) 98(57) 45(22) 39(14) 53(9) 99(8) 27(8) 69(6) 58(6) 59(5)
IR [cm^{-1}]: 3060 2920 2860 1530 1440 1380 1230 1160 1080 1040 850 810 690
Raman [cm^{-1}]: 3108 2923 1440 1352 1079 848 743 664 303 238
UV [nm]: 235(7943) 210(6026) hp
^{13}**C NMR [ppm]**: 15.0 123.4 125.7 127.2 139.4
^{1}**H NMR [ppm]**: 2.5 6.7 6.9 7.0 CDCl$_3$

398. 3-Methylthiophene

Syst. Name: Thiophene, 3-methyl-

CASRN: 616-44-4   **Beil Ref.:** 5-17-01-00331
Beil RN: 1300
MF: C_5H_6S **MP[°C]:** -69   **Den.[g/cm³]:** 1.0218^{20}
MW: 98.17   **BP[°C]:** 115.5 n_D: 1.5204^{20}
Sol.: H_2O 1; EtOH 5; eth
  5; ace 5; bz 5; chl 4
Vap. Press. [kPa]: 2.98^{25}; 9.78^{50}; 26.7^{75}; 63.1^{100}
Therm. Prop.[kJ/mol]: $\Delta_{fus}H$ = 10.53; $\Delta_{vap}H$ = 34.2; $\Delta_f H°$(l, 25°C) = 43.1
γ **[mN/m]:** 32.37^{20}
Elec. Prop.: μ = 0.95 D; IP = 8.40 eV
η **[mPa s]:** 0.635^{25}
MS: 97(100) 98(55) 45(27) 39(11) 53(9) 27(9) 99(8) 69(7) 71(5) 58(5)
IR [cm⁻¹]: 2940 1540 1450 1390 1240 1150 1080 990 940 860 830 760
UV [nm]: 235(6026) hx

---

399. 4-Methylvaleronitrile

Syst. Name: Pentanenitrile, 4-methyl-
Synonyms: Isoamyl cyanide; Isocapronitrile

CASRN: 542-54-1   **Beil Ref.:** 4-02-00-00946
Merck No.: 4999
Beil RN: 1698246
MF: $C_6H_{11}N$ **MP[°C]:** -51   **Den.[g/cm³]:** 0.8030^{20}
MW: 97.16   **BP[°C]:** 156.5 n_D: 1.4059^{20}
Sol.: H_2O 1; EtOH 3; eth
  5; ctc 2
Vap. Press. [kPa]: 3.82^{50}; 9.93^{75}; 22.7^{100}; 46.8^{125}; 88.4^{150}
γ **[mN/m]:** 26.53^{20}
Elec. Prop.: μ = (3.5 D); ε = 17.5^{22}
η **[mPa s]:** 0.980^{20}
MS: 55(100) 41(52) 43(46) 27(39) 39(29) 57(27) 54(26) 29(22) 82(13) 28(12)
IR [cm⁻¹]: 2940 2220 1470 1430 1390 1370 1180 1120 930 760
¹H NMR [ppm]: 1.0 1.6 2.3 CCl_4

400. Morpholine

Syst. Name: Morpholine
Synonyms: Tetrahydro-1,4-oxazine; Diethylene
oximide

CASRN: 110-91-8   **DOT No.**: 2054
Merck No.: 6194   **Beil Ref.**: 4-27-00-00015
Beil RN: 102549
MF: C$_4$H$_9$NO   **MP[°C]**: -4.9   **Den.[g/cm^3]**: 1.0005^{20}
MW: 87.12   **BP[°C]**: 128 n_D: 1.4548^{20}
Sol.: H$_2$O 5; EtOH 3; eth
  3; ace 3; bz 3; chl 2;
  os 3
Vap. Press. [kPa]: 1.34^{25}; 15.7^{75}; 40.4^{100}; 90.3^{125}
Therm. Prop.[kJ/mol]: $\Delta_{vap}H$ = 37.1
C_p **(liq.) [J/mol °C]**: 164.8^{25}
γ **[mN/m]**: 37.63^{20}
Elec. Prop.: μ = 1.55 D; ε = 7.42^{25}; IP = 8.20 eV
η **[mPa s]**: 2.02^{25}; 1.25^{50}; 0.850^{75}; 0.627^{100}
MS: 57(100) 29(100) 87(69) 28(69) 30(38) 56(33) 86(28) 31(28) 27(12) 15(7)
IR [cm^{-1}]: 3300 2940 2850 1440 1300 1230 1130 1080 1050 1020 880 820 780
Raman [cm^{-1}]: 3340 3300 2950 2900 2890 2840 2770 2740 2690 2630 1460
  1440 1380 1300 1270 1200 1150 1120 1090 1060 1030 1010 910 830 600
  540 480 440 420 270
^{1}H NMR [ppm]: 1.9 2.9 3.7 CDCl$_3$
Flammability: Flash pt. = 37°C; ign. temp. = 290°C; flam. lim. = 1-11%
TLV/TWA: 20 ppm (71 mg/m^3)

401. β-Myrcene
Syst. Name: 1,6-Octadiene, 7-methyl-3-methylene-

CASRN: 123-35-3   Beil Ref.: 4-01-00-01108
Merck No.: 6243
Beil RN: 1719990
MF: $C_{10}H_{16}$ BP[°C]: 167   Den.[g/cm^3]: 0.8013^{15}
MW: 136.24 n_D: 1.4722^{20}
Sol.: H_2O 1; EtOH 3; eth
    3; bz 3; chl 3; HOAc 3
Vap. Press. [kPa]: 0.280^{25}; 1.16^{50}; 3.83^{75}; 10.7^{100}; 25.8^{125}; 56.0^{150}
Therm. Prop.[kJ/mol]: $\Delta_f H°(l, 25°C) = 14.5$
Elec. Prop.: $\varepsilon = 2.3^{25}$
MS: 41(100) 93(87) 69(87) 39(34) 27(32) 53(17) 79(16) 77(16) 91(15) 94(14)
UV [nm]: 225(17378) EtOH

402. Naphthalene

Syst. Name: Naphthalene

CASRN: 91-20-3   **DOT No.**: 1334
Merck No.: 6289   **Beil Ref.**: 4-05-00-01640
Beil RN: 1421310
MF: $C_{10}H_8$ **MP[°C]**: 80.2   **Den.[g/cm³]**: 1.0253^{20}
MW: 128.17   **BP[°C]**: 217.9 n_D: 1.5898^{25}
Sol.: H_2O 1; EtOH 3; eth
  4; ace 4; bz 4; ctc 4;
  chl 4; CS_2 4; HOAc 3
Crit. Const.: $T_c = 475.3°C$; $P_c = 4.051$ MPa; $V_c = 407$ cm³
Vap. Press. [kPa]: 0.011^{25}; 0.768^{75}; 2.50^{100}; 6.84^{125}; 16.2^{150}
Therm. Prop.[kJ/mol]: $\Delta_{fus}H = 17.87$; $\Delta_{vap}H = 43.2$; $\Delta_f H°(s, 25°C) = 77.9$
γ **[mN/m]**: 31.8^{100}
Elec. Prop.: $\mu = 0$; $\varepsilon = 2.54^{90}$; IP = 8.14 eV
MS: 128(100) 127(12) 129(10) 51(8) 102(7) 64(7) 126(6) 63(5) 77(4) 75(4)
IR [cm⁻¹]: 3050 1590 1510 1390 1270 1130 1010 960 780
Raman [cm⁻¹]: 3060 3000 2970 1630 1580 1460 1450 1380 1240 1170 1150
  1020 780 760 720 510 390 120 110 80
UV [nm]: 334 320 311(239) 304(224) 301(294) 279(313) 286(3760) 283(3710)
  275(5530) 266(4990) 258(3470) 221(10600) MeOH
¹H NMR [ppm]: 7.4 7.7 CCl_4
Flammability: Flash pt. = 79°C; ign. temp. = 526°C; flam. lim. = 0.9-5.9%
TLV/TWA: 10 ppm (52 mg/m³)
Reg. Lists: CERCLA (RQ = 100 lb.); SARA 313 (1.0%); RCRA U165 (toxic)

403. Neopentane

Syst. Name: Propane, 2,2-dimethyl-
Synonyms: 2,2-Dimethylpropane;
  Tetramethylmethane

CASRN: 463-82-1   **DOT No.**: 2044
Merck No.: 6372   **Beil Ref.**: 4-01-00-00333
Beil RN: 1730722
MF: C_5H_{12} **MP[°C]**: -16.6   **Den.[g/cm^3]**: 0.5852^{25} (sat. press.)
MW: 72.15   **BP[°C]**: 9.4 n_D: 1.3476^6
Sol.: H_2O 1; EtOH 3; eth
  3; ctc 3
Crit. Const.: T_c = 160.7°C; P_c = 3.197 MPa; V_c = 307 cm^3
Vap. Press. [kPa]: 0.7^{-73}; 24.8^{-23}; 71.0^0; 171^{25}
Therm. Prop.[kJ/mol]: $\Delta_{fus}H$ = 3.10; $\Delta_{vap}H$ = 22.7; $\Delta_f H°$(g, 25°C) = -168.1
C_p **(liq.) [J/mol °C]**: 163.9^6
γ **[mN/m]**: 12.05^{20} (at saturation pressure)
Elec. Prop.: μ = 0; ε = 1.77^{23}; IP = 10.21 eV
η **[mPa s]**: 0.328^0 (at saturation pressure)
MS: 57(100) 41(42) 29(39) 27(15) 39(13) 15(5) 58(4) 56(4) 55(3) 43(2)
IR [cm^{-1}]: 2940 1470 1370 1250 720
^{13}C NMR [ppm]: 27.9 31.5
^{1}H NMR [ppm]: 0.9 CCl$_4$
Flammability: Flash pt. = -65°C; ign. temp. = 450°C; flam. lim. = 1.4-7.5%

404. 2-Nitroanisole
Syst. Name: Benzene, 1-methoxy-2-nitro-

CASRN: 91-23-6   **Beil Ref.:** 4-06-00-01249
Beil RN: 1868032
MF: $C_7H_7NO_3$ **MP[°C]:** 10.5   **Den.[g/cm^3]:** 1.2540^{20}
MW: 153.14   **BP[°C]:** 277 n_D: 1.5161^{20}
Sol.: H_2O 1; EtOH 5; eth
   5; ctc 3
Vap. Press. [kPa]: 2.42^{150}
γ [mN/m]: 45.7^{26}
Elec. Prop.: μ = (5.0 D); ε = 45.75^{20}
MS: 77(100) 106(87) 153(63) 92(62) 123(41) 63(38) 51(38) 64(35) 78(26)
   65(24)
IR [cm^{-1}]: 2940 1610 1490 1450 1350 1270 1160 1090 1040 1020 940 860 770
   740
UV [nm]: 321 258 211 209 MeOH
^{1}**H NMR [ppm]:** 3.9 7.3 CCl$_4$

405. Nitrobenzene

Syst. Name: Benzene, nitro-

CASRN: 98-95-3   **DOT No.**: 1662
Merck No.: 6509   **Beil Ref.**: 4-05-00-00708
Beil RN: 507540
MF: $C_6H_5NO_2$ **MP[°C]**: 5.7   **Den.[g/cm^3]**: 1.2037[20]
MW: 123.11   **BP[°C]**: 210.8 n_D: 1.5562[20]
Sol.: H_2O 2; EtOH 4; eth
   4; ace 4; bz 4; ctc 2
Vap. Press. [kPa]: 0.030[25]; 18.8[150]
Therm. Prop.[kJ/mol]: $\Delta_{fus}H$ = 11.59; $\Delta_f H°$(l, 25°C) = 12.5
C_p **(liq.) [J/mol °C]**: 185.8[25]
γ **[mN/m]**: 41.71[40]; $d\gamma/dT$ = 0.1157 mN/m °C
Elec. Prop.: μ = 4.22 D; ϵ = 35.6[20]; IP = 9.86 eV
η **[mPa s]**: 3.04[0]; 1.86[25]; 1.26[50]; 0.918[75]; 0.704[100]
MS: 77(100) 51(59) 123(42) 50(25) 30(15) 65(14) 39(10) 93(9) 74(7) 78(6)
IR [cm^{-1}]: 3100 2850 1620 1610 1530 1480 1350 1310 1170 1110 1070 1020
   930 850 790 700 680
Raman [cm^{-1}]: 3080 1590 1340 1110 1020 1000 850 610 390 180
UV [nm]: 259(7180) MeOH
^{13}C NMR [ppm]: 123.4 129.4 134.6 148.2 CDCl$_3$
^{1}H NMR [ppm]: 7.6 8.2 CCl$_4$
Flammability: Flash pt. = 88°C; ign. temp. = 482°C; flam. lim. = 1.8-9%
TLV/TWA: 1 ppm (5 mg/m^3)
Reg. Lists: CERCLA (RQ = 1000 lb.); SARA 313 (1.0%); RCRA U169 (toxic)

406. Nitroethane

Syst. Name: Ethane, nitro-

CASRN: 79-24-3   **DOT No.:** 2842
Merck No.: 6518   **Beil Ref.:** 4-01-00-00170
Beil RN: 1209324
MF: $C_2H_5NO_2$ **MP[°C]:** -89.5   **Den.[g/cm^3]:** 1.0448^{25}
MW: 75.07   **BP[°C]:** 114.0 n_D: 1.3917^{20}
Sol.: H_2O 2; EtOH 5; eth
  5; ace 3; chl 3; dil alk
  3
Vap. Press. [kPa]: 2.79^{25}; 9.95^{50}; 28.1^{75}; 66.6^{100}
Therm. Prop.[kJ/mol]: $\Delta_{fus}H$ = 9.85; $\Delta_{vap}H$ = 38.0; $\Delta_f H°$(l, 25°C) = -143.9
C_p **(liq.) [J/mol °C]:** 134.4^{25}
γ **[mN/m]:** 32.13^{25}; $d\gamma/dT$ = 0.1255 mN/m °C
Elec. Prop.: μ = 3.23 D; ε = 29.11^{15}; IP = 10.88 eV
η **[mPa s]:** 1.35^{-25}; 0.940^{0}; 0.688^{25}; 0.526^{50}; 0.415^{75}; 0.337^{100}
MS: 29(100) 30(12) 28(11) 26(9) 27(8) 43(5) 41(5) 14(5) 15(3) 46(2)
IR [cm^{-1}]: 3000 2950 2910 1550 1440 1390 1360 1250 1130 1100 990 870 810
  620 490 380
Raman [cm^{-1}]: 2960 2920 2750 1560 1440 1400 1370 1330 1270 1140 1100
  990 920 880 810 620 530 490 300
UV [nm]: 277 cyhex
^{13}C NMR [ppm]: 10.6 70.4
^{1}H NMR [ppm]: 1.6 4.3 $CDCl_3$
Flammability: Flash pt. = 28°C; ign. temp. = 414°C; flam. lim. = 3.4-17%
TLV/TWA: 100 ppm (307 mg/m^3)

407. Nitromethane

Syst. Name: Methane, nitro-
Synonyms: Nitrocarbol

CASRN: 75-52-5   **DOT No.**: 1261
Merck No.: 6532   **Beil Ref.**: 4-01-00-00100
Beil RN: 1698205
MF: CH_3NO_2 **MP[°C]**: -28.5   **Den.[g/cm^3]**: 1.1371^{20}
MW: 61.04   **BP[°C]**: 101.1 n_D: 1.3817^{20}
Sol.: H_2O 3; EtOH 3; eth
  3; ace 3; ctc 3; alk 3
Crit. Const.: T_c = 315°C; P_c = 5.87 MPa; V_c = 173 cm^3
Vap. Press. [kPa]: 4.79^{25}; 42.1^{75}; 97.6^{100}; 201^{125}
Therm. Prop.[kJ/mol]: $\Delta_{fus}H$ = 9.70; $\Delta_{vap}H$ = 34.0; $\Delta_f H°$(l, 25°C) = -113.1
C_p (liq.) [J/mol °C]: 106.6^{25}
γ [mN/m]: 36.53^{25}; $d\gamma/dT$ = 0.1678 mN/m °C
Elec. Prop.: μ = 3.46 D; ε = 37.27^{20}; IP = 11.02 eV
η [mPa s]: 1.31^{-25}; 0.875^{0}; 0.630^{25}; 0.481^{50}; 0.383^{75}; 0.317^{100}
MS: 30(100) 61(64) 46(39) 28(30) 45(8) 27(8) 44(7) 29(7) 60(5) 43(4)
IR [cm^{-1}]: 3030 2950 1560 1400 1370 1100 920 660
Raman [cm^{-1}]: 3060 2980 2780 1570 1410 1380 1310 1100 920 880 850 660
  610 520 480
UV [nm]: 277 cyhex
^{13}**C NMR [ppm]**: 57.3
^{1}**H NMR [ppm]**: 4.2 CCl_4
Flammability: Flash pt. = 35°C; ign. temp. = 418°C; flam. lim. = 7.3-22%
TLV/TWA: 100 ppm (250 mg/m^3)

408. 1-Nitropropane

Syst. Name: Propane, 1-nitro-

CASRN: 108-03-2   **DOT No.**: 2608
Merck No.: 6548   **Beil Ref.**: 4-01-00-00229
Beil RN: 506236
MF: $C_3H_7NO_2$ **MP[°C]**: -108   **Den.[g/cm^3]**: 0.9961^{25}
MW: 89.09   **BP[°C]**: 131.1 n_D: 1.4018^{20}
Sol.: H_2O 2; EtOH 5; eth
    5; chl 3
Vap. Press. [kPa]: 1.36^{25}; 15.3^{75}; 38.4^{100}; 84.4^{125}
Therm. Prop.[kJ/mol]: $\Delta_{vap}H$ = 38.5; $\Delta_f H°$(l, 25°C) = -167.2
C_p (liq.) [J/mol °C]: 175.3^{25}
γ [mN/m]: 30.64^{20}
Elec. Prop.: μ = 3.66 D; ε = 24.70^{15}; IP = 10.81 eV
η [mPa s]: 1.85^{-25}; 1.16^{0}; 0.798^{25}; 0.589^{50}; 0.460^{75}; 0.374^{100}
MS: 43(100) 27(93) 41(90) 39(34) 30(25) 44(20) 42(20) 26(20) 28(13) 54(12)
IR [cm^{-1}]: 2970 1550 1460 1430 1390 1230 1130 800
Raman [cm^{-1}]: 2940 2880 2750 1550 1460 1440 1380 1340 1290 1270 1220
    1130 1120 1090 1030 890 870 800 740 610 520 480 340 270 210
UV [nm]: 280 cyhex
^{1}H NMR [ppm]: 1.0 2.1 4.4 CDCl$_3$
Flammability: Flash pt. = 36°C; ign. temp. = 421°C; lower flam. lim. = 2.2%
TLV/TWA: 25 ppm (91 mg/m^3)

409. 2-Nitropropane

Syst. Name: Propane, 2-nitro-
Synonyms: Isonitropropane

CASRN: 79-46-9   **DOT No.:** 2608
Merck No.: 6549   **Beil Ref.:** 4-01-00-00230
Beil RN: 1740684
MF: $C_3H_7NO_2$ **MP[°C]:** -91.3   **Den.[g/cm^3]:** 0.9821^{25}
MW: 89.09   **BP[°C]:** 120.2 n_D: 1.3944^{20}
Sol.: H_2O 2; chl 3
Vap. Press. [kPa]: 2.30^{25}; 8.20^{50}; 23.2^{75}; 55.4^{100}; 116^{125}
Therm. Prop.[kJ/mol]: $\Delta_{vap}H$ = 36.8; $\Delta_f H°$(l, 25°C) = -180.3
C_p **(liq.) [J/mol °C]:** 170.3^{25}
γ **[mN/m]:** 29.29^{25}; $d\gamma/dT$ = 0.1158 mN/m °C
Elec. Prop.: μ = 3.73 D; ε = 26.74^{15}; IP = 10.71 eV
η **[mPa s]:** 0.721^{25}
MS: 43(100) 41(73) 27(71) 39(30) 30(18) 15(11) 42(9) 28(8) 26(8) 38(6)
IR [cm^{-1}]: 3000 2950 1550 1470 1400 1360 1300 1100 850
Raman [cm^{-1}]: 3010 2960 2940 2880 2800 2750 1560 1450 1400 1360 1310
    1180 1140 1100 950 900 850 720 620 530 350 310 280
UV [nm]: 278 cyhex
^{13}C NMR [ppm]: 19.1 79.1
Flammability: Flash pt. = 24°C; ign. temp. = 428°C; flam. lim. = 2.6-11.0%
TLV/TWA: 10 ppm (36 mg/m^3)
Reg. Lists: CERCLA (RQ = 10 lb.); SARA 313 (0.1%); RCRA U171 (toxic);
    confirmed or suspected carcinogen

410. Nonane

Syst. Name: Nonane

CASRN: 111-84-2   **DOT No.**: 1920
Beil RN: 1696917   **Beil Ref.**: 4-01-00-00447
MF: C_9H_{20} **MP[°C]**: -53.5   **Den.[g/cm^3]**: 0.7176^{20}
MW: 128.26   **BP[°C]**: 150.8 n_D: 1.4054^{20}
Sol.: H_2O 1; EtOH 4; eth
  4; ace 5; bz 5; chl 5;
  hp 5
Crit. Const.: T_c = 321.5°C; P_c = 2.29 MPa; V_c = 555 cm^3
Vap. Press. [kPa]: 0.570^{25}; 2.41^{50}; 7.85^{75}; 21.0^{100}; 48.5^{125}; 99.2^{150}
Therm. Prop.[kJ/mol]: $\Delta_{fus}H$ = 15.47; $\Delta_{vap}H$ = 36.9; $\Delta_f H°$(l, 25°C) = -274.7
C_p (liq.) [J/mol °C]: 284.4^{25}
γ [mN/m]: 22.38^{25}; dγ/dT = 0.0935 mN/m °C
Elec. Prop.: μ = 0; ε = 1.97^{20}; IP = 9.72 eV
η [mPa s]: 0.964^0; 0.665^{25}; 0.488^{50}; 0.375^{75}; 0.300^{100}
k [W/m °C]: 0.144^{-25}; 0.138^0; 0.131^{25}; 0.124^{50}; 0.118^{75}; 0.111^{100}
MS: 43(100) 57(75) 41(29) 84(26) 85(22) 29(22) 71(18) 56(16) 27(13) 42(12)
IR [cm^{-1}]: 2900 1460 1380 1340 1300 890 720
Raman [cm^{-1}]: 2962 2935 2893 2875 2860 2852 1446 1301 1078 890
^{13}**C NMR [ppm]**: 14.0 23.1 32.4 29.8 30.1
^{1}**H NMR [ppm]**: 0.9 1.3 CDCl$_3$
Flammability: Flash pt. = 31°C; ign. temp. = 205°C; flam. lim. = 0.8-2.9%
TLV/TWA: 200 ppm (1050 mg/m^3)

411. Nonanoic acid

Syst. Name: Nonanoic acid
Synonyms: Pelargonic acid; Nonylic acid

CASRN: 112-05-0   **Beil Ref.:** 4-02-00-01018
Merck No.: 7013
Beil RN: 1752351
MF: $C_9H_{18}O_2$ **MP[°C]:** 12.3   **Den.[g/cm^3]:** 0.9052^{20}
MW: 158.24   **BP[°C]:** 254.5 n_D: 1.4343^{19}
Sol.: H_2O 1; EtOH 3; eth
   3; chl 3
Crit. Const.: T_c = 438°C; P_c = 2.40 MPa
Vap. Press. [kPa]: 0.127^{100}; 0.637^{125}; 2.41^{150}
Therm. Prop.[kJ/mol]: $\Delta_{fus}H$ = 20.28; $\Delta_f H°$(l, 25°C) = -659.7
C_p **(liq.) [J/mol °C]:** 362.4^{25}
γ **[mN/m]:** 28.9^{27}
Elec. Prop.: μ = (0.8 D); ε = 2.48^{22}
η **[mPa s]:** 7.01^{25}; 3.71^{50}; 2.23^{75}; 1.48^{100}
MS: 60(100) 73(90) 57(58) 43(39) 41(38) 55(36) 115(20) 40(20) 69(18) 29(17)
IR [cm^{-1}]: 2940 2630 1720 1470 1430 1280 1240 1110 940 730
^{13}C NMR [ppm]: 14.1 22.7 24.8 29.2 29.3 31.9 34.2 180.6 CDCl$_3$
^{1}H NMR [ppm]: 0.9 1.3 2.4 10.5 CDCl$_3$

---

412. 1-Nonene

Syst. Name: 1-Nonene
Synonyms: 1-Nonylene

CASRN: 124-11-8   **Beil Ref.:** 4-01-00-00894
Beil RN: 1698439
MF: C_9H_{18} **MP[°C]:** -81.3   **Den.[g/cm^3]:** 0.7253^{25}
MW: 126.24   **BP[°C]:** 146.9 n_D: 1.4257^{20}
Vap. Press. [kPa]: 0.714^{25}; 2.90^{50}; 9.17^{75}; 24.0^{100}; 54.5^{125}; 110^{150}
Therm. Prop.[kJ/mol]: $\Delta_{fus}H$ = 18.08
γ **[mN/m]:** 22.56^{25}; $d\gamma/dT$ = 0.0938 mN/m °C
Elec. Prop.: μ = 0; ε = 2.18^{20}
η **[mPa s]:** 0.586^{25}
MS: 43(100) 41(83) 56(78) 55(66) 29(48) 27(41) 42(38) 70(36) 69(35) 39(32)
Raman [cm^{-1}]: 3081 3000 2982 2962 2900 2854 1641 1440 1417 1300
Flammability: Flash pt. = 26°C

413. Octane

Syst. Name: Octane

CASRN: 111-65-9   **DOT No.**: 1262
Merck No.: 6672   **Beil Ref.**: 4-01-00-00412
Beil RN: 1696875
MF: C_8H_{18} **MP[°C]**: -56.8   **Den.[g/cm^3]**: 0.6986^{25}
MW: 114.23   **BP[°C]**: 125.6 n_D: 1.3974^{20}
Sol.: H_2O 1; EtOH 5; eth
  3; ace 5; bz 5; chl 5;
  peth 5
Crit. Const.: T_c = 295.6°C; P_c = 2.49 MPa; V_c = 492 cm^3
Vap. Press. [kPa]: 0.058^{-25}; 0.386^0; 1.86^{25}; 6.71^{50}; 19.3^{75}; 46.8^{100}; 99.5^{125};
  190^{150}
Therm. Prop.[kJ/mol]: $\Delta_{fus}H$ = 20.65; $\Delta_{vap}H$ = 34.4; $\Delta_f H°$(l, 25°C) = -250.1
C_p (liq.) [J/mol °C]: 254.6^{25}
γ [mN/m]: 21.14^{25}; dγ/dT = 0.0951 mN/m °C
Elec. Prop.: ε = 1.948^{20}
η [mPa s]: 0.700^0; 0.508^{25}; 0.385^{50}; 0.302^{75}; 0.243^{100}
k [W/m °C]: 0.143^{-25}; 0.135^0; 0.128^{25}; 0.120^{50}; 0.113^{75}; 0.106^{100}
MS: 43(100) 57(30) 85(25) 41(25) 71(19) 29(17) 56(14) 70(10) 42(10) 27(10)
IR [cm^{-1}]: 2860 1470 1370 1140 1080 980 920 880 760 720
Raman [cm^{-1}]: 2970 2940 2900 2880 2860 2740 2680 1450 1300 1220 1190
  1160 1140 1080 1060 1040 1030 970 890 880 860 840 820 760 500 420 400
  380 350 280
^{13}C NMR [ppm]: 14.0 23.0 32.4 29.7
^{1}H NMR [ppm]: 0.9 1.3 CDCl$_3$
Flammability: Flash pt. = 13°C; ign. temp. = 206°C; flam. lim. = 1.0-6.5%
TLV/TWA: 300 ppm (1400 mg/m^3)

414. Octanenitrile

Syst. Name: Octanenitrile

CASRN: 124-12-9   **Beil Ref.:** 4-02-00-00993
Beil RN: 1744063
MF: $C_8H_{15}N$ **MP[°C]:** -45.6   **Den.[g/cm^3]:** 0.8136^{20}
MW: 125.21   **BP[°C]:** 205.2 n_D: 1.4203^{20}
Sol.: eth 4
Crit. Const.: $T_c = 401.3$°C; $P_c = 2.85$ MPa
Vap. Press. [kPa]: 3.28^{100}; 9.05^{125}; 21.7^{150}
Therm. Prop.[kJ/mol]: $\Delta_f H°$(l, 25°C) = -107.3
γ [mN/m]: 28.10^{20}
Elec. Prop.: ε = 13.90^{20}
η [mPa s]: 1.81^{15}
MS: 82(100) 41(63) 96(45) 54(41) 83(38) 69(36) 55(35) 43(24) 39(24) 97(20)

---

415. Octanoic acid

Syst. Name: Octanoic acid
Synonyms: Caprylic acid

CASRN: 124-07-2   **Beil Ref.:** 4-02-00-00982
Merck No.: 1765
Beil RN: 1747180
MF: $C_8H_{16}O_2$ **MP[°C]:** 16.3   **Den.[g/cm^3]:** 0.9106^{20}
MW: 144.21   **BP[°C]:** 239 n_D: 1.4285^{20}
Sol.: H_2O 2; EtOH 5; chl
  5; CH_3CN 5
Crit. Const.: $T_c = 422$°C; $P_c = 2.64$ MPa
Vap. Press. [kPa]: 1.44^{125}; 5.14^{150}
Therm. Prop.[kJ/mol]: $\Delta_{fus}H = 21.36$; $\Delta_{vap}H = 58.5$; $\Delta_f H°$(l, 25°C) = -636.0
C_p (liq.) [J/mol °C]: 297.9^{25}
γ [mN/m]: 28.7^{25}
Elec. Prop.: μ = (1.1 D); ε = 2.85^{15}
η [mPa s]: 5.02^{25}; 2.66^{50}; 1.65^{75}; 1.15^{100}
MS: 60(100) 73(66) 43(52) 32(39) 41(38) 55(37) 29(26) 85(17) 101(14) 84(14)
IR [cm^{-1}]: 2950 2860 2680 1700 1470 1410 1380 1290 1240 1210 1110 940
  730
Raman [cm^{-1}]: 2930 2910 2870 2860 2730 1650 1440 1410 1300 1170 1110
  1070 1060 1030 890 840 780 740 620 590 390 310 260
^{1}H NMR [ppm]: 0.9 1.3 2.4 11.5 CDCl$_3$

416. 1-Octanol

Syst. Name: 1-Octanol
Synonyms: Caprylic alcohol; Octyl alcohol

CASRN: 111-87-5   **Beil Ref.**: 4-01-00-01756
Merck No.: 6674
Beil RN: 1697461
MF: $C_8H_{18}O$ **MP[°C]**: -15.5   **Den.[g/cm³]**: 0.8262^{25}
MW: 130.23   **BP[°C]**: 195.1 n_D: 1.4295^{20}
Sol.: H_2O 1; EtOH 5; eth
    5; ctc 3
Crit. Const.: T_c = 379.4°C; P_c = 2.777 MPa; V_c = 490 cm³
Vap. Press. [kPa]: 0.010^{25}; 0.079^{50}; 0.554^{75}; 2.55^{100}; 8.72^{125}; 24.0^{150}
Therm. Prop.[kJ/mol]: $\Delta_{fus}H$ = 42.30; $\Delta_{vap}H$ = 46.9; $\Delta_fH°$(l, 25°C) = -426.5
C_p (liq.) [J/mol °C]: 305.2^{25}
γ [mN/m]: 27.10^{25}; dγ/dT = 0.0795 mN/m °C
Elec. Prop.: μ = (1.8 D); ϵ = 10.30^{20}
η [mPa s]: 7.29^{25}; 3.23^{50}; 1.68^{75}; 0.991^{100}
k [W/m °C]: 0.168^{0}; 0.161^{25}; 0.154^{50}; 0.147^{75}; 0.141^{100}
MS: 41(100) 56(85) 43(82) 55(81) 31(69) 27(69) 29(68) 42(62) 70(53) 69(48)
IR [cm⁻¹]: 3340 2930 2850 1470 1380 1060 730
Raman [cm⁻¹]: 3340 2930 2900 2870 2860 2730 1450 1440 1370 1300 1250
    1200 1180 1120 1080 1060 1050 1030 960 930 890 870 850 810 760 730 530
    480 450 400 340 260
UV [nm]: 197 gas
¹H NMR [ppm]: 0.9 1.3 3.5 3.9 CCl_4
Flammability: Flash pt. = 81°C

417. 2-Octanol

Syst. Name: 2-Octanol, (±)-
Synonyms: *sec*-Caprylic alcohol

CASRN: 4128-31-8   **Beil Ref.:** 4-01-00-01770
Merck No.: 6675
Beil RN: 1719325

MF: $C_8H_{18}O$	**MP[°C]:** -31.6	**Den.[g/cm^3]:** 0.8193^{20}
MW: 130.23	**BP[°C]:** 180	n_D: 1.4203^{20}

Sol.: H_2O 2; EtOH 3; eth
   3; ace 3
Crit. Const.: $T_c = 356.5°C$; $P_c = 2.754$ MPa
Vap. Press. [kPa]: 0.232^{50}; 1.40^{75}; 5.60^{100}; 41.9^{150}; 89^{175}
Therm. Prop.[kJ/mol]: $\Delta_{vap}H = 44.4$
C_p **(liq.) [J/mol °C]:** 330.1^{25}
γ **[mN/m]:** 26.28^{20}
Elec. Prop.: $\mu = (1.7$ D$)$; $\varepsilon = 8.13^{20}$
η **[mPa s]:** 6.49^{25}
MS: 45(100) 55(20) 43(13) 41(11) 97(9) 44(7) 84(7) 130(1)
IR [cm^{-1}]: 3350 2920 2850 1460 1370 1110 1060 940 840 720
Raman [cm^{-1}]: 2930 2910 2870 2860 2720 1450 1440 1360 1300 1180 1140
   1120 1070 1060 1030 970 930 880 840 810 740 480 440 400 340 250
UV [nm]: 175(316) hx
^{13}C NMR [ppm]: 14.1 22.7 23.4 25.9 29.5 32.0 39.4 67.8 CDCl$_3$
Flammability: Flash pt. = 88°C

418. 1-Octene

Syst. Name: 1-Octene
Synonyms: Caprylene

CASRN: 111-66-0   **Beil Ref.:** 4-01-00-00874
Merck No.: 1764
Beil RN: 1734497
MF: C_8H_{16} **MP[°C]:** -101.7   **Den.[g/cm^3]:** 0.7149^{20}
MW: 112.22   **BP[°C]:** 121.2 n_D: 1.4087^{20}
Sol.: H_2O 1; EtOH 5; eth
   3; ace 3; bz 3; ctc 2;
   chl 4; os 4
Crit. Const.: $T_c = 293.6°C$; $P_c = 2.675$ MPa; $V_c = 464$ cm^3
Vap. Press. [kPa]: 2.30^{25}; 8.07^{50}; 22.6^{75}; 53.7^{100}
Therm. Prop.[kJ/mol]: $\Delta_{fus}H = 15.57$; $\Delta_{vap}H = 34.1$; $\Delta_f H°$(l, 25°C) = -121.8
C_p (liq.) [J/mol °C]: 241.0^{25}
γ [mN/m]: 21.28^{25}; $d\gamma/dT = 0.0958$ mN/m °C
Elec. Prop.: $\mu = 0$; $\varepsilon = 2.11^{20}$; IP = 9.43 eV
η [mPa s]: 21.3^{25}
MS: 43(100) 41(82) 55(80) 56(67) 42(67) 70(54) 29(44) 27(31) 69(30) 39(29)
IR [cm^{-1}]: 3090 2950 2850 1840 1650 1470 1390 990 910 720
Raman [cm^{-1}]: 2990 2950 2900 2860 2850 2720 1640 1460 1440 1420 1300
   1210 1120 1080 1070 910 890 860 810 640 440 360 290
UV [nm]: 177(12589) hp
^{13}C NMR [ppm]: 14.1 22.7 29.0 29.1 31.9 34.0 114.1 139.0 CDCl$_3$
^{1}H NMR [ppm]: 0.9 1.3 2.1 4.8 4.9 5.7 CCl$_4$
Flammability: Flash pt. = 21°C; ign. temp. = 230°C

419. *cis*-2-Octene

Syst. Name: 2-Octene, (*Z*)-

CASRN: 7642-04-8   **Beil Ref.**: 4-01-00-00878
Beil RN: 1719496
MF: C_8H_{16} **MP[°C]**: -100.2   **Den.[g/cm³]**: 0.7243^{20}
MW: 112.22   **BP[°C]**: 125.6 n_D: 1.4150^{20}
Sol.: H_2O 1; EtOH 3; eth
   3; ace 3; bz 3; chl 3
Vap. Press. [kPa]: 19.3^{75}; 46.9^{100}; 99.5^{125}
Therm. Prop.[kJ/mol]: $\Delta_f H°$(l, 25°C) = -135.7
C_p **(liq.) [J/mol °C]**: 239^{25}
γ **[mN/m]**: 21.91^{20}
Elec. Prop.: μ = 0; ε = 2.06^{25}
MS: 55(100) 41(99) 29(64) 56(59) 42(51) 70(39) 27(36) 39(28) 69(25) 57(21)
IR [cm⁻¹]: 3000 2950 2910 2840 1640 1450 1380 960 720 690
UV [nm]: 179(14125) hx

---

420. *trans*-2-Octene

Syst. Name: 2-Octene, (*E*)-

CASRN: 13389-42-9   **Beil Ref.**: 4-01-00-00879
Beil RN: 1719497
MF: C_8H_{16} **MP[°C]**: -87.7   **Den.[g/cm³]**: 0.7199^{20}
MW: 112.22   **BP[°C]**: 125 n_D: 1.4132^{20}
Sol.: H_2O 1; EtOH 3; eth
   3; ace 3; bz 3; chl 4
Vap. Press. [kPa]: 19.6^{75}; 47.7^{100}; 101^{125}
Therm. Prop.[kJ/mol]: $\Delta_f H°$(l, 25°C) = -135.7
C_p **(liq.) [J/mol °C]**: 239^{25}
γ **[mN/m]**: 21.38^{20}
Elec. Prop.: μ = 0; ε = 2.00^{25}
MS: 55(100) 41(82) 29(50) 56(49) 42(42) 70(35) 27(30) 39(23) 69(22) 57(17)
UV [nm]: 183(12882) hx

421. Oleic acid

Syst. Name: 9-Octadecenoic acid (Z)-
Synonyms: *cis*-9-Octadecenoic acid

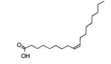

CASRN: 112-80-1   **Beil Ref.:** 4-02-00-01641
Merck No.: 6788
Beil RN: 1726542
MF: $C_{18}H_{34}O_2$ **MP[°C]:** 13.4   **Den.[g/cm³]:** 0.8935^{20}
MW: 282.47   **BP[°C]:** 360 n_D: 1.4582^{20}
Sol.: H_2O 1; EtOH 5; eth
   5; ace 5; bz 5; ctc 5;
   chl 5; MeOH 5
Vap. Press. [kPa]: 0.534^{200}; 3.97^{250}; 20.4^{300}; 79^{350}
Therm. Prop.[kJ/mol]: $\Delta_{vap}H = 67.4$
C_p (liq.) [J/mol °C]: 577^{50}
γ [mN/m]: 32.8^{20}
Elec. Prop.: $\mu = (1.2\ D)$; $\varepsilon = 2.34^{20}$
η [mPa s]: 38.8^{20}; 23.0^{25}; 4.85^{80}
MS: 55(100) 69(73) 43(59) 41(58) 83(53) 57(48) 73(47) 60(39) 97(38) 56(32)
IR [cm⁻¹]: 2940 2630 1700 1470 1410 1280 1240 940 720
UV [nm]: 185(6310) hx
¹³C NMR [ppm]: 14.1 22.7 24.7 27.2 29.1 29.4 29.6 29.7 32.0 34.1 129.6 129.9
   180.5 C
¹H NMR [ppm]: 0.9 1.3 2.0 2.3 5.3 11.9 CCl_4
Flammability: Flash pt. = 189°C; ign. temp. = 363°C

422. Pentachloroethane

Syst. Name: Ethane, pentachloro-
Synonyms: Pentalin; R 120

CASRN: 76-01-7   **DOT No.**: 1669
Merck No.: 7058   **Beil Ref.**: 4-01-00-00147
Beil RN: 1736845
MF: C_2HCl_5 **MP[°C]**: -29   **Den.[g/cm^3]**: 1.6796^{20}
MW: 202.29   **BP[°C]**: 159.8 n_D: 1.5025^{20}
Sol.: H_2O 1; EtOH 5; eth
   5
Vap. Press. [kPa]: 0.478^{25}; 1.99^{50}; 6.39^{75}; 16.9^{100}; 38.7^{125}; 78.6^{150}
Therm. Prop.[kJ/mol]: $\Delta_{fus}H$ = 11.34; $\Delta_{vap}H$ = 36.9; $\Delta_fH°$(l, 25°C) = -187.6
C_p **(liq.) [J/mol °C]**: 173.8^{25}
γ **[mN/m]**: 34.15^{25}; $d\gamma/dT$ = 0.1178 mN/m °C
Elec. Prop.: μ = 0.92 D; ε = 3.72^{25}; IP = 11.00 eV
η **[mPa s]**: 3.76^0; 2.25^{25}; 1.49^{50}; 1.06^{75}
MS: 167(100) 165(91) 117(90) 119(89) 83(58) 169(54) 130(43) 132(42) 60(40)
   85(37)
IR [cm^{-1}]: 3030 1250 1210 1050 1030 940 910 860 820 780 730
Raman [cm^{-1}]: 2980 1250 1210 1020 840 820 730 590 450 410 330 280 240
   230 180 90
Reg. Lists: CERCLA (RQ = 10 lb.); RCRA U184 (toxic)

423. Pentane

Syst. Name: Pentane

CASRN: 109-66-0   **DOT No.:** 1265
Merck No.: 7072   **Beil Ref.:** 4-01-00-00303
Beil RN: 969132
MF: C_5H_{12} **MP[°C]:** -129.7   **Den.[g/cm^3]:** 0.6262^{20}
MW: 72.15   **BP[°C]:** 36.0 n_D: 1.3575^{20}
Sol.: H_2O 2; EtOH 5; eth
   5; ace 5; bz 5; ctc 3;
   chl 5; hp 5
Crit. Const.: $T_c = 196.6°C$; $P_c = 3.370$ MPa; $V_c = 311$ cm^3
Vap. Press. [kPa]: 7.6^{-23}; 24.5^0; 68.3^{25}; 159^{50}
Therm. Prop.[kJ/mol]: $\Delta_{fus}H = 8.42$; $\Delta_{vap}H = 25.8$; $\Delta_f H°(l, 25°C) = -173.5$
C_p **(liq.) [J/mol °C]:** 167.2^{25}
γ **[mN/m]:** 15.49^{25}; $d\gamma/dT = 0.1102$ mN/m °C
Elec. Prop.: $\mu = 0$; $\varepsilon = 1.84^{20}$; IP = 10.35 eV
η **[mPa s]:** 0.351^{-25}; 0.274^0; 0.224^{25}
k **[W/m °C]:** 0.132^{-25}; 0.122^0; 0.113^{25}; 0.103^{50}; 0.095^{75}; 0.087^{100}
MS: 43(100) 42(55) 41(45) 27(42) 29(26) 39(19) 57(13) 28(9) 15(9) 72(8)
IR [cm^{-1}]: 2920 1460 1380 1340 1310 1260 1140 910 860 760 730
^{13}C NMR [ppm]: 13.7 22.6 34.5
^{1}H NMR [ppm]: 0.9 1.3 CCl$_4$
Flammability: Flash pt. < -40°C; ign. temp. = 260°C; flam. lim. = 1.5-7.8%
TLV/TWA: 600 ppm (1770 mg/m^3)

424. 1,5-Pentanediol

Syst. Name: 1,5-Pentanediol
Synonyms: Pentamethylene glycol

HO〜〜〜OH

CASRN: 111-29-5   **Beil Ref.**: 4-01-00-02540
Merck No.: 7073
Beil RN: 1560130
MF: $C_5H_{12}O_2$ **MP[°C]**: -18   **Den.[g/cm^3]**: 0.9914^{20}
MW: 104.15   **BP[°C]**: 239 n_D: 1.4494^{20}
Sol.: H_2O 3; EtOH 3; eth
   2; bz 2
Vap. Press. [kPa]: 3.36^{150}
Therm. Prop.[kJ/mol]: $\Delta_{vap}H = 60.7$; $\Delta_fH°(l, 25°C) = -531.5$
C_p **(liq.) [J/mol °C]**: 321.3^{25}
γ **[mN/m]**: 43.4^{20}
Elec. Prop.: $\mu = (2.5$ D$)$; $\epsilon = 26.2^{20}$
η **[mPa s]**: 115^{20}
MS: 31(100) 56(85) 41(67) 57(59) 55(51) 44(45) 29(37) 43(31) 68(29) 27(26)
IR [cm^{-1}]: 3330 2940 1450 1050 1030 970 910 890
^{13}C NMR [ppm]: 22.7 32.8 62.1
^{1}H NMR [ppm]: 1.5 3.6 D_2O
Flammability: Flash pt. = 129°C; ign. temp. = 335°C

425. Pentanenitrile

Syst. Name: Pentanenitrile
Synonyms: Valeronitrile; Butyl cyanide

CASRN: 110-59-8   **Beil Ref.**: 4-02-00-00875
Beil RN: 1736706
MF: C_5H_9N **MP[°C]**: -96.2   **Den.[g/cm^3]**: 0.8008^{20}
MW: 83.13   **BP[°C]**: 141.3 n_D: 1.3971^{20}
Sol.: eth 3; ace 3; bz 3;
    ctc 2
Crit. Const.: $T_c = 337.2°C$; $P_c = 3.58$ MPa
Vap. Press. [kPa]: 0.943^{25}; 11.2^{75}; 28.7^{100}; 63.8^{125}
Therm. Prop.[kJ/mol]: $\Delta_{fus}H = 4.73$; $\Delta_{vap}H = 36.1$; $\Delta_f H°(l, 25°C) = -33.1$
C_p **(liq.) [J/mol °C]**: 180
γ **[mN/m]**: 27.44^{20}
Elec. Prop.: $\mu = 4.12$ D; $\varepsilon = 20.04^{20}$
η **[mPa s]**: 0.693^{25}
MS: 41(100) 43(97) 54(54) 27(34) 55(21) 28(19) 29(16) 39(15) 42(5) 26(5)
IR [cm^{-1}]: 2940 2860 2220 1470 1430 1390 1110 930 910 740 730
UV [nm]: 341(126) cyhex
^{1}H NMR [ppm]: 1.0 1.6 2.3 CCl_4

426. Pentanoic acid

Syst. Name: Pentanoic acid
Synonyms: Valeric acid

CASRN: 109-52-4   **DOT No.:** 1760
Merck No.: 9815   **Beil Ref.:** 4-02-00-00868
Beil RN: 969454
MF: $C_5H_{10}O_2$ **MP[°C]:** -34   **Den.[g/cm^3]:** 0.9391^{20}
MW: 102.13   **BP[°C]:** 186.1 n_D: 1.4085^{20}
Sol.: H_2O 3; EtOH 3; eth
    3; ctc 2
Crit. Const.: T_c = 370°C; P_c = 3.58 MPa; V_c = 340 cm^3
Vap. Press. [kPa]: 0.002^0; 0.024^{25}; 0.172^{50}; 0.924^{75}; 3.86^{100}; 12.3^{125}; 32.5^{150}
Therm. Prop.[kJ/mol]: $\Delta_{fus}H$ = 14.16; $\Delta_{vap}H$ = 44.1; $\Delta_f H°$(l, 25°C) = -559.4
C_p **(liq.) [J/mol °C]:** 210.3^{25}
γ **[mN/m]:** 27.13^{20}; dγ/dT = 0.0887 mN/m °C
Elec. Prop.: μ = (1.6 D); ϵ = 2.66^{21}; IP = 10.53 eV
η **[mPa s]:** 1.97^{25}
MS: 60(100) 73(34) 27(33) 29(28) 41(21) 43(17) 45(16) 28(14) 42(12) 39(12)
IR [cm^{-1}]: 2940 2630 1720 1430 1280 1220 1110 940 750
Raman [cm^{-1}]: 2940 2920 2870 2740 1660 1440 1410 1300 1280 1230 1200
    1110 1050 1020 980 940 910 870 820 790 750 720 700 670 600 460 430 380
    350 300 250
^{1}H NMR [ppm]: 0.9 1.5 2.3 11.7 CCl$_4$
Flammability: Flash pt. = 96°C; ign. temp. = 400°C

427. 1-Pentanol

Syst. Name: 1-Pentanol
Synonyms: Amyl alcohol; Pentyl alcohol

CASRN: 71-41-0   **DOT No.**: 1105
Merck No.: 7074   **Beil Ref.**: 4-01-00-01640
Beil RN: 1730975
MF: $C_5H_{12}O$ **MP[°C]**: -78.9   **Den.[g/cm³]**: 0.8144^{20}
MW: 88.15   **BP[°C]**: 137.9 n_D: 1.4101^{20}
Sol.: H_2O 1; EtOH 5; eth
   5; ace 3; chl 3
Crit. Const.: $T_c = 315.0°C$; $P_c = 3.889$ MPa; $V_c = 326$ cm³
Vap. Press. [kPa]: 0.259^{25}; 1.66^{50}; 7.25^{75}; 24.0^{100}; 64.6^{125}; 149^{150}
Therm. Prop.[kJ/mol]: $\Delta_{fus}H = 9.83$; $\Delta_{vap}H = 44.4$; $\Delta_f H°(l, 25°C) = -351.6$
C_p **(liq.) [J/mol °C]**: 208.1^{25}
γ **[mN/m]**: 25.36^{25}; $d\gamma/dT = 0.0874$ mN/m °C
Elec. Prop.: $\mu = (1.7$ D); $\varepsilon = 15.13^{25}$; IP = 10.00 eV
η **[mPa s]**: 25.4^{-25}; 8.51^{0}; 3.62^{25}; 1.82^{50}; 1.04^{75}; 0.646^{100}
k **[W/m °C]**: 0.157^{0}; 0.153^{25}; 0.149^{50}; 0.145^{75}
MS: 42(100) 70(72) 55(65) 41(56) 31(47) 29(41) 27(26) 57(22) 28(22) 43(21)
IR [cm⁻¹]: 3340 2960 2860 1470 1380 1120 1050 1000 890 650
Raman [cm⁻¹]: 3300 3030 2950 2940 2910 2870 2750 1440 1360 1300 1270
   1230 1200 1110 1070 1050 1020 1010 980 880 850 830 760 510 500 430 390
   360 320
UV [nm]: 215(35) 178(251) undiluted
¹³C NMR [ppm]: 14.1 22.9 28.5 32.8 62.1
¹H NMR [ppm]: 0.9 1.4 3.5 4.4 CCl_4
Flammability: Flash pt. = 33°C; ign. temp. = 300°C; flam. lim. = 1.2-10.0%

428. 2-Pentanol

Syst. Name: 2-Pentanol
Synonyms: *sec*-Amyl alcohol; *sec*-Pentyl alcohol

CASRN: 6032-29-7   **Beil Ref.**: 3-01-00-01609
Merck No.: 7075
Beil RN: 1718819
MF: $C_5H_{12}O$ **MP[°C]**: -73   **Den.[g/cm^3]**: 0.8094^{20}
MW: 88.15   **BP[°C]**: 119.3 n_D: 1.4053^{20}
Sol.: H_2O 4; EtOH 3; eth
  3; ctc 3; chl 3
Crit. Const.: T_c = 287.3°C; P_c = 3.675 MPa
Vap. Press. [kPa]: 0.804^{25}; 4.35^{50}; 16.7^{75}; 49.9^{100}; 124^{125}
Therm. Prop.[kJ/mol]: $\Delta_{vap}H$ = 41.4; $\Delta_f H°$(l, 25°C) = -365.2
C_p (liq.) [J/mol °C]: 239.4^{25}
γ [mN/m]: 23.45^{25}; dγ/dT = 0.1004 mN/m °C
Elec. Prop.: μ = (1.7 D); ε = 13.71^{25}; IP = 9.78 eV
η [mPa s]: 3.47^{25}; 1.45^{50}; 0.761^{75}; 0.465^{100}
MS: 45(100) 43(20) 55(18) 27(17) 29(11) 41(9) 31(9) 15(9) 44(8) 39(8)
IR [cm^{-1}]: 3330 2940 1470 1370 1110 1020 940 890 830 760
Raman [cm^{-1}]: 3320 2970 2940 2910 2880 2730 1460 1370 1340 1300 1270
  1230 1150 1130 1060 1040 1000 950 900 870 850 840 810 750 610 500 470
  430 340
^{13}C NMR [ppm]: 14.3 19.4 23.6 41.9 67.3
^{1}H NMR [ppm]: 0.9 1.2 1.3 2.2 3.7 CDCl$_3$
Flammability: Flash pt. = 34°C; ign. temp. = 343°C; flam. lim. = 1.2-9.0%

429. 3-Pentanol

Syst. Name: 3-Pentanol

CASRN: 584-02-1   **DOT No.:** 2706
Merck No.: 7076   **Beil Ref.:** 4-01-00-01662
Beil RN: 1730964
MF: $C_5H_{12}O$ **MP[°C]:** -69   **Den.[g/cm^3]:** 0.8203^{20}
MW: 88.15   **BP[°C]:** 116.2 n_D: 1.4104^{20}
Sol.: H_2O 2; EtOH 3; eth
  3; ace 3; ctc 3
Crit. Const.: $T_c = 286.5$°C
Vap. Press. [kPa]: 1.10^{25}; 5.47^{50}; 19.9^{75}; 57.7^{100}; 141^{125}
Therm. Prop.[kJ/mol]: $\Delta_{vap}H = 43.5$; $\Delta_fH°(l, 25°C) = -368.9$
C_p **(liq.) [J/mol °C]:** 239.7^{25}
γ **[mN/m]:** 24.60^{20}
Elec. Prop.: $\mu = (1.6$ D); $\varepsilon = 13.35^{25}$; IP = 9.78 eV
η **[mPa s]:** 4.15^{25}; 1.47^{50}; 0.727^{75}; 0.436^{100}
MS: 59(100) 31(83) 41(42) 27(35) 29(34) 58(15) 43(15) 57(14) 39(12) 15(9)
IR [cm^{-1}]: 3330 2940 1470 1160 1110 1040 960 860
Raman [cm^{-1}]: 3360 2940 2880 2740 1470 1450 1370 1310 1280 1250 1150
  1130 1110 1060 1040 1020 960 920 860 830 780 760 610 510 480 410 360
  320 290 210
^{13}C NMR [ppm]: 10.1 30.0 74.1
^{1}H NMR [ppm]: 0.9 1.4 3.3 3.4 CCl_4
Flammability: Flash pt. = 41°C; ign. temp. = 435°C; flam. lim. = 1.2-9.0%

430. 1-Pentene

Syst. Name: 1-Pentene
Synonyms: α-Amylene

CASRN: 109-67-1   **Beil Ref.:** 4-01-00-00808
Merck No.: 7079
Beil RN: 1731629
MF: C_5H_{10} **MP[°C]:** -165.2   **Den.[g/cm³]:** 0.6405^{20}
MW: 70.13   **BP[°C]:** 29.9 n_D: 1.3715^{20}
Sol.: H_2O 1; EtOH 5; eth 5; bz 3; ctc 2; dil sulf 4
Crit. Const.: T_c = 191.63°C; P_c = 3.527 MPa; V_c = 293 cm³
Vap. Press. [kPa]: 31.4^0; 85.0^{25}; 193^{50}; 385^{75}; 695^{100}; 1162^{125}; 1829^{150}
Therm. Prop.[kJ/mol]: $\Delta_{fus}H$ = 5.81; $\Delta_{vap}H$ = 25.2; $\Delta_f H°$(l, 25°C) = -46.9
C_p (liq.) [J/mol °C]: 154.0^{25}
γ [mN/m]: 15.45^{25}; dγ/dT = 0.1099 mN/m °C
Elec. Prop.: μ = 0.5 D; ε = 2.01^{20}; IP = 9.52 eV
η [mPa s]: 0.313^{-25}; 0.241^0; 0.195^{25}
k [W/m °C]: 0.131^{-25}; 0.124^0; 0.116^{25}
MS: 42(100) 55(61) 41(50) 39(45) 27(40) 70(33) 29(27) 40(10) 15(10) 26(9)
IR [cm⁻¹]: 3030 2860 1820 1640 1450 1390 1000 910 760
Raman [cm⁻¹]: 3082 3000 2981 2966 2938 2908 2877 2844 1642 1302 1293 853
UV [nm]: 187(4467) 181(13804) 177(16596) gas
¹H NMR [ppm]: 0.9 1.3 2.0 4.9 5.0 5.7 CCl_4
Flammability: Flash pt. = -18°C; ign. temp. = 275°C; flam. lim. = 1.5-8.7%

431. *cis*-2-Pentene
Syst. Name: 2-Pentene, (Z)-
Synonyms: *cis*-β-Amylene

CASRN: 627-20-3   **Beil Ref.**: 4-01-00-00814
Merck No.: 7080
Beil RN: 1718794
MF: C_5H_{10} **MP[°C]**: -151.4   **Den.[g/cm^3]**: 0.6556^{20}
MW: 70.13   **BP[°C]**: 36.9 n_D: 1.3830^{20}
Sol.: H_2O 1; EtOH 5; eth
    5; bz 3; dil sulf 3
Crit. Const.: T_c = 202°C; P_c = 3.69 MPa
Vap. Press. [kPa]: 23.2^0; 66.0^{25}; 156^{50}; 318^{75}; 586^{100}; 994^{125}; 1580^{150}
Therm. Prop.[kJ/mol]: $\Delta_{fus}H$ = 7.12; $\Delta_{vap}H$ = 26.1; $\Delta_f H°$(l, 25°C) = -53.7
C_p (liq.) [J/mol °C]: 151.7^{25}
γ [mN/m]: 16.80^{25}; dγ/dT = 0.1172 mN/m °C
Elec. Prop.: μ = 0; IP = 9.04 eV
MS: 55(100) 42(46) 70(36) 39(35) 41(30) 29(30) 27(29) 53(8) 40(6) 15(6)
Raman [cm^{-1}]: 3016 2968 2937 2920 2879 2865 1658 1452 1266 860
UV [nm]: 205(891) 185(12589) 177(18197) 167(10715) gas
Flammability: Flash pt. < -20°C

432. *trans*-2-Pentene

Syst. Name: 2-Pentene, (*E*)-
Synonyms: *trans*-β-Amylene

CASRN: 646-04-8   **Beil Ref.:** 4-01-00-00814
Merck No.: 7080
Beil RN: 1718795
MF: C_5H_{10} **MP[°C]:** -140.2   **Den.[g/cm^3]:** 0.6431^{25}
MW: 70.13   **BP[°C]:** 36.3 n_D: 1.3793^{20}
Sol.: H_2O 1; EtOH 5; eth
  5; bz 3; dil sulf 4
Crit. Const.: T_c = 198°C; P_c = 3.52 MPa
Vap. Press. [kPa]: 23.9^0; 67.4^{25}; 159^{50}; 324^{75}; 597^{100}; 1012^{125}; 1607^{150}
Therm. Prop.[kJ/mol]: $\Delta_{fus}H$ = 8.36; $\Delta_{vap}H$ = 26.1; $\Delta_f H°$(l, 25°C) = -58.2
C_p **(liq.) [J/mol °C]:** 157.0^{25}
γ **[mN/m]:** 16.41^{25}; dγ/dT = 0.0997 mN/m °C
Elec. Prop.: μ = 0; IP = 9.04 eV
MS: 55(100) 42(44) 70(35) 39(33) 41(31) 29(28) 27(28) 53(8) 40(6) 15(6)
Raman [cm^{-1}]: 3000 2967 2937 2920 2895 2856 1673 1455 1308 1290 488
UV [nm]: 201(1288) 181(14125) 158(12023) gas
Flammability: Flash pt. < -20°C

433. Pentyl acetate

Syst. Name: Ethanoic acid, pentyl ester
Synonyms: Amyl acetate

CASRN: 628-63-7
Beil RN: 1744753
MF: $C_7H_{14}O_2$
MW: 130.19
Sol.: H_2O 2; EtOH 5; eth
    5; ctc 3

DOT No.: 1104
Beil Ref.: 4-02-00-00152
MP[°C]: -70.8   **Den.[g/cm^3]**: 0.8756^{20}
BP[°C]: 149.2 n_D: 1.4023^{20}

Vap. Press. [kPa]: 0.600^{25}; 2.33^{50}; 7.40^{75}; 20.1^{100}; 48.3^{125}; 105^{150}
Therm. Prop.[kJ/mol]: $\Delta_{vap}H = 41.0$
C_p **(liq.) [J/mol °C]**: 261.0^{25}
γ **[mN/m]**: 25.17^{25}; $d\gamma/dT = 0.0994$ mN/m °C
Elec. Prop.: $\mu = 1.75$ D; $\varepsilon = 4.79^{20}$
η **[mPa s]**: 0.862^{25}
MS: 43(100) 70(90) 42(52) 28(51) 61(50) 55(41) 73(21) 41(20) 29(14) 69(11)
IR [cm^{-1}]: 2960 2940 2880 1740 1470 1370 1240 1120 1040 970 950 920 820
    770 730
^{13}C NMR [ppm]: 14.1 20.5 22.8 28.6 28.9 64.3 170.1 diox
^{1}H NMR [ppm]: 0.9 1.4 2.0 4.0 CCl$_4$
Flammability: Flash pt. = 16°C; ign. temp. = 360°C; flam. lim. = 1.1-7.5%
TLV/TWA: 100 ppm (532 mg/m^3)
Reg. Lists: CERCLA (RQ = 5000 lb.)

434. Pentylamine

Syst. Name: 1-Pentanamine
Synonyms: Amylamine

CASRN: 110-58-7   **DOT No.:** 1106
Merck No.: 630   **Beil Ref.:** 4-04-00-00674
Beil RN: 505953
MF: $C_5H_{13}N$ **MP[°C]:** -55   **Den.[g/cm³]:** 0.7544^{20}
MW: 87.16   **BP[°C]:** 104.3 n_D: 1.448^{20}
Sol.: H_2O 5; EtOH 5; eth
   5; ace 4; bz 4; chl 2
Vap. Press. [kPa]: 4.00^{25}; 13.6^{50}; 37.5^{75}; 88.3^{100}; 184^{125}
Therm. Prop.[kJ/mol]: $\Delta_{vap}H = 34.0$
C_p **(liq.) [J/mol °C]:** 218.0^{25}
γ **[mN/m]:** 24.69^{25}; $d\gamma/dT = 0.1023$ mN/m °C
Elec. Prop.: $\varepsilon = 4.27^{20}$; IP = 8.67 eV
η **[mPa s]:** 1.03^{0}; 0.702^{25}; 0.493^{50}; 0.356^{75}
MS: 30(100) 87(8) 41(4) 28(4) 45(3) 42(3) 27(3) 56(2) 44(2) 43(2)
IR [cm⁻¹]: 3450 3330 2940 1610 1470 1390 1300 1220 1120 1080 980 820 730
¹³C NMR [ppm]: 14.3 23.0 29.7 34.0 42.5
¹H NMR [ppm]: 0.9 1.4 2.7 $CDCl_3$
Flammability: Flash pt. = -1°C; flam. lim. = 2.2-22%

---

435. Perfluorobutane

Syst. Name: Butane, decafluoro-
Synonyms: Decafluorobutane

CASRN: 355-25-9   **Beil Ref.:** 4-01-00-00245
Beil RN: 1777513
MF: C_4F_{10} **MP[°C]:** -128.2   **Den.[g/cm³]:** 1.6484^{25}
MW: 238.03   **BP[°C]:** -1.9
Sol.: bz 3; chl 3
Crit. Const.: $T_c = 113.3$°C; $P_c = 2.323$ MPa; $V_c = 378$ cm³
Vap. Press. [kPa]: 36.8^{25}; 111^{0}
Therm. Prop.[kJ/mol]: $\Delta_{vap}H = 22.9$
C_p **(liq.) [J/mol °C]:** 127.2^{20} (sat. press.)
γ **[mN/m]:** 10.6^{-8}
Elec. Prop.: $\mu = 0$
MS: 69(100) 119(18) 31(12) 131(8) 100(8) 50(4) 219(3) 150(3) 169(2) 93(1)

436. Perfluorocyclobutane

Syst. Name: Cyclobutane, octafluoro-
Synonyms: Octafluorocyclobutane; R C318

CASRN: 115-25-3   **DOT No.**: 1976
Merck No.: 6666   **Beil Ref.**: 4-05-00-00008
Beil RN: 1909266
MF: C_4F_8 **MP[°C]**: -40.1   **Den.[g/cm³]**: 1.500^{25} (sat. press.)
MW: 200.03   **BP[°C]**: -5.9
Sol.: eth 3
Crit. Const.: T_c = 115.31°C; P_c = 2.784 MPa; V_c = 324 cm³
Vap. Press. [kPa]: 42.8^{-25}; 129^0
Therm. Prop.[kJ/mol]: $\Delta_{fus}H$ = 2.77; $\Delta_{vap}H$ = 23.2; $\Delta_f H°$(g, 25°C) = -1543
C_p **(liq.) [J/mol °C]**: 209.8^{-6}
Elec. Prop.: μ = 0
η **[mPa s]**: 1.1^{-11}
MS: 100(100) 131(87) 31(54) 69(25) 50(13) 93(7) 132(3) 74(3) 101(2) 81(2)
IR [cm⁻¹]: 1410 1330 1280 1240 1150 1090 1040 960

437. β-Phellandrene

Syst. Name: Cyclohexene, 3-methylene-6-(1-methylethyl)-

Synonyms: *p*-Mentha-1(7),2-diene

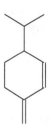

CASRN: 555-10-2   **Beil Ref.:** 2-05-00-00087
Merck No.: 7152
Beil RN: 2038351
MF: $C_{10}H_{16}$ **BP[°C]:** 171.5   **Den.[g/cm^3]:** 0.8520^{20}
MW: 136.24 n_D: 1.4788^{20}
Sol.: H_2O 1; EtOH 1; eth
   3
Vap. Press. [kPa]: 0.189^{25}; 0.843^{50}; 2.95^{75}
MS: 93(100) 77(27) 91(25) 136(18) 79(17) 94(15) 41(11) 80(9) 92(8) 39(8)
IR [cm^{-1}]: 3030 2940 1640 1590 1470 1390 1370 880 820 780
UV [nm]: 232(181970)
Flammability: Flash pt. = 49°C

438. Phenetole

Syst. Name: Benzene, ethoxy-
Synonyms: Ethyl phenyl ether

CASRN: 103-73-1   **Beil Ref.**: 4-06-00-00554
Merck No.: 7189
Beil RN: 636270
MF: $C_8H_{10}O$ **MP[°C]**: -29.5   **Den.[g/cm³]**: 0.9651^{20}
MW: 122.17   **BP[°C]**: 169.8 n_D: 1.5076^{20}
Sol.: H_2O 1; EtOH 3; eth
    3; ctc 3
Crit. Const.: T_c = 374°C; P_c = 3.42 MPa
Vap. Press. [kPa]: 0.204^{25}; 26.6^{125}; 58.6^{150}
Therm. Prop.[kJ/mol]: $\Delta_{vap}H$ = 40.7; $\Delta_fH°$(l, 25°C) = -152.6
C_p **(liq.) [J/mol °C]**: 228.5^{25}
γ **[mN/m]**: 32.41^{25}; $d\gamma/dT$ = 0.1104 mN/m °C
Elec. Prop.: μ = 1.45 D; ε = 4.22^{20}; IP = 8.13 eV
η **[mPa s]**: 1.20^{25}; 0.817^{50}; 0.594^{75}; 0.453^{100}
MS: 94(100) 122(39) 28(12) 66(11) 39(9) 77(8) 95(7) 65(7) 51(7) 29(6)
IR [cm⁻¹]: 3040 3000 2940 1600 1580 1500 1480 1390 1300 1240 1170 1120
    1050 920 880 800 750 690
Raman [cm⁻¹]: 3070 2980 2930 2880 2770 2720 1600 1580 1480 1450 1380
    1360 1330 1300 1240 1170 1150 1110 1080 1030 1000 920 810 790 760 610
    590 510 470 420 350 240 170
UV [nm]: 278(1190) 271(1400) 220(6400) MeOH
¹H NMR [ppm]: 1.3 3.9 6.9 CCl_4
Flammability: Flash pt. = 63°C

439. Phenol

Syst. Name: Phenol
Synonyms: Carbolic acid; Hydroxybenzene

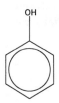

CASRN: 108-95-2
Merck No.: 7206
Beil RN: 969616
MF: C_6H_6O
MW: 94.11
Sol.: H_2O 3; EtOH 3; eth
  4; ace 5; bz 5; ctc 5;
  chl 3; CS_2 3

DOT No.: 1671
Beil Ref.: 4-06-00-00531

MP[°C]: 40.9
BP[°C]: 181.8

Den.[g/cm³]: 1.0545^{45}
n_D: 1.5408^{41}

Crit. Const.: $T_c = 421.1°C$; $P_c = 6.13$ MPa
Vap. Press. [kPa]: 0.055^{25}; 1.53^{75}; 5.47^{100}; 15.8^{125}; 38.8^{150}
Therm. Prop.[kJ/mol]: $\Delta_{fus}H = 11.29$; $\Delta_{vap}H = 45.7$; $\Delta_f H°(s, 25°C) = -165.1$
C_p (liq.) [J/mol °C]: 199.8^{41}
γ [mN/m]: 38.20^{50}; $d\gamma/dT = 0.1068$ mN/m °C
Elec. Prop.: $\mu = 1.224$ D; $\varepsilon = 12.40^{30}$; IP = 8.47 eV
η [mPa s]: 3.44^{50}; 1.78^{75}; 1.10^{100}
k [W/m °C]: 0.156^{50}; 0.153^{75}; 0.151^{100}
MS: 94(100) 66(25) 39(25) 65(21) 40(14) 55(10) 47(7) 38(7) 95(6) 63(5)
IR [cm⁻¹]: 3610 3360 3050 1930 1600 1500 1470 1340 1220 1070 1030 880
  810 750 690
Raman [cm⁻¹]: 3070 2950 1610 1600 1500 1480 1380 1320 1250 1170 1150
  1080 1030 1010 830 820 760 630 530 510 430 310 280 250
UV [nm]: 277 270 264 MeOH
¹³C NMR [ppm]: 115.4 121.0 129.7 154.9 $CDCl_3$
¹H NMR [ppm]: 6.1 6.9 $CDCl_3$
Flammability: Flash pt. = 79°C; ign. temp. = 715°C; flam. lim. = 1.8-8.6%
TLV/TWA: 5 ppm (19 mg/m³)
Reg. Lists: CERCLA (RQ = 1000 lb.); SARA 313 (1.0%); RCRA U188 (toxic)

440. Phenylethylamine

Syst. Name: Benzeneethanamine
Synonyms: Ethylphenylamine

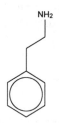

CASRN: 64-04-0   **Beil Ref.**: 4-12-00-02453
Merck No.: 7186
Beil RN: 507488
MF: $C_8H_{11}N$ **MP[°C]**: <0   **Den.[g/cm^3]**: 0.9580^{24}
MW: 121.18   **BP[°C]**: 197.5 n_D: 1.5290^{25}
Sol.: H_2O 3; eth 4; eth 4;
    ctc 3
MS: 30(100) 91(21) 92(8) 51(7) 121(6) 77(6) 65(5) 28(5) 63(4) 58(4)
IR [cm^{-1}]: 3333 3226 3030 2941 1613 1493 1449 1389 1075 1031 833 741 699
UV [nm]: 259 cyhex
^{1}H NMR [ppm]: 0.9 2.7 2.8 7.1 CCl_4

441. 2-Picoline

Syst. Name: Pyridine, 2-methyl-
Synonyms: 2-Methylpyridine

CASRN: 109-06-8   **Rel. CASRN**: 1333-41-1
Merck No.: 7372   **DOT No.**: 2313
Beil RN: 104581   **Beil Ref.**: 5-20-05-00464
MF: C_6H_7N **MP[°C]**: -66.7   **Den.[g/cm^3]**: 0.9443^{20}
MW: 93.13   **BP[°C]**: 129.3 n_D: 1.4957^{20}
Sol.: H_2O 4; EtOH 5; eth
   5; ace 4; ctc 3
Crit. Const.: T_c = 347.9°C; P_c = 4.60 MPa; V_c = 292 cm^3
Vap. Press. [kPa]: 1.50^{25}; 5.58^{50}; 16.5^{75}; 41.1^{100}; 89.4^{125}
Therm. Prop.[kJ/mol]: $\Delta_{fus}H$ = 9.72; $\Delta_{vap}H$ = 36.2; $\Delta_f H°$(l, 25°C) = 56.7
C_p **(liq.) [J/mol °C]**: 158.4^{25}
γ **[mN/m]**: 33.00^{25}; $d\gamma/dT$ = 0.1243 mN/m °C
Elec. Prop.: μ = 1.85 D; ε = 10.18^{20}; IP = 9.02 eV
η **[mPa s]**: 0.753^{25}
MS: 93(100) 66(41) 39(31) 92(20) 78(19) 51(19) 65(16) 38(13) 50(12) 52(11)
IR [cm^{-1}]: 3050 3000 1590 1560 1470 1420 1290 1140 1040 990 750 730
Raman [cm^{-1}]: 3140 3060 3050 3010 2920 2870 2730 1590 1570 1460 1430
   1380 1300 1240 1150 1100 1050 1000 970 810 800 630 550 470 360 210
UV [nm]: 268 261 256 MeOH
^{13}C NMR [ppm]: 24.3 120.6 123.0 135.9 149.4 158.6 diox
^{1}H NMR [ppm]: 2.5 6.9 7.4 8.4 CCl$_4$
Flammability: Flash pt. = 39°C; ign. temp. = 538°C
Reg. Lists: CERCLA (RQ = 5000 lb.); RCRA U191 (toxic)

442. 3-Picoline

Syst. Name: Pyridine, 3-methyl-
Synonyms: 3-Methylpyridine

CASRN: 108-99-6   **Rel. CASRN**: 1333-41-1
Merck No.: 7373   **DOT No.**: 2313
Beil RN: 1366   **Beil Ref.**: 5-20-05-00506
MF: C_6H_7N **MP[°C]**: -18.1   **Den.[g/cm^3]**: 0.9566^{20}
MW: 93.13   **BP[°C]**: 144.1 n_D: 1.5040^{20}
Sol.: H_2O 5; EtOH 5; eth
    5; ace 4; ctc 3
Crit. Const.: T_c = 371.9°C; P_c = 4.48 MPa; V_c = 288 cm^3
Vap. Press. [kPa]: 0.795^{25}; 3.18^{50}; 9.96^{75}; 26.0^{100}; 58.7^{125}
Therm. Prop.[kJ/mol]: $\Delta_{fus}H$ = 14.18; $\Delta_{vap}H$ = 37.4; $\Delta_f H°$(l, 25°C) = 61.9
C_p **(liq.) [J/mol °C]**: 158.7^{25}
γ **[mN/m]**: 35.06^{20}
Elec. Prop.: μ = (2.4 D); ϵ = 11.10^{30}; IP = 9.04 eV
η **[mPa s]**: 0.872^{25}
MS: 93(100) 39(51) 66(46) 92(31) 65(29) 40(19) 38(18) 67(13) 63(11) 51(11)
IR [cm^{-1}]: 3020 2920 1580 1480 1410 1380 1190 1130 1110 1030 790 720
Raman [cm^{-1}]: 3150 3130 3000 2920 2870 2750 1590 1560 1380 1230 1190
    1040 1030 810 790 630 540 340 220
UV [nm]: 269 262 257 MeOH
^{13}C NMR [ppm]: 18.9 124.1 133.8 137.3 147.8 150.5
^{1}H NMR [ppm]: 2.3 7.0 7.4 8.4 CCl$_4$

443. 4-Picoline

Syst. Name: Pyridine, 4-methyl-
Synonyms: 4-Methylpyridine

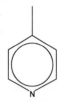

CASRN: 108-89-4
Merck No.: 7374
Beil RN: 104586
MF: C_6H_7N
MW: 93.13
Sol.: H_2O 5; EtOH 5; eth
  5; ace 3; ctc 3

Rel. CASRN: 1333-41-1
DOT No.: 2313
Beil Ref.: 5-20-05-00543
MP[°C]: 3.66   **Den.[g/cm^3]:** 0.9548^{20}
BP[°C]: 145.3 n_D: 1.5037^{20}

Crit. Const.: $T_c = 372.6°C$; $P_c = 4.70$ MPa; $V_c = 292$ cm^3
Vap. Press. [kPa]: 0.759^{25}; 3.04^{50}; 9.56^{75}; 25.0^{100}; 56.7^{125}
Therm. Prop.[kJ/mol]: $\Delta_{fus}H = 11.57$; $\Delta_{vap}H = 37.5$; $\Delta_f H°(l, 25°C) = 59.2$
C_p (liq.) [J/mol °C]: 159.0^{25}
γ [mN/m]: 35.45^{20}
Elec. Prop.: $\mu = 2.70$ D; $\varepsilon = 12.2^{20}$; IP = 9.04 eV
MS: 93(100) 39(44) 66(42) 65(26) 92(24) 40(17) 38(16) 51(13) 67(12) 63(10)
IR [cm^{-1}]: 3050 3010 2980 1600 1560 1490 1440 1410 1220 990 800 730
Raman [cm^{-1}]: 3150 3140 2980 2920 2870 2810 1600 1560 1490 1380 1330
  1220 1210 1070 990 800 670 510 480 340 210
UV [nm]: 262 255 MeOH
^{13}C NMR [ppm]: 21.5 125.8 147.7 150.1
^{1}H NMR [ppm]: 2.3 7.0 8.4 CCl_4
Flammability: Flash pt. = 57°C

444. α-Pinene

Syst. Name: Bicyclo[3.1.1]hept-2-ene, 2,6,6-
trimethyl-, (±)-
Synonyms: 2-Pinene

CASRN: 80-56-8   **DOT No.:** 2368
Merck No.: 7414   **Beil Ref.:** 2-05-00-00097
MF: $C_{10}H_{16}$ **MP[°C]:** -64   **Den.[g/cm³]:** 0.8539^{25}
MW: 136.24   **BP[°C]:** 156 n_D: 1.4632^{25}
Sol.: H_2O 1; EtOH 5; eth
   5; chl 5
Therm. Prop.[kJ/mol]: $\Delta_f H°(l, 25°C) = -16.4$
γ **[mN/m]:** 25.63^{25}; $d\gamma/dT = 0.0932$ mN/m °C
Elec. Prop.: $\varepsilon = 2.26^{30}$
η **[mPa s]:** 1.40^{20}
Flammability: Flash pt. = 35°C; ign. temp. = 275°C

---

445. β-Pinene

Syst. Name: Bicyclo[3.1.1]heptane, 6,6-dimethyl-2-
methylene-, (1R)-
Synonyms: 2(10)-Pinene; Nopinene

CASRN: 127-91-3   **Beil Ref.:** 4-05-00-00456
Merck No.: 7415
MF: $C_{10}H_{16}$ **MP[°C]:** -61.5   **Den.[g/cm³]:** 0.860^{25}
MW: 136.24   **BP[°C]:** 166 n_D: 1.4768^{25}
Sol.: H_2O 1; bz 3; EtOH
   3; eth 3; chl 3
γ **[mN/m]:** 26.82^{25}; $d\gamma/dT = 0.0900$ mN/m °C
η **[mPa s]:** 1.70^{20}
Flammability: Flash pt. = 38°C; ign. temp. = 275°C

446. Piperidine

Syst. Name: Piperidine
Synonyms: Hexahydropyridine; Azacyclohexane

CASRN: 110-89-4   **DOT No.:** 2401
Merck No.: 7438   **Beil Ref.:** 5-20-02-00003
Beil RN: 102438
MF: $C_5H_{11}N$ **MP[°C]:** -11.03   **Den.[g/cm^3]:** 0.8606^{20}
MW: 85.15   **BP[°C]:** 106.2 n_D: 1.4530^{20}
Sol.: H_2O 5; EtOH 5; eth
    3; ace 3; bz 3; chl 3
Crit. Const.: T_c = 321.0°C; P_c = 4.94 MPa; V_c = 288 cm^3
Vap. Press. [kPa]: 4.28^{25}; 13.5^{50}; 36.6^{75}; 84.1^{100}; 171^{125}
Therm. Prop.[kJ/mol]: $\Delta_{fus}H$ = 14.85; $\Delta_fH°$(l, 25°C) = -86.4
C_p **(liq.) [J/mol °C]:** 179.9^{25}
γ **[mN/m]:** 28.91^{25}; dγ/dT = 0.1153 mN/m °C
Elec. Prop.: μ = (1.2 D); ϵ = 4.33^{20}; IP = 8.05 eV
η **[mPa s]:** 1.57^{25}; 0.958^{50}; 0.649^{75}; 0.474^{100}
MS: 84(100) 85(53) 56(46) 57(43) 28(41) 29(37) 44(34) 42(30) 30(30) 43(25)
IR [cm^{-1}]: 3330 2860 1450 1320 1250 1190 1160 1140 1110 1050 1030 860
    750
Raman [cm^{-1}]: 3340 2930 2890 2850 2800 2730 2650 2620 1440 1350 1280
    1260 1160 1140 1050 1030 1000 900 850 810 740 550 440 430 400 250
UV [nm]: 190(4677) cyhex
^{13}**C NMR [ppm]:** 25.9 27.8 47.9 diox
Flammability: Flash pt. = 16°C; flam. lim. = 1-10%

447. Propanal

Syst. Name: Propanal
Synonyms: Propionaldehyde

CASRN: 123-38-6
Merck No.: 7835
Beil RN: 506010
MF: C_3H_6O
MW: 58.08
Sol.: H_2O 3; EtOH 5; eth
 5

DOT No.: 1275
Beil Ref.: 4-01-00-03165

MP[°C]: -80
BP[°C]: 48

Den.[g/cm^3]: 0.8657^{25}
n_D: 1.3636^{20}

Crit. Const.: $T_c = 231.3°C$; $P_c = 5.27$ MPa; $V_c = 204$ cm^3
Vap. Press. [kPa]: 13.5^0; 42.2^{25}
Therm. Prop.[kJ/mol]: $\Delta_{vap}H = 28.3$; $\Delta_f H°(l, 25°C) = -215.3$
C_p **(liq.) [J/mol °C]:** 137.2^{25}
Elec. Prop.: $\mu = 2.52$ D; $\varepsilon = 18.5^{17}$; IP = 9.95 eV
η **[mPa s]:** 0.321^{25}; 0.249^{50}
MS: 29(100) 58(59) 28(58) 27(39) 57(20) 31(4) 30(4) 42(3) 39(3) 59(2)
IR [cm^{-1}]: 2940 2780 2700 1750 1470 1410 1110 910 890 860 850
Raman [cm^{-1}]: 2980 2940 2900 2820 2730 1720 1690 1460 1390 1340 1250
 1100 1000 900 880 851 660 510 330 270
UV [nm]: 282(8) H_2O
^{1}H NMR [ppm]: 1.1 2.4 9.8 CDCl$_4$
Flammability: Flash pt. = -30°C; ign. temp. = 207°C; flam. lim. = 2.6-17%
Reg. Lists: CERCLA (RQ = 1 lb.); SARA 313 (1.0%)

448. Propane

Syst. Name: Propane
Synonyms: LPG

CASRN: 74-98-6 **Rel. CASRN**: 68476-85-7
Merck No.: 7809 **DOT No.**: 1978
Beil RN: 1730718 **Beil Ref.**: 4-01-00-00176
MF: C_3H_8 **MP[°C]**: -187.6 **Den.[g/cm^3]**: 0.493^{25} (sat. press.)
MW: 44.10 **BP[°C]**: -42.1
Sol.: H_2O 3; EtOH 3; eth
  4; ace 2; bz 4; chl 4
Crit. Const.: T_c = 96.68°C; P_c = 4.248 MPa; V_c = 200 cm^3
Vap. Press. [kPa]: 20.1^{-73}; 203^{-25}; 472^0; 939^{25}; 1665^{50}; 2702^{75}
Therm. Prop.[kJ/mol]: $\Delta_{fus}H$ = 3.53; $\Delta_{vap}H$ = 19.0; $\Delta_f H°$(g, 25°C) = -103.8
C_p **(liq.) [J/mol °C]**: 98.3^{-43}
γ **[mN/m]**: 13.59^{-50}; dγ/dT = 0.0874 mN/m °C
Elec. Prop.: μ = 0.084 D; ϵ = 1.67^{20}; IP = 10.95 eV
MS: 29(100) 28(59) 27(47) 44(27) 43(23) 39(21) 41(14) 26(12) 15(11) 42(6)
IR [cm^{-1}]: 2940 1470 1390 1160 1060 920 750
^{13}**C NMR [ppm]**: 15.6 16.1
Flammability: Flash pt. = -104°C; ign. temp. = 450°C; flam. lim. = 2.1-9.5%
TLV/TWA: 1000 ppm (1800 mg/m^3)

449. 1,2-Propanediol

Syst. Name: 1,2-Propanediol

Synonyms: Propylene glycol; 2-Hydroxypropanol

CASRN: 57-55-6   **Beil Ref.:** 3-01-00-02142

Merck No.: 7868

Beil RN: 1340498

MF: $C_3H_8O_2$ **MP[°C]:** -60   **Den.[g/cm^3]:** 1.0361^{20}

MW: 76.10   **BP[°C]:** 187.6 n_D: 1.4324^{20}

Sol.: H_2O 5; EtOH 5; eth 3; bz 3; chl 3

Vap. Press. [kPa]: 0.020^{25}; 0.175^{50}; 0.829^{75}; 3.15^{100}; 10.00^{125}; 27.5^{150}

Therm. Prop.[kJ/mol]: $\Delta_{vap}H$ = 52.4; $\Delta_f H°$(l, 25°C) = -485.7

C_p **(liq.) [J/mol °C]:** 190.8^{25}

γ **[mN/m]:** 36.51^{25}

Elec. Prop.: μ = (2.2 D); $\varepsilon = 27.5^{30}$

η **[mPa s]:** 248^{0}; 40.4^{25}; 11.3^{50}; 4.77^{75}; 2.75^{100}

k **[W/m °C]:** 0.202^{0}; 0.200^{25}; 0.199^{50}; 0.198^{75}; 0.197^{100}

MS: 45(100) 18(46) 29(21) 43(19) 31(18) 27(17) 28(11) 19(8) 44(6) 61(5)

IR [cm^{-1}]: 3330 2940 1450 1410 1370 1330 1280 1240 1140 1080 1040 990 940 920 830 800

Raman [cm^{-1}]: 3320 2970 2930 2880 2720 1460 1410 1350 1290 1240 1140 1080 1050 990 950 920 840 800 660 520 480 450 380 280

UV [nm]: 273(3802) sulf

^{13}C NMR [ppm]: 18.7 67.7 68.2 CDCl$_3$

^{1}H NMR [ppm]: 1.1 3.4 3.9 4.3 CDCl$_3$

Flammability: Flash pt. = 99°C; ign. temp. = 371°C; flam. lim. = 2.6-12.5%

450. 1,3-Propanediol

Syst. Name: 1,3-Propanediol
Synonyms: Trimethylene glycol; 1,3-
Dihydroxypropane

HO⌒⌒OH

CASRN: 504-63-2   **Beil Ref.**: 4-01-00-02493
Merck No.: 9629
Beil RN: 969155
MF: $C_3H_8O_2$ **MP[°C]**: -26.7   **Den.[g/cm^3]**: 1.0538^{20}
MW: 76.10   **BP[°C]**: 214.4 n_D: 1.4398^{20}
Sol.: H_2O 5; EtOH 5; eth
   4; bz 2
Vap. Press. [kPa]: 3.35^{125}; 10.0^{150}
Therm. Prop.[kJ/mol]: $\Delta_{vap}H$ = 57.9; $\Delta_f H°$(l, 25°C) = -464.9
γ **[mN/m]**: 45.62^{20}; dγ/dT = 0.0903 mN/m °C
Elec. Prop.: μ = (2.5 D); ε = 35.1^{20}
η **[mPa s]**: 46.6^{20}
MS: 28(100) 58(93) 31(76) 57(70) 29(40) 27(26) 45(24) 43(23) 19(18) 30(17)
IR [cm^{-1}]: 3360 2950 2900 1470 1420 1070 990 940 930 660
Raman [cm^{-1}]: 3260 2910 2880 1470 1440 1290 1240 1200 1060 990 920 870
   850 810 520 410 390 280
UV [nm]: 271(3715) sulf
^{1}H NMR [ppm]: 1.8 3.7 D_2O
Flammability: Ign. temp. = 400°C

451. Propanenitrile

Syst. Name: Propanenitrile
Synonyms: Ethyl cyanide

CASRN: 107-12-0   **DOT No.**: 2404
Merck No.: 7839   **Beil Ref.**: 4-02-00-00728
Beil RN: 773680
MF: C_3H_5N **MP[°C]**: -92.8   **Den.[g/cm^3]**: 0.7818^{20}
MW: 55.08   **BP[°C]**: 97.1 n_D: 1.3655^{20}
Sol.: H_2O 4; EtOH 3; eth
  3; ace 3; bz 3; ctc 3;
  os 3
Crit. Const.: T_c = 288.2°C; P_c = 4.26 MPa; V_c = 229 cm^3
Vap. Press. [kPa]: 0.276^{-25}; 1.66^0; 6.14^{25}
Therm. Prop.[kJ/mol]: $\Delta_{fus}H$ = 5.05; $\Delta_{vap}H$ = 31.8; $\Delta_f H°$(l, 25°C) = 15.5
C_p (liq.) [J/mol °C]: 119.3^{25}
γ [mN/m]: 26.75^{25}; $d\gamma/dT$ = 0.1153 mN/m °C
Elec. Prop.: μ = 4.05 D; ε = 29.7^{20}; IP = 11.84 eV
η [mPa s]: 0.294^{25}; 0.240^{50}; 0.202^{75}
MS: 28(100) 54(63) 26(20) 27(17) 52(11) 55(10) 51(9) 15(9) 53(7) 25(7)
IR [cm^{-1}]: 2860 2220 1430 1370 1320 1060 1000 780
Raman [cm^{-1}]: 3010 2960 2910 2860 2760 2640 2260 1470 1440 1390 1320
  1270 1080 1010 840 800 760 550 380 220
^{13}**C NMR [ppm]**: 10.6 10.8 120.8 CDCl$_3$
^{1}**H NMR [ppm]**: 1.3 2.3 CCl$_4$
Flammability: Flash pt. = 2°C; ign. temp. = 512°C; flam. lim. = 3.1-14%
Reg. Lists: CERCLA (RQ = 10 lb.); RCRA P101 (accutely hazardous)

452. Propanoic acid

Syst. Name: Propanoic acid
Synonyms: Propionic acid

CASRN: 79-09-4   **DOT No.:** 1848
Merck No.: 7837   **Beil Ref.:** 4-02-00-00695
Beil RN: 506071
MF: $C_3H_6O_2$ **MP[°C]:** -20.7   **Den.[g/cm³]:** 0.9930^{20}
MW: 74.08   **BP[°C]:** 141.1 n_D: 1.3809^{20}
Sol.: H_2O 5; EtOH 5; eth
   3; chl 2
Crit. Const.: T_c = 331°C; P_c = 4.53 MPa; V_c = 222 cm³
Vap. Press. [kPa]: 0.553^{25}; 2.21^{50}; 7.98^{75}; 23.7^{100}; 59.8^{125}; 132^{150}
Therm. Prop.[kJ/mol]: $\Delta_{fus}H$ = 10.66; $\Delta_{vap}H$ = 32.3; $\Delta_f H°$(l, 25°C) = -510.7
C_p (liq.) [J/mol °C]: 152.7^{25}
γ [mN/m]: 26.20^{25}; $d\gamma/dT$ = 0.0993 mN/m °C
Elec. Prop.: μ = 1.75 D; ε = 3.44^{25}; IP = 10.53 eV
η [mPa s]: 1.50^0; 1.03^{25}; 0.749^{50}; 0.569^{75}; 0.449^{100}
MS: 28(100) 29(84) 74(79) 27(62) 45(56) 73(48) 57(30) 26(21) 55(17) 56(16)
IR [cm⁻¹]: 2980 2540 1720 1460 1410 1320 1290 1240 1080 930 850
Raman [cm⁻¹]: 2990 2850 2900 2750 2640 1690 1670 1470 1430 1390 1260
   1080 1010 840 610 480 280
UV [nm]: 270 H_2O
¹H NMR [ppm]: 1.1 2.4 10.5 $CDCl_3$
Flammability: Flash pt. = 52°C; ign. temp. = 465°C; flam. lim. = 2.9-12.1%
TLV/TWA: 10 ppm (30 mg/m³)
Reg. Lists: CERCLA (RQ = 5000 lb.)

453. Propanoic anhydride

Syst. Name: Propanoic acid, anhydride
Synonyms: Propionic anhydride

CASRN: 123-62-6   **DOT No.**: 2496
Merck No.: 7838   **Beil Ref.**: 4-02-00-00722
Beil RN: 507066
MF: $C_6H_{10}O_3$ **MP[°C]**: -45   **Den.[g/cm^3]**: 1.0110^{20}
MW: 130.14   **BP[°C]**: 170 n_D: 1.4038^{20}
Sol.: eth 5; ctc 2
Vap. Press. [kPa]: 2.44^{50}; 8.87^{75}; 25.1^{100}; 59.1^{125}; 121^{150}
Therm. Prop.[kJ/mol]: $\Delta_{vap}H = 41.7$; $\Delta_f H°(l, 25°C) = -679.1$
C_p **(liq.) [J/mol °C]**: 235.0^{25}
γ **[mN/m]**: 29.70^{25}
Elec. Prop.: $\varepsilon = 18.30^{20}$
η **[mPa s]**: 1.06^{25}
MS: 57(100) 29(55) 28(20) 27(20) 74(19) 73(12) 45(11) 26(5) 30(4) 58(3)
IR [cm^{-1}]: 2990 2950 1820 1760 1470 1420 1350 1260 1040 800
Raman [cm^{-1}]: 2990 2950 2900 2750 1810 1750 1660 1470 1420 1390 1360 1260 1150 1080 1040 1010 990 970 890 860 830 780 700 630 560 450 340 290 250 170
^{13}C NMR [ppm]: 8.4 28.7 170.3 CDCl$_3$
^{1}H NMR [ppm]: 1.2 2.4 CCl$_4$
Flammability: Flash pt. = 63°C; ign. temp. = 285°C; flam. lim. = 1.3-9.5%
Reg. Lists: CERCLA (RQ = 5000 lb.)

454. 1-Propanol

Syst. Name: 1-Propanol
Synonyms: Propyl alcohol

CASRN: 71-23-8
Merck No.: 7854
Beil RN: 1098242
MF: C_3H_8O
MW: 60.10
Sol.: H_2O 5; EtOH 5; eth
5; ace 3; bz 4; chl 3

DOT No.: 1274
Beil Ref.: 4-01-00-01413

MP[°C]: -126.1
BP[°C]: 97.2

Den.[g/cm^3]: 0.8035^{20}
n_D: 1.3850^{20}

Crit. Const.: $T_c = 263.7$°C; $P_c = 5.169$ MPa; $V_c = 219$ cm^3
Vap. Press. [kPa]: 0.445^0; 2.76^{25}; 12.2^{50}; 40.9^{75}; 113^{100}; 265^{125}; 545^{150}
Therm. Prop.[kJ/mol]: $\Delta_{fus}H = 5.20$; $\Delta_{vap}H = 41.4$; $\Delta_f H°(l, 25$°C$) = -302.6$
C_p (liq.) [J/mol °C]: 143.7^{25}
γ [mN/m]: 23.32^{25}; dγ/d$T = 0.0777$ mN/m °C
Elec. Prop.: $\mu = 1.55$ D; $\varepsilon = 20.8^{20}$; IP = 10.22 eV
η [mPa s]: 8.65^{-25}; 3.82^0; 1.95^{25}; 1.11^{50}; 0.685^{75}
k [W/m °C]: 0.162^{-25}; 0.158^0; 0.154^{25}; 0.149^{50}; 0.145^{75}; 0.141^{100}
MS: 31(100) 27(19) 29(18) 59(11) 42(9) 60(7) 41(7) 28(7) 43(3) 32(3)
IR [cm^{-1}]: 3340 2950 2870 1450 1090 1050 1000 960
Raman [cm^{-1}]: 3300 2960 2940 2910 2880 2730 2670 1460 1350 1300 1270
1250 1100 1070 1060 1020 970 890 860 820 770 460 330
UV [nm]: 183(240) gas
^{13}C NMR [ppm]: 10.5 26.3 64.0 diox
^{1}H NMR [ppm]: 0.9 1.6 2.3 3.6 CDCl$_3$
Flammability: Flash pt. = 23°C; ign. temp. = 412°C; flam. lim. = 2.2-13.7%
TLV/TWA: 200 ppm (492 mg/m^3)

455. 2-Propanol

Syst. Name: 2-Propanol
Synonyms: Isopropyl alcohol

CASRN: 67-63-0
Merck No.: 5096
Beil RN: 635639

DOT No.: 1219
Beil Ref.: 4-01-00-01461

MF: C_3H_8O
MW: 60.10
Sol.: H_2O 5; EtOH 5; eth 5; ace 3; bz 4; chl 3

MP[°C]: -89.5
BP[°C]: 82.3

Den.[g/cm³]: 0.7855^{20}
n_D: 1.3776^{20}

Crit. Const.: $T_c = 235.2°C$; $P_c = 4.764$ MPa; $V_c = 220$ cm³
Vap. Press. [kPa]: 1.11^0; 6.02^{25}; 23.9^{50}; 75.3^{75}; 198^{100}; 452^{125}
Therm. Prop.[kJ/mol]: $\Delta_{fus}H = 5.37$; $\Delta_{vap}H = 39.9$; $\Delta_fH°(l, 25°C) = -318.1$
C_p (liq.) [J/mol °C]: 155.0^{25}
γ [mN/m]: 20.93^{25}; $d\gamma/dT = 0.0789$ mN/m °C
Elec. Prop.: $\mu = 1.56$ D; $\varepsilon = 20.18^{20}$; IP = 10.12 eV
η [mPa s]: 4.62^0; 2.04^{25}; 1.03^{50}; 0.576^{75}
k [W/m °C]: 0.146^{-25}; 0.141^0; 0.135^{25}; 0.129^{50}; 0.124^{75}; 0.118^{100}
MS: 45(100) 43(19) 27(17) 29(12) 41(7) 31(6) 19(6) 42(5) 44(4) 59(3)
IR [cm⁻¹]: 3340 2960 2880 1465 1410 1380 1340 1310 1160 1125 1105 950 810 480 420
Raman [cm⁻¹]: 3300 2970 2940 2920 2880 2760 2720 2650 1450 1410 1340 1300 1160 1130 1110 950 930 820 490 430 370 260
UV [nm]: 181(617) gas
¹³C NMR [ppm]: 25.4 63.7
¹H NMR [ppm]: 1.2 1.6 4.0 $CDCl_3$
Flammability: Flash pt. = 12°C; ign. temp. = 399°C; flam. lim. = 2.0-12.7%
TLV/TWA: 400 ppm (983 mg/m³)
Reg. Lists: SARA 313 (0.1%)

456. Propargyl acetate
Syst. Name: 2-Propyn-1-ol, ethanoate

CASRN: 627-09-8   **Beil Ref.**: 4-02-00-00197
Beil RN: 1742046
MF: $C_5H_6O_2$ **BP[°C]**: 121.5   **Den.[g/cm^3]**: 0.9982^{20}
MW: 98.10 n_D: 1.4187^{20}
Sol.: H_2O 2; EtOH 3; eth
    3
γ **[mN/m]**: 32.81^{20}
MS: 43(100) 39(18) 55(5) 70(4) 56(4) 44(3) 98(2)

---

457. Propargyl alcohol
Syst. Name: 2-Propyn-1-ol
Synonyms: 3-Hydroxy-1-propyne

CASRN: 107-19-7   **DOT No.**: 1986
Merck No.: 7819   **Beil Ref.**: 4-01-00-02214
Beil RN: 506003
MF: C_3H_4O **MP[°C]**: -51.8   **Den.[g/cm^3]**: 0.9478^{20}
MW: 56.06   **BP[°C]**: 113.6 n_D: 1.4322^{20}
Sol.: H_2O 3; EtOH 5; eth
    5; chl 3
γ **[mN/m]**: 35.2^{28}
Elec. Prop.: $\mu = 1.13$ D; $\varepsilon = 20.8^{20}$; IP = 10.51 eV
η **[mPa s]**: 1.68^{20}
MS: 55(100) 39(25) 28(20) 27(19) 29(16) 38(14) 26(11) 37(8) 53(6) 56(4)
IR [cm^{-1}]: 3230 2860 2080 1450 1410 1350 1300 1240 1030 980 920 680
Raman [cm^{-1}]: 3290 2930 2870 2110 1450 1360 1230 1030 980 910 670 550
    310 220
^{13}C NMR [ppm]: 50.0 73.8 83.0 diox
^{1}H NMR [ppm]: 2.5 2.8 4.3 $CDCl_3$
Flammability: Flash pt. = 36°C
TLV/TWA: 1 ppm (2.3 mg/m^3)
Reg. Lists: CERCLA (RQ = 1000 lb.); RCRA P102 (accutely hazardous)

458. Propyl acetate

Syst. Name: Ethanoic acid, propyl ester

CASRN: 109-60-4　　　　　　　　**DOT No.**: 1276
Merck No.: 7853　　　　　　　　　**Beil Ref.**: 4-02-00-00138
Beil RN: 1740764
MF: $C_5H_{10}O_2$　　**MP[°C]**: -93　　　**Den.[g/cm^3]**: 0.8878^{20}
MW: 102.13　　　**BP[°C]**: 101.5　　n_D: 1.3842^{20}
Sol.: H_2O 2; EtOH 5; eth
　5; ctc 3
Crit. Const.: T_c = 276.6°C; P_c = 3.36 MPa; V_c = 345 cm^3
Vap. Press. [kPa]: 4.49^{25}; 15.2^{50}; 41.6^{75}; 96.6^{100}; 198^{125}
Therm. Prop.[kJ/mol]: $\Delta_{vap}H$ = 33.9
C_p **(liq.) [J/mol °C]**: 196.2^{25}
γ **[mN/m]**: 23.80^{25}; $d\gamma/dT$ = 0.1120 mN/m °C
Elec. Prop.: μ = (1.8 D); ε = 5.62^{20}; IP = 10.04 eV
η **[mPa s]**: 0.768^0; 0.544^{25}; 0.406^{50}; 0.316^{75}; 0.255^{100}
MS: 43(100) 61(19) 31(18) 27(15) 42(11) 73(9) 41(9) 29(9) 59(5) 39(5)
IR [cm^{-1}]: 2960 2880 1740 1460 1390 1360 1240 1060 1040 1020 960 910 890
Raman [cm^{-1}]: 2940 2880 2740 1730 1450 1390 1370 1350 1280 1250 1150
　1130 1120 1100 1060 1050 1020 980 970 910 890 880 840 760 630 490 410
　350 310 260
UV [nm]: 203(51) H_2O 209(55) MeOH
^{1}H NMR [ppm]: 0.9 1.6 2.0 4.0 CCl_4
Flammability: Flash pt. = 13°C; ign. temp. = 450°C; flam. lim. = 1.7-8%
TLV/TWA: 200 ppm (835 mg/m^3)

459. Propylamine

Syst. Name: 1-Propanamine

CASRN: 107-10-8   **DOT No.:** 1277
Merck No.: 7855   **Beil Ref.:** 4-04-00-00464
Beil RN: 1098243
MF: C_3H_9N **MP[°C]:** -83   **Den.[g/cm^3]:** 0.7173^{20}
MW: 59.11   **BP[°C]:** 47.2 n_D: 1.3870^{20}
Sol.: H_2O 3; EtOH 4; eth
   4; ace 4; bz 3; ctc 2;
   chl 3
Crit. Const.: $T_c = 223.9°C$; $P_c = 4.72$ MPa
Vap. Press. [kPa]: 12.4^0; 42.1^{25}; 112^{50}; 251^{75}
Therm. Prop.[kJ/mol]: $\Delta_{fus}H = 10.97$; $\Delta_{vap}H = 29.6$; $\Delta_f H°(l, 25°C) = -101.5$
C_p **(liq.) [J/mol °C]:** 162.5^{25}
γ **[mN/m]:** 21.75^{25}; $d\gamma/dT = 0.1243$ mN/m °C
Elec. Prop.: $\mu = 1.17$ D; $\varepsilon = 5.08^{23}$; IP = 8.78 eV
η **[mPa s]:** 0.376^{25}
MS: 30(100) 28(13) 59(8) 27(7) 41(5) 42(3) 39(3) 29(3) 26(3) 18(3)
IR [cm^{-1}]: 3330 2940 1610 1470 1390 1080 900 810
^{13}**C NMR [ppm]:** 11.2 27.3 44.5
^{1}**H NMR [ppm]:** 0.9 1.5 2.1 2.7 CCl_4
Flammability: Flash pt. = -37°C; ign. temp. = 318°C; flam. lim. = 2.0-10.4%
Reg. Lists: CERCLA (RQ = 5000 lb.); RCRA U194 (toxic)

460. Propylbenzene

Syst. Name: Benzene, propyl-
Synonyms: Isocumene

CASRN: 103-65-1

Merck No.: 7856

Beil RN: 1903006

MF: C_9H_{12}

MW: 120.19

Sol.: H_2O 1; EtOH 5; eth
5; ace 5; bz 5; ctc 5;
peth 5

DOT No.: 2364

Beil Ref.: 4-05-00-00977

MP[°C]: -99.5

BP[°C]: 159.2

Den.[g/cm³]: 0.8620^{20}

n_D: 1.4920^{20}

Crit. Const.: $T_c = 365.20°C$; $P_c = 3.200$ MPa; $V_c = 440$ cm^3

Vap. Press. [kPa]: 16.6^{100}; 38.5^{125}; 79.5^{150}

Therm. Prop.[kJ/mol]: $\Delta_{fus}H = 9.27$; $\Delta_fH°(l, 25°C) = -38.3$

C_p **(liq.) [J/mol °C]:** 214.7^{25}

Elec. Prop.: $\mu = 0$; $\varepsilon = 2.37^{20}$; IP = 8.72 eV

η **[mPa s]:** 0.69^{38}

MS: 91(100) 120(21) 92(10) 38(10) 65(9) 78(6) 51(6) 27(5) 63(4) 105(3)

IR [cm⁻¹]: 2940 1590 1490 1450 1370 1090 1020 710

Raman [cm⁻¹]: 3069 3056 2936 2910 2876 2865 1607 1204 1033 1007 318

UV [nm]: 268(1730) 261(2180) 259(2140) 253(2090) 248(1940) MeOH

¹H NMR [ppm]: 0.9 1.6 2.6 CDCl$_3$

Flammability: Flash pt. = 30°C; ign. temp. = 450°C; flam. lim. = 0.8-6.0%

461. Propyl benzoate

Syst. Name: Benzoic acid, propyl ester
Synonyms: Propyl benzenecarboxylate

CASRN: 2315-68-6   **Beil Ref.:** 4-09-00-00288
Beil RN: 1863771
MF: $C_{10}H_{12}O_2$ **MP[°C]:** -51.6   **Den.[g/cm^3]:** 1.0230^{20}
MW: 164.20   **BP[°C]:** 211 n_D: 1.5000^{20}
Sol.: H_2O 1; EtOH 5; eth
   5
Vap. Press. [kPa]: 0.099^{50}; 0.420^{75}; 1.43^{100}; 4.11^{125}; 10.3^{150}
γ [mN/m]: 38.9^{23}
Elec. Prop.: ε = 5.78^{30}
IR [cm^{-1}]: 2940 1720 1610 1490 1450 1390 1320 1280 1180 1110 1020 940
   710

462. Propylene carbonate

Syst. Name: 1,3-Dioxolan-2-one, 4-methyl-
Synonyms: Propanediol-1,2-carbonate

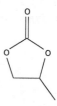

CASRN: 108-32-7   **Beil Ref.**: 5-19-04-00021
Beil RN: 107913
MF: $C_4H_6O_3$ **MP[°C]**: -48.8   **Den.[g/cm³]**: 1.2047^{20}
MW: 102.09   **BP[°C]**: 242 n_D: 1.4189^{20}
Sol.: H_2O 4; EtOH 4; eth
   4; ace 4; bz 4
Vap. Press. [kPa]: 0.131^{50}; 0.324^{75}; 0.710^{100}
Therm. Prop.[kJ/mol]: $\Delta_{fus}H = 9.62$; $\Delta_f H°(l, 25°C) = -613.2$
C_p **(liq.) [J/mol °C]**: 218.6^{25}
γ **[mN/m]**: 41.4^{25}
Elec. Prop.: $\mu = (4.9$ D); $\varepsilon = 66.14^{20}$
η **[mPa s]**: 2.53^{25}
MS: 28(100) 57(69) 43(66) 29(51) 27(45) 30(42) 26(19) 42(17) 31(16) 58(13)
IR [cm⁻¹]: 3000 2930 1800 1480 1390 1350 1180 1120 1050 780
¹H NMR [ppm]: 1.5 4.0 4.5 4.8 CDCl₃
Flammability: Flash pt. = 135°C

464. Propyl formate

Syst. Name: Methanoic acid, propyl ester

CASRN: 110-74-7   **DOT No.**: 1281
Merck No.: 7871   **Beil Ref.**: 4-02-00-00026
Beil RN: 1739272
MF: $C_4H_8O_2$ **MP[°C]**: -92.9   **Den.[g/cm^3]**: 0.9058^{20}
MW: 88.11   **BP[°C]**: 80.9 n_D: 1.3779^{20}
Sol.: H_2O 2; EtOH 5; eth
    5; ctc 2
Crit. Const.: T_c = 264.9°C; P_c = 4.06 MPa; V_c = 285 cm^3
Vap. Press. [kPa]: 2.76^0; 10.9^{25}; 33.4^{50}; 83.8^{75}; 180^{100}
Therm. Prop.[kJ/mol]: $\Delta_{vap}H$ = 33.6; $\Delta_f H°$(l, 25°C) = -500.3
C_p **(liq.) [J/mol °C]**: 171.4^{25}
γ **[mN/m]**: 24.53^{20}; $d\gamma/dT$ = 0.1119 mN/m °C
Elec. Prop.: μ = (1.9 D); ε = 6.92^{30}; IP = 10.52 eV
η **[mPa s]**: 0.669^0; 0.485^{25}; 0.370^{50}; 0.293^{75}
MS: 27(100) 29(92) 31(81) 42(60) 41(53) 43(29) 39(29) 26(28) 47(22) 30(16)
IR [cm^{-1}]: 2940 1720 1470 1390 1180 940 900
Raman [cm^{-1}]: 3100 2940 2900 2880 2740 2680 1720 1450 1380 1280 1180
    1150 1120 1050 1040 880 860 770 750 620 460 410 320 310 250 220
UV [nm]: 220(74) 216(79) 212(78) iso 212(71) MeOH
^{1}H NMR [ppm]: 1.0 1.7 4.1 7.9 CCl_4
Flammability: Flash pt. = -3°C; ign. temp. = 455°C

465. Pseudocumene

Syst. Name: Benzene, 1,2,4-trimethyl-
Synonyms: 1,2,4-Trimethylbenzene

CASRN: 95-63-6
Merck No.: 7929
Beil RN: 1903005
MF: C_9H_{12}
MW: 120.19
Sol.: H_2O 1; EtOH 5; eth 5; ace 5; bz 5; ctc 5; peth 5

Rel. CASRN: 25551-13-7, 15551-13-7
DOT No.: 1263
Beil Ref.: 4-05-00-01010
MP[°C]: -43.8   **Den.[g/cm³]**: 0.8758^{20}
BP[°C]: 169.3 n_D: 1.5048^{20}

Crit. Const.: T_c = 376.0°C; P_c = 3.232 MPa
Vap. Press. [kPa]: 0.300^{25}; 11.7^{100}; 28.3^{125}; 60.3^{150}
Therm. Prop.[kJ/mol]: $\Delta_{fus}H$ = 3.76; $\Delta_f H°$(l, 25°C) = -61.8
C_p (liq.) **[J/mol °C]**: 215.0^{25}
γ **[mN/m]**: 29.20^{25}; $d\gamma/dT$ = 0.1025 mN/m °C
Elec. Prop.: μ = 0; ε = 2.38^{20}; IP = 8.27 eV
η **[mPa s]**: 0.75^{38}
MS: 105(100) 120(56) 119(17) 77(15) 39(15) 51(11) 91(10) 27(10) 106(9) 79(7)
IR [cm⁻¹]: 3030 2941 1887 1724 1587 1493 1449 1389 1220 1149 1020 885 746
Raman [cm⁻¹]: 3040 3000 2970 2930 2860 2760 2730 2570 1620 1580 1450 1410 1380 1290 1250 1210 1160 1120 920 750 720 560 470 440 320 280 210 150
UV [nm]: 277 267 MeOH
¹³C NMR [ppm]: 19.1 19.5 20.8 126.4 129.5 130.4 133.1 135.0 136.1 $CDCl_3$
¹H NMR [ppm]: 2.2 6.8 CCl_4
Flammability: Flash pt. = 44°C; ign. temp. = 500°C; flam. lim. = 0.9-6.4%
TLV/TWA: 25 ppm (123 mg/m³)
Reg. Lists: SARA 313 (1.0%)

466. Pyridine

Syst. Name: Pyridine
Synonyms: Azine; Azabenzene

CASRN: 110-86-1
Merck No.: 7983
Beil RN: 103233
MF: C_5H_5N
MW: 79.10
Sol.: H_2O 5; EtOH 5; eth 5; ace 5; bz 5; chl 5

DOT No.: 1282
Beil Ref.: 5-20-05-00160

MP[°C]: -41.6
BP[°C]: 115.2

Den.[g/cm³]: 0.9819^{20}
n_D: 1.5095^{20}

Crit. Const.: $T_c = 346.9°C$; $P_c = 5.67$ MPa; $V_c = 243$ cm³
Vap. Press. [kPa]: 2.76^{25}; 9.57^{50}; 26.8^{75}; 63.7^{100}; 133^{125}; 252^{150}
Therm. Prop.[kJ/mol]: $\Delta_{fus}H = 8.28$; $\Delta_{vap}H = 35.1$; $\Delta_f H°(l, 25°C) = 100.2$
C_p (liq.) [J/mol °C]: 132.7^{25}
γ [mN/m]: 36.56^{25}; $d\gamma/dT = 0.1306$ mN/m °C
Elec. Prop.: $\mu = 2.215$ D; $\varepsilon = 13.26^{20}$; IP = 9.25 eV
η [mPa s]: 1.36^0; 0.879^{25}; 0.637^{50}; 0.497^{75}; 0.409^{100}
k [W/m °C]: 0.169^0; 0.165^{25}; 0.161^{50}; 0.158^{75}
MS: 79(100) 52(62) 51(31) 50(19) 78(11) 53(7) 39(7) 80(6) 27(3) 77(2)
IR [cm⁻¹]: 3080 3020 3000 1600 1580 1480 1440 1210 1140 1070 1030 990 740 700
Raman [cm⁻¹]: 3150 3090 3060 2990 2960 1600 1580 1570 1480 1220 1030 990 660 610
UV [nm]: 261 256 250 MeOH
¹³C NMR [ppm]: 123.6 135.7 149.8 $CDCl_3$
¹H NMR [ppm]: 7.0 7.6 8.6 $CDCl_3$
Flammability: Flash pt. = 20°C; ign. temp. = 482°C; flam. lim. = 1.8-12.4%
TLV/TWA: 5 ppm (16 mg/m³)
Reg. Lists: CERCLA (RQ = 1000 lb.); SARA 313 (1.0%); RCRA U196 (toxic)

467. Pyrrole

Syst. Name: 1*H*-Pyrrole
Synonyms: Azole; Imidole

CASRN: 109-97-7   **Beil Ref.**: 5-20-05-00003
Merck No.: 8025
Beil RN: 1159
MF: C_4H_5N **MP[°C]**: -23.4   **Den.[g/cm³]**: 0.9698^{20}
MW: 67.09   **BP[°C]**: 129.7 n_D: 1.5085^{20}
Sol.: H_2O 2; EtOH 3; eth
   3; ace 3; bz 3; chl 3
Crit. Const.: T_c = 366.6°C; P_c = 6.34 MPa; V_c = 200 cm³
Vap. Press. [kPa]: 1.10^{25}; 14.4^{75}; 38.2^{100}; 87.7^{125}; 179^{150}
Therm. Prop.[kJ/mol]: $\Delta_{fus}H$ = 7.91; $\Delta_{vap}H$ = 38.8; $\Delta_f H°$(l, 25°C) = 63.1
C_p **(liq.) [J/mol °C]**: 127.7^{25}
γ **[mN/m]**: 37.06^{25}; $d\gamma/dT$ = 0.1100 mN/m °C
Elec. Prop.: μ = 1.74 D; ϵ = 8.00^{20}; IP = 8.21 eV
η **[mPa s]**: 2.09^0; 1.23^{25}; 0.828^{50}; 0.612^{75}
MS: 67(100) 41(58) 39(58) 40(51) 28(42) 38(20) 37(12) 66(7) 68(5) 27(3)
IR [cm⁻¹]: 3450 3030 1430 1080 1050 1010 860 730
UV [nm]: 209(6730) MeOH
¹³C NMR [ppm]: 107.9 117.9 CDCl₃
¹H NMR [ppm]: 6.2 6.7 8.0 CDCl₃
Flammability: Flash pt. = 39°C

468. Pyrrolidine

Syst. Name: Pyrrolidine
Synonyms: Tetrahydropyrrole; Azacyclopentane

CASRN: 123-75-1   **DOT No.**: 1922
Merck No.: 8026   **Beil Ref.**: 5-20-01-00162
Beil RN: 102395
MF: C_4H_9N **MP[°C]**: -57.8   **Den.[g/cm³]**: 0.8586^{20}
MW: 71.12   **BP[°C]**: 86.5 n_D: 1.4431^{20}
Sol.: H_2O 5; EtOH 3; eth
  3; bz 2; chl 2
Crit. Const.: T_c = 295.1°C; P_c = 5.59 MPa; V_c = 238 cm³
Vap. Press. [kPa]: 8.40^{25}; 26.7^{50}; 69.0^{75}; 153^{100}
Therm. Prop.[kJ/mol]: $\Delta_{fus}H$ = 8.58; $\Delta_{vap}H$ = 33.0; $\Delta_f H°$(l, 25°C) = -41.0
C_p (liq.) [J/mol °C]: 156.6^{25}
γ [mN/m]: 29.23^{25}; $d\gamma/dT$ = 0.0900 mN/m °C
Elec. Prop.: μ = (1.6 D); ε = 8.30^{20}; IP = 8.00 eV
η [mPa s]: 1.91^{-25}; 1.07^0; 0.704^{25}; 0.512^{50}
MS: 43(100) 28(52) 70(33) 71(26) 42(22) 41(20) 27(16) 39(15) 29(10) 30(9)
IR [cm⁻¹]: 3330 2940 2860 1670 1560 1470 1430 1350 1300 1190 1110 1090
  1000 880
Raman [cm⁻¹]: 3348 3312 2956 2871 1481 1450 1284 1218 1105 900
UV [nm]: 196(1995) 171(2512) gas
¹³C NMR [ppm]: 25.7 47.1 CDCl₃
¹H NMR [ppm]: 1.5 2.4 2.7 bz
Flammability: Flash pt. = 3°C

469. 2-Pyrrolidone

Syst. Name: 2-Pyrrolidinone
Synonyms: γ-Butyrolactam

CASRN: 616-45-5   **Beil Ref.**: 5-21-06-00317
Merck No.: 8027
Beil RN: 105241
MF: C_4H_7NO **MP[°C]**: 25   **Den.[g/cm^3]**: 1.120^{20}
MW: 85.11   **BP[°C]**: 251 n_D: 1.4806^{30}
Sol.: H_2O 4; EtOH 4; eth
   4; bz 4; chl 4; CS_2 4
Vap. Press. [kPa]: 1.6^{133}
Therm. Prop.[kJ/mol]: $\Delta_f H°$(l, 25°C) = -286.2
C_p **(liq.) [J/mol °C]**: 169.4^{27}
Elec. Prop.: μ = (3.5 D); IP = 9.20 eV
η **[mPa s]**: 13.3^{25}
MS: 85(100) 42(43) 41(36) 28(33) 30(29) 56(16) 84(14) 40(12) 27(12) 29(9)
IR [cm^{-1}]: 3130 2940 2700 2560 1700 1490 1470 1430 1370 1280 1080 1000
   790 690
UV [nm]: 190(7079) H_2O
^{1}H NMR [ppm]: 2.2 3.4 7.7 CDCl$_3$
Flammability: Flash pt. = 129°C

470. Quinoline

Syst. Name: Quinoline
Synonyms: 1-Azanaphthalene; Benzopyridine

CASRN: 91-22-5   **DOT No.:** 2656
Merck No.: 8097   **Beil Ref.:** 5-20-07-00276
Beil RN: 107477
MF: C_9H_7N **MP[°C]:** -14.78   **Den.[g/cm^3]:** 1.0977^{15}
MW: 129.16   **BP[°C]:** 237.1 n_D: 1.6268^{20}
Sol.: H_2O 1; EtOH 5; eth
  5; ace 5; bz 5; ctc 3;
  CS_2 5
Crit. Const.: T_c = 509°C; P_c = 4.86 MPa; V_c = 437 cm^3
Vap. Press. [kPa]: 21.1^{175}; 41.9^{200}; 76.7^{225}
Therm. Prop.[kJ/mol]: $\Delta_{fus}H$ = 10.66; $\Delta_{vap}H$ = 49.7; $\Delta_f H°$(l, 25°C) = 141.2
C_p **(liq.) [J/mol °C]:** 194.9^{25}
γ **[mN/m]:** 42.59^{25}; $d\gamma/dT$ = 0.1063 mN/m °C
Elec. Prop.: μ = 2.29 D; ε = 9.16^{20}; IP = 8.62 eV
η **[mPa s]:** 3.34^{25}; 1.89^{50}; 1.20^{75}; 0.833^{100}
MS: 129(100) 51(28) 76(25) 128(24) 44(24) 50(20) 32(19) 75(18) 74(12)
  103(11)
IR [cm^{-1}]: 3050 1620 1600 1570 1500 1430 1390 1370 1310 1140 1120 1030
  940 810 790 740
Raman [cm^{-1}]: 3070 3010 1620 1600 1570 1500 1470 1460 1430 1400 1380
  1350 1320 1260 1230 1140 1120 1040 1020 990 960 780 760 520 490 190
UV [nm]: 313 306 300 294 288 277 231 226 205 MeOH
^{13}C NMR [ppm]: 120.8 126.3 127.6 128.0 129.2 135.7 148.1 150.0 CDCl$_3$
^{1}H NMR [ppm]: 7.1 7.8 8.8 CCl$_4$
Flammability: Ign. temp. = 480°C
Reg. Lists: CERCLA (RQ = 5000 lb.); SARA 313 (1.0%)

471. Salicylaldehyde

Syst. Name: Benzaldehyde, 2-hydroxy-
Synonyms: 2-Hydroxybenzaldehyde

CASRN: 90-02-8   **Beil Ref.**: 4-08-00-00176
Merck No.: 8295
Beil RN: 471388
MF: $C_7H_6O_2$ **MP[°C]**: -7   **Den.[g/cm³]**: 1.1674^{20}
MW: 122.12   **BP[°C]**: 197 n_D: 1.5740^{20}
Sol.: H_2O 2; EtOH 5; eth
  5; ace 4; bz 4; chl 2
Vap. Press. [kPa]: 4.42^{100}; 11.7^{125}; 27.3^{150}
Therm. Prop.[kJ/mol]: $\Delta_{vap}H = 38.2$
C_p **(liq.) [J/mol °C]**: 222^{18}
Elec. Prop.: $\mu = (2.9\ D)$; $\varepsilon = 18.35^{20}$
η **[mPa s]**: 2.90^{20}
MS: 28(100) 122(98) 121(92) 39(79) 65(52) 76(31) 32(31) 44(30) 93(26)
  38(23)
IR [cm⁻¹]: 3226 3030 2857 2778 2500 1667 1639 1587 1493 1471 1389 1333
  1282 1235 1205 1149 1124 1031 952 901 769 725 671
Raman [cm⁻¹]: 3070 2850 1660 1640 1620 1590 1580 1450 1390 1320 1280
  1230 1150 1110 1070 1030 880 860 770 660 560 450 410 290 270 210 150
UV [nm]: 328 259 cyhex
¹³C NMR [ppm]: 117.4 119.6 121.0 133.6 136.6 161.4 196.7 diox
¹H NMR [ppm]: 7.0 7.4 9.8 11.0 CDCl₃
Flammability: Flash pt. = 78°C

472. Styrene

Syst. Name: Benzene, ethenyl-
Synonyms: Vinylbenzene

CASRN: 100-42-5
Merck No.: 8830
Beil RN: 1071236
MF: C_8H_8
MW: 104.15
Sol.: H_2O 1; EtOH 3; eth
3; ace 3; bz 5; ctc 2;
peth 5; MeOH 3; CS_2
3

DOT No.: 2055
Beil Ref.: 4-05-00-01334

MP[°C]: -31
BP[°C]: 145

Den.[g/cm^3]: 0.9060^{20}
n_D: 1.5468^{20}

Vap. Press. [kPa]: 0.810^{25}; 3.18^{50}; 9.85^{75}; 25.5^{100}; 57.3^{125}
Therm. Prop.[kJ/mol]: $\Delta_{fus}H$ = 10.95; $\Delta_{vap}H$ = 38.7; $\Delta_f H°$(l, 25°C) = 103.8
C_p **(liq.) [J/mol °C]:** 182.0^{25}
γ **[mN/m]:** 32.3^{20}
Elec. Prop.: $\varepsilon = 2.47^{20}$; IP = 8.43 eV
η **[mPa s]:** 1.05^0; 0.695^{25}; 0.507^{50}; 0.390^{75}; 0.310^{100}
k **[W/m °C]:** 0.148^{-25}; 0.142^0; 0.137^{25}; 0.131^{50}; 0.126^{75}; 0.120^{100}
MS: 104(100) 103(41) 78(32) 51(28) 77(23) 105(12) 50(12) 52(11) 39(11)
102(10)
IR [cm^{-1}]: 3040 3030 3010 1630 1580 1500 1450 1410 1200 1080 1020 990
910 770 700
Raman [cm^{-1}]: 3090 3060 3010 2980 1670 1630 1610 1580 1560 1500 1450
1420 1340 1320 1300 1240 1210 1190 1160 1090 1040 1020 1010 910 850
820 780 620 560 450 240 220
UV [nm]: 289 281 272 246 cyhex
^{1}H NMR [ppm]: 5.1 5.6 6.6 7.2 CCl_4
Flammability: Flash pt. = 31°C; ign. temp. = 490°C; flam. lim. = 0.9-6.8%
TLV/TWA: 50 ppm (213 mg/m^3)
Reg. Lists: CERCLA (RQ = 1000 lb.); SARA 313 (0.1%); confirmed or
suspected carcinogen

473. Succinonitrile

Syst. Name: Butanedinitrile
Synonyms: 1,2-Dicyanoethane; Ethylene cyanide

CASRN: 110-61-2   **Beil Ref.**: 4-02-00-01923
Merck No.: 8843
Beil RN: 1098380
MF: $C_4H_4N_2$ **MP[°C]**: 54.5   **Den.[g/cm^3]**: 0.9867^{60}
MW: 80.09   **BP[°C]**: 266 n_D: 1.4173^{60}
Sol.: H_2O 4; EtOH 3; eth
   2; ace 3; bz 3; chl 3
Vap. Press. [kPa]: 0.001^{25}
Therm. Prop.[kJ/mol]: $\Delta_{fus}H = 3.70$; $\Delta_{vap}H = 48.5$
C_p **(liq.) [J/mol °C]**: 160.5^{62}
γ **[mN/m]**: 46.78^{60}
Elec. Prop.: $\mu = (3.7\ D)$; $\varepsilon = 62.6^{25}$; IP = 12.10 eV
η **[mPa s]**: 2.59^{60}
MS: 53(100) 28(90) 40(64) 79(33) 52(30) 51(20) 80(18) 27(14) 32(12) 38(11)
IR [cm^{-1}]: 3030 2220 1430 1330 1240 1190 1000 960 920 820 760
Raman [cm^{-1}]: 2980 2950 2840 2290 2250 2200 1430 1360 1330 1230 1200
   1030 1000 950 820 720 600 520 480 390 360 240 180
^{1}H NMR [ppm]: 2.8 CDCl$_3$
Flammability: Flash pt. = 132°C

474. Sulfolane
Syst. Name: Thiophene, tetrahydro-, 1,1-dioxide
Synonyms: Thiolane 1,1-dioxide

CASRN: 126-33-0   **Beil Ref.**: 5-17-01-00039
Merck No.: 8934
Beil RN: 107765
MF: $C_4H_8O_2S$ **MP[°C]**: 27.6   **Den.[g/cm³]**: 1.2723^{18}
MW: 120.17   **BP[°C]**: 287.3 n_D: 1.4833^{18}
Sol.: chl 3
Vap. Press. [kPa]: 1.83^{150}
Therm. Prop.[kJ/mol]: $\Delta_{fus}H = 1.43$
C_p **(liq.) [J/mol °C]**: 180^{30}
γ **[mN/m]**: 35.5^{30}
Elec. Prop.: $\mu = (4.8\ D)$; $\varepsilon = 43.26^{30}$; $IP = 9.80\ eV$
η **[mPa s]**: 6.28^{50}; 3.82^{75}; 2.56^{100}
MS: 41(100) 28(94) 56(82) 55(72) 120(37) 27(32) 39(19) 29(17) 26(11) 48(5)
IR [cm⁻¹]: 2960 1450 1420 1300 1250 1140 1110 1030 910 730 670 580 530
    440 390
¹³C NMR [ppm]: 22.8 51.5 diox
¹H NMR [ppm]: 2.2 3.0 CDCL₃
Flammability: Flash pt. = 177°C

475. α-Terpinene

Syst. Name: 1,3-Cyclohexadiene, 1-methyl-4-(1-methylethyl)-

Synonyms: 4-Isopropyl-1-methyl-1,3-cyclohexadiene

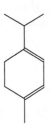

CASRN: 99-86-5   **Beil Ref.**: 4-05-00-00435
Beil RN: 1853379
MF: $C_{10}H_{16}$ **BP[°C]**: 174   **Den.[g/cm³]**: 0.8375^{19}
MW: 136.24 n_D: 1.477^{19}
Sol.: H_2O 1; EtOH 5; eth
   5
Elec. Prop.: ε = 2.45^{25}
MS: 121(100) 93(85) 136(43) 91(40) 77(34) 39(33) 27(33) 79(27) 41(26)
   43(18)
IR [cm⁻¹]: 2940 2860 1670 1470 1450 1390 1370 1330 1280 1240 1160 1090
   1080 1050 1020 990 890 830 760 730 680 670
UV [nm]: 265(7943) EtOH

476. α-Terpineol

Syst. Name: 3-Cyclohexene-1-methanol, α,α,4-
trimethyl-, (±)-

CASRN: 2438-12-2   **Beil Ref.**: 4-06-00-00252
Beil RN: 3195616
MF: $C_{10}H_{18}O$ **MP[°C]**: 40.5   **Den.[g/cm^3]**: 0.9337[20]
MW: 154.25   **BP[°C]**: 220 n_D: 1.4831[20]
Sol.: ace 4; bz 4; eth 4;
    EtOH 4
η **[mPa s]**: 59[20]; 26[25]
MS: 59(100) 93(85) 121(73) 43(72) 136(66) 28(54) 81(46) 154(6)
IR [cm^{-1}]: 3620 3340 3010 2960 2920 2830 1440 1380 1370 1250 1160 1130
    1100 1040 1020 940 930 920 840 800 780
Raman [cm^{-1}]: 3300 3050 3010 2970 2910 2870 2720 2670 1680 1610 1450
    1440 1370 1340 1310 1300 1250 1220 1170 1160 1140 1110 1080 1050 1020
    1000 980 950 920 870 840 800 780 760 730 710 650 640 600 570 550 510
    480 430 410 370 320 290 260 180
UV [nm]: 260(2) 215(178) EtOH
^{1}H NMR [ppm]: 1.1 1.7 5.3 CCl_4
Flammability: Flash pt. = 90°C

477. Terpinolene

Syst. Name: Cyclohexene, 1-methyl-4-(1-methylethylidene)-

Synonyms: *p*-Mentha-1,4(8)-diene; Isoterpinene

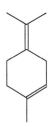

CASRN: 586-62-9   **DOT No.**: 2541
Beil RN: 1851203   **Beil Ref.**: 4-05-00-00437
MF: $C_{10}H_{16}$ **BP[°C]**: 186   **Den.[g/cm³]**: 0.8632^{15}
MW: 136.24 n_D: 1.4883^{20}
Sol.: H_2O 1; EtOH 5; eth
  5; bz 3; ctc 3
Vap. Press. [kPa]: 0.090^{25}; 0.438^{50}; 1.68^{75}; 5.31^{100}; 14.5^{125}; 34.9^{150}
Elec. Prop.: $\varepsilon = 2.29^{25}$
MS: 93(100) 121(93) 136(81) 41(65) 39(58) 91(56) 79(56) 27(54) 77(52)
  43(39)
IR [cm⁻¹]: 2960 2900 1440 1370 1250 1220 1200 1170 1150 1100 1050 1040
  920 910 790 760 730
Raman [cm⁻¹]: 3020 2960 2930 2880 2750 2730 1700 1680 1650 1610 1450
  1430 1370 1330 1310 1280 1250 1230 1210 1170 1140 1110 1080 1070 1040
  1020 980 960 930 920 900 860 850 810 800 790 770 730 700 650 620 580
  560 550 520 510 480 460 430 390 360 340 310 280 270 170
¹H NMR [ppm]: 1.7 2.0 2.3 2.7 5.3 CCl_4

478. 1,1,2,2-Tetrabromoethane

Syst. Name: Ethane, 1,1,2,2-tetrabromo-
Synonyms: Acetylene tetrabromide

CASRN: 79-27-6	**Rel. CASRN:** 25167-20-8
Merck No.: 9121	**DOT No.:** 2504
Beil RN: 1098321	**Beil Ref.:** 4-01-00-00162

MF: $C_2H_2Br_4$ **MP[°C]:** 0   **Den.[g/cm^3]:** 2.9655^{20}
MW: 345.65   **BP[°C]:** 243.5 n_D: 1.6353^{20}
Sol.: H_2O 1; EtOH 5; eth 5; ace 3; bz 3; ctc 2; chl 5; $PhNH_2$ 5; HOAc 5
Vap. Press. [kPa]: 0.003^{25}; 6.02^{150}
Therm. Prop.[kJ/mol]: $\Delta_{vap}H = 48.7$
C_p **(liq.) [J/mol °C]:** 165.7^{25}
γ **[mN/m]:** 48.65^{25}
Elec. Prop.: $\mu = (1.4\ D)$; $\varepsilon = 6.72^{30}$
η **[mPa s]:** 9.69^{25}
MS: 265(100) 267(99) 79(37) 186(36) 81(36) 263(35) 105(35) 107(33) 269(32) 184(19)
IR [cm^{-1}]: 2990 1240 1150 1130 1010 710 640 610 540
^{1}H NMR [ppm]: 6.0 CCl_4
Flammability: Ign. temp. = 335°C
TLV/TWA: 1 ppm (14 mg/m^3)

479. 1,1,2,2-Tetrachloro-1,2-difluoroethane

Syst. Name: Ethane, 1,1,2,2-tetrachloro-1,2-difluoro-
Synonyms: R 112; CFC 112

CASRN: 76-12-0　　　　　　**Beil Ref.**: 4-01-00-00146
Beil RN: 1740336
MF: $C_2Cl_4F_2$　　　　**MP[°C]**: 26　　　**Den.[g/cm^3]**: 1.6447^{25}
MW: 203.83　　　　　**BP[°C]**: 92.8　　n_D: 1.4130^{25}
Sol.: H_2O 1; EtOH 3; eth 3; chl 3
Crit. Const.: $T_c = 278°C$
Vap. Press. [kPa]: 1.73^0; 7.51^{25}; 23.5^{50}; 58.7^{75}; 124^{100}
Therm. Prop.[kJ/mol]: $\Delta_{fus}H = 3.70$; $\Delta_{vap}H = 35.0$
C_p **(liq.) [J/mol °C]**: 178.6^{27}
γ **[mN/m]**: 22.73^{30}
Elec. Prop.: $\varepsilon = 2.52^{35}$; IP = 11.30 eV
η **[mPa s]**: 1.21^{25}
MS: 101(100) 103(64) 167(54) 169(52) 117(19) 119(18) 171(17) 105(11) 31(11) 132(9)
IR [cm^{-1}]: 1140 1110 1090 1020 930 900 880 830 780 770 750
TLV/TWA: 500 ppm (4170 mg/m^3)

---

480. 1,1,1,2-Tetrachloro-2,2-difluoroethane

Syst. Name: Ethane, 1,1,1,2-tetrachloro-2,2-difluoro-
Synonyms: Tetrachloro-1,1-difluoroethane; R 112a

CASRN: 76-11-9　　　　　　**Beil Ref.**: 4-01-00-00146
Beil RN: 1699461
MF: $C_2Cl_4F_2$　　　　**MP[°C]**: 40.6　　**Den.[g/cm^3]**: 1.649^{25}
MW: 203.83　　　　　**BP[°C]**: 91.5
Sol.: H_2O 1; EtOH 3; eth 3; chl 3
Vap. Press. [kPa]: 7.36^{25}; 23.6^{50}; 60.2^{75}; 129^{100}
Therm. Prop.[kJ/mol]: $\Delta_{fus}H = 3.99$
MS: 167(100) 169(96) 117(85) 119(82) 171(31) 85(29) 121(26) 82(14) 47(14) 101(13)
IR [cm^{-1}]: 1160 1020 880 850 780 750
TLV/TWA: 500 ppm (4170 mg/m^3)

481. 1,1,1,2-Tetrachloroethane

Syst. Name: Ethane, 1,1,1,2-tetrachloro-

CASRN: 630-20-6   **Beil Ref.**: 4-01-00-00143
Beil RN: 1733216
MF: $C_2H_2Cl_4$ **MP[°C]**: -70.2   **Den.[g/cm^3]**: 1.5406^{20}
MW: 167.85   **BP[°C]**: 130.5 n_D: 1.4821^{20}
Sol.: H_2O 2; EtOH 5; eth
   5; ace 3; bz 3; chl 3
Vap. Press. [kPa]: 1.60^{25}; 5.83^{50}; 16.9^{75}; 41.2^{100}; 87.9^{125}; 169^{150}
C_p **(liq.) [J/mol °C]**: 153.8^{25}
Elec. Prop.: $\varepsilon = 9.22^{-66}$; IP = 11.10 eV
η **[mPa s]**: 3.66^{-25}; 2.20^{0}; 1.44^{25}; 1.01^{50}; 0.741^{75}; 0.570^{100}
MS: 131(100) 133(96) 117(76) 119(73) 95(34) 135(31) 121(23) 97(23) 61(19)
   60(18)
Flammability: Flash pt. = 47°C; flam. lim. = 5-12%
Reg. Lists: CERCLA (RQ = 100 lb.); RCRA U208 (toxic)

482. 1,1,2,2-Tetrachloroethane

Syst. Name: Ethane, 1,1,2,2-tetrachloro-
Synonyms: Acetylene tetrachloride; R 130

CASRN: 79-34-5   **Rel. CASRN**: 25322-20-7
Merck No.: 9125   **DOT No.**: 1702
Beil RN: 969206   **Beil Ref.**: 4-01-00-00144
MF: $C_2H_2Cl_4$ **MP[°C]**: -43.8   **Den.[g/cm³]**: 1.5953^{20}
MW: 167.85   **BP[°C]**: 146.5 n_D: 1.4940^{20}
Sol.: H_2O 2; EtOH 5; eth
   5; ace 3; bz 3; chl 3
Crit. Const.: T_c = 388.00°C
Vap. Press. [kPa]: 0.622^{25}; 2.78^{50}; 9.25^{75}; 24.9^{100}; 57.0^{125}
Therm. Prop.[kJ/mol]: $\Delta_{vap}H$ = 37.6; $\Delta_f H°$(l, 25°C) = -195.0
C_p **(liq.) [J/mol °C]**: 162.3^{25}
γ **[mN/m]**: 35.58^{25}; $d\gamma/dT$ = 0.1268 mN/m °C
Elec. Prop.: μ = 1.32 D; ϵ = 8.50^{20}; IP = 11.62 eV
η **[mPa s]**: 1.84^{15}
MS: 83(100) 85(63) 95(11) 87(10) 168(8) 133(8) 131(8) 96(8) 61(8) 60(8)
IR [cm⁻¹]: 2990 1280 1210 1020 800 740 650 550
Raman [cm⁻¹]: 2990 1310 1280 1250 1220 1200 1020 840 810 770 740 690
   660 590 550 410 370 360 330 290 250 230 180 90
¹H NMR [ppm]: 5.9 $CDCl_3$
Flammability: Flash pt. = 62°C; flam. lim. = 20-54%
TLV/TWA: 1 ppm (6.9 mg/m³)
Reg. Lists: CERCLA (RQ = 100 lb.); SARA 313 (0.1%); RCRA U209 (toxic);
   confirmed or suspected carcinogen

483. Tetrachloroethylene

Syst. Name: Ethene, tetrachloro-
Synonyms: Perclene

CASRN: 127-18-4   **DOT No.:** 1897
Merck No.: 9126   **Beil Ref.:** 4-01-00-00715
Beil RN: 1361721
MF: C_2Cl_4 **MP[°C]:** -22.3   **Den.[g/cm^3]:** 1.6227^{20}
MW: 165.83   **BP[°C]:** 121.3 n_D: 1.5053^{20}
Sol.: H_2O 1; EtOH 5; eth
    5; bz 5
Crit. Const.: T_c = 347.1°C
Vap. Press. [kPa]: 2.42^{25}; 8.27^{50}; 22.9^{75}; 54.2^{100}; 113^{125}
Therm. Prop.[kJ/mol]: $\Delta_{fus}H$ = 10.56; $\Delta_{vap}H$ = 34.7; $\Delta_f H°$(l, 25°C) = -50.6
C_p **(liq.) [J/mol °C]:** 143.4^{25}
γ **[mN/m]:** 31.30^{25}
Elec. Prop.: μ = 0; ε = 2.27^{30}; IP = 9.32 eV
η **[mPa s]:** 1.11^0; 0.844^{25}; 0.663^{50}; 0.535^{75}; 0.442^{100}
k **[W/m °C]:** 0.117^0; 0.110^{25}; 0.104^{50}; 0.097^{75}; 0.091^{100}
MS: 166(100) 164(82) 131(71) 129(71) 168(45) 94(38) 47(31) 96(24) 133(20)
    59(17)
IR [cm^{-1}]: 900 800 780 750
Raman [cm^{-1}]: *1570 1020 570 510 460 440 350 230
Flammability: Flash pt. = 45°C
TLV/TWA: 25 ppm (170 mg/m^3)
Reg. Lists: CERCLA (RQ = 100 lb.); SARA 313 (0.1%); RCRA U210 (toxic);
    confirmed or suspected carcinogen

484. Tetrachloromethane

Syst. Name: Methane, tetrachloro-
Synonyms: Carbon tetrachloride; R 10

CASRN: 56-23-5   **DOT No.**: 1846
Merck No.: 1822   **Beil Ref.**: 4-01-00-00056
Beil RN: 1098295
MF: CCl_4 **MP[°C]**: -23   **Den.[g/cm^3]**: 1.5940^{20}
MW: 153.82   **BP[°C]**: 76.8 n_D: 1.4601^{20}
Sol.: H_2O 1; EtOH 3; eth
  5; ace 3; bz 5; chl 5
Crit. Const.: T_c = 283.5°C; P_c = 4.516 MPa; V_c = 276 cm^3
Vap. Press. [kPa]: 0.959^{-25}; 4.49^0; 15.2^{25}; 41.6^{50}; 96.3^{75}; 197^{100}; 364^{125}
Therm. Prop.[kJ/mol]: $\Delta_{fus}H$ = 3.28; $\Delta_{vap}H$ = 29.8; $\Delta_f H°$(l, 25°C) = -128.2
C_p (liq.) [J/mol °C]: 130.7^{25}
γ [mN/m]: 26.43^{25}; dγ/dT = 0.1224 mN/m °C
Elec. Prop.: μ = 0; ε = 2.24^{20}; IP = 11.47 eV
η [mPa s]: 1.32^0; 0.908^{25}; 0.656^{50}; 0.494^{75}
k [W/m °C]: 0.104^0; 0.099^{25}; 0.093^{50}; 0.088^{75}
MS: 117(100) 119(98) 121(31) 82(24) 47(23) 84(16) 35(14) 49(8) 28(8) 36(6)
IR [cm^{-1}]: 1550 1250 1220 1000 970 780
Raman [cm^{-1}]: 790 760 460 310 220
^{13}C NMR [ppm]: 96.7
TLV/TWA: 5 ppm (31 mg/m^3)
Reg. Lists: CERCLA (RQ = 10 lb.); SARA 313 (0.1%); RCRA U211 (toxic);
  confirmed or suspected carcinogen

485. Tetraethylene glycol

Syst. Name: Ethanol, 2,2'-[oxybis(2,1-ethanediyloxy)]bis-

HO$\sim$O$\sim$O$\sim$O$\sim$OH

Synonyms: 3,6,9-Trioxa-1,11-undecanediol
CASRN: 112-60-7   **Beil Ref.:** 4-01-00-02403
Beil RN: 1634320
MF: $C_8H_{18}O_5$ **MP[°C]:** -6.2   **Den.[g/cm^3]:** 1.1285[15]
MW: 194.23   **BP[°C]:** 328 n_D: 1.4577[20]
Sol.: H_2O 4; EtOH 3; eth 3; ctc 3; diox 3
Vap. Press. [kPa]: 0.093[150]
Therm. Prop.[kJ/mol]: $\Delta_f H°$(l, 25°C) = -981.7
C_p **(liq.) [J/mol °C]:** 428.8[25]
Elec. Prop.: ε = 20.44[20]
MS: 45(100) 89(10) 44(8) 43(6) 31(6) 29(6) 27(6) 101(5) 75(5) 28(5)
IR [cm^{-1}]: 3400 2980 1720 1640 1450 1350 1290 1240 1120 1070 940 880 830
^{1}H NMR [ppm]: 3.5 3.6 CCl_4
Flammability: Flash pt. = 182°C

---

486. Tetraethylsilane

Syst. Name: Silane, tetraethyl-

CASRN: 631-36-7   **Beil Ref.:** 4-04-00-03895
Beil RN: 1732768
MF: $C_8H_{20}Si$ **BP[°C]:** 154.7   **Den.[g/cm^3]:** 0.7658[20]
MW: 144.33 n_D: 1.4268[20]
Sol.: H_2O 1
Crit. Const.: T_c = 330.6°C; P_c = 2.602 MPa
Vap. Press. [kPa]: 0.157[0]; 0.739[25]; 2.69[50]; 8.00[75]; 20.4[100]; 45.9[125]; 93.3[150]
Therm. Prop.[kJ/mol]: $\Delta_{fus}H$ = 13.01
C_p **(liq.) [J/mol °C]:** 298.1[25]
Elec. Prop.: μ = 0; ε = 2.09[20]; IP = 8.90 eV
η **[mPa s]:** 0.649[20]
MS: 87(100) 115(82) 59(35) 58(7) 116(6) 57(6) 31(6) 144(5) 88(5) 43(5)
IR [cm^{-1}]: 2940 1470 1430 1250 1080 1020 980 740
^{13}C NMR [ppm]: 2.0 6.2
^{1}H NMR [ppm]: 0.6 0.9 CCl_4

487. Tetrahydrofuran

Syst. Name: Furan, tetrahydro-
Synonyms: Oxolane; Tetramethylene oxide; 1,4-
Epoxybutane

CASRN: 109-99-9   **DOT No.**: 2056
Merck No.: 9144   **Beil Ref.**: 5-17-01-00027
Beil RN: 102391
MF: C_4H_8O **MP[°C]**: -108.3   **Den.[g/cm^3]**: 0.8892^{20}
MW: 72.11   **BP[°C]**: 65 n_D: 1.4050^{20}
Sol.: H_2O 3; EtOH 4; eth
   4; ace 4; bz 4; chl 3;
   os 4
Crit. Const.: T_c = 267.0°C; P_c = 5.19 MPa; V_c = 224 cm^3
Vap. Press. [kPa]: 21.6^{25}; 58.6^{50}; 135^{75}; 272^{100}
Therm. Prop.[kJ/mol]: $\Delta_{fus}H$ = 8.54; $\Delta_{vap}H$ = 29.8; $\Delta_f H°$(l, 25°C) = -216.2
C_p (liq.) [J/mol °C]: 124.1^{25}
γ [mN/m]: 26.4^{25}
Elec. Prop.: μ = 1.75 D; ϵ = 7.52^{22}; IP = 9.41 eV
η [mPa s]: 0.849^{-25}; 0.605^0; 0.456^{25}; 0.359^{50}
k [W/m °C]: 0.132^{-25}; 0.126^0; 0.120^{25}; 0.114^{50}
MS: 42(100) 41(52) 27(33) 72(29) 71(27) 39(24) 43(22) 29(22) 40(13) 15(10)
IR [cm^{-1}]: 2940 2700 1470 1370 1180 1060 910
Raman [cm^{-1}]: 2960 2940 2870 2710 2650 1480 1450 1360 1230 1170 1130
   1070 1020 910 580 280
^{13}C NMR [ppm]: 25.8 67.9 CDCl$_3$
^{1}H NMR [ppm]: 1.8 3.8 CDCl$_3$
Flammability: Flash pt. = -14°C; ign. temp. = 321°C; flam. lim. = 2-11.8%
TLV/TWA: 200 ppm (590 mg/m^3)
Reg. Lists: CERCLA (RQ = 1000 lb.); RCRA U213 (toxic)

488. Tetrahydrofurfuryl alcohol

Syst. Name: 2-Furanmethanol, tetrahydro-
Synonyms: Tetrahydro-2-furancarbinol

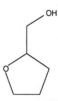

CASRN: 97-99-4   **Beil Ref.**: 5-17-03-00115
Merck No.: 9146
Beil RN: 102723
MF: $C_5H_{10}O_2$ **MP[°C]**: <-80   **Den.[g/cm^3]**: 1.0524^{20}
MW: 102.13   **BP[°C]**: 178 n_D: 1.4520^{20}
Sol.: ace 4; eth 4
Vap. Press. [kPa]: 0.100^{25}; 20.1^{125}; 45.8^{150}
Therm. Prop.[kJ/mol]: $\Delta_{vap}H$ = 45.2; $\Delta_f H°$(l, 25°C) = -435.7
C_p **(liq.) [J/mol °C]**: 181.2^{25}
γ **[mN/m]**: 37^{25}
Elec. Prop.: μ = (2.1 D); ϵ = 13.48^{30}
η **[mPa s]**: 6.24^{20}
MS: 71(100) 43(79) 41(44) 27(41) 29(36) 31(31) 42(22) 39(22) 44(12) 18(12)
Flammability: Flash pt. = 75°C; ign. temp. = 282°C; flam. lim. = 1.5-9.7%

489. 1,2,3,4-Tetrahydronaphthalene
Syst. Name: Naphthalene, 1,2,3,4-tetrahydro-
Synonyms: Tetralin; Benzocyclohexane

CASRN: 119-64-2   **Beil Ref.:** 4-05-00-01388
Merck No.: 9152
Beil RN: 1446407
MF: $C_{10}H_{12}$ **MP[°C]:** -35.7   **Den.[g/cm^3]:** 0.9660^{25}
MW: 132.21   **BP[°C]:** 207.6 n_D: 1.5413^{20}
Sol.: H_2O 1; EtOH 4; eth
4; chl 3; PhNH$_2$ 3
Crit. Const.: T_c = 447°C; P_c =3.65 MPa; V_c = 408 cm^3
Vap. Press. [kPa]: 0.050^{25}; 0.289^{50}; 1.11^{75}; 3.48^{100}; 9.21^{125}; 21.4^{150}
Therm. Prop.[kJ/mol]: $\Delta_{fus}H$ = 12.45; $\Delta_{vap}H$ = 43.9; $\Delta_f H°$(l, 25°C) = -29.2
C_p (liq.) [J/mol °C]: 217.5^{25}
γ [mN/m]: 33.17^{25}; dγ/dT = 0.0954 mN/m °C
Elec. Prop.: μ = 0; ϵ = 2.77^{25}; IP = 8.47 eV
η [mPa s]: 2.14^{25}
MS: 104(100) 132(53) 91(43) 51(17) 39(17) 131(15) 117(15) 115(14) 78(13)
77(13)
IR [cm^{-1}]: 3020 2930 1690 1600 1580 1500 1460 1440 1360 1280 1250 1110
1070 1040 950 810 750
Raman [cm^{-1}]: 3090 3060 2990 2950 2930 2900 2880 2840 2670 1600 1580
1460 1450 1430 1400 1380 1360 1340 1320 1300 1290 1240 1200 1160 1110
1080 1070 1040 1010 990 940 900 870 820 800 730 700 580 500 450 430
310 260 160
UV [nm]: 274(603) 267(550) EtOH
^{13}C NMR [ppm]: 23.6 29.5 125.5 129.0 136.8 diox
^{1}H NMR [ppm]: 1.8 2.8 7.1 CDCl$_3$
Flammability: Flash pt. = 71°C; ign. temp. = 385°C; flam. lim. = 0.8-5.0%

490. Tetrahydropyran

Syst. Name: 2H-Pyran, tetrahydro-
Synonyms: Oxane; Pentamethylene oxide

CASRN: 142-68-7   **Beil Ref.:** 5-17-01-00064
Merck No.: 9149
Beil RN: 102436
MF: $C_5H_{10}O$ **MP[°C]:** -45   **Den.[g/cm³]:** 0.8814^{20}
MW: 86.13   **BP[°C]:** 88 n_D: 1.4200^{20}
Sol.: EtOH 3; eth 3; bz
    3; ctc 3
Crit. Const.: T_c = 299.1°C; P_c = 4.77 MPa; V_c = 263 cm³
Vap. Press. [kPa]: 2.61^0; 9.54^{25}; 27.5^{50}
Therm. Prop.[kJ/mol]: $\Delta_{vap}H$ = 31.2; $\Delta_f H°$(l, 25°C) = -258.3
C_p **(liq.) [J/mol °C]:** 156.5^{25}
Elec. Prop.: μ = 1.74 D; ε = 5.66^{20}; IP = 9.25 eV
η **[mPa s]:** 0.764^{25}
MS: 41(100) 28(64) 56(57) 45(57) 29(51) 27(49) 85(47) 86(42) 39(28) 55(23)
IR [cm⁻¹]: 2940 2860 1450 1390 1350 1300 1280 1270 1210 1160 1090 1050
    1040 1020 970 870 860 820 780
Raman [cm⁻¹]: 2960 2930 2890 2850 2750 2710 2670 2640 1450 1440 1380
    1350 1300 1280 1260 1200 1170 1150 1050 1030 1010 870 850 820 500 470
    430 400 260
UV [nm]: 187(631) 175(1995) gas
¹³C NMR [ppm]: 24.2 27.2 68.6 diox
¹H NMR [ppm]: 1.6 3.6 CCl₄
Flammability: Flash pt. = -20°C

491. Tetrahydrothiophene
Syst. Name: Thiophene, tetrahydro-
Synonyms: Thiacyclopentane; Diethylene sulfide

CASRN: 110-01-0   **DOT No.**: 2412
Beil RN: 102392   **Beil Ref.**: 5-17-01-00036
MF: C_4H_8S **MP[°C]**: -96.1   **Den.[g/cm^3]**: 0.9987^{20}
MW: 88.17   **BP[°C]**: 121.0   **n_D**: 1.4871^{18}
Sol.: H_2O 1; EtOH 5; eth
  5; ace 5; bz 5; chl 3;
  os 5
Crit. Const.: T_c = 358.9°C
Vap. Press. [kPa]: 2.45^{25}; 8.32^{50}; 23.1^{75}; 54.5^{100}; 113^{125}; 214^{150}
Therm. Prop.[kJ/mol]: $\Delta_{vap}H$ = 34.7; $\Delta_f H°$(l, 25°C) = -72.9
γ **[mN/m]**: 35.0^{25}
Elec. Prop.: μ = (1.9 D); IP = 8.47 eV
η **[mPa s]**: 0.973^{25}; 0.912^{50}
MS: 60(100) 88(54) 45(37) 46(32) 47(26) 27(24) 59(18) 87(16) 39(14) 54(13)
IR [cm^{-1}]: 2940 1450 1300 1250 1190 1120 1050 1030 950 880 820 690
Raman [cm^{-1}]: 2950 2910 2860 1460 1440 1270 1260 1210 1130 1030 960 880
  820 690 520 470
UV [nm]: 210(1000) EtOH
^{13}C NMR [ppm]: 31.2 31.4 diox
^{1}H NMR [ppm]: 1.9 2.8 CDCl$_3$

492. 2,2,3,3-Tetramethylbutane

Syst. Name: Butane, 2,2,3,3-tetramethyl-

CASRN: 594-82-1   **DOT No.**: 1262
Beil RN: 1696864   **Beil Ref.**: 4-01-00-00447
MF: C_8H_{18} **MP[°C]**: 100.7   **Den.[g/cm^3]**: 0.8242[20]
MW: 114.23   **BP[°C]**: 106.4
Sol.: H_2O 1; eth 3; chl 3
Crit. Const.: T_c = 294.7°C; P_c = 2.87 MPa; V_c = 461 cm^3
Vap. Press. [kPa]: 0.555[0]; 2.76[25]; 10.6[50]; 32.6[75]; 84.5[100]; 165[125]
Therm. Prop.[kJ/mol]: $\Delta_{fus}H$ = 7.54; $\Delta_f H°$(s, 25°C) = -268.9
Elec. Prop.: μ = 0; IP = 9.80 eV
MS: 57(100) 41(28) 56(27) 43(18) 29(16) 39(7) 27(7) 99(6) 55(5) 15(5)
IR [cm^{-1}]: 2970 2740 1720 1480 1380 1180 1040 930 800
^{13}C NMR [ppm]: 25.6 35.0 diox
^{1}H NMR [ppm]: 1.0 CDCl$_3$

---

493. 2,2,3,3-Tetramethylpentane

Syst. Name: Pentane, 2,2,3,3-tetramethyl-

CASRN: 7154-79-2   **DOT No.**: 1920
Beil RN: 1730916   **Beil Ref.**: 4-01-00-00463
MF: C_9H_{20} **MP[°C]**: -9.8   **Den.[g/cm^3]**: 0.7530[25]
MW: 128.26   **BP[°C]**: 140.2 n_D: 1.4236[20]
Crit. Const.: T_c = 334.6°C; P_c = 2.741 MPa
Vap. Press. [kPa]: 13.0[75]; 31.5[100]; 67.3[125]
Therm. Prop.[kJ/mol]: $\Delta_{fus}H$ = 2.33; $\Delta_f H°$(l, 25°C) = -278.3
C_p (liq.) [J/mol °C]: 271.5[25]
Elec. Prop.: μ = 0
η [mPa s]: 0.61[38]
MS: 57(100) 43(63) 71(43) 41(31) 70(30) 29(22) 55(14) 27(13) 99(10) 56(9)
Raman [cm^{-1}]: * 1540 1478 1460 1239 1055 1015 919 860 658 467 363
^{13}C NMR [ppm]: 9.0 20.6 25.6 28.8 36.0 37.3 diox
Flammability: Flash pt. < 21°C; ign. temp. = 430°C; flam. lim. = 0.8-4.9%

494. 2,2,3,4-Tetramethylpentane

Syst. Name: Pentane, 2,2,3,4-tetramethyl-

CASRN: 1186-53-4   **DOT No.:** 1920
Beil RN: 1730911   **Beil Ref.:** 4-01-00-00463
MF: C_9H_{20} **MP[°C]:** -121.0   **Den.[g/cm^3]:** 0.7389[20]
MW: 128.26   **BP[°C]:** 133.0 n_D: 1.4147[20]
Crit. Const.: T_c = 319.6°C; P_c = 2.602 MPa
Vap. Press. [kPa]: 16.4[75]; 39.0[100]; 81.7[125]
Therm. Prop.[kJ/mol]: $\Delta_{fus}H$ = 0.50; $\Delta_f H°$(1, 25°C) = -277.7
Elec. Prop.: μ = 0
η **[mPa s]:** 0.60[38]
MS: 57(100) 43(67) 56(54) 41(34) 29(15) 71(12) 70(10) 27(10) 85(9) 55(9)
^{13}C NMR [ppm]: 11.6 17.3 24.5 27.4 28.2 34.0 47.9 diox
Flammability: Flash pt. < 21°C

---

495. 2,2,4,4-Tetramethylpentane

Syst. Name: Pentane, 2,2,4,4-tetramethyl-
Synonyms: Di-*tert*-butylmethane

CASRN: 1070-87-7   **DOT No.:** 1920
Beil RN: 1730923   **Beil Ref.:** 4-01-00-00464
MF: C_9H_{20} **MP[°C]:** -66.5   **Den.[g/cm^3]:** 0.7195[20]
MW: 128.26   **BP[°C]:** 122.2 n_D: 1.4069[20]
Sol.: H_2O 1; EtOH 4; bz
    4; os 4
Crit. Const.: T_c = 301.6°C; P_c = 2.485 MPa
Vap. Press. [kPa]: 8.75[50]; 23.4[75]; 53.7[100]
Therm. Prop.[kJ/mol]: $\Delta_{fus}H$ = 9.75; $\Delta_{vap}H$ = 32.5; $\Delta_f H°$(1, 25°C) = -280.0
C_p (liq.) [J/mol °C]: 266.3[25]
Elec. Prop.: μ = 0
η **[mPa s]:** 0.58[38]
MS: 57(100) 56(31) 41(23) 29(16) 43(8) 27(7) 39(6) 58(4) 55(4) 113(2)
^{13}C NMR [ppm]: 31.8 32.4 56.5 diox

496. 2,3,3,4-Tetramethylpentane
Syst. Name: Pentane, 2,3,3,4-tetramethyl-

CASRN: 16747-38-9   DOT No.: 1920
Beil RN: 1730918   Beil Ref.: 4-01-00-00464
MF: C_9H_{20} MP[°C]: -102.1   Den.[g/cm³]: 0.7547^{20}
MW: 128.26   BP[°C]: 141.5 n_D: 1.4222^{20}
Crit. Const.: T_c = 334.6°C; P_c = 2.716 MPa
Vap. Press. [kPa]: 12.4^{75}; 30.2^{100}; 64.8^{125}
Therm. Prop.[kJ/mol]: $\Delta_{fus}H$ = 9.00; $\Delta_f H°$(l, 25°C) = -277.9
Elec. Prop.: μ = 0
MS: 43(100) 85(60) 57(33) 41(30) 84(28) 69(15) 27(14) 39(8) 29(8) 55(7)
^{13}C NMR [ppm]: 17.2 18.9 33.6 37.1 diox

---

497. Tetramethylsilane
Syst. Name: Silane, tetramethyl-
Synonyms: TMS

CASRN: 75-76-3   DOT No.: 2749
Beil RN: 1696908   Beil Ref.: 4-04-00-03875
MF: $C_4H_{12}Si$ MP[°C]: -99.0   Den.[g/cm³]: 0.648^{19}
MW: 88.22   BP[°C]: 26.6 n_D: 1.3587^{20}
Sol.: H_2O 1; EtOH 4; eth
    4; sulf 1
Crit. Const.: T_c = 175.49°C; P_c = 2.821 MPa; V_c = 362 cm³
Vap. Press. [kPa]: 94.2^{25}; 215^{50}; 429^{75}; 769^{100}; 1271^{125}; 1964^{150}
Therm. Prop.[kJ/mol]: $\Delta_{fus}H$ = 6.88; $\Delta_f H°$(l, 25°C) = -264.0
C_p (liq.) [J/mol °C]: 204.1^{25}
Elec. Prop.: μ = 0; ε = 1.92^{20}; IP = 9.80 eV
MS: 73(100) 43(14) 45(12) 74(8) 29(7) 15(5) 75(4) 44(4) 42(4) 31(4)
IR [cm⁻¹]: 2980 2890 1420 1280 1240 850 690
^{1}H NMR [ppm]: 0.0 CCl_4

498. Tetramethylurea

Syst. Name: Urea, tetramethyl-

CASRN: 632-22-4   **Beil Ref.**: 4-04-00-00225
Merck No.: 9160
Beil RN: 773898
MF: $C_5H_{12}N_2O$ **MP[°C]**: -0.6   **Den.[g/cm³]**: 0.9687^{20}
MW: 116.16   **BP[°C]**: 176.5 n_D: 1.4496^{23}
Sol.: EtOH 2; eth 2; ctc
  2
Therm. Prop.[kJ/mol]: $\Delta_{fus}H = 14.10$
Elec. Prop.: $\mu = (3.5\ D)$; $\varepsilon = 23.10^{20}$
η **[mPa s]**: 139^{25}
MS: 72(100) 44(27) 116(24) 42(13) 15(13) 17(7) 28(5) 73(4) 56(4) 18(4)
IR [cm⁻¹]: 2950 2880 1640 1500 1370 1130 1050 910 780
UV [nm]: 217(1950) cyhex
¹³C NMR [ppm]: 38.5 165.4 CDCl₃
¹H NMR [ppm]: 2.8 CCl₄
Flammability: Flash pt. = 77°C

499. 2,2'-Thiodiethanol

Syst. Name: Ethanol, 2,2'-thiobis-
Synonyms: Bis(2-hydroxyethyl) sulfide

HO~~~S~~~OH

CASRN: 111-48-8   **Beil Ref.**: 4-01-00-02437
Merck No.: 9259
Beil RN: 1236325
MF: $C_4H_{10}O_2S$ **MP[°C]**: -10.2   **Den.[g/cm^3]**: 1.1793^{25}
MW: 122.19   **BP[°C]**: 282 n_D: 1.5211^{20}
Sol.: H_2O 5; EtOH 5; eth
   3; bz 2; chl 5; AcOEt
   5
Vap. Press. [kPa]: 0.726^{100}; 1.26^{125}; 2.04^{150}
γ [mN/m]: 53.8^{20}
Elec. Prop.: ε = 28.61^{20}
MS: 61(100) 45(68) 31(38) 104(36) 91(34) 47(26) 44(26) 27(24) 60(18) 43(17)
IR [cm^{-1}]: 3330 2940 1410 1280 1220 1160 1040 1010 940 770 690
^{1}H NMR [ppm]: 2.8 3.8 4.2 CDCl$_3$
Flammability: Flash pt. = 160°C; ign. temp. = 298°C

500. Thiophene

Syst. Name: Thiophene
Synonyms: Thiofuran

CASRN: 110-02-1
Merck No.: 9283
Beil RN: 103222
MF: C_4H_4S
MW: 84.14
Sol.: EtOH 5; eth 5; ace
5; bz 5; ctc 5; chl 2; os
5; hp 5; py 5; diox 5;
tol5

DOT No.: 2414
Beil Ref.: 5-17-01-00297

MP[°C]: -39.4
BP[°C]: 84.0

Den.[g/cm^3]: 1.0649^{20}
n_D: 1.5289^{20}

Crit. Const.: $T_c = 306.3°C$; $P_c = 5.69$ MPa; $V_c = 219$ cm^3
Vap. Press. [kPa]: 2.85^0; 10.6^{25}; 31.1^{50}; 75.8^{75}; 161^{100}; 307^{125}
Therm. Prop.[kJ/mol]: $\Delta_{fus}H = 5.09$; $\Delta_{vap}H = 31.5$; $\Delta_f H°(l, 25°C) = 80.2$
C_p **(liq.) [J/mol °C]**: 123.8^{25}
γ **[mN/m]**: 30.68^{25}; $d\gamma/dT = 0.1328$ mN/m °C
Elec. Prop.: $\mu = 0.55$ D; $\epsilon = 2.74^{20}$; IP = 8.87 eV
η **[mPa s]**: 0.613^{25}
k **[W/m °C]**: 0.199^{25}; 0.195^{50}; 0.191^{75}; 0.186^{100}
MS: 84(100) 58(65) 45(58) 39(29) 57(13) 38(8) 69(7) 37(7) 83(6) 50(6)
IR [cm^{-1}]: 3030 1590 1390 1250 1080 1030 900 870 830 710
Raman [cm^{-1}]: 3110 3090 3000 1410 1360 1080 1040 840 760 610 460
UV [nm]: 237 233 MeOH
^{13}C NMR [ppm]: 124.9 126.7 CDCl$_3$
^{1}H NMR [ppm]: 7.1 7.3 CDCl$_3$
Flammability: Flash pt. = -1°C

501. Toluene

Syst. Name: Benzene, methyl-
Synonyms: Methylbenzene; Phenylmethane

CASRN: 108-88-3
Merck No.: 9455
Beil RN: 635760
MF: C_7H_8
MW: 92.14
Sol.: H_2O 1; EtOH 5; eth 5; ace 3; bz 5; chl 3; CS_2 3; lig 3

DOT No.: 1294
Beil Ref.: 4-05-00-00766

MP[°C]: -94.9
BP[°C]: 110.6

Den.[g/cm^3]: 0.8669^{20}
n_D: 1.4961^{20}

Crit. Const.: $T_c = 318.60°C$; $P_c = 4.108$ MPa; $V_c = 316$ cm^3
Vap. Press. [kPa]: 3.79^{25}; 12.3^{50}; 32.6^{75}; 74.6^{100}; 149^{125}; 274^{150}
Therm. Prop.[kJ/mol]: $\Delta_{fus}H = 6.85$; $\Delta_{vap}H = 33.2$; $\Delta_f H°(l, 25°C) = 12.4$
C_p (liq.) [J/mol °C]: 157.0^{25}
γ [mN/m]: 27.93^{25}; $d\gamma/dT = 0.1189$ mN/m °C
Elec. Prop.: $\mu = 0.375$ D; $\varepsilon = 2.38^{23}$; IP = 8.82 eV
η [mPa s]: 1.17^{-25}; 0.778^{0}; 0.560^{25}; 0.424^{50}; 0.333^{75}; 0.270^{100}
k [W/m °C]: 0.1461^{-25}; 0.1386^{0}; 0.1311^{25}; 0.1236^{50}; 0.1161^{75}
MS: 91(100) 92(73) 39(20) 65(14) 63(11) 51(11) 50(7) 27(6) 93(5) 90(5)
IR [cm^{-1}]: 3040 2930 1610 1500 1460 1380 1180 1090 1040 900 730 700
Raman [cm^{-1}]: 3060 3000 2980 2910 2870 2730 1610 1580 1380 1210 1180 1150 1030 1000 780 620 520 220
UV [nm]: 268(222) 264(167) 261(238) 260(198) 255(175) 254(163) 249 208 MeOH
^{13}C NMR [ppm]: 21.2 125.5 128.3 129.1 137.7 diox
^{1}H NMR [ppm]: 2.3 7.2 CDCl$_3$
Flammability: Flash pt. = 4°C; ign. temp. = 480°C; flam. lim. = 1.1-7.1%
TLV/TWA: 50 ppm (188 mg/m^3)
Reg. Lists: CERCLA (RQ = 1000 lb.); SARA 313 (1.0%); RCRA U220 (toxic)

502. *o*-Toluidine

Syst. Name: Benzenamine, 2-methyl-
Synonyms: *o*-Methylaniline

CASRN: 95-53-4
Merck No.: 9462
Beil RN: 741981
MF: C$_7$H$_9$N   **MP[°C]**: -16.3   **Den.[g/cm^3]**: 0.9984^{20}
MW: 107.16   **BP[°C]**: 200.3 n_D: 1.5725^{20}
Sol.: H$_2$O 2; EtOH 5; eth
    5; ctc 5
Crit. Const.: T_c = 434°C; P_c = 4.37 MPa
Vap. Press. [kPa]: 0.043^{25}; 9.81^{125}; 23.9^{150}
Therm. Prop.[kJ/mol]: $\Delta_{vap}H$ = 44.6; $\Delta_fH°$(l, 25°C) = -6.3
C_p **(liq.) [J/mol °C]**: 209.6^{29}
γ **[mN/m]**: 40.03^{20}
Elec. Prop.: μ = (1.6 D); ε = 6.14^{25}; IP = 7.44 eV
η **[mPa s]**: 10.3^0; 3.82^{25}; 1.94^{50}; 1.20^{75}; 0.839^{100}
MS: 106(100) 107(83) 77(17) 79(13) 39(12) 53(10) 52(10) 54(9) 51(9) 28(9)
IR [cm^{-1}]: 3470 3380 3220 3020 2920 1620 1580 1500 1470 1440 1380 1300
    1270 1140 1060 1030 980 930 850 750 710
Raman [cm^{-1}]: 3370 3210 3040 2980 2940 2910 2880 2850 2740 2620 1620
    1600 1580 1490 1460 1440 1380 1310 1300 1270 1200 1150 1140 1080 1030
    980 850 740 580 530 510 440 420 310 260 180
UV [nm]: 285 233 cyhex
^{1}H NMR [ppm]: 2.0 3.2 6.7 CCl$_4$
Flammability: Flash pt. = 85°C; ign. temp. = 482°C
TLV/TWA: 2 ppm (8.8 mg/m^3)
Reg. Lists: CERCLA (RQ = 100 lb.); SARA 313 (0.1%); RCRA U328 (toxic);
    confirmed or suspected carcinogen

503. *m*-Toluidine

Syst. Name: Benzenamine, 3-methyl-
Synonyms: *m*-Methylaniline

CASRN: 108-44-1
Merck No.: 9462
Beil RN: 635944
DOT No.: 1708
Beil Ref.: 4-12-00-01813

MF: C_7H_9N **MP[°C]:** -31.2   **Den.[g/cm^3]:** 0.9889^{20}
MW: 107.16   **BP[°C]:** 203.3 n_D: 1.5681^{20}
Sol.: ace 4; bz 4; eth 4;
   EtOH 4
Crit. Const.: T_c = 434°C; P_c = 4.28 MPa
Vap. Press. [kPa]: 0.036^{25}; 8.65^{125}; 21.5^{150}
Therm. Prop.[kJ/mol]: $\Delta_{fus}H$ = 3.89; $\Delta_{vap}H$ = 44.9; $\Delta_f H°$(l, 25°C) = -8.1
C_p **(liq.) [J/mol °C]:** 227.0^{25}
γ **[mN/m]:** 38.02^{20}
Elec. Prop.: μ = (1.4 D); ε = 5.82^{25}; IP = 7.50 eV
η **[mPa s]:** 8.18^0; 3.31^{25}; 1.68^{50}; 1.01^{75}; 0.699^{100}
MS: 106(100) 107(84) 79(17) 77(17) 108(7) 78(6) 80(5) 89(4) 65(4) 53(4)
IR [cm^{-1}]: 3450 3360 3220 3040 2940 1620 1590 1500 1470 1310 1290 1170
   1000 930 870 850 770 690
Raman [cm^{-1}]: 3430 3350 3220 3040 3010 2910 2860 2730 1620 1580 1490
   1460 1370 1310 1290 1160 1110 1080 990 920 840 770 730 530 510 430 290
   230 210
UV [nm]: 286(1460) 236(8900) MeOH
^{1}H NMR [ppm]: 2.2 3.3 6.4 6.9 CCl$_4$
TLV/TWA: 2 ppm (8.8 mg/m^3)

504. *p*-Toluidine

Syst. Name: Benzenamine, 4-methyl-

Synonyms: *p*-Methylaniline

CASRN: 106-49-0   **DOT No.**: 1708

Merck No.: 9462   **Beil Ref.**: 4-12-00-01866

Beil RN: 471281

MF: C_7H_9N **MP[°C]**: 43.7   **Den.[g/cm^3]**: 0.9619^{20}

MW: 107.16   **BP[°C]**: 200.4 n_D: 1.5534^{45}

Sol.: H_2O 2; EtOH 4; eth 3; ace 3; ctc 3; py 4

Crit. Const.: T_c = 433°C; P_c = 4.58 MPa

Vap. Press. [kPa]: 0.191^{-25}; 0.638^0; 1.74^{25}; 4.07^{50}; 8.41^{75}; 15.8^{100}; 27.4^{125}; 44.4^{15}

Therm. Prop.[kJ/mol]: $\Delta_{fus}H$ = 18.22; $\Delta_{vap}H$ = 44.3; $\Delta_f H°$(s, 25°C) = -23.5

γ **[mN/m]**: 36.06^{45}

Elec. Prop.: μ = (1.5 D); ε = 5.06^{60}; IP = 7.24 eV

η **[mPa s]**: 1.94^{45}; 1.42^{60}

MS: 106(100) 107(68) 77(14) 79(9) 36(9) 28(9) 53(7) 52(6) 51(6) 108(5)

IR [cm^{-1}]: 3460 3380 3020 2920 1620 1510 1450 1260 1180 1120 1040 810

Raman [cm^{-1}]: 3440 3350 3230 3060 3040 3020 2930 2860 2740 1620 1380 1320 1280 1220 1180 850 720 650 470 420 340 260 130

UV [nm]: 294 291 288 237 cyhex

^{1}H NMR [ppm]: 2.2 3.3 6.3 6.8 CCl_4

Flammability: Flash pt. = 87°C; ign. temp. = 482°C

TLV/TWA: 2 ppm (8.8 mg/m^3)

Reg. Lists: CERCLA (RQ = 100 lb.); RCRA U353 (toxic); confirmed or suspected carcinogen

505. α-Tolunitrile

Syst. Name: Benzeneacetonitrile
Synonyms: Benzyl cyanide

CASRN: 140-29-4   **DOT No.:** 2470
Merck No.: 1145   **Beil Ref.:** 4-09-00-01663
Beil RN: 385941
MF: C_8H_7N **MP[°C]:** -23.8   **Den.[g/cm³]:** 1.0205^{15}
MW: 117.15   **BP[°C]:** 233.5 n_D: 1.5211^{25}
Vap. Press. [kPa]: 0.318^{75}; 1.13^{100}; 3.38^{125}; 8.77^{150}
γ [mN/m]: 41.36^{20}
Elec. Prop.: μ = (3.5 D); ε = 17.87^{26}
η [mPa s]: 2.16^{20}
MS: 117(100) 90(43) 116(35) 89(22) 51(13) 39(11) 63(10) 118(9) 91(8) 50(8)
IR [cm⁻¹]: 3060 3030 2250 1600 1500 1450 1420 1080 1030 1010 940 730 700
620 470 430
Raman [cm⁻¹]: 3080 3070 3020 2920 2260 1600 1590 1420 1300 1190 1150
1030 1010 810 800 740 620 430 350 220 120
UV [nm]: 263(149) 257(194) 251(168) 246(135) MeOH
¹H NMR [ppm]: 0.8 0.9 1.6 2.3 CCl₄
Flammability: Flash pt. = 113°C

506. Triacetin

Syst. Name: 1,2,3-Propanetriol, triethanoate
Synonyms: 1,2,3-Propanetriol triacetate; Glycerol
  triacetate

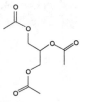

CASRN: 102-76-1   **Beil Ref.**: 4-02-00-00253
Merck No.: 9504
Beil RN: 1792353
MF: $C_9H_{14}O_6$ **MP[°C]**: -78   **Den.[g/cm^3]**: 1.1583^{20}
MW: 218.21   **BP[°C]**: 259 n_D: 1.4301^{20}
Sol.: H_2O 2; EtOH 5; eth
  5; ace 4; bz 5; chl 5;
  lig 2; CS_2 2
Vap. Press. [kPa]: 0.003^{50}
Therm. Prop.[kJ/mol]: $\Delta_{vap}H$ = 57.8; $\Delta_f H°$(l, 25°C) = -1331
C_p (liq.) [J/mol °C]: 384.7^{25}
Elec. Prop.: ε = 7.11^{20}
η [mPa s]: 16^{25}
MS: 43(100) 103(44) 145(34) 116(17) 115(13) 44(10) 86(9) 28(8) 73(7) 42(7)
IR [cm^{-1}]: 2960 1740 1430 1360 1210 1040
^{1}H NMR [ppm]: 2.1 4.2 4.3 5.2 CDCl$_3$
Flammability: Flash pt. = 138°C; ign. temp. = 433°C; lower flam. lim. = 1.0%

507. Tribromomethane

Syst. Name: Methane, tribromo-
Synonyms: Bromoform

CASRN: 75-25-2   **DOT No.**: 2515
Merck No.: 1407   **Beil Ref.**: 4-01-00-00082
Beil RN: 1731048
MF: $CHBr_3$ **MP[°C]**: 8.0   **Den.[g/cm^3]**: 2.899^{15}
MW: 252.73   **BP[°C]**: 149.1 n_D: 1.6005^{15}
Sol.: H_2O 2; EtOH 5; eth
  5; bz 3; ctc 2; chl 3;
  lig 3
Vap. Press. [kPa]: 0.726^{25}; 2.83^{50}; 8.77^{75}; 22.7^{100}; 51.2^{125}; 103^{150}
Therm. Prop.[kJ/mol]: $\Delta_{vap}H$ = 39.7; $\Delta_f H°(l, 25°C)$ = -28.5
C_p **(liq.) [J/mol °C]**: 130.7^{25}
γ **[mN/m]**: 44.87^{25}; $d\gamma/dT$ = 0.1308 mN/m °C
Elec. Prop.: μ = 0.99 D; ε = 4.40^{10}; IP = 10.48 eV
η **[mPa s]**: 1.86^{25}; 1.37^{50}; 1.03^{75}
MS: 173(100) 171(50) 175(49) 93(22) 91(22) 79(18) 81(17) 94(13) 92(13)
  254(11)
IR [cm^{-1}]: 3010 1140 690 650
Raman [cm^{-1}]: 3010 1290 1140 650 530 270 220 150
UV [nm]: 224(2138) 205(2138) hp
^{13}C NMR [ppm]: 12.3
^{1}H NMR [ppm]: 6.8 CCl_4
Flammability: Flash pt. = 83°C
TLV/TWA: 0.5 ppm (5.2 mg/m^3)
Reg. Lists: CERCLA (RQ = 100 lb.); SARA 313 (1.0%); RCRA U225 (toxic)

508. Tributylamine

Syst. Name: 1-Butanamine, *N,N*-dibutyl-

Synonyms: *N,N*-Dibutyl-1-butanamine

CASRN: 102-82-9   **DOT No.**: 2542

Merck No.: 9530   **Beil Ref.**: 4-04-00-00554

Beil RN: 1698872

MF: $C_{12}H_{27}N$ **MP[°C]**: -70   **Den.[g/cm³]**: 0.7770^{20}

MW: 185.35   **BP[°C]**: 216.5 n_D: 1.4299^{20}

Sol.: H_2O 2; EtOH 4; eth 4; ace 3; bz 3; ctc 2

Vap. Press. [kPa]: 0.010^{25}; 0.881^{75}; 2.69^{100}; 7.12^{125}; 16.8^{150}

Therm. Prop.[kJ/mol]: $\Delta_{vap}H$ = 46.9; $\Delta_f H°$(l, 25°C) = -281.6

γ [mN/m]: 24.39^{25}; dγ/dT = 0.0831 mN/m °C

Elec. Prop.: μ = (0.8 D); ε = 2.34^{20}; IP = 7.40 eV

η [mPa s]: 1.60^{15}

MS: 142(100) 100(19) 143(11) 29(8) 185(7) 57(6) 44(6) 41(6) 30(5) 86(4)

IR [cm⁻¹]: 2940 2860 1470 1370 1300 1270 1180 1090 940 900 780 730

Raman [cm⁻¹]: 2940 2910 2880 2800 2730 1460 1450 1380 1300 1230 1180 1110 1050 980 960 900 880 860 830 810 780 740 570 510 480 430 400 350 270

UV [nm]: 242(1000) $CHCl_3$

¹³C NMR [ppm]: 14.1 20.8 29.5 54.1 $CDCl_3$

¹H NMR [ppm]: 0.9 1.3 2.3 CCl_4

Flammability: Flash pt. = 86°C; flam. lim. = 1-5%

509. Tributyl borate

Syst. Name: Boric acid, tributyl ester
Synonyms: Butyl borate

CASRN: 688-74-4   **Beil Ref.**: 4-01-00-01544
Beil RN: 1703865
MF: $C_{12}H_{27}BO_3$ **MP[°C]**: <-70   **Den.[g/cm^3]**: 0.8567^{20}
MW: 230.16   **BP[°C]**: 234 n_D: 1.4106^{18}
Sol.: EtOH 3; eth 4; bz
  3; MeOH 4
Therm. Prop.[kJ/mol]: $ɔ45_{vap}H = 56.1$
γ **[mN/m]**: 23.00^{25}
Elec. Prop.: $\mu = (0.8\ D)$; $\varepsilon = 2.23^{20}$
η **[mPa s]**: 1.64^{25}
MS: 57(100) 56(43) 28(30) 41(28) 55(26) 29(22) 101(20) 117(16) 187(12)
  131(12)
IR [cm^{-1}]: 2960 2870 1490 1420 1370 1300 1260 1230 1070 1030 970 850 690
  660
Flammability: Flash pt. = 93°C

510. Tributyl phosphate

Syst. Name: Phosphoric acid, tributyl ester
Synonyms: Butyl phosphate

CASRN: 126-73-8   **Beil Ref.**: 4-01-00-01531
Merck No.: 9531
Beil RN: 1710584
MF: $C_{12}H_{27}O_4P$ **BP[°C]**: 289   **Den.[g/cm³]**: 0.9727^{25}
MW: 266.32 n_D: 1.4224^{25}
Sol.: H_2O 3; EtOH 5; eth
    3; bz 3; CS_2 3
Vap. Press. [kPa]: 18.7^{225}; 38.0^{250}; 72.4^{275}
γ **[mN/m]**: 27.55^{20}
Elec. Prop.: $\mu = (3.1\ D)$; $\varepsilon = 8.34^{20}$
η **[mPa s]**: 3.39^{25}; 1.61^{65}
MS: 99(100) 155(32) 211(30) 57(15) 29(10) 125(9) 137(8) 41(7) 55(6) 28(6)
IR [cm⁻¹]: 2940 1470 1390 1280 1030 910 810 780 740
Raman [cm⁻¹]: 2980 2940 2920 2880 2740 1450 1440 1390 1300 1260 1230
    1150 1120 1060 1040 980 950 900 880 830 810 770 720 550 500 470 360
    280 250
¹³C NMR [ppm]: 13.5 18.7 32.5 67.3 $CDCl_3$
¹H NMR [ppm]: 0.9 1.5 4.0 CCl_4
Flammability: Flash pt. = 146°C
TLV/TWA: 0.2 ppm (2.2 mg/m³)

511. 1,1,1-Trichloroethane

Syst. Name: Ethane, 1,1,1-trichloro-
Synonyms: Methyl chloroform; R140a

CASRN: 71-55-6   **DOT No.:** 2831
Merck No.: 9549   **Beil Ref.:** 4-01-00-00138
Beil RN: 1731614
MF: $C_2H_3Cl_3$ **MP[°C]:** -30.4   **Den.[g/cm^3]:** 1.3390^{20}
MW: 133.40   **BP[°C]:** 74.0 n_D: 1.4379^{20}
Sol.: H_2O 2; EtOH 3; eth
   5; chl 3
Crit. Const.: T_c = 272°C; P_c = 4.30 MPa
Vap. Press. [kPa]: 4.80^0; 16.5^{25}; 45.1^{50}; 104^{75}; 211^{100}
Therm. Prop.[kJ/mol]: $\Delta_{fus}H$ = 2.73; $\Delta_{vap}H$ = 29.9; $\Delta_f H°$(l, 25°C) = -177.4
C_p **(liq.) [J/mol °C]:** 144.3^{25}
γ **[mN/m]:** 25.18^{25}; dγ/dT = 0.1242 mN/m °C
Elec. Prop.: μ = 1.755 D; ε = 7.24^{20}; IP = 11.00 eV
η **[mPa s]:** 1.85^{-25}; 1.16^0; 0.793^{25}; 0.578^{50}; 0.428^{75}
k **[W/m °C]:** 0.106^0; 0.101^{25}; 0.096^{50}
MS: 97(100) 99(64) 61(58) 26(31) 27(24) 117(19) 63(19) 119(18) 35(17)
   62(11)
IR [cm^{-1}]: 2940 1450 1390 1250 1120 1090 880 710
^{13}**C NMR [ppm]:** 46.3 96.2
^{1}**H NMR [ppm]:** 2.7 CCl_4
Flammability: Flash pt. = -1°C; ign. temp. = 537°C; flam. lim. = 7.5-12.5%
TLV/TWA: 350 ppm (1910 mg/m^3)
Reg. Lists: CERCLA (RQ = 1000 lb.); SARA 313 (1.0%); RCRA U226 (toxic)

512. 1,1,2-Trichloroethane

Syst. Name: Ethane, 1,1,2-trichloro-
Synonyms: Trichloroethane; R 140

CASRN: 79-00-5   **Beil Ref.**: 4-01-00-00139
Merck No.: 9550
Beil RN: 1731726
MF: $C_2H_3Cl_3$ **MP[°C]**: -36.6   **Den.[g/cm^3]**: 1.4397^{20}
MW: 133.40   **BP[°C]**: 113.8 n_D: 1.4714^{20}
Sol.: H_2O 1; EtOH 3; eth
   3; chl 3
Vap. Press. [kPa]: 3.10^{25}; 10.1^{50}; 28.2^{75}; 66.6^{100}
Therm. Prop.[kJ/mol]: $\Delta_{fus}H$ = 11.54; $\Delta_{vap}H$ = 34.8; $\Delta_f H°$(l, 25°C) = -191.5
C_p **(liq.) [J/mol °C]**: 150.9^{25}
γ **[mN/m]**: 34.02^{25}; $d\gamma/dT$ = 0.1351 mN/m °C
Elec. Prop.: μ = (1.4 D); ε = 7.19^{25}; IP = 11.00 eV
η **[mPa s]**: 1.19^{20}
MS: 97(100) 83(95) 99(62) 85(60) 61(58) 26(23) 96(21) 63(19) 27(17) 98(15)
IR [cm^{-1}]: 3030 1430 1300 1270 1240 1210 1160 1050 930 770 730 690 670
Raman [cm^{-1}]: 3000 2970 1430 1310 1260 1210 1160 1050 1010 930 780 730
   700 670 640 530 390 330 280 260 190 170 110
^{13}C NMR [ppm]: 50.1 70.4 $CDCl_3$
^{1}H NMR [ppm]: 4.0 5.8 $CDCl_3$
Flammability: Flash pt. = 32°C; ign. temp. = 460°C; flam. lim. = 6-28%
TLV/TWA: 10 ppm (55 mg/m^3)
Reg. Lists: CERCLA (RQ = 100 lb.); SARA 313 (1.0%); RCRA U227 (toxic);
   confirmed or suspected carcinogen

513. Trichloroethylene

Syst. Name: Ethene, trichloro-
Synonyms: R 1120

CASRN: 79-01-6   **DOT No.**: 1710
Merck No.: 9552   **Beil Ref.**: 4-01-00-00712
Beil RN: 1736782
MF: C_2HCl_3 **MP[°C]**: -84.7   **Den.[g/cm^3]**: 1.4642^{20}
MW: 131.39   **BP[°C]**: 87.2 n_D: 1.4773^{20}
Sol.: H_2O 2; EtOH 5; eth
  5; ace 3; ctc 2; chl 3
Crit. Const.: T_c = 271.1°C; P_c = 5.02 MPa
Vap. Press. [kPa]: 2.73^0; 9.91^{25}; 28.6^{50}; 69.3^{75}
Therm. Prop.[kJ/mol]: $\Delta_{vap}H$ = 31.4; $\Delta_f H°$(l, 25°C) = -43.6
C_p **(liq.) [J/mol °C]**: 124.4^{25}
γ **[mN/m]**: 28.8^{25}
Elec. Prop.: μ = (0.8 D); ε = 3.39^{28}; IP = 9.47 eV
η **[mPa s]**: 0.703^0; 0.545^{25}; 0.444^{50}; 0.376^{75}
k **[W/m °C]**: 0.133^{-25}; 0.124^0; 0.116^{25}; 0.108^{50}; 0.100^{75}
MS: 95(100) 130(90) 132(85) 60(65) 97(64) 35(40) 134(27) 47(26) 62(21)
  59(13)
IR [cm^{-1}]: 3040 1580 1550 1240 930 840 780 630 450
Raman [cm^{-1}]: 3080 1850 1670 1580 1550 1240 1230 930 900 830 780 660
  630 450 410 380 270 210 170
^{1}H NMR [ppm]: 6.5 CCl_4
Flammability: Flash pt. = 32°C; ign. temp. = 420°C; flam. lim. = 8-10.5%
TLV/TWA: 50 ppm (269 mg/m^3)
Reg. Lists: CERCLA (RQ = 100 lb.); SARA 313 (1.0%); RCRA U228 (toxic);
  confirmed or suspected carcinogen

514. Trichloroethylsilane

Syst. Name: Silane, trichloroethyl-

CASRN: 115-21-9   **DOT No.**: 1196
Beil RN: 1361384   **Beil Ref.**: 4-04-00-04227
MF: $C_2H_5Cl_3Si$ **MP[°C]**: -105.6   **Den.[g/cm^3]**: 1.2373^{20}
MW: 163.51   **BP[°C]**: 100.5 n_D: 1.4256^{20}
Sol.: ctc 3
Vap. Press. [kPa]: 6.29^{25}; 19.0^{50}; 47.7^{75}; 104^{100}
Therm. Prop.[kJ/mol]: $\Delta_{fus}H$ = 6.96
Elec. Prop.: μ = (2.0 D)
MS: 135(100) 133(100) 126(67) 128(46) 137(37) 98(22) 63(21) 35(16) 100(14)
    127(11)
IR [cm^{-1}]: 2940 1470 1430 1390 1250 1220 1010 960 700
^{1}H NMR [ppm]: 1.3 CCl$_4$
Flammability: Flash pt. = 22°C

515. Trichlorofluoromethane

Syst. Name: Methane, trichlorofluoro-
Synonyms: Fluorotrichloromethane; R 11; CFC 11

CASRN: 75-69-4   **Beil Ref.**: 4-01-00-00054
Merck No.: 9553
Beil RN: 1732469
MF: CCl_3F **MP[°C]**: -111.1
MW: 137.37   **BP[°C]**: 23.7
Crit. Const.: T_c = 198.1°C; P_c = 4.41 MPa; V_c = 248 cm^3
Vap. Press. [kPa]: 0.4^{-73}; 12.1^{-25}; 40.3^0; 106^{25}; 237^{50}; 824^{100}
Therm. Prop.[kJ/mol]: $\Delta_{fus}H$ = 6.90; $\Delta_{vap}H$ = 25.1; $\Delta_f H°$(g, 25°C) = -268.3
C_p **(liq.) [J/mol °C]**: 121.6^{25} (sat. press.)
γ [mN/m]: 19.09^{15}
Elec. Prop.: μ = 0.46 D; $\varepsilon = 3.00^{20}$; IP = 11.77 eV
η [mPa s]: 0.740^{-25}; 0.539^0; 0.421^{25}
MS: 101(100) 103(66) 66(13) 105(11) 35(11) 47(9) 31(8) 82(4) 68(4) 37(4)
IR [cm^{-1}]: 2130 1670 1390 1350 1250 1160 1090 930 840 750
TLV/TWA: 1000 ppm (5620 mg/m^3)
Reg. Lists: CERCLA (RQ = 5000 lb.); SARA 313 (1.0%); RCRA U121 (toxic)

516. Trichloromethane

Syst. Name: Methane, trichloro-
Synonyms: Chloroform; R 20

CASRN: 67-66-3
Merck No.: 2141
Beil RN: 1731042
MF: $CHCl_3$
MW: 119.38
Sol.: H_2O 2; EtOH 5; eth 5; ace 3; bz 5; ctc 3; lig 5

DOT No.: 1888
Beil Ref.: 4-01-00-00042

MP[°C]: -63.6
BP[°C]: 61.1

Den.[g/cm³]: 1.4832^{20}
n_D: 1.4459^{20}

Crit. Const.: $T_c = 263.3°C$; $P_c = 5.47$ MPa; $V_c = 239$ cm³
Vap. Press. [kPa]: 1.75^{-25}; 8.02^0; 26.2^{25}; 69.3^{50}; 156^{75}; 308^{100}
Therm. Prop.[kJ/mol]: $\Delta_{fus}H = 8.80$; $\Delta_{vap}H = 29.2$; $\Delta_f H°(l, 25°C) = -134.5$
C_p **(liq.) [J/mol °C]**: 114.2^{25}
γ **[mN/m]**: 26.67^{25}; $d\gamma/dT = 0.1295$ mN/m °C
Elec. Prop.: $\mu = 1.04$ D; $\varepsilon = 4.81^{20}$; IP = 11.37 eV
η **[mPa s]**: 0.988^{-25}; 0.706^0; 0.537^{25}; 0.427^{50}
k **[W/m °C]**: 0.127^{-25}; 0.122^0; 0.117^{25}; 0.112^{50}; 0.107^{75}; 0.102^{100}
MS: 83(100) 85(64) 47(35) 35(19) 48(16) 49(12) 87(10) 37(6) 50(5) 84(4)
IR [cm⁻¹]: 3020 1210 760 670
Raman [cm⁻¹]: 3010 1210 750 660 360 260
¹³C NMR [ppm]: 77.7
¹H NMR [ppm]: 7.2 CCl_4
TLV/TWA: 10 ppm (49 mg/m³)
Reg. Lists: CERCLA (RQ = 10 lb.); SARA 313 (0.1%); RCRA U044 (toxic); confirmed or suspected carcinogen

517. (Trichloromethyl)benzene

Syst. Name: Benzene, (trichloromethyl)-
Synonyms: Benzotrichloride

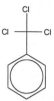

CASRN: 98-07-7   **DOT No.:** 2226
Merck No.: 1120   **Beil Ref.:** 4-05-00-00820
Beil RN: 508152
MF: $C_7H_5Cl_3$ **MP[°C]:** -5   **Den.[g/cm^3]:** 1.3723^{20}
MW: 195.47   **BP[°C]:** 221 n_D: 1.5580^{20}
Sol.: H_2O 1; EtOH 3; eth
    3; bz 3
Vap. Press. [kPa]: 0.180^{50}; 0.721^{75}; 2.36^{100}; 6.59^{125}; 16.1^{150}
γ [mN/m]: 23.39^{20}
Elec. Prop.: μ = (2.0 D); ε = 6.9^{21}; IP _ 9.60 eV
MS: 159(100) 161(64) 89(14) 163(11) 28(10) 63(9) 160(8) 123(8) 62(8) 124(6)
IR [cm^{-1}]: 3070 1490 1450 1190 1180 870 800 720 690 630
Raman [cm^{-1}]: 3170 3070 1600 1490 1450 1370 1290 1190 1180 1160 1130
    1040 1010 870 800 790 730 630 510 410 360 320 270 240 180 120
UV [nm]: 274(450) 267(558) 261(499) 225(6910) MeOH
^{1}H NMR [ppm]: 7.3 7.8 CCl_4
Flammability: Flash pt. = 127°C; ign. temp. = 211°C
Reg. Lists: CERCLA (RQ = 10 lb.); SARA 313 (0.1%); RCRA U023 (toxic);
    confirmed or suspected carcinogen

518. Trichloromethylsilane
Syst. Name: Silane, trichloromethyl-

CASRN: 75-79-6   **DOT No.**: 1250
Beil RN: 1361381   **Beil Ref.**: 4-04-00-04212
MF: CH₃Cl₃Si   **MP[°C]**: -90   **Den.[g/cm³]**: 1.273^{20}
MW: 149.48   **BP[°C]**: 65.6 n_D: 1.4106^{20}
Sol.: H₂O 6; EtOH 6
Crit. Const.: $T_c = 244°C$; $P_c = 3.28$ MPa; $V_c = 348$ cm³
Vap. Press. [kPa]: 22.5^{25}; 59.1^{50}; 132^{75}
Therm. Prop.[kJ/mol]: $\Delta_{fus}H = 8.94$
γ [mN/m]: 19.30^{25}
Elec. Prop.: μ = (1.9 D); IP = 11.36 eV
MS: 135(100) 133(99) 137(35) 113(20) 63(20) 150(13) 148(13) 115(13) 65(7)
    152(5)
IR [cm⁻¹]: 2940 2900 1400 1270 1020 800 760 560 450
Raman [cm⁻¹]: 2980 2920 1410 1270 800 760 690 570 450 240 170
Flammability: Flash pt. = -9°C; ign. temp. = >404°C; flam. lim. = 7.6->20%

519. 1,2,3-Trichloropropane

Syst. Name: Propane, 1,2,3-trichloro-
Synonyms: Allyl trichloride; Trichlorohydrin

CASRN: 96-18-4   **Beil Ref.**: 4-01-00-00199
Beil RN: 1732068
MF: $C_3H_5Cl_3$ **MP[°C]**: -14.7   **Den.[g/cm³]**: 1.3889^{20}
MW: 147.43   **BP[°C]**: 157 n_D: 1.4852^{20}
Sol.: H_2O 2; EtOH 3; eth
   3; ctc 2; chl 4
Vap. Press. [kPa]: 0.492^{25}; 6.64^{75}; 17.8^{100}; 41.1^{125}; 84.6^{150}
Therm. Prop.[kJ/mol]: $\Delta_{vap}H$ = 37.1; $\Delta_f H°$(l, 25°C) = -230.6
C_p **(liq.) [J/mol °C]**: 183.6^{25}
γ **[mN/m]**: 37.05^{25}
Elec. Prop.: ε = 7.5^{20}
η **[mPa s]**: 2.50^{20}
MS: 75(100) 39(58) 49(42) 110(38) 61(34) 77(33) 112(22) 27(16) 97(15)
   38(15)
IR [cm⁻¹]: 2940 1450 1330 1280 1250 1220 1180 1150 1100 990 930 910 870
   780 750 710 670   660
Raman [cm⁻¹]: 3010 2960 2860 2840 1440 1420 1340 1290 1280 1250 1200
   1140 1090 1020 990 920 900 860 800 750 730 720 660 630 520 410 370 350
   280 220 190 170 140
¹³C NMR [ppm]: 45.3 59.0 diox
¹H NMR [ppm]: 3.9 4.2 CCl₄
Flammability: Flash pt. = 71°C; flam. lim. = 3.2-12.6%
TLV/TWA: 10 ppm (60 mg/m³)
Reg. Lists: Confirmed or suspected carcinogen

520. 1,1,2-Trichlorotrifluoroethane

Syst. Name: Ethane, 1,1,2-trichloro-1,2,2-trifluoro-
Synonyms: R 113; CFC 113

CASRN: 76-13-1   **Beil Ref.**: 4-01-00-00142
Beil RN: 1740335
MF: $C_2Cl_3F_3$ **MP[°C]**: -35   **Den.[g/cm^3]**: 1.5635^{25}
MW: 187.38   **BP[°C]**: 47.7 n_D: 1.3557^{25}
Sol.: H_2O 1; EtOH 3; eth
   5; bz 5
Crit. Const.: $T_c = 214.2°C$; $P_c = 3.42$ MPa; $V_c = 325$ cm^3
Vap. Press. [kPa]: 3.88^{-25}; 15.1^0; 44.8^{25}; 110^{50}; 232^{75}; 438^{100}; 757^{125}; 1225^{150}
Therm. Prop.[kJ/mol]: $\Delta_{fus}H = 2.47$; $\Delta_{vap}H = 27.0$; $\Delta_fH°(l, 25°C) = -805.8$
C_p (liq.) [J/mol °C]: 170.1^{25}
γ [mN/m]: 17.75^{20}
Elec. Prop.: $\varepsilon = 2.41^{25}$; IP = 11.99 eV
η [mPa s]: 1.47^{-25}; 0.945^0; 0.656^{25}; 0.481^{50}
MS: 101(100) 151(68) 103(64) 85(45) 31(45) 153(44) 35(20) 66(19) 47(18)
   87(14)
IR [cm^{-1}]: 1210 1160 1100 1040 1030 890 810
TLV/TWA: 1000 ppm (7670 mg/m^3)
Reg. Lists: SARA 313 (1.0%)

521. Tri-*o*-cresyl phosphate

Syst. Name: Phosphoric acid, tris(2-methylphenyl) ester

Synonyms: *o*-Tolyl phosphate

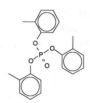

CASRN: 78-30-8

Merck No.: 9676

Beil RN: 1892885

MF: $C_{21}H_{21}O_4P$

MW: 368.37

Sol.: H_2O 1; EtOH 4; eth 4; ctc 4; tol 4; HOAc 3

Rel. CASRN: 1330-78-5

DOT No.: 2574

Beil Ref.: 4-06-00-01979

MP[°C]: 11

BP[°C]: 410

Den.[g/cm³]: 1.1955^{20}

n_D: 1.5575^{20}

Vap. Press. [kPa]: 0.007^{150}; 0.095^{200}; 0.788^{250}; 4.49^{300}; 19.4^{350}; 67^{400}

Elec. Prop.: $\mu = (2.9\ D)$; $\varepsilon = 6.7^{25}$

η [mPa s]: 86.6^{25}

MS: 165(100) 91(62) 179(54) 181(48) 368(37) 107(34) 277(33) 180(33) 77(33) 65(32)

IR [cm⁻¹]: 3050 2940 1590 1490 1410 1300 1220 1160 1100 1040 960 810 760 710

¹H NMR [ppm]: 2.2 7.1 CCl_4

Flammability: Flash pt. = 225°C; ign. temp. = 385°C

TLV/TWA: 0.1 mg/m³

522. Tri-*m*-cresyl phosphate

Syst. Name: Phosphoric acid, tris(3-methylphenyl)
ester
Synonyms: *m*-Tolyl phosphate

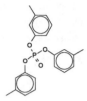

CASRN: 563-04-2   **Rel. CASRN**: 1330-78-5
Beil RN: 2063138   **DOT No.**: 2574
                                        Beil Ref.: 4-06-00-02057
MF: $C_{21}H_{21}O_4P$ **MP[°C]**: 25.5   **Den.[g/cm^3]**: 1.150^{25}
MW: 368.37 n_D: 1.5575^{20}
Sol.: H_2O 1; EtOH 2; eth
   3; ctc 4; tol 4; HOAc 3
Vap. Press. [kPa]: 0.001^{150}; 0.008^{175}; 0.048^{200}; 0.229^{225}; 0.935^{250}
Elec. Prop.: $\mu = (3.0\ D)$
MS: 368(100) 91(85) 165(75) 367(61) 65(60) 77(50) 90(38) 243(32) 89(30)
   79(28)
IR [cm^{-1}]: 3030 2920 1610 1590 1490 1450 1300 1240 1140 1020 970 880 780
   680
UV [nm]: 270(944) 263(1080) MeOH
^{1}H NMR [ppm]: 2.3 7.1 $CDCl_3$
TLV/TWA: 0.1 mg/m^3

523. Tri-*p*-cresyl phosphate

Syst. Name: Phosphoric acid, tris(4-methylphenyl) ester

Synonyms: *p*-Tolyl phosphate

CASRN: 78-32-0

Beil RN: 1891089

Rel. CASRN: 1330-78-5

DOT No.: 2574

Beil Ref.: 4-06-00-02130

MF: $C_{21}H_{21}O_4P$

MP[°C]: 77.5

Den.[g/cm^3]: 1.247^{25}

MW: 368.37

Sol.: EtOH 3; eth 3; bz 3; chl 3; HOAc 3

Vap. Press. [kPa]: 0.002^{150}

Elec. Prop.: $\mu = (3.2\ D)$

MS: 368(100) 367(68) 107(50) 77(43) 91(35) 108(33) 165(30) 65(24) 79(23) 369(18)

IR [cm^{-1}]: 3030 2920 1890 1600 1500 1450 1300 1190 1160 1100 1020 960 820 720 700

Raman [cm^{-1}]: 3070 3020 2990 2920 2860 2740 1600 1450 1380 1300 1230 1210 1180 1160 960 930 830 800 710 690 650 560 540 510 490 470 410 350 340 320 250 210

UV [nm]: 269 262 209 MeOH

^{1}H NMR [ppm]: 2.3 7.1 CDCl$_3$

TLV/TWA: 0.1 mg/m^3

524. Tricyclene

Syst. Name: Tricyclo[2.2.1.02,6]heptane, 1,7,7-trimethyl-
Synonyms: Cyclene

CASRN: 508-32-7   **Beil Ref.**: 4-05-00-00468
Beil RN: 1901153
MF: $C_{10}H_{16}$ **MP[°C]**: 67.5   **Den.[g/cm^3]**: 0.8668^{80}
MW: 136.24   **BP[°C]**: 152.5 n_D: 1.4296^{80}
MS: 93(100) 41(26) 121(25) 39(24) 136(21) 27(20) 79(19) 92(18) 91(16) 77(16)

---

525. Tridecane

Syst. Name: Tridecane

CASRN: 629-50-5   **Beil Ref.**: 4-01-00-00512
Beil RN: 1733089
MF: $C_{13}H_{28}$ **MP[°C]**: -5.3   **Den.[g/cm^3]**: 0.7564^{20}
MW: 184.37   **BP[°C]**: 235.4 n_D: 1.4256^{20}
Sol.: H_2O 1; EtOH 4; eth 4; ctc 3
Crit. Const.: $T_c = 402°C$; $P_c = 1.68$ MPa
Vap. Press. [kPa]: 0.005^{25}; 8.31^{150}
Therm. Prop.[kJ/mol]: $\Delta_{fus}H = 28.50$; $\Delta_{vap}H = 45.7$
C_p **(liq.) [J/mol °C]**: 406.7^{25}
γ **[mN/m]**: 25.55^{25}; $d\gamma/dT = 0.0872$ mN/m °C
Elec. Prop.: $\mu = 0$; $\varepsilon = 2.02^{20}$
η **[mPa s]**: 2.91^{0}; 1.72^{25}; 1.13^{50}; 0.796^{75}; 0.594^{100}
k **[W/m °C]**: 0.137^{25}; 0.132^{50}; 0.127^{75}; 0.122^{100}
MS: 57(100) 43(91) 71(51) 41(34) 85(24) 29(23) 56(14) 55(12) 27(11) 42(10)
IR [cm^{-1}]: 2920 2850 1470 1380 1300 720
Raman [cm^{-1}]: 2961 2890 2851 2729 1437 1300 1132 1078 1064 890
^{13}C NMR [ppm]: 14.2 23.0 29.8 30.1 32.4 diox
^{1}H NMR [ppm]: 0.9 1.3 CCl_4
Flammability: Flash pt. = 79°C

526. 1-Tridecene

Syst. Name: 1-Tridecene

CASRN: 2437-56-1   **Beil Ref.**: 4-01-00-00921
Beil RN: 1744660
MF: $C_{13}H_{26}$ **MP[°C]**: -13   **Den.[g/cm^3]**: 0.7658^{20}
MW: 182.35   **BP[°C]**: 232.8 n_D: 1.4340^{20}
Sol.: H_2O 1; EtOH 4; eth
   4; bz 3
Vap. Press. [kPa]: 1.36^{105}
Therm. Prop.[kJ/mol]: $\Delta_{fus}H = 22.83$
C_p **(liq.) [J/mol °C]**: 391.8^{25}
γ **[mN/m]**: 25.80^{25}; $d\gamma/dT = 0.0884$ mN/m °C
Elec. Prop.: $\mu = 0$; $\varepsilon = 2.14^{20}$
η **[mPa s]**: 1.50^{25}
MS: 41(100) 43(90) 55(82) 57(60) 69(59) 56(59) 83(58) 70(56) 29(54) 97(39)
Raman [cm^{-1}]: 3080 2999 2981 2960 2893 2851 1642 1438 1415 1300
Flammability: Flash pt. = 79°C

527. Triethanolamine

Syst. Name: Ethanol, 2,2',2"-nitrilotris-
Synonyms: Tris(2-hydroxyethyl)amine

CASRN: 102-71-6   **Beil Ref.**: 4-04-00-01524
Merck No.: 9581
Beil RN: 1699263
MF: $C_6H_{15}NO_3$ **MP[°C]**: 20.5   **Den.[g/cm³]**: 1.1242^{20}
MW: 149.19   **BP[°C]**: 335.4 n_D: 1.4852^{20}
Sol.: H_2O 5; EtOH 5; eth
   2; bz 2; chl 3; lig 2
Vap. Press. [kPa]: 11.2^{260}; 21.4^{280}; 39.2^{300}
Therm. Prop.[kJ/mol]: $\Delta_{fus}H$ = 27.20
C_p **(liq.) [J/mol °C]**: 389.0^{25}
Elec. Prop.: μ = (3.6 D); ε = 29.36^{25}; IP = 7.90 eV
η **[mPa s]**: 609^{25}; 114^{50}; 31.5^{75}; 11.7^{100}
MS: 118(100) 56(69) 45(60) 42(56) 44(27) 43(25) 41(14) 116(8) 57(8) 86(7)
IR [cm⁻¹]: 3350 2940 2880 2820 1450 1410 1360 1230 1150 1070 1030 880
   730
Raman [cm⁻¹]: 3300 2950 2880 2830 2700 1460 1380 1330 1300 1250 1150
   1060 1030 900 880 770 730 530 440 350 310 260
¹H NMR [ppm]: 2.7 3.6 D_2O
Flammability: Flash pt. = 179°C; flam. lim. = 1-10%
TLV/TWA: 5 mg/m³

528. Triethylamine

Syst. Name: Ethanamine, *N,N*-diethyl-
Synonyms: *N,N*-Diethylethanamine

CASRN: 121-44-8   **DOT No.**: 1296
Merck No.: 9582   **Beil Ref.**: 4-04-00-00322
Beil RN: 605283
MF: $C_6H_{15}N$ **MP[°C]**: -114.7   **Den.[g/cm^3]**: 0.7275^{20}
MW: 101.19   **BP[°C]**: 89 n_D: 1.4010^{20}
Sol.: H_2O 3; EtOH 3; eth
   3; ace 4; bz 4; ctc 3;
   chl 4
Crit. Const.: $T_c = 262.5°C$; $P_c = 3.032$ MPa; $V_c = 389$ cm^3
Vap. Press. [kPa]: 7.70^{25}; 25.8^{50}; 65.8^{75}
Therm. Prop.[kJ/mol]: $\Delta_{vap}H = 31.0$; $\Delta_f H°(l, 25°C) = -127.7$
C_p (liq.) [J/mol °C]: 219.9^{25}
γ [mN/m]: 20.22^{25}; $d\gamma/dT = 0.0992$ mN/m °C
Elec. Prop.: $\mu = 0.66$ D; $\epsilon = 2.42^{20}$; IP = 7.50 eV
η [mPa s]: 0.455^0; 0.347^{25}; 0.273^{50}; 0.221^{75}
MS: 86(100) 30(68) 58(37) 28(24) 29(23) 27(19) 44(18) 101(17) 42(16) 56(8)
IR [cm^{-1}]: 3030 2940 2780 1470 1390 1300 1210 1140 1080 1000 810 750
Raman [cm^{-1}]: 2970 2930 2880 2800 2760 2740 2710 1460 1370 1290 1210
   1150 1080 1070 1030 1000 920 900 810 740 540 470 440 340 290
UV [nm]: 196(5012) hp
^{13}C NMR [ppm]: 13.8 58.2
^{1}H NMR [ppm]: 1.0 2.4 CCl$_4$
Flammability: Flash pt. = -7°C; ign. temp. = 249°C; flam. lim. = 1.2-8.0%
TLV/TWA: 10 ppm (41 mg/m^3)
Reg. Lists: CERCLA (RQ = 5000 lb.)

529. Triethylene glycol

Syst. Name: Ethanol, 2,2'-[1,2-
ethanediylbis(oxy)]bis-

Synonyms: 1,2-Bis(2-hydroxyethoxy)ethane;
Triglycol

CASRN: 112-27-6   **Beil Ref.**: 4-01-00-02400

Merck No.: 9585

Beil RN: 969357

MF: $C_6H_{14}O_4$ **MP[°C]**: -7   **Den.[g/cm³]**: 1.1274^{15}

MW: 150.17   **BP[°C]**: 285 n_D: 1.4531^{20}

Sol.: H_2O 5; EtOH 5; eth
2; bz 5; chl 2; tol 5;
peth 1

Vap. Press. [kPa]: 0.247^{125}; 0.885^{150}

Therm. Prop.[kJ/mol]: $\Delta_{vap}H$ = 71.4; $\Delta_f H°$(l, 25°C) = -804.2

C_p **(liq.) [J/mol °C]**: 327.6^{25}

γ **[mN/m]**: 45.2^{20}

Elec. Prop.: $\varepsilon = 23.69^{20}$

η **[mPa s]**: 49^{20}; 8.5^{60}

MS: 45(100) 58(11) 89(9) 31(8) 29(8) 75(7) 44(7) 43(7) 27(7) 28(5)

IR [cm⁻¹]: 3380 2870 1450 1350 1250 1120 1070 940 890

¹H NMR [ppm]: 3.5 3.7 $CDCl_3$

Flammability: Flash pt. = 177°C; ign. temp. = 371°C; flam. lim. = 0.9-9.2%

530. Triethyl phosphate

Syst. Name: Phosphoric acid, triethyl ester
Synonyms: Ethyl phosphate

CASRN: 78-40-0 **Beil Ref.**: 4-01-00-01339
Merck No.: 9589
Beil RN: 1705772
MF: $C_6H_{15}O_4P$ **MP[°C]**: -56.4 **Den.[g/cm^3]**: 1.0695^{20}
MW: 182.16 **BP[°C]**: 215.5 n_D: 1.4053^{20}
Sol.: H_2O 3; EtOH 4; eth
   3; bz 3; chl 2
γ **[mN/m]**: 30.2^{18}
Elec. Prop.: μ = (3.1 D); ε = 13.20^{25}
η **[mPa s]**: 2.15^{25}
MS: 99(100) 81(71) 155(56) 82(45) 45(45) 109(44) 127(41) 43(24) 125(16)
   111(14)
IR [cm^{-1}]: 2990 2910 1480 1450 1400 1380 1280 1160 1040 970 820 750 540
Raman [cm^{-1}]: 2990 2950 2900 2880 2780 2730 1480 1460 1440 1370 1280
   1170 1100 1040 950 810 740 520 470 320 190
^{13}C NMR [ppm]: 15.9 63.4
^{1}H NMR [ppm]: 1.4 4.1 CDCl$_3$
Flammability: Flash pt. = 115°C; ign. temp. = 454°C

531. Trifluoroacetic acid

Syst. Name: Ethanoic acid, trifluoro-

CASRN: 76-05-1   **DOT No.**: 2699
Merck No.: 9595   **Beil Ref.**: 4-02-00-00458
Beil RN: 742035
MF: $C_2HF_3O_2$ **MP[°C]**: -15.2   **Den.[g/cm^3]**: 1.5351^{25}
MW: 114.02   **BP[°C]**: 73
Sol.: H_2O 3; EtOH 3; eth
   3; ace 3
Crit. Const.: $T_c = 218.2$°C; $P_c = 3.258$ MPa; $V_c = 204$ cm^3
Vap. Press. [kPa]: 4.00^0; 15.1^{25}; 45.1^{50}; 113^{75}
Therm. Prop.[kJ/mol]: $\Delta_{vap}H = 33.3$; $\Delta_fH°(l, 25°C) = -1070$
γ **[mN/m]**: 13.53^{25}; $d\gamma/dT = 0.0844$ mN/m °C
Elec. Prop.: $\mu = 2.28$ D; $\epsilon = 8.42^{20}$; IP = 11.46 eV
η **[mPa s]**: 0.808^{25}; 0.571^{50}
MS: 45(100) 69(70) 51(36) 28(28) 50(15) 44(11) 43(7) 97(5) 31(5) 29(5)
IR [cm^{-1}]: 3587 3134 2992 2597 1830 1792 1466 1415 1240 1199 1180 1122
   903 781 706 665
^{13}C NMR [ppm]: 115.0 163.0 CDCl$_3$
^{1}H NMR [ppm]: 11.3 CDCl$_3$

532. 1,1,1-Trifluoroethane

Syst. Name: Ethane, 1,1,1-trifluoro-
Synonyms: Methylfluoroform; R 143a

CASRN: 420-46-2   **DOT No.**: 2035
Beil RN: 1731552   **Beil Ref.**: 4-01-00-00122
MF: $C_2H_3F_3$ **MP[°C]**: -111.3
MW: 84.04   **BP[°C]**: -47.5
Sol.: eth 3; chl 3
Crit. Const.: T_c = 73.2°C; P_c = 3.76 MPa; V_c = 194 cm^3
Vap. Press. [kPa]: 264^{-25}; 627^0; 1267^{25}
Therm. Prop.[kJ/mol]: $\Delta_{fus}H$ = 6.19; $\Delta_f H°$(g, 25°C) = -744.6
C_p **(liq.) [J/mol °C]**: 109.7^{-53}
Elec. Prop.: μ = 2.347 D; IP = 12.90 eV
MS: 69(100) 65(38) 15(17) 45(12) 31(10) 64(9) 33(9) 14(6) 44(4) 13(3)
IR [cm^{-1}]: 3000 1470 1420 1280 1230 1190 1090 980 910 830 600

---

533. 2,2,2-Trifluoroethanol

Syst. Name: Ethanol, 2,2,2-trifluoro-

CASRN: 75-89-8   **Beil Ref.**: 4-01-00-01370
Beil RN: 1733203
MF: $C_2H_3F_3O$ **MP[°C]**: -43.5   **Den.[g/cm^3]**: 1.3842^{20}
MW: 100.04   **BP[°C]**: 74 n_D: 1.2907^{22}
Sol.: EtOH 4; eth 3; ace
  3; bz 3; chl 3
Vap. Press. [kPa]: 9.87^{25}; 37.0^{50}
Therm. Prop.[kJ/mol]: $\Delta_f H°$(l, 25°C) = -932.4
Elec. Prop.: ϵ = 27.68^{20}
η **[mPa s]**: 1.99^{20}
MS: 31(100) 33(24) 61(19) 29(19) 51(16) 69(9) 32(6) 49(5) 83(4) 81(4)
IR [cm^{-1}]: 3330 2940 1470 1430 1370 1280 1160 1090 940 830 660
^{1}H NMR [ppm]: 3.4 3.9 CDCl$_3$

534. (Trifluoromethyl)benzene

Syst. Name: Benzene, (trifluoromethyl)-
Synonyms: Benzotrifluoride

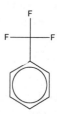

CASRN: 98-08-8   **DOT No.**: 2338
Merck No.: 1121   **Beil Ref.**: 4-05-00-00802
Beil RN: 1906908
MF: $C_7H_5F_3$ **MP[°C]**: -29.1   **Den.[g/cm³]**: 1.1884^{20}
MW: 146.11   **BP[°C]**: 102.1 n_D: 1.4146^{20}
Sol.: EtOH 5; eth 5; ace
    5; bz 5; ctc 5
Vap. Press. [kPa]: 5.14^{25}; 16.2^{50}; 42.4^{75}; 95.4^{100}; 190^{125}
Therm. Prop.[kJ/mol]: $\Delta_{fus}H = 13.46$; $\Delta_{vap}H = 32.6$
C_p **(liq.) [J/mol °C]**: 188.4^{25}
γ **[mN/m]**: 23.41^{20}
Elec. Prop.: $\mu = 2.86$ D; $\varepsilon = 9.22^{25}$; IP = 9.69 eV
η **[mPa s]**: 0.574^{20}
MS: 146(100) 145(40) 127(34) 96(28) 77(10) 51(10) 147(8) 75(6) 50(6) 128(3)
IR [cm⁻¹]: 3080 3040 1610 1460 1320 1170 1130 1070 1020 920 770 690 660
Raman [cm⁻¹]: 3080 3030 3000 1610 1590 1370 1330 1190 1170 1120 1070
    1050 1030 1010 990 850 810 770 660 620 490 410 340 150
UV [nm]: 266 259 253 249 cyhex
¹³C NMR [ppm]: 124.6 125.4 128.9 131.1 131.9 $CDCl_3$
¹H NMR [ppm]: 7.5 CCl_4
Flammability: Flash pt. = 12°C

535. Trimethylamine

Syst. Name: Methanamine, *N,N*-dimethyl-
Synonyms: *N,N*-Dimethylmethanamine

CASRN: 75-50-3   **DOT No.:** 1083
Merck No.: 9625   **Beil Ref.:** 4-04-00-00134
Beil RN: 956566
MF: C_3H_9N **MP[°C]:** -117.1   **Den.[g/cm^3]:** 0.627^{25} (sat. press.)
MW: 59.11   **BP[°C]:** 2.8 n_D: 1.3631^0
Sol.: H_2O 4; EtOH 3; eth
   3; bz 3; chl 4; tol 4
Crit. Const.: T_c = 159.64°C; P_c = 4.087 MPa; V_c = 254 cm^3
Vap. Press. [kPa]: 30.5^{-25}; 90.7^0; 215^{25}
Therm. Prop.[kJ/mol]: $\Delta_{fus}H$ = 6.55; $\Delta_{vap}H$ = 22.9; $\Delta_fH°$(g, 25°C) = -23.7
C_p **(liq.) [J/mol °C]:** 137.9^{25} (sat. press.)
γ **[mN/m]:** 13.41^{25} (at saturation press.); dγ/dT = 0.1133 mN/m °C
Elec. Prop.: μ = 0.612 D; ϵ = 2.44^{25}; IP = 7.82 eV
η **[mPa s]:** 0.177^{25} (at saturation pressure)
k **[W/m °C]:** 0.143^{-25}; 0.133^0
MS: 58(100) 59(47) 30(29) 42(26) 44(17) 15(14) 28(10) 18(10) 43(8) 57(7)
IR [cm^{-1}]: 2940 2780 1470 1280 1210 1190 1110 1060 1040 850 830
UV [nm]: 227(891) 19(3890) 161(2512) gas
^{13}**C NMR [ppm]:** 47.5
Flammability: Flash pt. = -7°C; ign. temp. = 190°C; flam. lim. = 2.0-11.6%
TLV/TWA: 5 ppm (12 mg/m^3)
Reg. Lists: CERCLA (RQ = 100 lb.)

536. 1,2,3-Trimethylbenzene
Syst. Name: Benzene, 1,2,3-trimethyl-
Synonyms: Hemimellitene

CASRN: 526-73-8
Beil RN: 1903410

Rel. CASRN: 25551-13-7, 15551-13-7
DOT No.: 2325
Beil Ref.: 4-05-00-01007

MF: C_9H_{12}
MW: 120.19
Sol.: H_2O 1; EtOH 5; eth 5; ace 5; bz 5; ctc 5; peth 5

MP[°C]: -25.4
BP[°C]: 176.1

Den.[g/cm^3]: 0.8944^{20}
n_D: 1.5139^{20}

Crit. Const.: T_c = 391.4°C; P_c = 3.454 MPa
Vap. Press. [kPa]: 9.44^{100}; 23.2^{125}; 50.3^{150}
Therm. Prop.[kJ/mol]: $\Delta_{fus}H$ = 8.37; $\Delta_f H°$(l, 25°C) = -58.5
C_p (liq.) [J/mol °C]: 216.4^{25}
Elec. Prop.: μ = 0; ε = 2.66^{20}; IP = 8.42 eV
η [mPa s]: 0.74^{38}
MS: 105(100) 120(47) 39(22) 77(17) 91(14) 51(14) 27(14) 79(12) 119(11) 106(9)
IR [cm^{-1}]: 2940 1590 1470 1390 1150 1080 1000 770
Raman [cm^{-1}]: 3080 3040 3000 2920 2870 2730 1600 1580 1480 1440 1380 1250 1190 1160 1090 990 820 660 540 480 270 230
UV [nm]: 269(303) 261(374) MeOH
^{1}H NMR [ppm]: 2.2 2.3 7.0 CDCl$_3$
Flammability: Flash pt. = 44°C; ign. temp. = 470°C; flam. lim. = 0.8-6.6%
TLV/TWA: 25 ppm (123 mg/m^3)

537. 2,2,3-Trimethylbutane

Syst. Name: Butane, 2,2,3-trimethyl-
Synonyms: Triptane

CASRN: 464-06-2
Beil RN: 1730756
MF: C_7H_{16}
MW: 100.20
Sol.: H_2O 1; EtOH 3; eth
   3; ace 4; bz 4; ctc 4;
   peth 4

DOT No.: 1206
Beil Ref.: 4-01-00-00410
MP[°C]: -25   **Den.[g/cm³]:** 0.6901^{20}
BP[°C]: 80.8 n_D: 1.3864^{20}

Crit. Const.: T_c = 258.1°C; P_c = 2.954 MPa; V_c = 398 cm³
Vap. Press. [kPa]: 9.13^0; 28.5^{25}; 72.2^{50}; 157^{75}
Therm. Prop.[kJ/mol]: $\Delta_{fus}H$ = 2.20; $\Delta_{vap}H$ = 28.9; $\Delta_f H°$(l, 25°C) = -236.5
C_p (liq.) [J/mol °C]: 213.5^{25}
Elec. Prop.: μ = 0; ε = 1.93^{20}
η [mPa s]: 0.48^{38}
MS: 57(100) 43(71) 56(63) 41(53) 85(30) 29(26) 27(18) 39(14) 15(6) 55(5)
IR [cm⁻¹]: 2940 2860 1470 1370 1220 1160 1110 1090 1000 920 890 830
Raman [cm⁻¹]: 2960 2910 2870 2800 2710 1470 1460 1330 1250 1220 1080
   1000 920 830 690 530 390 360
¹³C NMR [ppm]: 15.9 27.2 32.9 38.1
¹H NMR [ppm]: 0.8 1.3 CCl_4
Flammability: Flash pt. < 0°C; ign. temp. = 412°C

538. 2,2,5-Trimethylhexane

Syst. Name: Hexane, 2,2,5-trimethyl-

CASRN: 3522-94-9   **DOT No.**: 1920
Beil RN: 1730924   **Beil Ref.**: 4-01-00-00460
MF: C_9H_{20} **MP[°C]**: -105.7   **Den.[g/cm³]**: 0.7072^{20}
MW: 128.26   **BP[°C]**: 124.0 n_D: 1.3997^{20}
Sol.: H_2O 1; EtOH 4; eth
    4; ace 4; bz 4; ctc 3;
    lig 5; os 4
Crit. Const.: T_c = 295°C
Vap. Press. [kPa]: 2.21^{25}; 7.62^{50}; 21.2^{75}; 50.0^{100}; 104^{125}
Therm. Prop.[kJ/mol]: $\Delta_{fus}H$ = 6.19; $\Delta_{vap}H$ = 33.7; $\Delta_f H°$(l, 25°C) = -293.3
C_p **(liq.) [J/mol °C]**: 207.9^{25}
γ **[mN/m]**: 19.59^{25}; $d\gamma/dT$ = 0.0879 mN/m °C
Elec. Prop.: μ = 0
MS: 57(100) 56(35) 71(18) 41(17) 43(14) 29(8) 70(4) 58(4) 113(3) 55(3)
IR [cm⁻¹]: 2960 2880 1470 1400 1390 1370 1250 1200 920
Raman [cm⁻¹]: 3010 2960 2940 2910 2870 2780 2760 2720 1470 1450 1370
    1340 1320 1280 1250 1210 1170 1120 1100 1060 1020 960 930 920 830 800
    750 710 550 500 480 440 410 380 340 300 250 130
¹³C NMR [ppm]: 22.5 28.9 29.3 30.1 33.9 42.0 diox
¹H NMR [ppm]: 0.9 1.2 CCl_4
Flammability: Flash pt. = 13°C

539. 2,3,5-Trimethylhexane

Syst. Name: Hexane, 2,3,5-trimethyl-

CASRN: 1069-53-0   **DOT No.**: 1920
Beil RN: 1696923   **Beil Ref.**: 4-01-00-00461
MF: C_9H_{20} **MP[°C]**: -127.9   **Den.[g/cm^3]**: 0.7218[20]
MW: 128.26   **BP[°C]**: 131.4 n_D: 1.4051[20]
Vap. Press. [kPa]: 1.57[25]; 5.68[50]; 16.4[75]; 39.9[100]; 85.1[125]; 163[150]
Therm. Prop.[kJ/mol]: $\Delta_{fus}H$ = 10.00; $\Delta_{vap}H$ = 34.4; $\Delta_f H°$(l, 25°C) = -284.0
Elec. Prop.: $\mu = 0$
MS: 43(100) 85(32) 57(24) 84(17) 41(17) 71(13) 56(9) 69(8) 27(7) 42(6)
^{13}**C NMR [ppm]**: 15.3 17.8 20.0 21.9 23.5 25.7 32.4 36.2 43.9 diox

540. 2,2,3-Trimethylpentane

Syst. Name: Pentane, 2,2,3-trimethyl-
Synonyms: 2-*tert*-Butylbutane

CASRN: 564-02-3   **DOT No.**: 1262
Beil RN: 1696866   **Beil Ref.**: 4-01-00-00438
MF: C_8H_{18} **MP[°C]**: -112.2   **Den.[g/cm^3]**: 0.7161[20]
MW: 114.23   **BP[°C]**: 110 n_D: 1.4030[20]
Sol.: H_2O 1; EtOH 5; eth
    5; ace 5; bz 3; chl 5;
    hp 5
Crit. Const.: T_c = 290.4°C; P_c = 2.730 MPa; V_c = 436 cm^3
Vap. Press. [kPa]: 4.28[25]; 13.4[50]; 34.5[75]; 76.5[100]; 151[125]; 273[150]
Therm. Prop.[kJ/mol]: $\Delta_{fus}H$ = 8.62; $\Delta_{vap}H$ = 31.9; $\Delta_f H°$(l, 25°C) = -256.9
C_p **(liq.) [J/mol °C]**: 245.5[25]
γ **[mN/m]**: 20.22[25]; dγ/dT = 0.0895 mN/m °C
Elec. Prop.: $\mu = 0$; $\varepsilon = 1.96$[20]
η **[mPa s]**: 0.598[20]
MS: 57(100) 56(58) 41(33) 43(23) 29(23) 27(11) 39(8) 55(6) 58(4) 99(3)
IR [cm^{-1}]: 2950 2880 1470 1400 1380 1360 1250 1220 1200 1160 1080 1030
    1000 970 960
^{13}**C NMR [ppm]**: 13.0 13.3 24.4 27.1 33.0 45.4 diox
Flammability: Flash pt. < 21°C; ign. temp. = 346°C

541. 2,2,4-Trimethylpentane

Syst. Name: Pentane, 2,2,4-trimethyl-
Synonyms: Isooctane

CASRN: 540-84-1   **DOT No.**: 1262
Merck No.: 5079   **Beil Ref.**: 4-01-00-00439
Beil RN: 1696876
MF: C_8H_{18} **MP[°C]**: -107.3   **Den.[g/cm³]**: 0.6877^{25}
MW: 114.23   **BP[°C]**: 99.2 n_D: 1.3915^{20}
Sol.: H_2O 1; EtOH 5; eth
   3; ace 5; bz 5; ctc 3;
   chl 5; hp 5
Crit. Const.: T_c = 270.9°C; P_c = 2.568 MPa; V_c = 468 cm³
Vap. Press. [kPa]: 1.73^0; 6.50^{25}; 19.5^{50}; 48.8^{75}; 103^{100}; 198^{125}; 351^{150}
Therm. Prop.[kJ/mol]: $\Delta_{fus}H$ = 9.04; $\Delta_{vap}H$ = 30.8; $\Delta_f H°$(l, 25°C) = -259.2
C_p **(liq.) [J/mol °C]**: 239.1^{25}
γ [mN/m]: 18.33^{25}; dγ/dT = 0.0888 mN/m °C
Elec. Prop.: μ = 0; ε = 1.94^{20}; IP = 9.86 eV
η [mPa s]: 0.504^{20}
MS: 57(100) 41(31) 56(28) 43(24) 29(16) 27(9) 39(7) 58(4) 55(4) 99(2)
IR [cm⁻¹]: 2960 2900 2875 1470 1395 1370 1250 1200 1170 975
Raman [cm⁻¹]: 2960 2940 2920 2880 2790 2770 2720 1460 1390 1350 1290
   1250 1210 1170 1110 1100 1020 980 950 930 900 830 750 510 450 420 350
   300 200
¹³C NMR [ppm]: 24.9 25.5 30.2 31.2 53.4 diox
¹H NMR [ppm]: 0.9 1.0 1.1 1.7 CCl₄
Flammability: Flash pt. = -12°C; ign. temp. = 418°C
Reg. Lists: CERCLA (RQ = 1 lb.)

542. 2,3,3-Trimethylpentane

Syst. Name: Pentane, 2,3,3-trimethyl-

CASRN: 560-21-4

Beil RN: 1696865

MF: C_8H_{18}

MW: 114.23

DOT No.: 1262

Beil Ref.: 4-01-00-00445

MP[°C]: -100.9   **Den.[g/cm³]**: 0.7262^{20}

BP[°C]: 114.8 n_D: 1.4075^{20}

Sol.: H_2O 1; EtOH 4; eth 5; ace 5; bz 5; chl 5; hp 5; os 4

Crit. Const.: T_c = 300.5°C; P_c = 2.820 MPa; V_c = 455 cm³

Vap. Press. [kPa]: 3.60^{25}; 11.4^{50}; 29.7^{75}; 66.6^{100}; 133^{125}; 241^{150}

Therm. Prop.[kJ/mol]: $\Delta_{fus}H$ = 0.86; $\Delta_{vap}H$ = 32.1; $\Delta_fH°$(l, 25°C) = -253.5

C_p **(liq.) [J/mol °C]**: 245.6^{25}

Elec. Prop.: μ = 0; ϵ = 1.98^{20}

η **[mPa s]**: 0.51^{38}

MS: 43(100) 71(45) 70(36) 57(36) 41(29) 85(25) 27(18) 55(16) 29(16) 39(11)

IR [cm⁻¹]: 2970 1470 1380 1370 1190 1160 1090 1000

Raman [cm⁻¹]: 2980 2940 2900 2780 2720 1470 1450 1400 1320 1290 1230 1210 1190 1150 1110 1090 1050 1010 990 960 930 900 830 780 720 690 680 540 470 430 370 300 230

¹³C NMR [ppm]: 7.9 17.1 23.3 32.6 34.9 35.1 diox

¹H NMR [ppm]: 0.8 0.8 1.4 CCl_4

Flammability: Flash pt. < 21°C; ign. temp. = 425°C

543. 2,3,4-Trimethylpentane
Syst. Name: Pentane, 2,3,4-trimethyl-

CASRN: 565-75-3　　　　　　　　**DOT No.**: 1262
Beil RN: 1696869　　　　　　　　**Beil Ref.**: 4-01-00-00446
MF: C_8H_{18}　　　**MP[°C]**: -109.2　　**Den.[g/cm^3]**: 0.7191[20]
MW: 114.23　　　**BP[°C]**: 113.5　　　n_D: 1.4042[20]
Sol.: H_2O 1; EtOH 4; eth
　5; ace 5; bz 5; ctc 2;
　chl 5; hp 5; tol 4; os 4
Crit. Const.: T_c = 293.4°C; P_c = 2.730 MPa; V_c = 461 cm^3
Vap. Press. [kPa]: 11.6[50]; 30.4[75]; 68.7[100]
Therm. Prop.[kJ/mol]: $\Delta_{fus}H$ = 9.27; $\Delta_{vap}H$ = 32.4; $\Delta_f H°$(l, 25°C) = -255.0
C_p (liq.) [J/mol °C]: 247.3[25]
Elec. Prop.: μ = 0;　ε = 1.97[20]
η **[mPa s]**: 0.49[38]
MS: 43(100) 71(62) 70(41) 41(25) 27(18) 55(17) 57(16) 29(14) 39(10) 42(7)
IR [cm^{-1}]: 2940 2860 1460 1380 1110 1030
^{13}C NMR [ppm]: 10.4 18.1 21.4 29.8 45.3 diox
^{1}H NMR [ppm]: 0.8 1.9 CCl_4

544. Trimethyl phosphate

Syst. Name: Phosphoric acid, trimethyl ester
Synonyms: Methyl phosphate

CASRN: 512-56-1   **Beil Ref.**: 4-01-00-01259
Beil RN: 1071731
MF: $C_3H_9O_4P$ **MP[°C]**: -46   **Den.[g/cm^3]**: 1.2144^{20}
MW: 140.08   **BP[°C]**: 197.2 n_D: 1.3967^{20}
Sol.: H_2O 4; EtOH 2; eth
    3
Vap. Press. [kPa]: 0.110^{25}; 0.505^{50}; 1.82^{75}; 5.41^{100}; 13.8^{125}; 31.3^{150}
γ **[mN/m]**: 37.8^{17}
Elec. Prop.: μ = (3.2 D); ε = 20.6^{20}
η **[mPa s]**: 2.03^{25}
MS: 110(100) 109(35) 79(34) 95(25) 80(23) 15(20) 140(18) 47(10) 31(7)
    139(5)
IR [cm^{-1}]: 2940 2860 1470 1280 1190 1040 850 750 740
UV [nm]: 262(9) 229(55) EtOH 263(11) 230(62) H_2O
^{13}C NMR [ppm]: 54.2
Flammability: Flash pt. = 107°C

545. 2,4,6-Trimethylpyridine
Syst. Name: Pyridine, 2,4,6-trimethyl-
Synonyms: 2,4,6-Collidine

CASRN: 108-75-8   **Beil Ref.**: 5-20-06-00093
Merck No.: 9632
Beil RN: 107283
MF: $C_8H_{11}N$ **MP[°C]**: -46   **Den.[g/cm³]**: 0.9166^{22}
MW: 121.18   **BP[°C]**: 170.6 n_D: 1.4959^{25}
Sol.: H_2O 3; EtOH 3; eth
  3; ace 3; ctc 3
Therm. Prop.[kJ/mol]: $\Delta_{fus}H$ = 9.53; $\Delta_{vap}H$ = 39.9
Elec. Prop.: μ = (2.1 D); ε = 7.81^{25}; IP _ 8.9 eV
MS: 121(100) 39(27) 79(26) 120(24) 106(17) 27(16) 77(13) 51(11) 42(11)
  122(10)
IR [cm⁻¹]: 2940 1610 1560 1410 1370 1220 1030 990 920 840 730
Raman [cm⁻¹]: 3040 2920 2860 2730 1600 1570 1440 1410 1380 1380 1320
  1150 1100 1060 1030 880 590 540 510 280 230 200
UV [nm]: 264 208 MeOH
¹³C NMR [ppm]: 21.7 24.7 121.8 147.7 157.9
¹H NMR [ppm]: 2.2 2.4 6.6 CCl_4

546. Trinonafluorobutylamine

Syst. Name: 1-Butanamine, 1,1,2,2,3,3,4,4,4-
nonafluoro-*N*,*N*-bis(nonafluorobutyl)-
Synonyms: Heptacosafluorotributylamine

CASRN: 311-89-7   **Beil Ref.**: 4-02-00-00819
Beil RN: 1813883
MF: $C_{12}F_{27}N$ **BP[°C]**: 178   **Den.[g/cm^3]**: 1.884^{25}
MW: 671.10   **n_D**: 1.291^{25}
Sol.: ace 3
Vap. Press. [kPa]: 0.073^{25}; 0.441^{50}; 1.92^{75}; 6.51^{100}; 18.3^{125}; 44.3^{150}
Therm. Prop.[kJ/mol]: $_{vap}H = 46.4$
C_p (liq.) [J/mol °C]: 418.4^{25}
γ [mN/m]: 16^{25}
Elec. Prop.: $\varepsilon = 2.15^{20}$
η [mPa s]: 5.3^{25}
MS: 219(100) 69(77) 131(69) 220(60) 114(59) 614(57) 100(49) 671(0)
IR [cm^{-1}]: 1300 1230 1150 950 790 730

547. 1-Undecene

Syst. Name: 1-Undecene

CASRN: 821-95-4   **Beil Ref.**: 4-01-00-00910
Beil RN: 1740044
MF: $C_{11}H_{22}$ **MP[°C]**: -49.2   **Den.[g/cm^3]**: 0.7503^{20}
MW: 154.30   **BP[°C]**: 192.7   **n_D**: 1.4261^{20}
Sol.: H_2O 1; eth 3; chl 3;
   lig 3
Vap. Press. [kPa]: 1.55^{75}; 5.00^{100}; 13.4^{125}; 31.1^{150}
Therm. Prop.[kJ/mol]: $\Delta_{fus}H = 16.99$
C_p (liq.) [J/mol °C]: 344.9^{25}
Elec. Prop.: $\mu = 0$; $\varepsilon = 2.14^{20}$
η [mPa s]: 0.81^{38}
MS: 41(100) 43(87) 55(80) 70(67) 56(67) 69(55) 29(55) 83(51) 57(50) 27(46)
IR [cm^{-1}]: 3130 2940 1640 1470 1370 990 910 720
Raman [cm^{-1}]: 3081 2999 2981 2961 2895 2852 1641 1437 1415 1299
^{1}H NMR [ppm]: 0.9 1.3 2.0 4.8 4.9 5.6 CCl_4
Flammability: Flash pt. = 71°C

548. Veratrole

Syst. Name: Benzene, 1,2-dimethoxy-
Synonyms: 1,2-Dimethoxybenzene

CASRN: 91-16-7   **Beil Ref.**: 4-06-00-05564
Merck No.: 9857
Beil RN: 1364621
MF: $C_8H_{10}O_2$ **MP[°C]**: 22.5   **Den.[g/cm^3]**: 1.0810^{25}
MW: 138.17   **BP[°C]**: 206 n_D: 1.5827^{21}
Sol.: H_2O 2; EtOH 3; eth
   3; ctc 3
Therm. Prop.[kJ/mol]: $\Delta_{fus}H$ = 16.04; $\Delta_f H°$(l, 25°C) = -290.3
γ **[mN/m]**: 26.0^{131}
Elec. Prop.: μ = (1.3 D); ε = 4.45^{20}
η **[mPa s]**: 3.28^{25}
MS: 138(100) 95(65) 77(48) 123(44) 52(42) 41(33) 65(30) 51(29) 39(19)
   63(17)
IR [cm^{-1}]: 2941 2857 1587 1493 1449 1333 1235 1163 1124 1031 901 806 741
Raman [cm^{-1}]: 3180 3080 3000 2930 2840 1590 1500 1450 1330 1300 1260
   1180 1160 1050 1030 810 760 580 540 480 380 330 210 170
UV [nm]: 275(3920) 225(11700) MeOH
^{1}H NMR [ppm]: 3.7 6.8 CCl_4

549. Vinyl acetate

Syst. Name: Ethanoic acid, ethenyl ester

CASRN: 108-05-4   **DOT No.:** 1301
Merck No.: 9896   **Beil Ref.:** 4-02-00-00176
Beil RN: 1209327
MF: $C_4H_6O_2$ **MP[°C]:** -93.2   **Den.[g/cm³]:** 0.9317^{20}
MW: 86.09   **BP[°C]:** 72.5 n_D: 1.3959^{20}
Sol.: H_2O 1; EtOH 5; eth
  3; ace 3; bz 3; ctc 3;
  chl 3; os 3
Vap. Press. [kPa]: 15.4^{25}; 45.0^{50}
Therm. Prop.[kJ/mol]: $\Delta_{vap}H = 34.6$; $\Delta_f H°(l, 25°C) = -280.1$
C_p **(liq.) [J/mol °C]:** 169.5^{25}
γ **[mN/m]:** 23.95^{20}
Elec. Prop.: $\mu = (1.8\ D)$; IP = 9.19 eV
η **[mPa s]:** 0.421^{20}
MS: 43(100) 28(45) 42(26) 44(24) 86(20) 31(10) 32(7) 29(7) 45(2) 41(2)
IR [cm⁻¹]: 3100 2980 1760 1650 1470 1290 1220 1130 1020 970 950 880 850
  790
Raman [cm⁻¹]: 3130 3100 3050 3000 2940 2850 2730 1760 1690 1650 1620
  1430 1390 1380 1300 1220 1140 980 960 880 850 710 640 470 410 230
UV [nm]: 258(1) hx
¹³C NMR [ppm]: 20.2 96.8 141.8 167.6 diox
¹H NMR [ppm]: 2.1 4.6 4.9 7.3 $CDCl_3$
Flammability: Flash pt. = -8°C; ign. temp. = 402°C; flam. lim. = 2.6-13.4%
TLV/TWA: 10 ppm (35 mg/m³)
Reg. Lists: CERCLA (RQ = 5000 lb.); SARA 313 (1.0%)

550. *o*-Xylene

Syst. Name: Benzene, 1,2-dimethyl-
Synonyms: 1,2-Dimethylbenzene

CASRN: 95-47-6
Merck No.: 9988
Beil RN: 1815558
MF: C_8H_{10}
MW: 106.17
Sol.: H_2O 1; EtOH 5; eth
5; ace 5; bz 5; ctc 5;
peth 5

Rel. CASRN: 1330-20-7
DOT No.: 1307
Beil Ref.: 4-05-00-00917
MP[°C]: -25.2   **Den.[g/cm^3]**: 0.8802^{10}
BP[°C]: 144.5 n_D: 1.5055^{20}

Crit. Const.: $T_c = 357.2°C$; $P_c = 3.732$ MPa; $V_c = 370$ cm^3
Vap. Press. [kPa]: 0.880^{25}; 3.40^{50}; 10.4^{75}; 26.5^{100}; 58.9^{125}
Therm. Prop.[kJ/mol]: $\Delta_{fus}H = 13.61$; $\Delta_{vap}H = 36.2$; $\Delta_f H°(l, 25°C) = -24.4$
C_p (liq.) [J/mol °C]: 186.1^{25}
γ [mN/m]: 29.76^{25}; $d\gamma/dT = 0.1101$ mN/m °C
Elec. Prop.: $\mu = 0.640$ D; $\varepsilon = 2.56^{20}$; IP = 8.56 eV
η [mPa s]: 1.08^{0}; 0.760^{25}; 0.561^{50}; 0.432^{75}; 0.345^{100}
k [W/m °C]: 0.131^{25}; 0.126^{50}; 0.120^{75}; 0.114^{100}
MS: 91(100) 106(40) 39(21) 105(17) 51(17) 77(15) 27(12) 65(10) 92(8) 79(8)
IR [cm^{-1}]: 3030 2940 1590 1470 1390 1210 1120 1060 1030 750
Raman [cm^{-1}]: 3080 3050 2980 2920 2880 2860 2730 1610 1580 1450 1380
1220 1160 1050 980 740 580 510 260 180
UV [nm]: 270(211) 262(254) MeOH
^{1}H NMR [ppm]: 2.2 7.1 CDCl$_3$
Flammability: Flash pt. = 32°C; ign. temp. = 463°C; flam. lim. = 0.9-6.7%
TLV/TWA: 100 ppm (434 mg/m^3)
Reg. Lists: CERCLA (RQ = 1000 lb.); SARA 313 (1.0%); RCRA U239 (toxic)

551. *m*-Xylene

Syst. Name: Benzene, 1,3-dimethyl-
Synonyms: 1,3-Dimethylbenzene

CASRN: 108-38-3   **Rel. CASRN:** 1330-20-7
Merck No.: 9988   **DOT No.:** 1307
Beil RN: 605441   **Beil Ref.:** 4-05-00-00932
MF: C_8H_{10} **MP[°C]:** -47.8   **Den.[g/cm^3]:** 0.8642^{20}
MW: 106.17   **BP[°C]:** 139.1 n_D: 1.4972^{10}
Sol.: H_2O 1; EtOH 5; eth
   5; ace 5; bz 5; chl 3;
   os 5; peth 5
Crit. Const.: $T_c = 343.9$°C; $P_c = 3.541$ MPa; $V_c = 375$ cm^3
Vap. Press. [kPa]: 1.13^{25}; 4.15^{50}; 12.4^{75}; 31.1^{100}; 68.3^{125}
Therm. Prop.[kJ/mol]: $\Delta_{fus}H = 11.55$; $\Delta_{vap}H = 35.7$; $\Delta_f H°(l, 25°C) = -25.4$
C_p **(liq.) [J/mol °C]:** 183.0^{25}
γ **[mN/m]:** 28.47^{25}; $d\gamma/dT = 0.1104$ mN/m °C
Elec. Prop.: $\mu = 0$; $\varepsilon = 2.36^{20}$; IP = 8.56 eV
η **[mPa s]:** 0.795^0; 0.581^{25}; 0.445^{50}; 0.353^{75}; 0.289^{100}
k **[W/m °C]:** 0.130^{25}; 0.124^{50}; 0.118^{75}; 0.113^{100}
MS: 91(100) 106(65) 105(29) 39(18) 51(15) 77(14) 27(10) 92(8) 79(8) 78(8)
IR [cm^{-1}]: 3030 2940 1590 1470 1370 1160 1100 1050 890
Raman [cm^{-1}]: 3060 3010 2920 2860 2730 1620 1590 1380 1260 1250 1170
   1090 1040 1000 830 640 620 280 230 210
UV [nm]: 265 cyhex
^{13}C NMR [ppm]: 21.2 126.2 128.3 130.0 137.6 diox
^{1}H NMR [ppm]: 2.3 7.0 CDCl$_3$
Flammability: Flash pt. = 27°C; ign. temp. = 527°C; flam. lim. = 1.1-7.0%
TLV/TWA: 100 ppm (434 mg/m^3)
Reg. Lists: CERCLA (RQ = 1000 lb.); SARA 313 (1.0%); RCRA U239 (toxic)

552. *p*-Xylene

Syst. Name: Benzene, 1,4-dimethyl-
Synonyms: 1,4-Dimethylbenzene

CASRN: 106-42-3
Merck No.: 9988
Beil RN: 1901563
MF: C_8H_{10}
MW: 106.17

Rel. CASRN: 1330-20-7
DOT No.: 1307
Beil Ref.: 4-05-00-00951
MP[°C]: 13.2   **Den.[g/cm³]**: 0.8611^{20}
BP[°C]: 138.3 n_D: 1.4958^{20}

Sol.: H_2O 1; EtOH 5; eth 5; ace 5; bz 5; chl 3; os 5; peth 5

Crit. Const.: T_c = 343.1°C; P_c = 3.511 MPa; V_c = 378 cm³

Vap. Press. [kPa]: 1.19^{25}; 4.34^{50}; 12.8^{75}; 32.1^{100}; 69.9^{125}

Therm. Prop.[kJ/mol]: $\Delta_{fus}H$ = 16.81; $\Delta_{vap}H$ = 35.7; $\Delta_f H°$(l, 25°C) = -24.4

C_p (liq.) [J/mol °C]: 181.5^{25}

γ [mN/m]: 28.01^{25}; $d\gamma/dT$ = 0.1074 mN/m °C

Elec. Prop.: μ = 0; ϵ = 2.27^{20}; IP = 8.44 eV

η [mPa s]: 0.603^{25}; 0.457^{50}; 0.359^{75}; 0.290^{100}

k [W/m °C]: 0.130^{25}; 0.124^{50}; 0.118^{75}; 0.112^{100}

MS: 91(100) 106(62) 105(30) 51(16) 39(16) 77(13) 27(11) 92(7) 78(7) 65(7)

IR [cm⁻¹]: 2940 2860 1890 1790 1640 1490 1430 1390 1210 1110 1020 780

Raman [cm⁻¹]: 3060 3030 3010 2910 2860 2720 1620 1580 1450 1380 1310 1210 1180 830 810 650 460 410

UV [nm]: 275 212 cyhex

¹H NMR [ppm]: 2.3 7.1 $CDCl_3$

Flammability: Flash pt. = 27°C; ign. temp. = 528°C; flam. lim. = 1.1-7.0%

TLV/TWA: 100 ppm (434 mg/m³)

Reg. Lists: CERCLA (RQ = 1000 lb.); SARA 313 (1.0%); RCRA U239 (toxic)

553. 2,3-Xylenol

Syst. Name: Phenol, 2,3-dimethyl-
Synonyms: 2,3-Dimethylphenol

CASRN: 526-75-0   **Rel. CASRN**: 1300-71-6
Merck No.: 9989   **DOT No.**: 2261
Beil RN: 1906267   **Beil Ref.**: 4-06-00-03096
MF: $C_8H_{10}O$ **MP[°C]**: 72.8 n_D: 1.5420^{20}
MW: 122.17   **BP[°C]**: 216.9
Sol.: H_2O 2; EtOH 3; eth
   3
Crit. Const.: T_c = 449.7°C
Vap. Press. [kPa]: 1.60^{100}; 5.11^{125}; 13.6^{150}
Therm. Prop.[kJ/mol]: $\Delta_{fus}H$ = 21.02; $\Delta_f H°$(s, 25°C) = -241.1
Elec. Prop.: ε = 4.81^{70}; IP = 8.26 eV
MS: 107(100) 122(83) 121(28) 77(28) 79(19) 32(17) 91(15) 39(14) 28(14)
   78(11)
IR [cm^{-1}]: 1390 1320 1280 1220 1160 1090 1060 990 900 820 770 710
UV [nm]: 273(1445) MeOH 278(1549) 273(1413) 269(1380) hx 280(1549)
   cyhex
Reg. Lists: CERCLA (RQ = 1000 lb.)

554. 2,4-Xylenol

Syst. Name: Phenol, 2,4-dimethyl-
Synonyms: 2,4-Dimethylphenol

CASRN: 105-67-9   **Rel. CASRN**: 1300-71-6
Merck No.: 9989   **DOT No.**: 2261
Beil RN: 636244   **Beil Ref.**: 4-06-00-03126
MF: $C_8H_{10}O$ **MP[°C]**: 24.5   **Den.[g/cm^3]**: 0.9650^{20}
MW: 122.17   **BP[°C]**: 210.9 n_D: 1.5420^{14}
Sol.: H_2O 2; EtOH 5; eth
    5; ctc 3
Crit. Const.: T_c = 434.5°C
Vap. Press. [kPa]: 0.022^{25}; 1.93^{100}; 6.12^{125}; 16.2^{150}
Therm. Prop.[kJ/mol]: $\Delta_{vap}H$ = 47.1; $\Delta_f H°$(l, 25°C) = -228.7
γ **[mN/m]**: 31.10^{40}; $d\gamma/dT$ = 0.0869 mN/m °C
Elec. Prop.: μ = (1.4 D); ε = 5.06^{30}; IP = 8.00 eV
MS: 122(100) 107(92) 121(54) 77(29) 39(28) 27(20) 91(19) 51(19) 79(15)
    53(13)
IR [cm^{-1}]: 3448 3030 2941 1613 1493 1351 1266 1205 1124 813 775
Raman [cm^{-1}]: 3060 3020 2920 2870 2740 1620 1520 1450 1420 1380 1270
    1180 1150 1120 1040 980 940 890 820 780 730 580 490 450 350 280 220
UV [nm]: 292 MeOH
^{1}H NMR [ppm]: 2.1 2.2 5.5 6.5 6.7 6.8 CCl_4
Reg. Lists: CERCLA (RQ = 1000 lb.); SARA 313 (1.0%); RCRA U101 (toxic)

555. 2,5-Xylenol

Syst. Name: Phenol, 2,5-dimethyl-
Synonyms: 2,5-Dimethylphenol

CASRN: 95-87-4   **Rel. CASRN**: 1300-71-6
Merck No.: 9989   **DOT No.**: 2261
Beil RN: 1099260   **Beil Ref.**: 4-06-00-03164
MF: $C_8H_{10}O$ **MP[°C]**: 74.8
MW: 122.17   **BP[°C]**: 211.1
Sol.: H_2O 3; EtOH 3; eth
  4; chl 2
Crit. Const.: T_c = 433.8°C
Vap. Press. [kPa]: 0.022^{25}; 1.96^{100}; 6.17^{125}; 16.2^{150}
Therm. Prop.[kJ/mol]: $\Delta_{fus}H$ = 23.38; $\Delta_{vap}H$ = 46.9; $\Delta_f H°$(s, 25°C) = -246.6
γ [mN/m]: 29.49^{85}; $d\gamma/dT$ = 0.0850 mN/m °C
Elec. Prop.: μ = (1.4 D); ε = 5.36^{65}
η [mPa s]: 1.55^{80}
MS: 107(100) 122(93) 28(67) 121(38) 77(37) 39(21) 91(19) 79(19) 36(19)
  27(18)
IR [cm^{-1}]: 3600 3470 3030 2920 2850 1630 1590 1540 1520 1460 1420 1270
  1230 1170 1150 1110 990 930 850 800 730
Raman [cm^{-1}]: 3300 3060 3030 2960 2920 2860 2730 1620 1590 1530 1450
  1380 1370 1290 1270 1250 1200 1150 1120 1040 1000 940 870 800 760 730
  580 520 500 450 330 260 240 150 100
UV [nm]: 276(1810) MeOH
^{1}H NMR [ppm]: 2.2 5.0 6.5 6.6 6.9 CDCl$_3$
Reg. Lists: CERCLA (RQ = 1000 lb.)

556. 2,6-Xylenol

Syst. Name: Phenol, 2,6-dimethyl-
Synonyms: 2,6-Dimethylphenol

CASRN: 576-26-1
Merck No.: 9989
Beil RN: 1446677
MF: $C_8H_{10}O$
MW: 122.17
Sol.: H_2O 3; EtOH 3; eth
  3; ctc 3

Rel. CASRN: 1300-71-6
DOT No.: 2261
Beil Ref.: 4-06-00-03112
MP[°C]: 45.7
BP[°C]: 201.0

Crit. Const.: T_c = 427.9°C
Vap. Press. [kPa]: 0.019^{25}; 3.41^{100}; 9.69^{125}; 23.6^{150}
Therm. Prop.[kJ/mol]: $\Delta_{fus}H$ = 18.90; $\Delta_{vap}H$ = 44.5; $\Delta_f H°$(s, 25°C) = -237.4
Elec. Prop.: μ = (1.4 D); ε = 4.90^{40}; IP = 8.05 eV
MS: 107(100) 122(98) 121(37) 77(34) 28(34) 79(23) 91(22) 78(19) 39(19)
  38(17)
IR [cm^{-1}]: 3630 3050 3030 2990 2920 2860 1600 1480 1430 1380 1320 1260
  1200 1090 1020 910 830 770 730
Raman [cm^{-1}]: 3050 2980 2950 2920 2860 2840 2740 1620 1600 1480 1440
  1430 1380 1320 1270 1210 1200 1160 1100 1040 1000 910 900 830 770 750
  720 680 590 560 500 480 300 250
UV [nm]: 272(1480) MeOH 278(1560) 240(1800) MeOH: KOH
^{1}H NMR [ppm]: 2.2 4.4 6.5 7.0 CCl_4
Reg. Lists: CERCLA (RQ = 1000 lb.)

557. 3,4-Xylenol

Syst. Name: Phenol, 3,4-dimethyl-
Synonyms: 3,4-Dimethylphenol

CASRN: 95-65-8
Merck No.: 9989
Beil RN: 1099267
MF: $C_8H_{10}O$
MW: 122.17

Rel. CASRN: 1300-71-6
DOT No.: 2261
Beil Ref.: 4-06-00-03099
MP[°C]: 60.8   **Den.[g/cm^3]:** 0.9830^{20}
BP[°C]: 227

Sol.: H_2O 2; EtOH 3; eth
   5; ctc 3
Crit. Const.: T_c = 456.7°C
Vap. Press. [kPa]: 3.16^{125}; 9.14^{150}; 22.5^{175}; 49.1^{200}; 96.5^{225}
Therm. Prop.[kJ/mol]: $\Delta_{fus}H$ = 18.13; $\Delta_{vap}H$ = 49.7; $\Delta_f H°$(s, 25°C) = -242.3
γ **[mN/m]:** 28.92^{75}; $d\gamma/dT$ = 0.0910 mN/m °C
Elec. Prop.: μ = (1.6 D); ε = 9.02^{60}; IP = 8.09 eV
η **[mPa s]:** 3.00^{80}
MS: 107(100) 122(80) 44(49) 121(45) 28(32) 77(24) 91(17) 32(17) 39(11)
   123(10)
IR [cm^{-1}]: 3333 3030 2941 1613 1493 1470 1333 1299 1266 1149 1124 1020
   943 855 833
Raman [cm^{-1}]: 3050 2990 2930 2880 2760 1620 1600 1500 1450 1390 1340
   1300 1270 1230 1200 1160 1120 1050 1030 1000 950 890 870 810 760 730
   700 580 560 510 470 350 320 300 230
UV [nm]: 270(1890) 218(6230) MeOH
^{1}H NMR [ppm]: 2.1 5.4 6.5 6.8 CCl_4
Reg. Lists: CERCLA (RQ = 1000 lb.)

558. 3,5-Xylenol

Syst. Name: Phenol, 3,5-dimethyl-
Synonyms: 3,5-Dimethylphenol

CASRN: 108-68-9
Merck No.: 9989
Beil RN: 774117
MF: $C_8H_{10}O$
MW: 122.17
Sol.: H_2O 3; EtOH 3; ctc 3

Rel. CASRN: 1300-71-6
DOT No.: 2261
Beil Ref.: 4-06-00-03141
MP[°C]: 63.6   **Den.[g/cm³]**: 0.9680^{20}
BP[°C]: 221.7

Crit. Const.: $T_c = 442.5°C$
Vap. Press. [kPa]: 3.83^{125}; 10.8^{150}; 26.3^{175}; 56.6^{200}; 110.1^{225}
Therm. Prop.[kJ/mol]: $\Delta_{fus}H = 18.00$; $\Delta_{vap}H = 49.3$; $\Delta_fH°(s, 25°C) = -244.4$
γ [mN/m]: 28.04^{75}; $d\gamma/dT = 0.0801$ mN/m °C
Elec. Prop.: $\mu = (1.6$ D); $\varepsilon = 9.06^{50}$
η [mPa s]: 2.42^{80}
MS: 93(100) 39(48) 44(47) 66(40) 28(34) 43(24) 92(23) 65(23) 38(17) 51(16)
IR [cm⁻¹]: 3448 3030 2941 1613 1493 1316 1220 1163 1042 962 847 699
Raman [cm⁻¹]: 3040 2980 2920 2880 2850 2820 2730 1620 1590 1500 1450 1390 1340 1290 1270 1200 1160 1120 1050 1030 990 950 870 810 760 730 700 580 560 500 470 350 230
UV [nm]: 282 276 203 MeOH
¹H NMR [ppm]: 2.2 5.7 6.3 6.5 CCl_4
Reg. Lists: CERCLA (RQ = 1000 lb.)

559. 2,3-Xylidine

Syst. Name: Benzenamine, 2,3-dimethyl-
Synonyms: 2,3-Dimethylaniline

CASRN: 87-59-2
Beil RN: 742174

Rel. CASRN: 1300-73-8
DOT No.: 1711
Beil Ref.: 4-12-00-02497

MF: $C_8H_{11}N$ **MP[°C]**: <-15   **Den.[g/cm^3]**: 0.9931[20]
MW: 121.18   **BP[°C]**: 221.5 n_D: 1.5684[20]
Sol.: H_2O 2; EtOH 4; eth 4; ctc 3
MS: 106(100) 121(95) 120(56) 77(19) 91(13) 103(11) 79(10) 78(10) 122(9) 107(8)
IR [cm^{-1}]: 3448 3333 3030 2857 1613 1587 1471 1429 1370 1282 1099 990 775 714 654
UV [nm]: 286 236 cyhex
^{13}C NMR [ppm]: 12.4 20.3 113.1 120.4 120.6 125.9 136.8 144.5 CDCl$_3$
^{1}H NMR [ppm]: 2.0 2.2 3.4 6.5 6.8 CCl$_4$
Flammability: Flash pt. = 97°C; lower flam. lim. = 1.0%
TLV/TWA: 0.5 ppm (2.5 mg/m^3)
Reg. Lists: Confirmed or suspected carcinogen

560. 2,4-Xylidine

Syst. Name: Benzenamine, 2,4-dimethyl-
Synonyms: 2,4-Dimethylaniline

CASRN: 95-68-1
Beil RN: 636243

Rel. CASRN: 1300-73-8
DOT No.: 1711
Beil Ref.: 4-12-00-02545

MF: $C_8H_{11}N$ **MP[°C]**: -14.3 **Den.[g/cm^3]**: 0.9723^{20}
MW: 121.18 **BP[°C]**: 214 **n_D**: 1.5569^{20}
Sol.: H_2O 2; EtOH 3; eth
   3; bz 3; ctc 2
Vap. Press. [kPa]: 1.82^{100}; 5.66^{125}; 15.0^{150}
γ [mN/m]: 36.75^{25}
Elec. Prop.: $\mu = (1.4$ D$)$; $\varepsilon = 4.9^{20}$
MS: 121(100) 120(78) 106(57) 77(15) 28(15) 91(12) 122(9) 18(8) 93(6) 118(5)
IR [cm^{-1}]: 3448 3333 3030 2941 1613 1515 1493 1471 1370 1266 1235 1149
   1031 1000 870 806 725 680
UV [nm]: 287 235 MeOH
^{1}H NMR [ppm]: 2.0 2.2 3.4 6.4 6.7 CCl_4
TLV/TWA: 0.5 ppm (2.5 mg/m^3)
Reg. Lists: Confirmed or suspected carcinogen

561. 2,5-Xylidine

Syst. Name: Benzenamine, 2,5-dimethyl-
Synonyms: 2,5-Dimethylaniline

CASRN: 95-78-3
Beil RN: 2205178

Rel. CASRN: 1300-73-8
DOT No.: 1711
Beil Ref.: 4-12-00-02567

MF: $C_8H_{11}N$ **MP[°C]**: 15.5 **Den.[g/cm^3]**: 0.9790^{21}
MW: 121.18 **BP[°C]**: 214 **n_D**: 1.5591^{21}
Sol.: H_2O 2; eth 3; ctc 3
MS: 121(100) 106(91) 120(78) 77(25) 91(19) 78(12) 103(10) 79(10) 122(9) 93(9)
IR [cm^{-1}]: 3330 2940 1610 1560 1490 1450 1390 1300 1000 860 790
UV [nm]: 288 237 cyhex
^{1}H NMR [ppm]: 2.0 2.2 3.2 6.3 6.8 CCl$_4$
TLV/TWA: 0.5 ppm (2.5 mg/m^3)
Reg. Lists: Confirmed or suspected carcinogen

562. 2,6-Xylidine

Syst. Name: Benzenamine, 2,6-dimethyl-
Synonyms: 2,6-Dimethylaniline

CASRN: 87-62-7   **Rel. CASRN**: 1300-73-8
Beil RN: 636332   **DOT No.**: 1711
                                      Beil Ref.: 4-12-00-02521
MF: $C_8H_{11}N$ **MP[°C]**: 11.2   **Den.[g/cm³]**: 0.9842^{20}
MW: 121.18   **BP[°C]**: 215 n_D: 1.5610^{20}
Sol.: eth 4; EtOH 4
Vap. Press. [kPa]: 2.38^{100}; 6.36^{125}; 15.1^{150}
C_p **(liq.) [J/mol °C]**: 238.9^{25}
Elec. Prop.: $\mu = (1.6\ D)$
MS: 121(100) 106(59) 120(56) 77(18) 91(12) 122(9) 78(9) 93(8) 103(7) 79(6)
IR [cm⁻¹]: 3450 3330 3030 2940 1610 1490 1450 1390 1320 1280 1240 1090
  760 740
UV [nm]: 283(2020) 233(7490) MeOH
¹H NMR [ppm]: 2.0 3.2 6.6 CCl_4
Flammability: Flash pt. = 96°C
TLV/TWA: 0.5 ppm (2.5 mg/m³)
Reg. Lists: Confirmed or suspected carcinogen

563. 3,4-Xylidine

Syst. Name: Benzenamine, 3,4-dimethyl-
Synonyms: 3,4-Dimethylaniline

CASRN: 95-64-7   **Rel. CASRN**: 1300-73-8
Beil RN: 507414   **DOT No.**: 1711
                                                  Beil Ref.: 4-12-00-02502
MF: $C_8H_{11}N$ **MP[°C]**: 51   **Den.[g/cm^3]**: 1.076^{18}
MW: 121.18   **BP[°C]**: 228
Sol.: H_2O 2; eth 3; chl 2;
   lig 4
MS: 121(100) 106(94) 120(92) 77(24) 91(17) 122(9) 93(8) 107(7) 79(7) 78(7)
IR [cm^{-1}]: 3448 3333 3226 3030 2941 1613 1587 1515 1449 1370 1299 1266
   1163 1020 877 820 709
UV [nm]: 293 238 cyhex
^{1}H NMR [ppm]: 2.2 3.4 6.5 6.9 CDCl$_3$
TLV/TWA: 0.5 ppm (2.5 mg/m^3)
Reg. Lists: Confirmed or suspected carcinogen

564. 3,5-Xylidine

Syst. Name: Benzenamine, 3,5-dimethyl-
Synonyms: 3,5-Dimethylaniline

CASRN: 108-69-0   **Rel. CASRN**: 1300-73-8
Beil RN: 507051   **DOT No.**: 1711
                                         Beil Ref.: 4-12-00-02561
MF: $C_8H_{11}N$ **MP[°C]**: 9.8   **Den.[g/cm^3]**: 0.9706^{20}
MW: 121.18   **BP[°C]**: 220.5 n_D: 1.5581^{20}
Sol.: H_2O 2; eth 3; ctc 3
MS: 121(100) 106(79) 120(54) 77(20) 91(13) 79(10) 122(9) 93(8) 39(8) 107(7)
IR [cm^{-1}]: 3448 3333 3226 3030 2941 1613 1471 1370 1316 1176 1031 943
826 685
UV [nm]: 288(2420) 239(11000) 207(35900) MeOH
^{1}H NMR [ppm]: 2.2 3.3 6.1 6.3 CCl_4
TLV/TWA: 0.5 ppm (2.5 mg/m^3)
Reg. Lists: Confirmed or suspected carcinogen

Indexes

Name Index

This index includes all primary names, systematic names, and synonyms. The number following the name is the ID for the solvent. Names are alphabetized without regard to locant numbers, spaces, punctuation, or descriptors such as cis, tran, etc.

(Trifluoromethyl)benzene: 534
Triglycol: 529
1,2,6-Trihydroxyhexane: 292
1,2,3-Trihydroxypropane: 280
Trimethylamine: 535
1,3,5-Trimethylbenzene: 333
1,2,4-Trimethylbenzene: 465
1,2,3-Trimethylbenzene: 536
1,7,7-Trimethylbicyclo[2.2.1]hepten-2-one: 78
2,2,3-Trimethylbutane: 537
Trimethylcarbinol: 392
3,5,5-Trimethyl-2-cyclohexen-1-one: 325
Trimethylene glycol: 450
2,2,5-Trimethylhexane: 538
2,3,5-Trimethylhexane: 539
2,2,3-Trimethylpentane: 540
2,2,4-Trimethylpentane: 541
2,3,3-Trimethylpentane: 542
2,3,4-Trimethylpentane: 543
Trimethyl phosphate: 544
2,4,6-Trimethylpyridine: 545
Trinonafluorobutylamine: 546
3,6,9-Trioxa-1,11-undecanediol: 485
Triptane: 537
Tris(dimethylamino)phosphine oxide: 289
Tris(2-hydroxyethyl)amine: 527
1-Undecene: 547
Urea, tetramethyl-: 498

Valeric acid: 426
Valeronitrile: 425
Veratrole: 548
Vinyl acetate: 549
Vinylbenzene: 472
Vinyl bromide: 40
Vinyl butyl ether: 75
Vinylcarbinol: 15
Vinyl chloride: 88
Vinyl cyanide: 12
Vinylene carbonate: 244
Vinyl ethyl ether: 270
Vinylidene chloride: 149
Vinylidene fluoride: 178
o-Xylene: 550
m-Xylene: 551
p-Xylene: 552
2,3-Xylenol: 553
2,4-Xylenol: 554
2,5-Xylenol: 555
2,6-Xylenol: 556
3,4-Xylenol: 557
3,5-Xylenol: 558
2,3-Xylidine: 559
2,4-Xylidine: 560
2,5-Xylidine: 561
2,6-Xylidine: 562
3,4-Xylidine: 563
3,5-Xylidine: 564

Molecular Formula Index

This index is by molecular formula in the Hill order. The number following the name is the ID for the solvent.

C_3H_5ClO Epichlorohydrin: 228
$C_3H_5Cl_3$ 1,2,3-Trichloropropane: 519
C_3H_5N Propanenitrile: 451
C_3H_5NO Hydracrylonitrile: 306
$C_3H_6Br_2$ 1,2-Dibromopropane: 134
$C_3H_6Cl_2$ 1,2-Dichloropropane: 154
C_3H_6O 1,2-Propylene oxide: 463
C_3H_6O Propanal: 447
C_3H_6O Acetone: 6
C_3H_6O Allyl alcohol: 15
$C_3H_6O_2$ Methyl acetate: 338
$C_3H_6O_2$ 1,3-Dioxolane: 220
$C_3H_6O_2$ Ethyl formate: 258
$C_3H_6O_2$ Propanoic acid: 452
C_3H_7Br 1-Bromopropane: 45
C_3H_7Br 2-Bromopropane: 46
C_3H_7Cl 1-Chloropropane: 98
C_3H_7Cl 2-Chloropropane: 99
C_3H_7I 1-Iodopropane: 313
C_3H_7I 2-Iodopropane: 314
C_3H_7N Allylamine: 16
C_3H_7NO *N*-Methylacetamide: 337
C_3H_7NO *N,N*-Dimethylformamide: 198
$C_3H_7NO_2$ 1-Nitropropane: 408
$C_3H_7NO_2$ 2-Nitropropane: 409
C_3H_8 Propane: 448
C_3H_8O 1-Propanol: 454
C_3H_8O 2-Propanol: 455
$C_3H_8O_2$ Ethylene glycol monomethyl
                ether: 256
$C_3H_8O_2$ Methylal: 342
$C_3H_8O_2$ 1,2-Propanediol: 449
$C_3H_8O_2$ 1,3-Propanediol: 450
$C_3H_8O_3$ Glycerol: 280
C_3H_9N Trimethylamine: 535
C_3H_9N Propylamine: 459
C_3H_9N Isopropylamine: 327
$C_3H_9O_4P$ Trimethyl phosphate: 544
C_4F_8 Perfluorocyclobutane: 436
C_4F_{10} Perfluorobutane: 435
$C_4H_4N_2$ Succinonitrile: 473
C_4H_4O Furan: 277
C_4H_4S Thiophene: 500
C_4H_5N Methylacrylonitrile: 341
C_4H_5N Pyrrole: 467
$C_4H_5NO_2$ Methyl cyanoacetate: 351
C_4H_6O *trans*-Crotonaldehyde: 109
$C_4H_6O_2$ Vinyl acetate: 549
$C_4H_6O_2$ *trans*-Crotonic acid: 110
$C_4H_6O_2$ γ-Butyrolactone: 76
$C_4H_6O_2$ Methyl acrylate: 340
$C_4H_6O_2$ Methacrylic acid: 335
$C_4H_6O_3$ Acetic anhydride: 5
$C_4H_6O_3$ Propylene carbonate: 462

C_4H_7N Butanenitrile: 53
C_4H_7N 2-Methylpropanenitrile: 389
C_4H_7NO 2-Pyrrolidone: 469
$C_4H_8Cl_2O$ Bis(2-chloroethyl) ether: 30
C_4H_8O Butanal: 48
C_4H_8O Methyl ethyl ketone: 361
C_4H_8O *trans*-Crotonyl alcohol: 112
C_4H_8O *cis*-Crotonyl alcohol: 111
C_4H_8O Ethyl vinyl ether: 270
C_4H_8O Tetrahydrofuran: 487
C_4H_8O Isobutanal: 315
$C_4H_8O_2$ Ethyl acetate: 232
$C_4H_8O_2$ Butanoic acid: 55
$C_4H_8O_2$ *trans*-2-Butene-1,4-diol: 60
$C_4H_8O_2$ *cis*-2-Butene-1,4-diol: 59
$C_4H_8O_2$ 1,4-Dioxane: 219
$C_4H_8O_2$ Propyl formate: 464
C_4H_8O 1,2-Epoxybutane: 229
$C_4H_8O_2$ 2-Methylpropanoic acid: 390
$C_4H_8O_2S$ Sulfolane: 474
C_4H_8S Tetrahydrothiophene: 491
C_4H_9Br 1-Bromobutane: 33
C_4H_9Br 2-Bromobutane: 34
C_4H_9Br 2-Bromo-2-methylpropane: 42
C_4H_9Cl 1-Chlorobutane: 83
C_4H_9Cl 2-Chlorobutane: 84
C_4H_9Cl 1-Chloro-2-methylpropane: 92
C_4H_9Cl 2-Chloro-2-methylpropane: 93
C_4H_9I 1-Iodobutane: 308
C_4H_9I 2-Iodobutane: 309
C_4H_9I 1-Iodo-2-methylpropane: 312
C_4H_9N Pyrrolidine: 468
C_4H_9NO *N,N*-Dimethylacetamide: 185
C_4H_9NO Morpholine: 400
C_4H_9NO *N*-Methylpropanamide: 388
C_4H_{10} Butane: 49
C_4H_{10} Isobutane: 316
$C_4H_{10}O$ 1-Butanol: 57
$C_4H_{10}O$ 2-Butanol: 58
$C_4H_{10}O$ Diethyl ether: 170
$C_4H_{10}O$ 2-Methyl-1-propanol: 391
$C_4H_{10}O$ 2-Methyl-2-propanol: 392
$C_4H_{10}O_2$ 1,3-Butanediol: 50
$C_4H_{10}O_2$ 1,4-Butanediol: 51
$C_4H_{10}O_2$ 2,3-Butanediol: 52
$C_4H_{10}O_2$ Ethylene glycol dimethyl ether:
                251
$C_4H_{10}O_2$ Ethylene glycol monoethyl
                ether: 255
$C_4H_{10}O_2S$ 2,2'-Thiodiethanol: 499
$C_4H_{10}O_3$ Diethylene glycol: 162
$C_4H_{10}S$ 1-Butanethiol: 54
$C_4H_{10}S$ Diethyl sulfide: 176
$C_4H_{11}N$ Butylamine: 63

$C_6H_{12}O_2$ Isobutyl acetate: 317
$C_6H_{12}O_2$ Ethyl butanoate: 238
$C_6H_{12}O_2$ Hexanoic acid: 293
$C_6H_{12}O_2$ Diacetone alcohol: 128
$C_6H_{12}O_3$ Ethylene glycol ethyl ether acetate: 252
$C_6H_{13}N$ Cyclohexylamine: 118
C_6H_{14} Hexane: 290
C_6H_{14} 2-Methylpentane: 377
C_6H_{14} 3-Methylpentane: 378
C_6H_{14} 2,2-Dimethylbutane: 189
C_6H_{14} 2,3-Dimethylbutane: 190
$C_6H_{14}O$ Butyl ethyl ether: 69
$C_6H_{14}O$ 3,3-Dimethyl-1-butanol: 191
$C_6H_{14}O$ 2-Ethyl-1-butanol: 239
$C_6H_{14}O$ 2,3-Dimethyl-2-butanol: 192
$C_6H_{14}O$ 3,3-Dimethyl-2-butanol: 193
$C_6H_{14}O$ 1-Hexanol: 294
$C_6H_{14}O$ 2-Hexanol: 295
$C_6H_{14}O$ 3-Hexanol: 296
$C_6H_{14}O$ 2-Methyl-1-pentanol: 379
$C_6H_{14}O$ 3-Methyl-1-pentanol: 380
$C_6H_{14}O$ 2-Methyl-2-pentanol: 381
$C_6H_{14}O$ 3-Methyl-2-pentanol: 382
$C_6H_{14}O$ 4-Methyl-2-pentanol: 383
$C_6H_{14}O$ 2-Methyl-3-pentanol: 384
$C_6H_{14}O$ 3-Methyl-3-pentanol: 385
$C_6H_{14}O$ Dipropyl ether: 225
$C_6H_{14}O$ Diisopropyl ether: 183
$C_6H_{14}O_2$ Acetal: 1
$C_6H_{14}O_2$ Ethylene glycol diethyl ether: 250
$C_6H_{14}O_2$ Ethylene glycol monobutyl ether: 254
$C_6H_{14}O_2$ Hexylene glycol: 304
$C_6H_{14}O_3$ Diethylene glycol dimethyl ether: 165
$C_6H_{14}O_3$ Diethylene glycol monoethyl ether: 166
$C_6H_{14}O_3$ 1,2,6-Hexanetriol: 292
$C_6H_{14}O_4$ Triethylene glycol: 529
$C_6H_{15}N$ Triethylamine: 528
$C_6H_{15}N$ Dipropylamine: 224
$C_6H_{15}N$ Diisopropylamine: 182
$C_6H_{15}NO_3$ Triethanolamine: 527
$C_6H_{15}O_4P$ Triethyl phosphate: 530
$C_6H_{18}N_3OP$
          Hexamethylphosphoric triamide: 289
$C_7H_5Cl_3$ (Trichloromethyl)benzene: 517
$C_7H_5F_3$ (Trifluoromethyl)benzene: 534
C_7H_5N Benzonitrile: 24
$C_7H_6Cl_2$ Benzal chloride: 20
$C_7H_6Cl_2$ 3,4-Dichlorotoluene: 158

$C_7H_6Cl_2$ 2,4-Dichlorotoluene: 157
C_7H_6O Benzaldehyde: 21
$C_7H_6O_2$ Salicylaldehyde: 471
C_7H_7Cl Benzyl chloride: 28
C_7H_7Cl o-Chlorotoluene: 101
C_7H_7Cl m-Chlorotoluene: 102
C_7H_7Cl p-Chlorotoluene: 103
C_7H_7F o-Fluorotoluene: 272
C_7H_7F m-Fluorotoluene: 273
C_7H_7F p-Fluorotoluene: 274
$C_7H_7NO_3$ 2-Nitroanisole: 404
C_7H_8 Toluene: 501
C_7H_8O Anisole: 19
C_7H_8O Benzyl alcohol: 26
C_7H_8O o-Cresol: 106
C_7H_8O m-Cresol: 107
C_7H_8O p-Cresol: 108
C_7H_9N o-Toluidine: 502
C_7H_9N m-Toluidine: 503
C_7H_9N p-Toluidine: 504
C_7H_9N N-Methylaniline: 344
C_7H_9N 2,4-Lutidine: 331
C_7H_9N 2,6-Lutidine: 332
$C_7H_{12}O_4$ Dimethyl glutarate: 199
$C_7H_{12}O_4$ Diethyl malonate: 172
C_7H_{14} Methylcyclohexane: 352
C_7H_{14} 1-Heptene: 285
C_7H_{14} trans-2-Heptene: 287
C_7H_{14} cis-2-Heptene: 286
$C_7H_{14}O$ Methyl pentyl ketone: 387
$C_7H_{14}O$ Methyl isopentyl ketone: 370
$C_7H_{14}O$ Ethyl butyl ketone: 240
$C_7H_{14}O$ Diisopropyl ketone: 184
$C_7H_{14}O$ 1-Methylcyclohexanol: 353
$C_7H_{14}O$ cis-2-Methylcyclohexanol: 354
$C_7H_{14}O$ trans-2-Methylcyclohexanol: 355
$C_7H_{14}O$ cis-3-Methylcyclohexanol: 356
$C_7H_{14}O$ trans-3-Methylcyclohexanol: 357
$C_7H_{14}O$ cis-4-Methylcyclohexanol: 358
$C_7H_{14}O$ trans-4-Methylcyclohexanol: 359
$C_7H_{14}O_2$ Pentyl acetate: 433
$C_7H_{14}O_2$ Isopentyl acetate: 323
$C_7H_{14}O_2$ Ethyl isovalerate: 264
C_7H_{16} Heptane: 281
C_7H_{16} 2-Methylhexane: 367
C_7H_{16} 3-Methylhexane: 368
C_7H_{16} 2,2-Dimethylpentane: 210
C_7H_{16} 2,3-Dimethylpentane: 211
C_7H_{16} 2,4-Dimethylpentane: 212
C_7H_{16} 3,3-Dimethylpentane: 213
C_7H_{16} 3-Ethylpentane: 268

$C_{10}H_7Br$ 1-Bromonaphthalene: 43
$C_{10}H_7Cl$ 1-Chloronaphthalene: 94
$C_{10}H_8$ Naphthalene: 402
$C_{10}H_{10}O_4$ Dimethyl phthalate: 214
$C_{10}H_{12}$ 1,2,3,4-Tetrahydronaphthalene: 489
$C_{10}H_{12}O_2$ Propyl benzoate: 461
$C_{10}H_{14}$ Butylbenzene: 66
$C_{10}H_{14}$ *sec*-Butylbenzene: 67
$C_{10}H_{14}$ *tert*-Butylbenzene: 68
$C_{10}H_{14}$ Isobutylbenzene: 319
$C_{10}H_{14}$ *p*-Cymene: 123
$C_{10}H_{16}$ Camphene: 77
$C_{10}H_{16}$ α-Terpinene: 475
$C_{10}H_{16}$ β-Phellandrene: 437
$C_{10}H_{16}$ Dipentene: 221
$C_{10}H_{16}$ *d*-Limonene: 329
$C_{10}H_{16}$ *l*-Limonene: 330
$C_{10}H_{16}$ Terpinolene: 477
$C_{10}H_{16}$ β-Myrcene: 401
$C_{10}H_{16}$ Tricyclene: 524
$C_{10}H_{16}$ β-Pinene: 445
$C_{10}H_{16}$ α-Pinene: 444
$C_{10}H_{16}O$ (+)-Camphor: 78
$C_{10}H_{18}$ *cis*-Decahydronaphthalene: 124
$C_{10}H_{18}$ *trans*-Decahydronaphthalene: 125
$C_{10}H_{18}O$ α-Terpineol: 476
$C_{10}H_{18}O$ Cineole: 105
$C_{10}H_{20}$ 1-Decene: 127
$C_{10}H_{20}O_2$ 2-Ethylhexyl acetate: 262
$C_{10}H_{20}O_2$ Isopentyl isopentanoate: 324
$C_{10}H_{21}Br$ 1-Bromodecane: 38
$C_{10}H_{22}$ Decane: 126
$C_{10}H_{22}O$ Diisopentyl ether: 181

$C_{10}H_{22}O$ Dipentyl ether: 222
$C_{10}H_{22}O_2$ Ethylene glycol dibutyl ether: 249
$C_{11}H_{10}$ 1-Methylnaphthalene: 372
$C_{11}H_{12}O_2$ Ethyl *trans*-cinnamate: 241
$C_{11}H_{16}$ *p-tert*-Butyltoluene: 74
$C_{11}H_{22}$ 1-Undecene: 547
$C_{12}F_{27}N$ Trinonafluorobutylamine: 546
$C_{12}H_{10}O$ Diphenyl ether: 223
$C_{12}H_{12}$ 1,2-Dimethylnaphthalene: 208
$C_{12}H_{12}$ 1,6-Dimethylnaphthalene: 209
$C_{12}H_{16}$ Cyclohexylbenzene: 119
$C_{12}H_{20}O_4$ Dibutyl maleate: 138
$C_{12}H_{24}$ 1-Dodecene: 227
$C_{12}H_{26}$ Dodecane: 226
$C_{12}H_{26}O_3$ Diethylene glycol dibutyl ether: 163
$C_{12}H_{27}BO_3$ Tributyl borate: 509
$C_{12}H_{27}N$ Tributylamine: 508
$C_{12}H_{27}O_4P$ Tributyl phosphate: 510
$C_{13}H_{26}$ 1-Tridecene: 526
$C_{13}H_{28}$ Tridecane: 525
$C_{14}H_{12}O_2$ Benzyl benzoate: 27
$C_{14}H_{14}O$ Dibenzyl ether: 129
$C_{16}H_{22}O_4$ Dibutyl phthalate: 139
$C_{18}H_{34}O_2$ Oleic acid: 421
$C_{18}H_{34}O_4$ Dibutyl sebacate: 140
$C_{19}H_{36}O_2$ Methyl oleate: 376
$C_{21}H_{21}O_4P$ Tri-*o*-cresyl phosphate: 521
$C_{21}H_{21}O_4P$ Tri-*m*-cresyl phosphate: 522
$C_{21}H_{21}O_4P$ Tri-*p*-cresyl phosphate: 523
$C_{22}H_{42}O_2$ Butyl oleate: 72
$C_{22}H_{44}O_2$ Butyl stearate: 73
$C_{24}H_{38}O_4$ Bis(2-ethylhexyl) phthalate: 31

CAS Registry Number Index

This Index is ordered by Chemical Abstracts Service Registry Number. Some Registry Numbers are generic in nature; e.g., 1319-77-3 refers to "cresols (mixed isomers)". Thus some compounds appear under two or more Registry Numbers, the CAS RN for the specific isomer and the CAS RN for the generic compound(s). The number following the name is the ID for the solvent.

78-75-1	1,2-Dibromopropane: 134
78-76-2	2-Bromobutane: 34
78-78-4	Isopentane: 322
78-81-9	Isobutylamine: 318
78-82-0	2-Methylpropanenitrile: 389
78-83-1	2-Methyl-1-propanol: 391
78-84-2	Isobutanal: 315
78-87-5	1,2-Dichloropropane: 154
78-92-2	2-Butanol: 58
78-93-3	Methyl ethyl ketone: 361
79-00-5	1,1,2-Trichloroethane: 512
79-01-6	Trichloroethylene: 513
79-09-4	Propanoic acid: 452
79-10-7	Acrylic acid: 11
79-16-3	*N*-Methylacetamide: 337
79-20-9	Methyl acetate: 338
79-24-3	Nitroethane: 406
79-27-6	1,1,2,2-Tetrabromoethane: 478
79-29-8	2,3-Dimethylbutane: 190
79-31-2	2-Methylpropanoic acid: 390
79-34-5	1,1,2,2-Tetrachloroethane: 482
79-35-6	1,1-Dichloro-2,2-difluoroethylene: 145
79-41-4	Methacrylic acid: 335
79-46-9	2-Nitropropane: 409
79-92-5	Camphene: 77
80-56-8	α-Pinene: 444
80-62-6	Methyl methacrylate: 371
84-74-2	Dibutyl phthalate: 139
87-59-2	2,3-Xylidine: 559
87-62-7	2,6-Xylidine: 562
90-02-8	Salicylaldehyde: 471
90-11-9	1-Bromonaphthalene: 43
90-12-0	1-Methylnaphthalene: 372
90-13-1	1-Chloronaphthalene: 94
91-16-7	Veratrole: 548
91-20-3	Naphthalene: 402
91-22-5	Quinoline: 470
91-23-6	2-Nitroanisole: 404
93-58-3	Methyl benzoate: 345
93-89-0	Ethyl benzoate: 237
94-96-2	2-Ethyl-1,3-hexanediol: 260
95-47-6	*o*-Xylene: 550
95-48-7	*o*-Cresol: 106
95-49-8	*o*-Chlorotoluene: 101
95-50-1	*o*-Dichlorobenzene: 142
95-51-2	*o*-Chloroaniline: 81
95-52-3	*o*-Fluorotoluene: 272
95-53-4	*o*-Toluidine: 502
95-63-6	Pseudocumene: 465
95-64-7	3,4-Xylidine: 563
95-65-8	3,4-Xylenol: 557
95-68-1	2,4-Xylidine: 560
95-73-8	2,4-Dichlorotoluene: 157
95-75-0	3,4-Dichlorotoluene: 158
95-78-3	2,5-Xylidine: 561
95-87-4	2,5-Xylenol: 555
95-92-1	Diethyl oxalate: 173
96-14-0	3-Methylpentane: 378
96-18-4	1,2,3-Trichloropropane: 519
96-22-0	Diethyl ketone: 171
96-33-3	Methyl acrylate: 340
96-37-7	Methylcyclopentane: 360
96-47-9	2-Methyltetrahydrofuran: 396
96-48-0	γ-Butyrolactone: 76
96-49-1	Ethylene carbonate: 244
97-85-8	Isobutyl isobutanoate: 321
97-95-0	2-Ethyl-1-butanol: 239
97-99-4	Tetrahydrofurfuryl alcohol: 488
98-00-0	Furfuryl alcohol: 279
98-01-1	Furfural: 278
98-06-6	*tert*-Butylbenzene: 68
98-07-7	(Trichloromethyl)benzene: 517
98-08-8	(Trifluoromethyl)benzene: 534
98-51-1	*p-tert*-Butyltoluene: 74
98-82-8	Cumene: 113
98-86-2	Acetophenone: 8
98-87-3	Benzal chloride: 20
98-95-3	Nitrobenzene: 405
99-86-5	α-Terpinene: 475
99-87-6	*p*-Cymene: 123
100-41-4	Ethylbenzene: 236
100-42-5	Styrene: 472
100-44-7	Benzyl chloride: 28
100-47-0	Benzonitrile: 24
100-51-6	Benzyl alcohol: 26
100-52-7	Benzaldehyde: 21
100-61-8	*N*-Methylaniline: 344
100-66-3	Anisole: 19
101-84-8	Diphenyl ether: 223
102-71-6	Triethanolamine: 527
102-76-1	Triacetin: 506
102-82-9	Tributylamine: 508
103-09-3	2-Ethylhexyl acetate: 262
103-50-4	Dibenzyl ether: 129
103-65-1	Propylbenzene: 460
103-73-1	Phenetole: 438
104-51-8	Butylbenzene: 66
104-75-6	2-Ethylhexylamine: 263
104-76-7	2-Ethyl-1-hexanol: 261
105-30-6	2-Methyl-1-pentanol: 379
105-34-0	Methyl cyanoacetate: 351
105-37-3	Ethyl propanoate: 269
105-45-3	Methyl acetoacetate: 339
105-46-4	*sec*-Butyl acetate: 62

Compound Class Index

In this index, solvents are sorted by chemical class. Within each class, compounds are ordered first by number of carbon atoms, second by number of hydrogen atoms, and then by the other elements in alphabetical order (the Hill Convention). The number following the name is the ID for the solvent.

Hydrocarbons

Propane: 448
Butane: 49
Isobutane: 316
Cyclopentene: 122
Cyclopentane: 120
1-Pentene: 430
trans-2-Pentene: 432
cis-2-Pentene: 431
Isopentane: 322
Pentane: 423
Neopentane: 403
Benzene: 22
Cyclohexene: 117
Cyclohexane: 114
Methylcyclopentane: 360
1-Hexene: 297
trans-2-Hexene: 299
cis-2-Hexene: 298
trans-3-Hexene: 301
cis-3-Hexene: 300
Hexane: 290
2-Methylpentane: 377
3-Methylpentane: 378
2,2-Dimethylbutane: 189
2,3-Dimethylbutane: 190
Toluene: 501
Methylcyclohexane: 352
1-Heptene: 285
trans-2-Heptene: 287
cis-2-Heptene: 286
Heptane: 281
2-Methylhexane: 367
3-Methylhexane: 368
2,2-Dimethylpentane: 210
2,3-Dimethylpentane: 211
2,4-Dimethylpentane: 212
3,3-Dimethylpentane: 213
3-Ethylpentane: 268
2,2,3-Trimethylbutane: 537
Styrene: 472
o-Xylene: 550
m-Xylene: 551

p-Xylene: 552
Ethylbenzene: 236
cis-1,2-Dimethylcyclohexane: 194
trans-1,2-Dimethylcyclohexane: 195
Ethylcyclohexane: 243
1-Octene: 418
trans-2-Octene: 420
cis-2-Octene: 419
2,2,3,3-Tetramethylbutane: 492
2-Methylheptane: 364
3-Methylheptane: 365
4-Methylheptane: 366
2,2-Dimethylhexane: 201
2,3-Dimethylhexane: 202
2,5-Dimethylhexane: 204
3,3-Dimethylhexane: 205
3,4-Dimethylhexane: 206
3-Ethylhexane: 259
Octane: 413
3-Ethyl-2-methylpentane: 266
3-Ethyl-3-methylpentane: 267
2,2,3-Trimethylpentane: 540
2,2,4-Trimethylpentane: 541
2,3,3-Trimethylpentane: 542
2,3,4-Trimethylpentane: 543
2,4-Dimethylhexane: 203
Cumene: 113
Propylbenzene: 460
1,2,3-Trimethylbenzene: 536
Pseudocumene: 465
Mesitylene: 333
1-Nonene: 412
Nonane: 410
2-Methyloctane: 373
3-Methyloctane: 374
4-Methyloctane: 375
2,2-Dimethylheptane: 200
2,2,5-Trimethylhexane: 538
2,3,5-Trimethylhexane: 539
2,2,3,3-Tetramethylpentane: 493
2,2,3,4-Tetramethylpentane: 494
2,2,4,4-Tetramethylpentane: 495
2,3,3,4-Tetramethylpentane: 496
2,3-Diethylpentane: 174

Halogenated hydrocarbons

Alcohols (mono- and polyhydroxy)

Ethers

Dimethyl ether: 197
1,2-Propylene oxide: 463
1,3-Dioxolane: 220
Furan: 277
Ethyl vinyl ether: 270
Tetrahydrofuran: 487
1,4-Dioxane: 219
1,2-Epoxybutane: 229
Diethyl ether: 170
Ethylene glycol dimethyl ether: 251
2-Methyltetrahydrofuran: 396
Tetrahydropyran: 490
Butyl vinyl ether: 75
Butyl ethyl ether: 69
Dipropyl ether: 225
Diisopropyl ether: 183
Ethylene glycol diethyl ether: 250
Diethylene glycol dimethyl ether: 165
Anisole: 19
Phenetole: 438
Veratrole: 548
Dibutyl ether: 137
Diethylene glycol diethyl ether: 164
Benzyl ethyl ether: 29
Cineole: 105
Diisopentyl ether: 181
Dipentyl ether: 222
Ethylene glycol dibutyl ether: 249
Diphenyl ether: 223
Diethylene glycol dibutyl ether: 163
Dibenzyl ether: 129

Aldehydes and ketones

Acetaldehyde: 2
Acrolein: 10
Propanal: 447
Acetone: 6
trans-Crotonaldehyde: 109
Butanal: 48
Methyl ethyl ketone: 361
Isobutanal: 315
Cyclopentanone: 121
Acetylacetone: 9
Methyl propyl ketone: 393
Diethyl ketone: 171
Cyclohexanone: 116
Mesityl oxide: 334
4-Methyl-4-penten-2-one: 386
Butyl methyl ketone: 71
Methyl isobutyl ketone: 369

Benzaldehyde: 21
Methyl pentyl ketone: 387
Methyl isopentyl ketone: 370
Ethyl butyl ketone: 240
Diisopropyl ketone: 184
Acetophenone: 8
Hexyl methyl ketone: 305
Isophorone: 325
Diisobutyl ketone: 180
(+)-Camphor: 78

Acids and acid anhydrides

Formic acid: 276
Acetic acid: 4
Acrylic acid: 11
Propanoic acid: 452
trans-Crotonic acid: 110
Methacrylic acid: 335
Acetic anhydride: 5
Butanoic acid: 55
2-Methylpropanoic acid: 390
3-Methylbutanoic acid: 346
Pentanoic acid: 426
Propanoic anhydride: 453
Hexanoic acid: 293
Butanoic anhydride: 56
Octanoic acid: 415
Nonanoic acid: 411
Oleic acid: 421

Esters

Methyl formate: 363
Ethylene carbonate: 244
Methyl acetate: 338
Ethyl formate: 258
Trimethyl phosphate: 544
Vinyl acetate: 549
γ-Butyrolactone: 76
Methyl acrylate: 340
Propylene carbonate: 462
Ethyl acetate: 232
Propyl formate: 464
Propargyl acetate: 456
Allyl acetate: 14
Ethyl acrylate: 234
Methyl methacrylate: 371
Isopropyl acetate: 326
Propyl acetate: 458
Butyl formate: 70
Isobutyl formate: 320

Nitrogen compounds

α-Tolunitrile: 505
2,3-Xylidine: 559
2,4-Xylidine: 560
2,5-Xylidine: 561
2,6-Xylidine: 562
3,4-Xylidine: 563
3,5-Xylidine: 564
N,N-Dimethylaniline: 188
Phenylethylamine: 440
2,4,6-Trimethylpyridine: 545
Octanenitrile: 414
Dibutylamine: 136
2-Ethylhexylamine: 263
Isoquinoline: 328
Quinoline: 470
Tributylamine: 508

Sulfur compounds

Carbon disulfide: 80
Dimethyl sulfoxide: 218
Dimethyl sulfide: 217
Dimethyl disulfide: 196
Thiophene: 500
Sulfolane: 474
Tetrahydrothiophene: 491
1-Butanethiol: 54
Diethyl sulfide: 176
2-Methylthiophene: 397
3-Methylthiophene: 398
Benzenethiol: 23
Dibutyl sulfide: 141

Silicon compounds

Trichloromethylsilane: 518
Trichloroethylsilane: 514
Tetramethylsilane: 497
Tetraethylsilane: 486

Compounds with more than one characteristic group

Trifluoroacetic acid: 531
2,2,2-Trifluoroethanol: 533
Ethylene chlorohydrin: 245
Ethanolamine: 231
Epichlorohydrin: 228
Hydracrylonitrile: 306
Ethylene glycol monomethyl ether: 256
Methylal: 342
Methyl cyanoacetate: 351
Bis(2-chloroethyl) ether: 30
Morpholine: 400
Ethylene glycol monoethyl ether: 255
2,2'-Thiodiethanol: 499
Diethylene glycol: 162
2-Aminoisobutanol: 17
Diethanolamine: 159
Furfural: 278
Furfuryl alcohol: 279
Ethyl cyanoacetate: 242
Methyl acetoacetate: 339
Tetrahydrofurfuryl alcohol: 488
Ethylene glycol momomethyl ether acetate: 253
Ethyl lactate: 265
Diethylene glycol monomethyl ether: 168
o-Chloroaniline: 81
Ethyl acetoacetate: 233
Diacetone alcohol: 128
Ethylene glycol ethyl ether acetate: 252
Acetal: 1
Ethylene glycol monobutyl ether: 254
Diethylene glycol monoethyl ether: 166
Triethylene glycol: 529
Triethanolamine: 527
Hexamethylphosphoric triamide: 289
Salicylaldehyde: 471
2-Nitroanisole: 404
Methyl salicylate: 395
Diethylene glycol monoethyl ether acetate: 167
Tetraethylene glycol: 485
Trinonafluorobutylamine: 546

Boiling Point Index

This index is arranged by normal boiling point in °C (at 760 mmHg pressure). The number following the name is the ID for the solvent.

67.9	*trans*-2-Hexene: 299	91.2	2-Bromobutane: 34
68.0	Bromochloromethane: 37	91.5	1,1,1,2-Tetrachloro-2,2-difluoroethane: 480
68.2	2-Chlorobutane: 84	92	3-Methylhexane: 368
68.5	1-Chloro-2-methylpropane: 92	92.1	Diethyl sulfide: 176
68.5	Diisopropyl ether: 183	92.3	Butyl ethyl ether: 69
68.7	Hexane: 290	92.5	1,2-Dibromo-1,1-difluoroethane: 130
68.8	*cis*-2-Hexene: 298	92.8	1,1,2,2-Tetrachloro-1,2-difluoroethane: 479
71.1	1-Bromopropane: 45	93.5	3-Ethylpentane: 268
71.8	Methylcyclopentane: 360	93.6	1-Heptene: 285
72.5	Iodoethane: 310	94	Butyl vinyl ether: 75
72.5	Vinyl acetate: 549	96.4	1,2-Dichloropropane: 154
73	Trifluoroacetic acid: 531	97.0	Allyl alcohol: 15
73.3	2-Bromo-2-methylpropane: 42	97	Dibromomethane: 133
74.0	1,1,1-Trichloroethane: 511	97.1	Propanenitrile: 451
74	2,2,2-Trifluoroethanol: 533	97.2	1-Propanol: 454
74.8	Butanal: 48	98	*trans*-2-Heptene: 287
76.8	Tetrachloromethane: 484	98.2	Isobutyl formate: 320
77.0	Butylamine: 63	98.4	*cis*-2-Heptene: 286
77.1	Ethyl acetate: 232	98.5	1-Butanethiol: 54
77.3	Acrylonitrile: 12	98.5	Heptane: 281
78	1,3-Dioxolane: 220	98.9	1-Chloro-3-methylbutane: 90
78	2-Methyltetrahydrofuran: 396	99.1	Ethyl propanoate: 269
78.2	Ethanol: 230	99.2	2,2,4-Trimethylpentane: 541
78.6	1-Chlorobutane: 83	99.4	Ethyl acrylate: 234
79.2	2,2-Dimethylpentane: 210	99.5	2-Butanol: 58
79.5	Methyl ethyl ketone: 361	100.5	Methyl methacrylate: 371
80.0	Benzene: 22	100.5	Trichloroethylsilane: 514
80.2	Hexafluorobenzene: 288	100.9	Methylcyclohexane: 352
80.4	2,4-Dimethylpentane: 212	101	Formic acid: 276
80.7	Cyclohexane: 114	101.1	Nitromethane: 407
80.7	Methyl acrylate: 340	101.5	1,4-Dioxane: 219
80.8	2,2,3-Trimethylbutane: 537	101.5	Propyl acetate: 458
80.9	Propyl formate: 464	101.6	1-Bromobutane: 33
81.6	Acetonitrile: 7	101.9	Diethyl ketone: 171
82.3	2-Propanol: 455	102.1	(Trifluoromethyl)benzene: 534
82.4	2-Methyl-2-propanol: 392	102.2	Acetal: 1
82.9	Cyclohexene: 117	102.2	*trans*-Crotonaldehyde: 109
83.5	1,2-Dichloroethane: 148	102.2	Methyl propyl ketone: 393
83.9	Diisopropylamine: 182	102.4	2-Methyl-2-butanol: 349
84.0	Thiophene: 500	102.6	1-Iodopropane: 313
84.7	Fluorobenzene: 271	103.5	Allyl acetate: 14
85	Ethylene glycol dimethyl ether: 251	103.9	2-Methylpropanenitrile: 389
86.0	3,3-Dimethylpentane: 213	104.3	Pentylamine: 434
86.5	Pyrrolidine: 468	106.1	Butyl formate: 70
87.2	Trichloroethylene: 513	106.2	Piperidine: 446
88	Tetrahydropyran: 490	106.4	2,2,3,3-Tetramethylbutane: 492
88.6	Isopropyl acetate: 326	106.8	2,2-Dimethylhexane: 201
89	Triethylamine: 528	107	1-Bromo-2-chloroethane: 35
89.5	2-Iodopropane: 314	107.8	1-Chloropentane: 97
89.7	2,3-Dimethylpentane: 211	107.8	2-Methyl-1-propanol: 391
90.0	Dipropyl ether: 225	109.1	2,5-Dimethylhexane: 204
90.0	2-Methylhexane: 367		
90.3	Methylacrylonitrile: 341		

141.5	2,3,3,4-Tetramethylpentane: 496
141.9	1,2-Dibromopropane: 134
142.4	4-Methyloctane: 375
142.5	Isopentyl acetate: 323
143	3,3-Dimethyl-1-butanol: 191
143	Ethylene glycol momomethyl ether acetate: 253
143.2	2-Methyloctane: 373
144	Methyl isopentyl ketone: 370
144.1	2,6-Lutidine: 332
144.1	3-Picoline: 442
144.2	3-Methyloctane: 374
144.5	*o*-Xylene: 550
145	Styrene: 472
145.3	4-Picoline: 443
146.3	3,3-Diethylpentane: 175
146.5	1,1,2,2-Tetrachloroethane: 482
146.9	1-Nonene: 412
147	2-Ethyl-1-butanol: 239
147	Ethyl butyl ketone: 240
147.5	*sec*-Hexyl acetate: 303
148	*N*-Methylpropanamide: 388
148.6	Isobutyl isobutanoate: 321
149	2-Methyl-1-pentanol: 379
149.1	Tribromomethane: 507
149.2	Pentyl acetate: 433
150.8	Nonane: 410
151.0	Methyl pentyl ketone: 387
152.4	Cumene: 113
152.5	Tricyclene: 524
153	*N,N*-Dimethylformamide: 198
153	3-Methyl-1-pentanol: 380
153.7	Anisole: 19
154.4	2-Methylpropanoic acid: 390
154.5	Ethyl lactate: 265
154.7	Tetraethylsilane: 486
155	1-Methylcyclohexanol: 353
155.4	Cyclohexanone: 116
156.0	Bromobenzene: 32
156	α-Pinene: 444
156.4	Ethylene glycol ethyl ether acetate: 252
156.5	4-Methylvaleronitrile: 399
157	3-Heptanol: 284
157	1,2,3-Trichloropropane: 519
157.6	1-Hexanol: 294
158.5	Camphene: 77
158.5	2,4-Lutidine: 331
159.0	*o*-Chlorotoluene: 101
159	2-Heptanol: 283
159.2	Propylbenzene: 460
159.6	Dibutylamine: 136
159.8	Pentachloroethane: 422
160.8	Cyclohexanol: 115

161.7	Furfural: 278
161.8	*m*-Chlorotoluene: 102
162	Diethylene glycol dimethyl ether: 165
162.4	*p*-Chlorotoluene: 103
162.5	Methacrylic acid: 335
163.6	Hexanenitrile: 291
163.7	Butanoic acid: 55
164.7	Mesitylene: 333
165	*N,N*-Dimethylacetamide: 185
165	*cis*-2-Methylcyclohexanol: 354
165.5	2-Aminoisobutanol: 17
166	β-Pinene: 445
167	*trans*-3-Methylcyclohexanol: 357
167	β-Myrcene: 401
167.5	*trans*-2-Methylcyclohexanol: 355
167.9	Diacetone alcohol: 128
168	*cis*-3-Methylcyclohexanol: 356
168.4	Ethylene glycol monobutyl ether: 254
169.1	Benzenethiol: 23
169.1	*tert*-Butylbenzene: 68
169.2	2-Ethylhexylamine: 263
169.3	Pseudocumene: 465
169.4	Diisobutyl ketone: 180
169.8	Phenetole: 438
170	Propanoic anhydride: 453
170.5	1-Decene: 127
170.6	2,4,6-Trimethylpyridine: 545
171	Ethanolamine: 231
171	Furfuryl alcohol: 279
171.5	Hexyl acetate: 302
171.5	β-Phellandrene: 437
171.7	Methyl acetoacetate: 339
172	3-(Chloromethyl)heptane: 91
172.5	Diisopentyl ether: 181
172.5	Hexyl methyl ketone: 305
172.7	Isobutylbenzene: 319
173	*m*-Dichlorobenzene: 143
173	*cis*-4-Methylcyclohexanol: 358
173.3	*sec*-Butylbenzene: 67
174	*p*-Dichlorobenzene: 144
174	*trans*-4-Methylcyclohexanol: 359
174	α-Terpinene: 475
174.1	Decane: 126
176.1	1,2,3-Trimethylbenzene: 536
176.4	Cineole: 105
176.4	1-Heptanol: 282
176.5	3-Methylbutanoic acid: 346
176.5	Tetramethylurea: 498
177.1	*p*-Cymene: 123

221	Hydracrylonitrile: 306	256	Diethylene glycol dibutyl ether: 163
221	(Trichloromethyl)benzene: 517		
221.5	2,3-Xylidine: 559	258.0	Diphenyl ether: 223
221.7	3,5-Xylenol: 558	259	1-Chloronaphthalene: 94
222.0	Acetamide: 3	259	Triacetin: 506
222.9	Methyl salicylate: 395	264	1,6-Dimethylnaphthalene: 209
227	3,4-Xylenol: 557	266	Succinonitrile: 473
228	3,4-Xylidine: 563	266.5	1,2-Dimethylnaphthalene: 208
232.5	Hexamethylphosphoric triamide: 289	268.8	Diethanolamine: 159
		270	ε-Caprolactam: 79
232.8	1-Tridecene: 526	271.5	Ethyl *trans*-cinnamate: 241
233.5	α-Tolunitrile: 505	277	2-Nitroanisole: 404
234	Tributyl borate: 509	280	Dibutyl maleate: 138
235	1,4-Butanediol: 51	281	1-Bromonaphthalene: 43
235	*cis*-2-Butene-1,4-diol: 59	282	2,2'-Thiodiethanol: 499
235.4	Tridecane: 525	283.7	Dimethyl phthalate: 214
237.1	Quinoline: 470	285	Triethylene glycol: 529
239	Octanoic acid: 415	287.3	Sulfolane: 474
239	1,5-Pentanediol: 424	289	Tributyl phosphate: 510
240.1	Cyclohexylbenzene: 119	290	Glycerol: 280
240.6	1-Bromodecane: 38	295	Adiponitrile: 13
242	Propylene carbonate: 462	298	Dibenzyl ether: 129
243.2	Isoquinoline: 328	323.5	Benzyl benzoate: 27
243.5	1,1,2,2-Tetrabromoethane: 478	328	Tetraethylene glycol: 485
244	2-Ethyl-1,3-hexanediol: 260	335.4	Triethanolamine: 527
244.7	1-Methylnaphthalene: 372	340	Dibutyl phthalate: 139
245.8	Diethylene glycol: 162	343	Butyl stearate: 73
248	Ethylene carbonate: 244	344.5	Dibutyl sebacate: 140
251	2-Pyrrolidone: 469	360	Oleic acid: 421
254.5	Nonanoic acid: 411	384	Bis(2-ethylhexyl) phthalate: 31
		410	Tri-*o*-cresyl phosphate: 521

Melting Point Index

Melting points are given in °C. The number following the name is the ID for the solvent.

<-80	Dibutyl maleate: 138	-126.1	Dipropyl ether: 225
<-80	Tetrahydrofurfuryl alcohol: 488	-126.1	1-Propanol: 454
<-75	Methyl acrylate: 340	-126	2-Bromopropene: 47
<-72	*sec*-Butylamine: 64	-124	Butyl ethyl ether: 69
<-70	Tributyl borate: 509	-123.8	2,2-Dimethylpentane: 210
<-50	1,3-Butanediol: 50	-123.1	1-Chlorobutane: 83
<-30	*trans*-Crotonyl alcohol: 112	-123	Acetaldehyde: 2
<-15	2-Ethyl-1-butanol: 239	-122.8	1-Chloropropane: 98
<-15	2,3-Xylidine: 559	-122.5	1,1-Dichloroethylene: 149
<0	Phenylethylamine: 440	-121.1	2,2-Dimethylhexane: 201
-187.6	Propane: 448	-121	4-Methylheptane: 366
-181	Chlorotrifluoromethane: 104	-121.0	2,2,3,4-Tetramethylpentane: 494
-165.2	1-Pentene: 430	-120	3-Methylheptane: 365
-162.9	3-Methylpentane: 378	-119.9	2,4-Dimethylpentane: 212
-159.9	Isopentane: 322	-119.7	1-Heptene: 285
-158	Dichlorodifluoromethane: 146	-119.4	3-Methylhexane: 368
-157.4	Chlorodifluoromethane: 86	-118.6	Bromoethane: 39
-153.7	Chloroethylene: 88	-118.6	3-Ethylpentane: 268
-153.7	2-Methylpentane: 377	-118.2	2-Methylhexane: 367
-151.4	*cis*-2-Pentene: 431	-117.2	2-Chloropropane: 99
-150	1,2-Epoxybutane: 229	-117.2	3-Methyl-1-butanol: 348
-144	1,1-Difluoroethylene: 178	-117.1	Trimethylamine: 535
-142.5	Methylcyclopentane: 360	-117	1,1-Difluoroethane: 177
-141.5	Dimethyl ether: 197	-116.3	Diethyl ether: 170
-141.1	*cis*-2-Hexene: 298	-116	1,1-Dichloro-2,2-difluoroethane: 145
-140.2	*trans*-2-Pentene: 432		
-139.7	1-Hexene: 297	-115.8	Ethyl vinyl ether: 270
-138.7	Chloroethane: 87	-115.7	1-Butanethiol: 54
-138.3	Isobutane: 316	-115.4	*trans*-3-Hexene: 301
-138.2	Butane: 49	-115	Bromochlorofluoromethane: 36
-137.8	Bromoethylene: 40	-114.9	3-Ethyl-2-methylpentane: 266
-137.8	*cis*-3-Hexene: 300	-114.7	2-Butanol: 58
-135.1	Cyclopentene: 122	-114.7	Triethylamine: 528
-135	Dichlorofluoromethane: 152	-114.1	Ethanol: 230
-134.9	3,3-Dimethylpentane: 213	-113.3	4-Methyloctane: 375
-134.5	3-Chloropropene: 100	-113	2,2-Dimethylheptane: 200
-133	*trans*-2-Hexene: 299	-112.4	1-Bromobutane: 33
-131.3	2-Chlorobutane: 84	-112.2	2,2,3-Trimethylpentane: 540
-130.8	1-Chloro-1,1-difluoroethane: 85	-112	1,2-Propylene oxide: 463
-130.3	1-Chloro-2-methylpropane: 92	-111.9	2-Bromobutane: 34
-129.7	Pentane: 423	-111.9	Butanenitrile: 53
-129	Allyl alcohol: 15	-111.5	Carbon disulfide: 80
-128.8	2,3-Dimethylbutane: 190	-111.3	Ethylcyclohexane: 243
-128.2	Perfluorobutane: 435	-111.3	1,1,1-Trifluoroethane: 532
-127.9	2,3,5-Trimethylhexane: 539	-111.1	Iodoethane: 310
-126.6	Methylcyclohexane: 352	-111.1	Trichlorofluoromethane: 515
-126.1	3,3-Dimethylhexane: 205		

-110.4	1,2-Dibromotetrafluoroethane: 135	-95	1,3-Dioxolane: 220
-110	1-Bromopropane: 45	-94.9	Ethylbenzene: 236
-109.5	*trans*-2-Heptene: 287	-94.9	Toluene: 501
-109.2	2,3,4-Trimethylpentane: 543	-94.8	Acetone: 6
-108.9	2-Methylheptane: 364	-94	1,2-Dichlorotetrafluoroethane: 156
-108.3	Tetrahydrofuran: 487	-93.8	Cyclopentane: 120
-108	2-Methyl-1-propanol: 391	-93.7	Bromomethane: 41
-108	1-Nitropropane: 408	-93.4	Methylamine: 343
-107.6	3-Methyloctane: 374	-93.2	Vinyl acetate: 549
-107.3	2,2,4-Trimethylpentane: 541	-93	Propyl acetate: 458
-105.7	2,2,5-Trimethylhexane: 538	-92.9	Propyl formate: 464
-105.6	Trichloroethylsilane: 514	-92.8	Propanenitrile: 451
-104.8	Methylal: 342	-92.2	Dimethylamine: 187
-104.4	1-Chloro-3-methylbutane: 90	-92	Butyl vinyl ether: 75
-104.2	2-Iodobutane: 309	-91.5	Butyl formate: 70
-103.9	Diethyl sulfide: 176	-91.3	2-Nitropropane: 409
-103.5	Cyclohexene: 117	-91	2,5-Dimethylhexane: 204
-103	1-Iodobutane: 308	-90.9	3-Ethyl-3-methylpentane: 267
-103	2-Methyl-2-pentanol: 381	-90.6	Heptane: 281
-102.1	2,3,3,4-Tetramethylpentane: 496	-90	*trans*-1,2-Dimethylcyclohexane: 195
-101.7	1-Octene: 418		
-101.3	1-Iodopropane: 313	-90	2-Iodopropane: 314
-100.9	2,3,3-Trimethylpentane: 542	-90	4-Methyl-2-pentanol: 383
-100.4	1,2-Dichloropropane: 154	-90	Trichloromethylsilane: 518
-100.2	*cis*-2-Octene: 419	-89.8	1-Butanol: 57
-100	Acetal: 1	-89.5	Nitroethane: 406
-99.5	Propylbenzene: 460	-89.5	2-Propanol: 455
-99.4	Chloropentafluoroethane: 96	-89	2-Bromopropane: 46
-99.3	Ethyl isovalerate: 264	-88.2	Allylamine: 16
-99	Butanal: 48	-87.9	Bromochloromethane: 37
-99	1-Chloropentane: 97	-87.9	Butylbenzene: 66
-99	2,2-Dimethylbutane: 189	-87.7	Acrolein: 10
-99	Methyl formate: 363	-87.7	*trans*-2-Octene: 420
-99.0	Tetramethylsilane: 497	-87	*m*-Fluorotoluene: 273
-98.9	*sec*-Butyl acetate: 62	-86.8	Diisopropyl ether: 183
-98.8	Isobutyl acetate: 317	-86.7	Isobutylamine: 318
-98.3	Dimethyl sulfide: 217	-86.6	Methyl ethyl ketone: 361
-98	Ethyl butanoate: 238	-85.6	Furan: 277
-98	Methyl acetate: 338	-85.1	Ethylene glycol monomethyl ether: 256
-97.7	Chloromethane: 89		
-97.6	Methanol: 336	-85	Dimethyl disulfide: 196
-97	*d*-Limonene: 329	-84.7	Trichloroethylene: 513
-96.9	1,1-Dichloroethane: 147	-84	Methyl isobutyl ketone: 369
-96.2	Pentanenitrile: 425	-83.6	Ethyl acetate: 232
-96.1	Tetrahydrothiophene: 491	-83.5	Acrylonitrile: 12
-96.0	Cumene: 113	-83	Propylamine: 459
-95.8	Isobutyl formate: 320	-82.7	*sec*-Butylbenzene: 67
-95.5	Dipentene: 221	-81.3	1-Nonene: 412
-95.3	Hexane: 290	-80.9	Hexyl acetate: 302
-95.2	Dibutyl ether: 137	-80.7	Isobutyl isobutanoate: 321
-95.1	Dichloromethane: 153	-80.5	Ethylamine: 235
-95.1	Isopropylamine: 327	-80.3	Hexanenitrile: 291
-95	1-Bromopentane: 44	-80.3	2-Methyloctane: 373

-46	Trimethyl phosphate: 544	-31	Styrene: 472
-46	2,4,6-Trimethylpyridine: 545	-30.9	*N*-Methylpropanamide: 388
-45.6	Octanenitrile: 414	-30.6	Bromobenzene: 32
-45.2	Chlorobenzene: 82	-30.4	1-Methylnaphthalene: 372
-45	Benzyl chloride: 28	-30.4	1,1,1-Trichloroethane: 511
-45	Diethylene glycol diethyl ether: 164	-30.3	*trans*-Decahydronaphthalene: 125
-45	Ethyl acetoacetate: 233	-29.7	Decane: 126
-45	Propanoic anhydride: 453	-29.5	Phenetole: 438
-45	Tetrahydropyran: 490	-29.3	3-Methylbutanoic acid: 346
-44.7	Mesitylene: 333	-29.2	1-Bromodecane: 38
-44.6	1-Hexanol: 294	-29.1	(Trifluoromethyl)benzene: 534
-44	Diacetone alcohol: 128	-29	Pentachloroethane: 422
-43.8	Acetonitrile: 7	-28.5	Nitromethane: 407
-43.8	Pseudocumene: 465	-26.7	1,3-Propanediol: 450
-43.8	1,1,2,2-Tetrachloroethane: 482	-26.4	Butyl oleate: 72
-43.5	2,2,2-Trifluoroethanol: 533	-26	Benzaldehyde: 21
-43.3	γ-Butyrolactone: 76	-26	2-Chloro-2-methylpropane: 93
-43	Diethyl carbonate: 161	-26	Epichlorohydrin: 228
-42.9	*cis*-Decahydronaphthalene: 124	-26	Ethyl lactate: 265
-42.5	Dimethyl glutarate: 199	-25.4	1,2,3-Trimethylbenzene: 536
-42.2	Fluorobenzene: 271	-25.2	*o*-Xylene: 550
-41.6	Pyridine: 466	-25	Diethylene glycol monoethyl ether acetate: 167
-41.5	Diisobutyl ketone: 180		
-40.6	Diethyl oxalate: 173	-25	2,2,3-Trimethylbutane: 537
-40.1	Perfluorocyclobutane: 436	-24.8	*m*-Dichlorobenzene: 143
-40	2-Ethyl-1,3-hexanediol: 260	-24	*N*-Methyl-2-pyrrolidone: 394
-39.4	Thiophene: 500	-23.8	α-Tolunitrile: 505
-39	Diethylenetriamine: 169	-23.6	3-Methyl-3-pentanol: 385
-39	Diethyl ketone: 171	-23.4	Pyrrole: 467
-39	Ethyl butyl ketone: 240	-23	Acetylacetone: 9
-37.5	Anisole: 19	-23	Tetrachloromethane: 484
-36.6	1,1,2-Trichloroethane: 512	-22.5	Ethyl cyanoacetate: 242
-36.5	Furfural: 278	-22.5	Methyl cyanoacetate: 351
-35.8	Methylacrylonitrile: 341	-22.3	Tetrachloroethylene: 483
-35.7	1,2,3,4-Tetrahydronaphthalene: 489	-20.7	Propanoic acid: 452
		-20	*N,N*-Dimethylacetamide: 185
-35.6	*o*-Chlorotoluene: 101	-19.9	Methyl oleate: 376
-35.5	1,2-Dichloroethane: 148	-19	Dimethyl maleate: 207
-35.2	1-Dodecene: 227	-18.1	3-Picoline: 442
-35	Dibutyl phthalate: 139	-18	1,5-Pentanediol: 424
-35	Methyl pentyl ketone: 387	-17.7	Cyclohexylamine: 118
-35	1,1,2-Trichlorotrifluoroethane: 520	-17	Benzal chloride: 20
		-16.9	1,6-Dimethylnaphthalene: 209
-34	Ethyl benzoate: 237	-16.7	1-Bromo-2-chloroethane: 35
-34	1-Heptanol: 282	-16.7	*o*-Dichlorobenzene: 142
-34	Pentanoic acid: 426	-16.6	Neopentane: 403
-33.1	3,3-Diethylpentane: 175	-16.3	*o*-Toluidine: 502
-31.6	2-Octanol: 417	-16.2	2-Bromo-2-methylpropane: 42
-31.3	Iodobenzene: 307	-16	Hexyl methyl ketone: 305
-31.2	*m*-Toluidine: 503	-15.5	1-Octanol: 416
-31	Cyclohexanone: 116	-15.2	Benzyl alcohol: 26
-31	Ethylene glycol diacetate: 248	-15.2	3,4-Dichlorotoluene: 158
-31	Furfuryl alcohol: 279	-15.2	Trifluoroacetic acid: 531

-15	Methyl benzoate: 345	6.1	Diiodomethane: 179
-14.9	Benzenethiol: 23	6.6	Cyclohexane: 114
-14.78	Quinoline: 470	7	*cis*-2-Methylcyclohexanol: 354
-14.7	1,2,3-Trichloropropane: 519	7.3	Cyclohexylbenzene: 119
-14.3	2,4-Xylidine: 560	7.5	*p*-Chlorotoluene: 103
-14	*o*-Chloroaniline: 81	7.6	2,3-Butanediol: 52
-14	2,3-Dimethyl-2-butanol: 192	8.0	Tribromomethane: 507
-13.5	2,4-Dichlorotoluene: 157	8.3	Formic acid: 276
-13	Ethylene glycol: 247	9.8	3,5-Xylidine: 564
-13	1-Tridecene: 526	9.9	1,2-Dibromoethane: 131
-12.7	Benzonitrile: 24	10	Ethyl *trans*-cinnamate: 241
-11.03	Piperidine: 446	10.3	Dimethyl adipate: 186
-10.4	Diethylene glycol: 162	10.5	Ethanolamine: 231
-10.2	2,2'-Thiodiethanol: 499	10.5	2-Nitroanisole: 404
-10	Dibutyl sebacate: 140	11	Tri-*o*-cresyl phosphate: 521
-9.8	2,2,3,3-Tetramethylpentane: 493	11.1	Ethylenediamine: 246
-9.6	Dodecane: 226	11.2	2,6-Xylidine: 562
-9.2	*cis*-4-Methylcyclohexanol: 358	11.8	*m*-Cresol: 107
-8.8	2-Methyl-2-butanol: 349	11.8	1,4-Dioxane: 219
-8.1	Isophorone: 325	12.3	Acrylic acid: 11
-8	Methyl salicylate: 395	12.3	Nonanoic acid: 411
-7	Salicylaldehyde: 471	13.2	*p*-Xylene: 552
-7	Triethylene glycol: 529	13.4	Oleic acid: 421
-6.2	Tetraethylene glycol: 485	15.5	2,5-Xylidine: 561
-6.1	2,6-Lutidine: 332	16	Methacrylic acid: 335
-6.0	Aniline: 18	16.3	Octanoic acid: 415
-5.7	Butanoic acid: 55	16.6	Acetic acid: 4
-5.5	*cis*-3-Methylcyclohexanol: 356	18.2	Glycerol: 280
-5.3	Tridecane: 525	18.5	Dimethyl sulfoxide: 218
-5	(Trichloromethyl)benzene: 517	19	Dimethyl succinate: 216
-4.9	Morpholine: 400	20	Acetophenone: 8
-3.8	*N*-Methylformamide: 362	20.1	1,4-Butanediol: 51
-3	Hexanoic acid: 293	20.5	Triethanolamine: 527
-2.5	1-Chloronaphthalene: 94	21	Benzyl benzoate: 27
-2.0	*trans*-2-Methylcyclohexanol: 355	22.5	Veratrole: 548
		24.5	2,4-Xylenol: 554
-0.9	1-Bromonaphthalene: 43	25	*trans*-2-Butene-1,4-diol: 60
-0.6	Tetramethylurea: 498	25	1-Methylcyclohexanol: 353
-0.5	*trans*-3-Methylcyclohexanol: 357	25	2-Pyrrolidone: 469
		25.4	Cyclohexanol: 115
0	1,1,2,2-Tetrabromoethane: 478	25.5	2-Aminoisobutanol: 17
0.8	Cineole: 105	25.5	Tri-*m*-cresyl phosphate: 522
0.8	1,2-Dimethylnaphthalene: 208	25.8	2-Methyl-2-propanol: 392
1	Adiponitrile: 13	26	1,1,2,2-Tetrachloro-1,2-difluoroethane: 479
1.8	Dibenzyl ether: 129		
2.0	*cis*-2-Butene-1,4-diol: 59	26.47	Isoquinoline: 328
2.4	*N,N*-Dimethylaniline: 188	26.8	Diphenyl ether: 223
2.55	Formamide: 275	27	Butyl stearate: 73
3.66	4-Picoline: 443	27.5	Methyl acetoacetate: 339
5.3	Hexafluorobenzene: 288	27.6	Sulfolane: 474
5.5	Benzene: 22	28	Diethanolamine: 159
5.5	Dimethyl phthalate: 214	28	*N*-Methylacetamide: 337
5.6	3,3-Dimethyl-2-butanol: 193	29.8	*o*-Cresol: 106
5.7	Nitrobenzene: 405	35.5	*p*-Cresol: 108

36.4	Ethylene carbonate: 244
40.5	α-Terpineol: 476
40.6	1,1,1,2-Tetrachloro-2,2-difluoroethane: 480
40.9	Phenol: 439
43.7	*p*-Toluidine: 504
45.7	2,6-Xylenol: 556
51	3,4-Xylidine: 563
51.5	Camphene: 77
52.5	2,2-Dimethyl-1-propanol: 215
52.7	*p*-Dichlorobenzene: 144
54.5	Succinonitrile: 473

60.8	3,4-Xylenol: 557
63.6	3,5-Xylenol: 558
67.5	Tricyclene: 524
69.3	ε-Caprolactam: 79
72	*trans*-Crotonic acid: 110
72.8	2,3-Xylenol: 553
74.8	2,5-Xylenol: 555
77.5	Tri-*p*-cresyl phosphate: 523
80.2	Naphthalene: 402
81	Acetamide: 3
100.7	2,2,3,3-Tetramethylbutane: 492
178.8	(+)-Camphor: 78

Density Index

This index is arranged by density in g/cm^3. The superscript gives the temperature in °C. The number following the name is the ID for the solvent.

0.493^{25}	Propane: 448	0.7075^{16}	3-Methylheptane: 365
0.5510^{25}	Isobutane: 316	0.708^{20}	*cis*-2-Heptene: 286
0.573^{25}	Butane: 49	0.7095^{25}	2-Methyloctane: 373
0.5852^{25}	Neopentane: 403	0.7100^{20}	3,3-Dimethylhexane: 205
0.6201^{20}	Isopentane: 322	0.7105^{20}	2,2-Dimethylheptane: 200
0.6262^{20}	Pentane: 423	0.7136^{20}	3-Ethylhexane: 259
0.627^{25}	Trimethylamine: 535	0.7138^{20}	Diethyl ether: 170
0.6405^{20}	1-Pentene: 430	0.7149^{20}	1-Octene: 418
0.6431^{25}	*trans*-2-Pentene: 432	0.7151^{25}	3,4-Dimethylhexane: 206
0.6444^{25}	2,2-Dimethylbutane: 189	0.7153^{20}	Diisopropylamine: 182
0.648^{19}	Tetramethylsilane: 497	0.716^{25}	4-Methyloctane: 375
0.650^{25}	2-Methylpentane: 377	0.7161^{20}	2,2,3-Trimethylpentane: 540
0.6548^{25}	Hexane: 290	0.717^{25}	3-Methyloctane: 374
0.6556^{20}	*cis*-2-Pentene: 431	0.7173^{20}	Propylamine: 459
0.656^{25}	Methylamine: 343	0.7176^{20}	Nonane: 410
0.6598^{25}	3-Methylpentane: 378	0.7191^{20}	2,3,4-Trimethylpentane: 543
0.6616^{20}	2,3-Dimethylbutane: 190	0.7193^{20}	3-Ethyl-2-methylpentane: 266
0.6727^{20}	2,4-Dimethylpentane: 212	0.7195^{20}	2,2,4,4-Tetramethylpentane: 495
0.6731^{20}	1-Hexene: 297	0.7199^{20}	*trans*-2-Octene: 420
0.6732^{25}	*trans*-2-Hexene: 299	0.7218^{20}	2,3,5-Trimethylhexane: 539
0.6739^{20}	2,2-Dimethylpentane: 210	0.724^{25}	Isobutylamine: 318
0.677^{25}	Ethylamine: 235	0.7241^{20}	Diisopropyl ether: 183
0.6772^{20}	*trans*-3-Hexene: 301	0.7243^{20}	*cis*-2-Octene: 419
0.6787^{20}	2-Methylhexane: 367	0.7246^{20}	*sec*-Butylamine: 64
0.6796^{20}	*cis*-3-Hexene: 300	0.7253^{25}	1-Nonene: 412
0.6804^{0}	Dimethylamine: 187	0.7262^{20}	2,3,3-Trimethylpentane: 542
0.6837^{20}	Heptane: 281	0.7274^{20}	3-Ethyl-3-methylpentane: 267
0.6869^{20}	*cis*-2-Hexene: 298	0.7275^{20}	Triethylamine: 528
0.687^{21}	3-Methylhexane: 368	0.7300^{20}	Decane: 126
0.6877^{25}	2,2,4-Trimethylpentane: 541	0.7389^{20}	2,2,3,4-Tetramethylpentane: 494
0.6891^{20}	Isopropylamine: 327	0.7400^{20}	Dipropylamine: 224
0.6901^{25}	2,5-Dimethylhexane: 204	0.7408^{20}	1-Decene: 127
0.6901^{20}	2,2,3-Trimethylbutane: 537	0.7414^{20}	Butylamine: 63
0.6912^{25}	2,3-Dimethylhexane: 202	0.7420^{20}	2,3-Diethylpentane: 174
0.6936^{20}	3,3-Dimethylpentane: 213	0.7457^{20}	Cyclopentane: 120
0.6951^{20}	2,3-Dimethylpentane: 211	0.7466^{20}	Dipropyl ether: 225
0.6953^{20}	2,2-Dimethylhexane: 201	0.7486^{20}	Methylcyclopentane: 360
0.6958^{20}	*tert*-Butylamine: 65	0.7487^{20}	Dodecane: 226
0.6962^{25}	2,4-Dimethylhexane: 203	0.7495^{20}	Butyl ethyl ether: 69
0.6970^{20}	1-Heptene: 285	0.7503^{20}	1-Undecene: 547
0.6980^{20}	2-Methylheptane: 364	0.7530^{25}	2,2,3,3-Tetramethylpentane: 493
0.6982^{20}	3-Ethylpentane: 268	0.7536^{20}	3,3-Diethylpentane: 175
0.6986^{25}	Octane: 413	0.7544^{20}	Pentylamine: 434
0.7012^{20}	*trans*-2-Heptene: 287	0.7547^{20}	2,3,3,4-Tetramethylpentane: 496
0.7046^{20}	4-Methylheptane: 366	0.7564^{20}	Tridecane: 525
0.7056^{20}	Diethylamine: 160	0.758^{20}	Allylamine: 16
0.7072^{20}	2,2,5-Trimethylhexane: 538	0.7584^{20}	1-Dodecene: 227

0.7589^{20}	Ethyl vinyl ether: 270
0.7658^{20}	Tetraethylsilane: 486
0.7658^{20}	1-Tridecene: 526
0.7670^{20}	Dibutylamine: 136
0.7684^{20}	Dibutyl ether: 137
0.7694^{20}	Methylcyclohexane: 352
0.7704^{20}	2-Methylpropanenitrile: 389
0.7720^{20}	Cyclopentene: 122
0.7760^{20}	*trans*-1,2-Dimethylcyclohexane: 195
0.7770^{20}	Tributylamine: 508
0.7777^{20}	Diisopentyl ether: 181
0.7785^{20}	Cyclohexane: 114
0.7818^{20}	Propanenitrile: 451
0.7833^{20}	Dipentyl ether: 222
0.7834^{18}	Acetaldehyde: 2
0.7855^{20}	2-Propanol: 455
0.7857^{20}	Acetonitrile: 7
0.7880^{20}	Ethylcyclohexane: 243
0.7887^{20}	2-Methyl-2-propanol: 392
0.7888^{20}	Butyl vinyl ether: 75
0.7891^{20}	Isobutanal: 315
0.7893^{20}	Ethanol: 230
0.7899^{20}	Acetone: 6
0.7914^{20}	Methanol: 336
0.7936^{20}	Butanenitrile: 53
0.7963^{20}	*cis*-1,2-Dimethylcyclohexane: 194
0.7978^{20}	Methyl isobutyl ketone: 369
0.8001^{20}	Methylacrylonitrile: 341
0.8008^{20}	Pentanenitrile: 425
0.8013^{15}	β-Myrcene: 401
0.8016^{20}	Butanal: 48
0.8018^{20}	2-Methyl-1-propanol: 391
0.8030^{20}	4-Methylvaleronitrile: 399
0.8035^{20}	1-Propanol: 454
0.8051^{20}	Hexanenitrile: 291
0.8054^{20}	Methyl ethyl ketone: 361
0.8060^{20}	Acrylonitrile: 12
0.8062^{20}	Diisobutyl ketone: 180
0.8063^{20}	2-Butanol: 58
0.8075^{20}	4-Methyl-2-pentanol: 383
0.809^{20}	Methyl propyl ketone: 393
0.8094^{20}	2-Pentanol: 428
0.8096^{20}	2-Methyl-2-butanol: 349
0.8098^{20}	1-Butanol: 57
0.8104^{20}	3-Methyl-1-butanol: 348
0.8108^{20}	Diisopropyl ketone: 184
0.8110^{20}	Cyclohexene: 117
0.8111^{20}	Methyl pentyl ketone: 387
0.8113^{20}	Butyl methyl ketone: 71
0.812^{20}	2,2-Dimethyl-1-propanol: 215
0.8122^{25}	3,3-Dimethyl-2-butanol: 193
0.8136^{20}	1-Hexanol: 294

0.8136^{20}	Octanenitrile: 414
0.8144^{20}	1-Pentanol: 427
0.8150^{25}	2-Methyl-1-butanol: 347
0.8159^{20}	2-Hexanol: 295
0.816^{19}	Diethyl ketone: 171
0.8167^{20}	2-Heptanol: 283
0.8180^{20}	3-Methyl-2-butanol: 350
0.8182^{20}	3-Hexanol: 296
0.8183^{20}	Ethyl butyl ketone: 240
0.8191^{20}	Cyclohexylamine: 118
0.8193^{20}	2-Octanol: 417
0.820^{20}	Hexyl methyl ketone: 305
0.8203^{20}	3-Pentanol: 429
0.8219^{20}	1-Heptanol: 282
0.8227^{20}	3-Heptanol: 284
0.8236^{20}	2,3-Dimethyl-2-butanol: 192
0.8242^{20}	3-Methyl-1-pentanol: 380
0.8242^{20}	2,2,3,3-Tetramethylbutane: 492
0.8243^{20}	2-Methyl-3-pentanol: 384
0.8254^{20}	Acetal: 1
0.8262^{25}	1-Octanol: 416
0.8263^{20}	2-Methyl-1-pentanol: 379
0.8286^{20}	3-Methyl-3-pentanol: 385
0.8297^{20}	1,2-Epoxybutane: 229
0.8307^{20}	3-Methyl-2-pentanol: 382
0.8319^{25}	Ethylene glycol dibutyl ether: 249
0.8319^{25}	2-Ethyl-1-hexanol: 261
0.832^{25}	Ethyleneimine: 257
0.8326^{20}	2-Ethyl-1-butanol: 239
0.8350^{16}	2-Methyl-2-pentanol: 381
0.8362^{20}	Diethyl sulfide: 176
0.8375^{19}	α-Terpinene: 475
0.8386^{20}	Dibutyl sulfide: 141
0.840^{20}	Acrolein: 10
0.8402^{21}	Dipentene: 221
0.8411^{20}	*d*-Limonene: 329
0.8411^{20}	4-Methyl-4-penten-2-one: 386
0.8416^{20}	1-Butanethiol: 54
0.8420^{20}	2-Chloro-2-methylpropane: 93
0.843^{20}	*l*-Limonene: 330
0.844^{15}	3,3-Dimethyl-1-butanol: 191
0.8483^{20}	Dimethyl sulfide: 217
0.8484^{20}	Ethylene glycol diethyl ether: 250
0.8516^{20}	*trans*-Crotonaldehyde: 109
0.8520^{20}	β-Phellandrene: 437
0.8521^{20}	*trans*-Crotonyl alcohol: 112
0.8532^{20}	Isobutylbenzene: 319
0.8539^{25}	α-Pinene: 444
0.8540^{20}	Allyl alcohol: 15
0.854^{25}	Butyl stearate: 73
0.8542^{20}	Isobutyl isobutanoate: 321
0.8552^{20}	2-Methyltetrahydrofuran: 396

0.8567^{20} Tributyl borate: 509
0.8573^{20} *p*-Cymene: 123
0.8583^{19} Isopentyl isopentanoate: 324
0.8586^{20} Pyrrolidine: 468
0.859^{0} 1,2-Propylene oxide: 463
0.8593^{20} Methylal: 342
0.860^{25} β-Pinene: 445
0.8601^{20} Butylbenzene: 66
0.8606^{20} Piperidine: 446
0.8611^{20} *p*-Xylene: 552
0.8612^{20} *p-tert*-Butyltoluene: 74
0.8617^{20} 2-Chloropropane: 99
0.8618^{20} Cumene: 113
0.8620^{20} Propylbenzene: 460
0.8621^{20} *sec*-Butylbenzene: 67
0.8632^{15} Terpinolene: 477
0.8642^{20} *m*-Xylene: 551
0.8652^{20} Mesitylene: 333
0.8653^{20} Mesityl oxide: 334
0.8656^{20} Ethyl isovalerate: 264
0.8657^{25} Propanal: 447
0.8662^{20} *cis*-Crotonyl alcohol: 111
0.8665^{20} *tert*-Butylbenzene: 68
0.8668^{80} Tricyclene: 524
0.8669^{20} Toluene: 501
0.8670^{20} Ethylbenzene: 236
0.8691^{20} Ethylene glycol dimethyl ether: 251
0.8699^{20} *trans*-Decahydronaphthalene: 125
0.8704^{15} Butyl oleate: 72
0.8712^{20} Isobutyl acetate: 317
0.8718^{20} 2-Ethylhexyl acetate: 262
0.8718^{20} Isopropyl acetate: 326
0.8732^{20} 2-Chlorobutane: 84
0.8738^{20} 1-Chlorooctane: 95
0.8739^{20} Methyl oleate: 376
0.8748^{20} *sec*-Butyl acetate: 62
0.8750^{20} 1-Chloro-3-methylbutane: 90
0.8756^{20} Pentyl acetate: 433
0.8758^{20} Pseudocumene: 465
0.876^{15} Isopentyl acetate: 323
0.8765^{20} Benzene: 22
0.8769^{20} 3-(Chloromethyl)heptane: 91
0.8773^{20} 1-Chloro-2-methylpropane: 92
0.8776^{20} Isobutyl formate: 320
0.8779^{15} Hexyl acetate: 302
0.879^{20} Camphene: 77
0.8802^{10} *o*-Xylene: 550
0.8805^{25} *sec*-Hexyl acetate: 303
0.8814^{20} Tetrahydropyran: 490
0.8820^{20} 1-Chloropentane: 97
0.8825^{20} Butyl acetate: 61
0.8844^{15} Ethyl butanoate: 238

0.885^{25} Diethylene glycol dibutyl ether: 163
0.8862^{20} 1-Chlorobutane: 83
0.8878^{20} Propyl acetate: 458
0.888^{20} Methyl isopentyl ketone: 370
0.8885^{20} Butyl formate: 70
0.8892^{20} Tetrahydrofuran: 487
0.8899^{20} 1-Chloropropane: 98
0.8902^{25} Chloroethane: 87
0.8917^{20} Ethyl propanoate: 269
0.8935^{20} Oleic acid: 421
0.8944^{20} 1,2,3-Trimethylbenzene: 536
0.896^{20} 1,1-Difluoroethane: 177
0.8965^{20} *cis*-Decahydronaphthalene: 124
0.8979^{20} Ethylenediamine: 246
0.9003^{20} Ethyl acetate: 232
0.9015^{20} Ethylene glycol monobutyl ether: 254
0.9052^{20} Nonanoic acid: 411
0.9058^{20} Propyl formate: 464
0.9060^{20} Styrene: 472
0.9063^{20} Diethylene glycol diethyl ether: 164
0.9106^{20} Chloroethylene: 88
0.9106^{20} Octanoic acid: 415
0.911^{25} Chloromethane: 89
0.9118^{21} *trans*-4-Methylcyclohexanol: 359
0.9155^{20} *cis*-3-Methylcyclohexanol: 356
0.9166^{22} 2,4,6-Trimethylpyridine: 545
0.9168^{20} Ethyl formate: 258
0.9170^{20} *cis*-4-Methylcyclohexanol: 358
0.9194^{20} 1-Methylcyclohexanol: 353
0.9214^{30} *trans*-3-Methylcyclohexanol: 357
0.9226^{20} 2,6-Lutidine: 332
0.923^{15} Hexylene glycol: 304
0.9234^{20} Ethyl acrylate: 234
0.9247^{20} *trans*-2-Methylcyclohexanol: 355
0.9255^{20} Isophorone: 325
0.9267^{20} Cineole: 105
0.9274^{20} Hexanoic acid: 293
0.9275^{20} Allyl acetate: 14
0.9297^{20} Ethylene glycol monoethyl ether: 255
0.9305^{25} *N*-Methylpropanamide: 388
0.9309^{20} 2,4-Lutidine: 331
0.931^{20} 3-Methylbutanoic acid: 346
0.9317^{20} Vinyl acetate: 549
0.9325^{22} 2-Ethyl-1,3-hexanediol: 260
0.9337^{20} α-Terpineol: 476
0.934^{20} 2-Aminoisobutanol: 17
0.9342^{20} Methyl acetate: 338

0.9360^{20} *cis*-2-Methylcyclohexanol: 354
0.9366^{25} *N,N*-Dimethylacetamide: 185
0.9371^{25} *N*-Methylacetamide: 337
0.9376^{20} 3-Chloropropene: 100
0.9387^{20} Diacetone alcohol: 128
0.9391^{20} Pentanoic acid: 426
0.9405^{15} Dibutyl sebacate: 140
0.9427^{20} Cyclohexylbenzene: 119
0.9434^{20} Diethylene glycol dimethyl
        ether: 165
0.944^{25} *N,N*-Dimethylformamide: 198
0.9440^{20} Methyl methacrylate: 371
0.9443^{20} 2-Picoline: 441
0.9478^{20} Benzyl ethyl ether: 29
0.9478^{20} Cyclohexanone: 116
0.9478^{20} Propargyl alcohol: 457
0.9487^{20} Cyclopentanone: 121
0.9514^{20} Furan: 277
0.9535^{20} Methyl acrylate: 340
0.9548^{20} 4-Picoline: 443
0.9557^{20} *N,N*-Dimethylaniline: 188
0.9566^{20} 3-Picoline: 442
0.9569^{20} Diethylenetriamine: 169
0.9577^{20} Butanoic acid: 55
0.9580^{24} Phenylethylamine: 440
0.9604^{77} *trans*-Crotonic acid: 110
0.9619^{20} *p*-Toluidine: 504
0.9624^{20} Cyclohexanol: 115
0.9647^{20} Ethylene glycol monomethyl
        ether: 256
0.9650^{20} 2,4-Xylenol: 554
0.9651^{20} Phenetole: 438
0.9660^{25} 1,2,3,4-Tetrahydronaphthalene:
        489
0.9668^{20} Butanoic anhydride: 56
0.9676^{20} Adiponitrile: 13
0.9680^{20} 3,5-Xylenol: 558
0.9681^{20} 2-Methylpropanoic acid: 390
0.9687^{20} Tetramethylurea: 498
0.9698^{20} Pyrrole: 467
0.9706^{20} 3,5-Xylidine: 564
0.9721^{25} Acetylacetone: 9
0.9723^{20} 2,4-Xylidine: 560
0.9727^{25} Tributyl phosphate: 510
0.9740^{20} Ethylene glycol ethyl ether
        acetate: 252
0.9742^{20} Methyl formate: 363
0.9752^{20} Diethyl carbonate: 161
0.9790^{21} 2,5-Xylidine: 561
0.981^{25} Bis(2-ethylhexyl) phthalate: 31
0.9819^{20} Pyridine: 466
0.9821^{25} 2-Nitropropane: 409
0.9830^{20} 3,4-Xylenol: 557
0.9842^{20} 2,6-Xylidine: 562

0.9867^{60} Succinonitrile: 473
0.9885^{20} Diethylene glycol monoethyl
        ether: 166
0.9889^{20} *m*-Toluidine: 503
0.9891^{20} *N*-Methylaniline: 344
0.990^{25} (+)-Camphor: 78
0.9914^{20} 1,5-Pentanediol: 424
0.9930^{20} Propanoic acid: 452
0.9931^{20} 2,3-Xylidine: 559
0.9940^{20} Anisole: 19
0.9961^{25} 1-Nitropropane: 408
0.9974^{20} *m*-Fluorotoluene: 273
0.9975^{20} *p*-Fluorotoluene: 274
0.9982^{20} Propargyl acetate: 456
0.9984^{20} *o*-Toluidine: 502
0.9986^{85} Acetamide: 3
0.9987^{20} Tetrahydrothiophene: 491
1.0005^{20} Morpholine: 400
1.0021^{20} 1,6-Dimethylnaphthalene: 209
1.0033^{20} 2,3-Butanediol: 52
1.0041^{13} *o*-Fluorotoluene: 272
1.0053^{20} 1,3-Butanediol: 50
1.0074^{19} Ethylene glycol momomethyl
        ether acetate: 253
1.0093^{15} Benzonitrile: 24
1.0096^{20} Diethylene glycol monoethyl
        ether acetate: 167
1.011^{19} *N*-Methylformamide: 362
1.0110^{20} Propanoic anhydride: 453
1.0153^{20} Methacrylic acid: 335
1.0171^{20} 1,4-Butanediol: 51
1.0179^{20} 1,2-Dimethylnaphthalene: 208
1.0180^{20} Ethanolamine: 231
1.0185^{40} *p*-Cresol: 108
1.0193^{20} 2-Methylthiophene: 397
1.0202^{20} 1-Methylnaphthalene: 372
1.0205^{15} α-Tolunitrile: 505
1.0217^{20} Aniline: 18
1.0218^{20} 3-Methylthiophene: 398
1.0225^{20} Fluorobenzene: 271
1.0230^{25} *N*-Methyl-2-pyrrolidone: 394
1.0230^{20} Propyl benzoate: 461
1.0253^{20} Naphthalene: 402
1.0281^{20} Acetophenone: 8
1.03^{20} Hexamethylphosphoric triamide:
        289
1.0328^{20} Ethyl lactate: 265
1.0337^{20} 1,4-Dioxane: 219
1.0341^{20} *m*-Cresol: 107
1.035^{20} Diethylene glycol monomethyl
        ether: 168
1.0361^{20} 1,2-Propanediol: 449
1.0368^{10} Ethyl acetoacetate: 233
1.0404^{25} Hydracrylonitrile: 306

1.0415^{10} Benzaldehyde: 21
1.0419^{24} Benzyl alcohol: 26
1.0428^{20} Dibenzyl ether: 129
1.0448^{25} Nitroethane: 406
1.0465^{20} Dibutyl phthalate: 139
1.0491^{20} Ethyl *trans*-cinnamate: 241
1.0492^{20} Acetic acid: 4
1.0511^{20} Acrylic acid: 11
1.0511^{15} Ethyl benzoate: 237
1.0524^{20} Tetrahydrofurfuryl alcohol: 488
1.0538^{20} 1,3-Propanediol: 450
1.0545^{45} Phenol: 439
1.0550^{20} Benzyl acetate: 25
1.0551^{20} Diethyl malonate: 172
1.0600^{20} Dimethyl adipate: 186
1.060^{20} 1,3-Dioxolane: 220
1.0625^{20} Dimethyl disulfide: 196
1.0649^{20} Thiophene: 500
1.0654^{20} Ethyl cyanoacetate: 242
1.0661^{30} Diphenyl ether: 223
1.0695^{20} Triethyl phosphate: 530
1.0697^{20} *p*-Chlorotoluene: 103
1.0698^{20} *cis*-2-Butene-1,4-diol: 59
1.0700^{20} *trans*-2-Butene-1,4-diol: 60
1.0702^{20} 1-Bromodecane: 38
1.075^{20} *m*-Chlorotoluene: 102
1.076^{18} 3,4-Xylidine: 563
1.0762^{20} Methyl acetoacetate: 339
1.0775^{20} Benzenethiol: 23
1.0785^{20} Diethyl oxalate: 173
1.0810^{25} Veratrole: 548
1.082^{20} Acetic anhydride: 5
1.0825^{20} *o*-Chlorotoluene: 101
1.0876^{20} Dimethyl glutarate: 199
1.0910^{30} Isoquinoline: 328
1.0933^{15} Methyl benzoate: 345
1.0966^{20} Diethanolamine: 159
1.0977^{15} Quinoline: 470
1.1004^{20} Benzyl chloride: 28
1.1014^{20} Dimethyl sulfoxide: 218
1.1043^{20} Ethylene glycol diacetate: 248
1.1049^{20} 1,2,6-Hexanetriol: 292
1.1058^{20} Chlorobenzene: 82
1.107^{25} 1-Chloro-1,1-difluoroethane: 85
1.1088^{20} Ethylene glycol: 247
1.1121^{25} Benzyl benzoate: 27
1.1197^{15} Diethylene glycol: 162
1.1198^{20} Dimethyl succinate: 216
1.120^{20} 2-Pyrrolidone: 469
1.1225^{25} Methyl cyanoacetate: 351
1.1242^{20} Triethanolamine: 527
1.1274^{15} Triethylene glycol: 529
1.1284^{16} γ-Butyrolactone: 76
1.1285^{15} Tetraethylene glycol: 485

1.1296^{20} Furfuryl alcohol: 279
1.1334^{20} Formamide: 275
1.135^{25} *o*-Cresol: 106
1.1371^{20} Nitromethane: 407
1.150^{25} Tri-*m*-cresyl phosphate: 522
1.1560^{20} 1,2-Dichloropropane: 154
1.1583^{20} Triacetin: 506
1.1594^{20} Furfural: 278
1.1606^{20} Dimethyl maleate: 207
1.1674^{20} Salicylaldehyde: 471
1.1757^{20} 1,1-Dichloroethane: 147
1.1793^{25} 2,2'-Thiodiethanol: 499
1.181^{25} Methyl salicylate: 395
1.1812^{20} Epichlorohydrin: 228
1.1884^{20} (Trifluoromethyl)benzene: 534
1.1905^{20} Dimethyl phthalate: 214
1.1938^{20} 1-Chloronaphthalene: 94
1.1955^{20} Tri-*o*-cresyl phosphate: 521
1.2019^{20} Ethylene chlorohydrin: 245
1.2037^{20} Nitrobenzene: 405
1.2047^{20} Propylene carbonate: 462
1.213^{20} 1,1-Dichloroethylene: 149
1.2144^{20} Trimethyl phosphate: 544
1.2182^{20} 1-Bromopentane: 44
1.22^{20} Bis(2-chloroethyl) ether: 30
1.220^{20} Formic acid: 276
1.2351^{20} 1,2-Dichloroethane: 148
1.2373^{20} Trichloroethylsilane: 514
1.247^{25} Tri-*p*-cresyl phosphate: 523
1.2475^{55} *p*-Dichlorobenzene: 144
1.2476^{20} 2,4-Dichlorotoluene: 157
1.2540^{20} 2-Nitroanisole: 404
1.2564^{20} 3,4-Dichlorotoluene: 158
1.2565^{20} *trans*-1,2-Dichloroethylene: 151
1.2585^{20} 2-Bromobutane: 34
1.26^{25} Benzal chloride: 20
1.2613^{20} Glycerol: 280
1.2632^{20} Carbon disulfide: 80
1.2723^{18} Sulfolane: 474
1.273^{20} Trichloromethylsilane: 518
1.2758^{20} 1-Bromobutane: 33
1.2837^{20} *cis*-1,2-Dichloroethylene: 150
1.2884^{20} *m*-Dichlorobenzene: 143
1.3059^{20} *o*-Dichlorobenzene: 142
1.3140^{20} 2-Bromopropane: 46
1.3214^{39} Ethylene carbonate: 244
1.3266^{20} Dichloromethane: 153
1.3390^{20} 1,1,1-Trichloroethane: 511
1.3537^{20} 1-Bromopropane: 45
1.3723^{20} (Trichloromethyl)benzene: 517
1.3842^{20} 2,2,2-Trifluoroethanol: 533
1.3889^{20} 1,2,3-Trichloropropane: 519
1.3965^{16} 2-Bromopropene: 47
1.405^{9} Dichlorofluoromethane: 152

1.4278[20]   2-Bromo-2-methylpropane: 42
1.4397[20]   1,1,2-Trichloroethane: 512
1.455[25]   1,1-Dichlorotetrafluoroethane:
            155
1.455[25]   1,2-Dichlorotetrafluoroethane:
            156
1.4604[20]   Bromoethane: 39
1.4642[20]   Trichloroethylene: 513
1.4785[20]   1-Bromonaphthalene: 43
1.4832[20]   Trichloromethane: 516
1.4909[-69] Chlorodifluoromethane: 86
1.4933[20]   Bromoethylene: 40
1.4950[20]   Bromobenzene: 32
1.500[25]   Perfluorocyclobutane: 436
1.5351[25]   Trifluoroacetic acid: 531
1.5406[20]   1,1,1,2-Tetrachloroethane: 481
1.555[-20]   1,1-Dichloro-2,2-
            difluoroethylene: 145
1.5635[25]   1,1,2-Trichlorotrifluoroethane:
            520
1.5678[-42] Chloropentafluoroethane: 96
1.5920[20]   2-Iodobutane: 309
1.5940[20]   Tetrachloromethane: 484
1.5953[20]   1,1,2,2-Tetrachloroethane: 482
1.6035[20]   1-Iodo-2-methylpropane: 312
1.6154[20]   1-Iodobutane: 308
1.6184[20]   Hexafluorobenzene: 288
1.6227[20]   Tetrachloroethylene: 483

1.6447[25]   1,1,2,2-Tetrachloro-1,2-
            difluoroethane: 479
1.6484[25]   Perfluorobutane: 435
1.649[25]   1,1,1,2-Tetrachloro-2,2-
            difluoroethane: 480
1.6755[20]   Bromomethane: 41
1.6796[20]   Pentachloroethane: 422
1.7042[20]   2-Iodopropane: 314
1.7392[20]   1-Bromo-2-chloroethane: 35
1.7489[20]   1-Iodopropane: 313
1.8308[20]   Iodobenzene: 307
1.884[25]   Trinonafluorobutylamine: 546
1.9324[20]   1,2-Dibromopropane: 134
1.9344[20]   Bromochloromethane: 37
1.9358[20]   Iodoethane: 310
1.9771[0]   Bromochlorofluoromethane: 36
2.149[25]   1,2-Dibromotetrafluoroethane:
            135
2.1791[20]   1,2-Dibromoethane: 131
2.2238[20]   1,2-Dibromo-1,1-difluoroethane:
            130
2.279[20]   Iodomethane: 311
2.421[20]   Dibromofluoromethane: 132
2.4969[20]   Dibromomethane: 133
2.899[15]   Tribromomethane: 507
2.9655[20]   1,1,2,2-Tetrabromoethane: 478
3.3212[20]   Diiodomethane: 179

Dielectric Constant Index

This index is arranged by value of the dielectric constant. The applicable temperature in °C is given as a superscript. The number following the name is the ID for the solvent.

2.37^{25} l-Limonene: 330
2.38^{23} Toluene: 501
2.38^{20} Cumene: 113
2.38^{20} Pseudocumene: 465
2.38^{25} Dipentene: 221
2.39^{55} p-Dichlorobenzene: 144
2.41^{25} 1,1,2-Trichlorotrifluoroethane: 520
2.42^{20} Triethylamine: 528
2.44^{25} Trimethylamine: 535
2.45^{20} Ethylbenzene: 236
2.45^{25} α-Terpinene: 475
2.47^{20} Styrene: 472
2.48^{0} 1,2-Dichlorotetrafluoroethane: 156
2.48^{22} Nonanoic acid: 411
2.52^{25} 1,1,1,2-Tetrachloro-2,2-difluoroethane: 480
2.52^{35} 1,1,2,2-Tetrachloro-1,2-difluoroethane: 479
2.54^{90} Naphthalene: 402
2.56^{20} o-Xylene: 550
2.58^{20} 2-Methylpropanoic acid: 390
2.60^{25} Hexanoic acid: 293
2.61^{25} 1,2-Dimethylnaphthalene: 208
2.63^{20} Carbon disulfide: 80
2.64^{20} Methylal: 342
2.66^{20} 1,2,3-Trimethylbenzene: 536
2.66^{21} Pentanoic acid: 426
2.73^{20} 1,6-Dimethylnaphthalene: 209
2.74^{20} Thiophene: 500
2.77^{20} Dibutylamine: 136
2.77^{25} 1,2,3,4-Tetrahydronaphthalene: 489
2.80^{25} Dipentyl ether: 222
2.82^{20} Diisopentyl ether: 181
2.82^{24} Diethyl carbonate: 161
2.85^{15} Octanoic acid: 415
2.88^{4} Furan: 277
2.92^{20} 1-Methylnaphthalene: 372
2.98^{14} Butanoic acid: 55
3.00^{20} Trichlorofluoromethane: 515
3.01^{-150} Chlorotrifluoromethane: 104
3.07^{20} Dipropylamine: 224
3.08^{20} Dibutyl ether: 137
3.12^{25} Butyl stearate: 73
3.21^{20} Methyl oleate: 376
3.38^{24} Dipropyl ether: 225
3.39^{28} Trichloroethylene: 513
3.44^{25} Propanoic acid: 452
3.50^{-150} Dichlorodifluoromethane: 146
3.68^{20} Diethylamine: 160
3.72^{25} Pentachloroethane: 422
3.73^{10} Diphenyl ether: 223

3.80^{25} Acetal: 1
3.81^{30} Diisopropyl ether: 183
3.82^{20} Dibenzyl ether: 129
3.90^{25} Benzyl ethyl ether: 29
3.90^{20} Ethylene glycol diethyl ether: 250
4.00^{25} Butyl oleate: 72
4.22^{20} Phenetole: 438
4.23^{25} o-Fluorotoluene: 272
4.26^{30} Benzenethiol: 23
4.27^{20} Diethyl ether: 170
4.27^{20} Pentylamine: 434
4.29^{25} Dibutyl sulfide: 141
4.30^{21} Anisole: 19
4.32^{20} 3-Methyl-3-pentanol: 385
4.33^{20} Piperidine: 446
4.39^{15} Isopentyl isopentanoate: 324
4.40^{10} Tribromomethane: 507
4.42^{20} Hexyl acetate: 302
4.43^{21} Isobutylamine: 318
4.44^{25} 1-Bromodecane: 38
4.45^{20} Veratrole: 548
4.54^{20} Dibutyl sebacate: 140
4.55^{20} Cyclohexylamine: 118
4.57^{25} Cineole: 105
4.59^{20} Iodobenzene: 307
4.60^{20} 1,1-Dichloroethylene: 149
4.60^{10} 1,2-Dibromopropane: 134
4.71^{20} Butylamine: 63
4.71^{20} Ethyl isovalerate: 264
4.72^{20} o-Chlorotoluene: 101
4.72^{20} Isopentyl acetate: 323
4.77^{25} 1-Bromonaphthalene: 43
4.79^{20} Pentyl acetate: 433
4.81^{20} Trichloromethane: 516
4.81^{70} 2,3-Xylenol: 553
4.9^{20} 2,4-Xylidine: 560
4.90^{25} N,N-Dimethylaniline: 188
4.90^{40} 2,6-Xylenol: 556
4.96^{20} 1,2-Dibromoethane: 131
5.02^{20} m-Dichlorobenzene: 143
5.04^{25} 1-Chloronaphthalene: 94
5.05^{25} 1-Chlorooctane: 95
5.06^{60} p-Toluidine: 504
5.06^{30} 2,4-Xylenol: 554
5.07^{20} Butyl acetate: 61
5.07^{20} Isobutyl acetate: 317
5.08^{23} Propylamine: 459
5.14^{20} sec-Butyl acetate: 62
5.18^{28} Ethyl butanoate: 238
5.20^{15} 1-Butanethiol: 54
5.26^{30} Benzyl benzoate: 27
5.26^{25} Dimethylamine: 187
5.3^{20} Bis(2-ethylhexyl) phthalate: 31

9.02^{60} 3,4-Xylenol: 557
9.06^{50} 3,5-Xylenol: 558
9.16^{20} Quinoline: 470
9.20^{25} *cis*-1,2-Dichloroethylene: 150
9.20^{15} Methyl formate: 363
9.22^{25} (Trifluoromethyl)benzene: 534
9.22^{-66} 1,1,1,2-Tetrachloroethane: 481
9.30^{25} Ethylene glycol monobutyl ether: 254
9.39^{28} 3,4-Dichlorotoluene: 158
9.45^{20} Chloroethane: 87
9.46^{20} 2-Bromopropane: 46
9.51^{20} Hexyl methyl ketone: 305
9.6^{25} Dimethyl disulfide: 196
9.60^{20} 2,4-Lutidine: 331
9.66^{25} 3-Hexanol: 296
9.66^{20} 2-Chloro-2-methylpropane: 93
9.71^{3} Bromomethane: 41
9.72^{21} 2-Heptanol: 283
9.91^{20} Diisobutyl ketone: 180
10.0^{22} Chloromethane: 89
10.10^{25} 1,1-Dichloroethane: 147
10.12^{20} *o*-Dichlorobenzene: 142
10.18^{20} 2-Picoline: 441
10.30^{20} 1-Octanol: 416
10.42^{20} 1,2-Dichloroethane: 148
10.98^{20} 2-Bromo-2-methylpropane: 42
11.0^{25} Isoquinoline: 328
11.06^{25} 2-Hexanol: 295
11.10^{30} 3-Picoline: 442
11.75^{20} 1-Heptanol: 282
11.92^{30} Benzyl alcohol: 26
11.95^{20} Methyl pentyl ketone: 387
12.1^{25} 3-Methyl-2-butanol: 350
12.2^{20} 4-Picoline: 443
12.40^{30} Phenol: 439
12.44^{25} *m*-Cresol: 107
12.47^{25} 2-Methyl-2-propanol: 392
12.62^{20} Diethylenetriamine: 169
12.7^{20} Ethyl butyl ketone: 240
12.8^{20} Butanoic anhydride: 56
13.03^{20} 1-Hexanol: 294
13.05^{25} *p*-Cresol: 108
13.11^{20} Methyl isobutyl ketone: 369
13.20^{25} Triethyl phosphate: 530
13.26^{20} Pyridine: 466
13.35^{25} 3-Pentanol: 429
13.38^{25} Ethylene glycol monoethyl ether: 255
13.40^{20} *o*-Chloroaniline: 81
13.45^{25} Butanal: 48
13.48^{30} Tetrahydrofurfuryl alcohol: 488
13.53^{20} Methyl isopentyl ketone: 370
13.58^{25} Cyclopentanone: 121

13.71^{25} 2-Pentanol: 428
13.82^{20} Ethylenediamine: 246
13.90^{20} Octanenitrile: 414
14.0^{20} Ethyl acetoacetate: 233
14.56^{20} Butyl methyl ketone: 71
15.13^{25} 1-Pentanol: 427
15.4^{30} Ethyl lactate: 265
15.45^{20} Methyl propyl ketone: 393
15.6^{0} Mesityl oxide: 334
15.63^{25} 2-Methyl-1-butanol: 347
15.63^{20} 3-Methyl-1-butanol: 348
16.05^{20} *cis*-3-Methylcyclohexanol: 356
16.1^{20} Cyclohexanone: 116
16.40^{20} Cyclohexanol: 115
16.7^{-58} Methylamine: 343
16.85^{25} Furfuryl alcohol: 279
17.00^{20} Diethyl ketone: 171
17.2^{25} Ethylene glycol monomethyl ether: 256
17.26^{20} 2-Butanol: 58
17.26^{25} Hexanenitrile: 291
17.44^{25} Acetophenone: 8
17.5^{22} 4-Methylvaleronitrile: 399
17.84^{20} 1-Butanol: 57
17.85^{20} Benzaldehyde: 21
17.87^{26} α-Tolunitrile: 505
17.93^{20} 2-Methyl-1-propanol: 391
18.2^{25} Diacetone alcohol: 128
18.3^{25} Ethyleneimine: 257
18.30^{20} Propanoic anhydride: 453
18.35^{20} Salicylaldehyde: 471
18.5^{17} Propanal: 447
18.56^{20} Methyl ethyl ketone: 361
18.73^{20} 2-Ethyl-1,3-hexanediol: 260
19.7^{20} Allyl alcohol: 15
20.04^{20} Pentanenitrile: 425
20.18^{20} 2-Propanol: 455
20.44^{20} Tetraethylene glycol: 485
20.6^{20} Trimethyl phosphate: 544
20.8^{20} 1-Propanol: 454
20.8^{20} Propargyl alcohol: 457
21.0^{18} Acetaldehyde: 2
21.01^{20} Acetone: 6
21.20^{20} Bis(2-chloroethyl) ether: 30
22.45^{20} Acetic anhydride: 5
22.6^{22} Epichlorohydrin: 228
23.10^{20} Tetramethylurea: 498
23.4^{30} Hexylene glycol: 304
23.69^{20} Triethylene glycol: 529
24.42^{20} 2-Methylpropanenitrile: 389
24.70^{15} 1-Nitropropane: 408
24.83^{20} Butanenitrile: 53
25.3^{20} Ethanol: 230
25.75^{20} Diethanolamine: 159